Modern Environments
and Human Health

*To my mother, Joan Byrne, without whose influence
I would never have become an anthropologist, and to my endlessly
supportive husband and partner, Derek Anderson.*

Modern Environments and Human Health

Revisiting the Second Epidemiologic Transition

Edited by

Molly K. Zuckerman
Department of Anthropology and Middle Eastern Cultures
Mississippi State University, Starkville, MS

WILEY Blackwell

Copyright © 2014 by John Wiley & Sons, Inc. All rights reserved

Published by John Wiley & Sons, Inc., Hoboken, New Jersey

Published simultaneously in Canada

No part of this publication may be reproduced, stored in a retrieval system, or transmitted in any form or by any means, electronic, mechanical, photocopying, recording, scanning, or otherwise, except as permitted under Section 107 or 108 of the 1976 United States Copyright Act, without either the prior written permission of the Publisher, or authorization through payment of the appropriate per-copy fee to the Copyright Clearance Center, Inc., 222 Rosewood Drive, Danvers, MA 01923, (978) 750-8400, fax (978) 750-4470, or on the web at www.copyright.com. Requests to the Publisher for permission should be addressed to the Permissions Department, John Wiley & Sons, Inc., 111 River Street, Hoboken, NJ 07030, (201) 748-6011, fax (201) 748-6008, or online at http://www.wiley.com/go/permission.

Limit of Liability/Disclaimer of Warranty: While the publisher and author have used their best efforts in preparing this book, they make no representations or warranties with respect to the accuracy or completeness of the contents of this book and specifically disclaim any implied warranties of merchantability or fitness for a particular purpose. No warranty may be created or extended by sales representatives or written sales materials. The advice and strategies contained herein may not be suitable for your situation. You should consult with a professional where appropriate. Neither the publisher nor author shall be liable for any loss of profit or any other commercial damages, including but not limited to special, incidental, consequential, or other damages.

For general information on our other products and services or for technical support, please contact our Customer Care Department within the United States at (800) 762-2974, outside the United States at (317) 572-3993 or fax (317) 572-4002.

Wiley also publishes its books in a variety of electronic formats. Some content that appears in print may not be available in electronic formats. For more information about Wiley products, visit our web site at www.wiley.com.

Library of Congress Cataloging-in-Publication data has been applied for

ISBN 978-1-118-50420-8 (hardback)

Printed and bound in Malaysia by Vivar Printing Sdn Bhd

10 9 8 7 6 5 4 3 2 1

Contents

Contributors vii
Acknowledgments ix

1. Introduction: Interdisciplinary Approaches to the Second Epidemiologic Transition — 1
   Molly K. Zuckerman

Part 1 Causes of the Second Epidemiologic Transition

2. Infectious Disease in Philadelphia, 1690–1807: An Ecological Perspective — 17
   Gilda M. Anroman

3. Modeling the Second Epidemiologic Transition in London: Patterns of Mortality and Frailty during Industrialization — 35
   Sharon N. DeWitte

4. The Wider Background of the Second Transition in Europe: Information from Skeletal Material — 55
   Nikola Koepke

5. The Epidemiological Transition in Practice: Consumption, Phthisis, and TB in the 19th Century — 81
   Jeffrey K. Beemer

Part 2 Epidemic Infectious Disease and the Second Epidemiologic Transition

6. Agent-Based Modeling and the Second Epidemiologic Transition — 105
   Carolyn Orbann, Jessica Dimka, Erin Miller and Lisa Sattenspiel

7. Does Exposure to Influenza Very Early in Life Affect Mortality Risk during a Subsequent Outbreak? The 1890 and 1918 Pandemics in Canada — 123
   Stacey Hallman and Alain Gagnon

Part 3 Regional and Temporal Variation in the Second Epidemiologic Transition

8. The Second Epidemiologic Transition in Western Poland — 139
   Alicja Budnik

9 The Timing of the Second Epidemiologic Transition in Small US Towns
and Cities: Evidence from Local Cemeteries 163
Lisa Sattenspiel and Rebecca S. Lander

10 Industrialization and the Changing Mortality Environment in an English
Community during the Industrial Revolution 179
Peter M. Kitson

Part 4 Marginalized and Underrepresented Communities in the Second Epidemiologic Transition

11 Short Women and Their Stagnating Growth: A Study of Biological
Welfare and Inequality of Women in Postcolonial India 201
Aravinda Meera Guntupalli

12 Tracking the Second Epidemiologic Transition Using Bioarchaeological
Data on Infant Morbidity and Mortality 225
Megan A. Perry

13 The Biological Effects of Urbanization and In-Migration on
19th-Century-Born African Americans and Euro-Americans of Low
Socioeconomic Status: An Anthropological and Historical Approach 243
Carlina de la Cova

Part 5 The Environment and the Second Epidemiologic Transition

14 Reassessing the Good and Bad of Modern Environments: Developing
a More Comprehensive Approach to Health Trend Assessment 267
Lawrence M. Schell

15 Childhood Lead Exposure in the British Isles during the Industrial Revolution 279
*Andrew Millard, Janet Montgomery, Mark Trickett, Julia Beaumont,
Jane Evans, and Simon Chenery*

16 The Hygiene Hypothesis and the Second Epidemiologic Transition 301
Molly K. Zuckerman and George J. Armelagos

17 Comparative Parasitological Perspectives on Epidemiologic Transitions:
The Americas and Europe 321
Karl J. Reinhard and Elisa Pucu de Araújo

Part 6 Epilogue

18 The Second Epidemiologic Transition, Adaptation, and the Evolutionary Paradigm 339
George J. Armelagos

19 The Second Epidemiologic Transition from an Epidemiologist's Perspective 353
Nancy L. Fleischer and Robert E. McKeown

20 Methodological Perspectives on the Second Epidemiologic Transition:
Current and Future Research 369
Richard H. Steckel

21 The Current State of Knowledge on the Industrial Epidemiologic
Transition: Where Do We Go from Here? 377
Timothy B. Gage

Index 393

Contributors

Gilda M. Anroman
School of Pharmacy,
Notre Dame of Maryland University,
Baltimore, MD

George J. Armelagos
Department of Anthropology,
Emory University,
Atlanta, GA

Julia Beaumont
Division of Archaeological,
Geographical and
Environmental Sciences,
University of Bradford,
Bradford, UK

Jeffrey K. Beemer
Department of Sociology,
University of Massachusetts Amherst,
Amherst, MA

Alicja Budnik
Instytut Antropologii,
Uniwersytet Im. Adama Mickiewicza,
Poznań, Poland

Simon Chenery
British Geological Survey,
Keyworth, UK

Carlina de la Cova
Department of Anthropology,
University of South Carolina,
Columbia, SC

Sharon N. DeWitte
Department of Anthropology,
University of South Carolina,
Columbia, SC

Jessica Dimka
Department of Anthropology,
University of Missouri,
Columbia, MO

Jane Evans
NERC Isotope Geoscience Laboratory,
Keyworth, UK

Nancy L. Fleischer
Department of Epidemiology and
Biostatistics, Arnold School of Public
Health, University of South Carolina,
Columbia, SC

Timothy B. Gage
Department of Anthropology
and the Department of Epidemiology,
University of Albany,
State University of New York,
Albany, NY

Alain Gagnon
Département de Démographie,
Université de Montréal, Montreal,
Quebec, Canada

Aravinda Meera Guntupalli
Centre for Research on Ageing,
University of Southampton,
Southampton, UK

Stacey Hallman
Department of Sociology,
Western University, London,
Ontario, Canada

Peter M. Kitson
Cambridge Group for the History
of Population and Social Structure
and the Faculty of History,
University of Cambridge,
Cambridge, UK

Nikola Koepke
Departament d'Història i
Institucions Econòmiques,
Universitat de Barcelona,
Barcelona, Spain

Robert E. McKeown
Department of Epidemiology
and Biostatistics, Arnold
School of Public Health,
University of South Carolina,
Columbia, SC

Andrew Millard
Department of Archaeology,
Durham University,
Durham, UK

Erin Miller
Department of Anthropology,
University of Missouri,
Columbia, MO

Janet Montgomery
Department of Archaeology,
Durham University,
Durham, UK

Carolyn Orbann
Department of Anthropology,
University of Missouri,
Columbia, MO

Megan A. Perry
Department of Anthropology,
East Carolina University,
Greenville, NC

Elisa Pucu de Araújo
Manter Laboratory of Parasitology,
University of Nebraska at Lincoln,
Lincoln, NE

Karl J. Reinhard
School of Natural Resources,
University of Nebraska at Lincoln,
Lincoln, NE

Lisa Sattenspiel
Department of Anthropology,
University of Missouri,
Columbia, MO

Lawrence M. Schell
Department of Epidemiology and
Biostatistics and Department of
Anthropology, University of Albany,
State University of New York,
Albany, NY

Rebecca S. Lander
Department of Anthropology,
University of Missouri,
Columbia, MO

Richard H. Steckel
Department of Economics,
The Ohio State University,
Columbus, OH

Mark Trickett
Department of Archaeology,
Thomas Jefferson's Monticello,
Charlottesville, VA

Molly K. Zuckerman
Department of Anthropology
and Middle Eastern Cultures,
Mississippi State University,
Starkville, MS

Acknowledgments

This volume and the conference that inspired it would have been impossible without the help of several individuals and institutions. The volume represents the contributions of scholars who attended the third annual Postdoctoral Fellow's Conference, "Moving the Middle to the Foreground: Re-Visiting the Second Epidemiological Transition," which was held on April 18 and 19, 2011, at the University of South Carolina (USC). Thanks go to the South Carolina Institute for Archaeology and Anthropology (SCIAA) for sponsoring the conference and the postdoctoral fellowship program and to the Departments of History, Anthropology, and Epidemiology and Biostatistics and the College of Arts and Sciences at the University of South Carolina for supporting the conference. Thanks also to Charlie Cobb and to Nena Rice for their invaluable assistance in orchestrating the conference. Thanks also go to the Historic Columbia Foundation, the Inn at USC, and private donations from Ernest Helms III, MD, and from Francis and Mary Neuffer, MDs, for supporting the conference. Support was provided during the preparation of this volume by SCIAA's postdoctoral fellowship program and by Mississippi State University. Thanks are due to all of the authors for their contributions to the conference and to the volume, and for their invaluable internal reviewing of each other's works.

They must also be thanked for their patience in awaiting this text's appearance in its final, published form. Thanks are also in order to the anonymous reviewers of the manuscript in its early stages at Wiley-Blackwell for their comments. Jayson Zoino provided assistance with the references included in the volume, and the editorial and production staff at Wiley-Blackwell provided excellent guidance in bringing this volume to press.

Lastly, thank you to Casey Bouskill and Leslie Jo Weaver Meek for their thoughts about the utility of epidemiologic transition theory in the modern era, to George Armelagos, my PhD advisor, for getting me interested in epidemiologic transitions, and to Derek Anderson for patiently listening to my many musings about this subject over the past few years.

Chapter 1

Introduction: Interdisciplinary Approaches to the Second Epidemiologic Transition

Molly K. Zuckerman
Department of Anthropology and Middle Eastern Cultures, Mississippi State University, Starkville, MS

Introduction

The concept of an epidemiologic transition has served as a guiding principle for much of the discussion about economic growth and human health over the past several decades (Barrett et al., 1998). It constitutes a trend wherein a high burden of mortality from infectious disease, largely acute, epidemic *childhood* diseases, like smallpox or measles, is replaced by one of chronic and non-communicable diseases (NCDs), such as cardiovascular disease, cancer, and diabetes. The *classic* model was originally formulated to capture the changes in cause-specific mortality that followed the industrial revolution in the United States and Western Europe, but, in a modified form, the transition is ongoing in many developing low- and middle-income countries (LMICs). Recent scholarship has also placed the classic transition within an expanded evolutionary framework, recognizing a *first* transition coincident with the Neolithic and the intensification of agriculture, and a *third* of emerging and reemerging infectious diseases in the modern era (Barrett et al., 1998). This positions the classic transition as the *second epidemiologic transition*, as it is known in this volume.

Epidemiologic transition theory provides a model for understanding the relationships between economic, demographic, ecological, and social factors and the evolution and spread of disease (Armelagos and Barnes, 1999). As such, it has become a paradigmatic theoretical framework in public health (Caldwell, 2001; Szreter, 2002; Girard, 2005; Huicho et al., 2009), demography (McKeown and Record, 1962; Kirk, 1996; Salomon and Murray, 2002), biological and medical anthropology (Armelagos et al., 2005; Munoz-Tuduri et al., 2006), economics (Morand, 2004), and, to a much lesser extent, epidemiology (Harper and Armelagos, 2010; Fleischer and McKeown, this volume). The transition and the economic and social processes that drove it constitute one of the greatest social and environmental transformations in history (Armelagos et al., 2005). Industrialization and urbanization ushered in the chronic conditions and NCDs that not only plague high-income,

---

Modern Environments and Human Health: Revisiting the Second Epidemiologic Transition,
First Edition. Edited by Molly K. Zuckerman.
© 2014 John Wiley & Sons, Inc. Published 2014 by John Wiley & Sons, Inc.

developed countries but also increasingly afflict LMICs, which still also carry a high burden of mortality and morbidity from infectious conditions (Marinho et al., 2013).

Because of this, understanding the causes of the transition, particularly the decline in infectious disease, its scope, and how it played out differently across different regions, communities, and time periods is critical to many concerns. These issues are fundamental to planning health services and to public health and economic policy in both high-income nations and LMICs (Caldwell, 2001a). When the health consequences of industrialization specifically are added to the model, they are also significant for modern governmental policy on central control versus free markets (Lindert, 1994). They are important for continuing debates about the allocation of funds for public health and the role of individual responsibility, broad-based efforts to redistribute social, political, and economic resources, or targeted, well-funded interventions in raising living standards and promoting good health, especially in an era of limited funds. Additionally, they are germane to the discussions about the relationship between nutrition, food supply, economic development, and population growth that have engaged the natural and social sciences since Malthus (Colgrove, 2002). Most broadly, they are critical to understanding how humans interact with their environments and the impact of environmental change on human health and adaptability. In the face of climate change, ecological instability, and a world in which more than 50% of humans live in urban environments (United Nations, 2012), it is increasingly important to understand the dynamics between human health and modern environments.

Despite its significance, controversy surrounds the second transition. As contributions in this volume discuss, fundamental questions persist about its causes, especially declining infectious disease mortality (Gage, 2005). Scholars have critiqued the model's focus on mortality and neglect of morbidity and more holistic components of health (Johansson, 1992; Riley, 1992; Riley and Alter, 1989). They have also highlighted the nearly exclusive use of national-level data from the United States and Western Europe in studies of the transition, which obscures potential variation across regions, populations, and over time as well as by factors such as age, sex, race, or ethnicity (Barrett et al., 1998; McKeown, 2009; Fleishcher and McKeown, this volume).

Most of the controversy arises from the fundamental limitations of the material traditionally used to study total and cause-specific mortality (Gage, 2005). Total mortality is reconstructed from census and vital statistics records. However, vital registration systems only started in the 18th century in Northern Europe and after 1800 in the US and Western Europe; data from other countries, especially LMICs, exists only for much later periods or not at all. Cause-specific mortality is even more poorly documented. Reporting only started in England and Wales in 1830, and little data from other countries predates this. Documentary data on morbidity is even more limited and largely unavailable for any country before 1960 (Vallin, 1991; Gage, 2005).

The contributions to this volume—and the scholarship it is intended to spur—aim to end this stalemate. Both in sum and independently, these studies propose and demonstrate a novel approach to the transition that brings in theoretical approaches, methods, evidence types, and data sets from multiple disciplines. Some are more traditional for this topic, such as demography and economic history. Others, such as archaeoparasitology, bioarchaeology, and geochemistry, are less so. Studies of historical trends in health tend to systematically use and uncritically privilege documentary evidence, producing a one-sided view of history that is mostly reflective of textual sources (Perry, 2007). However, a more balanced, dynamic interplay between sources can be achieved by equally weighting each data type, analyzing them separately, and allowing them to extend and inform each other. Consistencies, nonconformities, and ambiguities between them can be identified and used to generate new research questions (see Buikstra et al., 2000). Ultimately, such a strategy can generate more information than can separate consideration of the sources (Swedlund and Herring, 2003). This volume applies this approach to studies of the transition, both explicitly and implicitly. Several studies in this volume are explicitly interdisciplinary (e.g., Anroman; de la Cova; Koepke; all this volume). Others studies represent a single disciplinary approach, but grouped, approach a single issue from multiple perspectives (e.g., Orbann et al.; Hallman and Gagnon; both this volume).

The Volume

This volume is not comprehensive; it does not represent the full scope of research on the transition or all of the disciplinary approaches that can be brought to bear on it. It is not the final say. Instead, it demonstrates how interdisciplinary approaches can elucidate several specific neglected or poorly understood areas of the transition, such as the experiences of women, children, and ethnic minorities, and regional and temporal variation in the mode and pace of the transition. It also demonstrates how they can be used to investigate unexplored aspects of the transition, such as chronic disease from industrial toxicants during the industrial revolution, which are wholly inaccessible through traditional materials but can yield powerful insights for modern populations.

Each author was encouraged to be explicit about the interdisciplinary aspects of their research, their theoretical approach and methodology, and the rationales for their use. They were also encouraged to discuss the benefits, limitations, and interpretive issues posed by the type of data they employed and, overall, the advantages and disadvantages of their approach. Accordingly, these studies should foster novel, productive approaches to the transition and similar epidemiologic and demographic processes, encourage researchers to explore the unique vantage points of other disciplines for their research, and unite across these divisions.

The volume is divided into sections that reflect not the disciplinary traditions or types of evidence employed by different authors, but the larger issues that each addresses. The studies thus come into play with each other on a series of central uniting themes that are key to research on the second epidemiologic transition. Part 1 addresses standing conflicts and controversies about its causes, focusing on the decline in infectious disease. In Part 2, the authors address the 1918 Spanish influenza pandemic, the last great burst of epidemic crisis mortality, and its role in causing and contributing to the transition. Part 3 discusses largely unanswered questions about how the transition played out across different communities and population subgroups, such as women and ethnic minorities, and the interpretive issues involved for historically underrepresented groups. Part 4 attends to the poorly understood issues of regional and temporal variation in the transition and the implications of this variation for models of the second transition and epidemiologic transition theory. Lastly, in Part 5, the authors address an understudied aspect of the transition: environmental quality and the role of parasites and industrial pollutants in changing patterns of infectious and chronic disease. These sections are not exclusive; nearly all of the studies in Parts 3–5, for instance, shed some light into causes of the transition and, by presenting data from understudied sources, time periods, or geographical regions, into regional and temporal variation. Importantly, many of the studies also defy categorization into a single section by their overarching focus on defining and investigating health as something more than longevity and mortality.

The volume concludes with four epilogues written by senior scholars charged with evaluating where research on the transition currently stands and where it should proceed from here. Armelagos places research in the volume within an evolutionary context and ties it into human adaptability and environmental change. Fleischer and McKeown discuss the place of epidemiologic transition theory in epidemiology and the creation of a space for dialogue on health transitions between epidemiologists and scholars from other disciplines. Steckel sums the current state of affairs in terms of methodology and highlights important topics for future research and the evidence and data types required for those advances. Lastly, Gage provides an overview of current knowledge on the transition and recommends a framework for future research.

The Demographic and Epidemiologic Transitions

The epidemiologic transition models the causes of death that accompanied the demographic transition, the secular declines in mortality and fertility, and the consequent population growth that accompanied industrialization; several of the contributions to this volume address aspects of both transitions. The theory is a simplified, descriptive model encompassing multiple phases (Thompson, 1929). Stage I is largely preindustrial and is characterized by high mortality and fertility, variable but overall low life expectancy, and slow but stable population growth punctuated by bursts of crisis mortality. Stage II, beginning in the mid-19th century in Europe, featured declining normal mortality and continued high fertility. Stage III continues the mortality trend but features declining fertility, increased life expectancy, and more sustained population growth. Declining normal mortality began in the mid-19th century in Western and Northern Europe and the US and around 1920 in LMICs, like Chile. Crisis mortality continued into the 20th century, perhaps ending only with the 1918–1919 Spanish influenza. Mortality declines completed in the mid-20th century in high-income nations but continue in LMICs. Stage IV features low, still declining mortality, as observed in developed countries after WWII (Vallin, 1991). Mortality declines continue, especially among the elderly, but slowly, likely due to lower mortality among infants, children, and young adults (Gage 2005).

The epidemiologic transition incorporates consistent patterns in cause of death accompanying the demographic transition (Omran, 1977). The original formulation involves several stages. Stage I, the *Age of Pestilence and Famine*, is preindustrial and features high frequencies of epidemic infectious disease and crisis mortality. Stages II and III, the *Age of Receding Pandemics*, witness a shift from epidemic to endemic infectious disease and declining crisis mortality. In Stage IV, the *Age of Degenerative and Man-Made Diseases* [sic], degenerative diseases like diabetes and stroke replace infectious disease, ushered in by age-related degenerative processes and anthropogenic factors, namely the environmental hazards and nutritional and behavioral patterns associated with industrialization and urban living. This neat dichotomy is blurred by the chronic nature of some infectious diseases, such as tuberculosis, as well as increasing recognition of the role of infectious disease and inflammatory processes in many chronic conditions (e.g., HPV and cervical cancer). Rather than precluding use of transition theory, this issue highlights the importance of historical relationships between humans and pathogens for understanding the current patterns of human health (see Zuckerman and Armelagos, this volume).

As contributors discuss, this model also presents a simplification of reality; epidemiologic evidence clearly shows that the timing and pace of the transition varied between nations. As Fleischer and McKeown (this volume) discuss, Omran accounted for variants on the classic model, and several scholars have added additional stages. These variants suggest that socioeconomic and local conditions, such as the degree of urbanization, seem to greatly influence the extent, nature, and timing of the transition (Dobson, 1997; Woods, 2000), both now and historically. For instance, due to structural inequalities, rapid urbanization, and inadequate public health infrastructure in many LMICs, such as India (Guntupalli, this volume), populations seem to be undergoing concurrent second and third transitions, with more affluent segments of the population suffering from chronic and NCDs and poorer segments still carrying a high burden of infectious disease. Importantly, the industrial transition also is not yet complete in high-income nations; scholars do not yet know whether postindustrial populations will return to Stage I-type mortality and fertility regimes or not (Gage and DeWitte, 2009).

Some scholars have also proposed the idea of a "health transition" (Riley, 1989), which further elaborates on the demographic and epidemiologic transition models by incorporating the secular trends in morbid states that accompany the transitions. This model summarizes the finding that the prevalence—but not necessarily the incidence—of morbidity increases as the demographic transition progresses and longevity increases (Riley and Alter, 1989; Roos et al., 1993), with attendant impacts on quality of life (Johansson, 1992; Riley, 1992). Secular changes in morbidity are difficult to document due to complex relationships between the incidence and prevalence of morbid conditions, and because the accuracy of diagnoses of morbid states have improved over time amongst medical practitioners and examiners (see Gage, 2005). Relationships between morbidity and mortality are also difficult to reconstruct because there is not a straightforward, direct relationship between them (Gage, 1989; Crimmins et al., 1994; Usher, 2000); as Gage and DeWitte (2009: 650) note, this is because mortality is an "absorbing state": it removes individuals from a population. If individuals in poor health are more likely to die than those in better health, there will often be an inverse relationship between morbidity and mortality in a given population (Riley, 1992). Reconstructions of this dynamic also depend on the exact definition of "health" employed in a given study. These range from a focus on mortality, such as the definition of life expectancy or age-standardized death rates employed by Gage (2005), to a biomedical definition premised on the absence of disease, to the more holistic one currently favored in epidemiology and anthropology: a "complete physical, mental, and social well-being" (WHO, 1948). Studies of the health transition typically define health as the ability to perform "activities of daily life" and occasionally the capacity for more vigorous physical measures, but as Gage and DeWitte (2009) note, these capacities are culturally relative. Furthermore, documentary data on such aspects of "health," as well as morbidity, is limited to the 20th century, meaning that traditional studies of health and the health transition are largely limited to the current era and western, highly documented populations. However, as several of the studies in this volume demonstrate, skeletal data, particularly frequencies of various pathological lesions and skeletal stress indicators, can be used to reconstruct morbidity in past populations, therefore providing a venue for evaluating health transitions. While interpretive biases must be taken into account (e.g., Wood et al., 1992), skeletal data can give direct insights into patterns of morbidity surrounding the second transition. For instance, Perry (this volume) supplements her analysis of patterns of infant mortality preceding the transition with skeletal data on multiple indicators of quality of life and health, such as frequencies of metabolic deficiencies and general infection. DeWitte (this volume) employ frequencies of various skeletal lesions to reconstruct levels of morbidity prior to the epidemiologic transition in London, and assess relationships between morbidity and mortality.

Part 1: Causes of the Second Epidemiologic Transition

Despite decades of research, the cause of the transition, specifically the decline in infectious disease mortality, remains contentious. The most influential proposal came from McKeown (1976), who proposed that a rising standard of living, namely, improved per capita nutritional consumption from improved economic conditions, bolstered resistance to infectious disease. In other words, rather than medical advances or public health interventions, the invisible hand of economic forces precipitated the transformation of human demographic and epidemiologic patterns (Colgrove, 2002; see Kitson, this volume). The McKeown thesis has ongoing resonance due to its public health and economic policy implications (Bynum, 2008) but it has ultimately been discredited; scholars now understand that public measures played a major role in reducing infectious disease mortality (Szreter, 1988), with medical advances such as vaccination, the germ theory of disease, general education, and improved hygiene exerting a delayed, supplemental effect (Cutler and Miller, 2005; see Reinhard and Pucu de Araújo, this volume). However, these insights have not resolved the controversy; a convincing and more comprehensive explanation of the decline remains unavailable (Gage, 2005).

This volume does not attempt such an explanation, but studies within demonstrate that an interdisciplinary approach can reduce ambiguity and isolate causal factors in the decline. They also reveal that causes of the decline were in many cases condition-specific rather than universal, with clear implications for public health and economic policy. For instance, due to the temporal limitations of documentary evidence, *why* and *when* the transition began in a particular context is often unclear. As DeWitte (this volume) emphasizes, this is a substantial issue. The reasons underlying the transition—why some portions of the population begin to succumb to some causes but not others, while others are able to survive to older ages—are ultimately of the greatest interest to researchers. With an eye on specifics rather than universalities, this can be accomplished by looking beyond traditional historical methodologies and available documentary materials. For instance, Anroman (this volume) brings together methods, paradigms, and evidence types from multiple disciplines to knit together an understanding of how individual's relationships and interactions *with* (rather than *within*) their total environment—their physical, biological, psychological, cultural, political, and socioeconomic universe—precipitated patterns of declining crisis mortality in 17th- to early 19th-century Philadelphia. Other contributors employ skeletal evidence, particularly taking advantage of the insights that large, aggregated databases of skeletal data can grant into population-level processes. While skeletal material is subject to a raft of theoretical and material interpretive issues (e.g., Wood et al., 1992), it provides direct evidence of the biological experiences of past populations and exists for most regions and time periods (see Steckel, this volume). Perplexingly though, it has scarcely been applied to studies of the demographic or second epidemiologic transition. In this volume, DeWitte employs skeletal data from the Wellcome Osteological Research Database,[1] which includes thousands of skeletons from London from the Roman era into the mid-19th century, to model relationships between age, sex, experiences of physiological stress (e.g., malnutrition), and patterns of frailty and mortality from the 16th century into the mid-19th and assess how these patterns may have caused the transition in London. Importantly, by analyzing skeletal data from such a well-documented urban center, DeWitte also generates paleodemographic signatures that can be used as baseline expectations for studies of the transitions in less well-documented samples. Koepke applies econometric history methods to data on stature, a sensitive measure of environmental quality and net nutrition, from nearly 20,000 skeletons from across Europe, dating from the 8th century BC to the 18th century AD, to reconstruct the biological standard of living and therefore the conditions that preceded and precipitated the transition in Europe. This data set allows detection of key regional differences that highlight why the transition began first in Western Europe.

Lastly, interdisciplinary approaches can also highlight and engage with issues that are fundamental to understanding the causes of the decline yet are inaccessible to or neglected

---

[1] http://www.museumoflondon.org.uk/Collections-Research/LAARC/Centre-for-Human-Bioarchaeology/Database/

by traditional methods and national-level data. Beemer (this volume) engages with an issue highlighted by Gage (2005) and others: that much of the controversy is attributable to the cause-of-death data and the ways in which diseases and causes of death are identified, grouped, disaggregated, and interpreted. Rather than exploring this issue on the national level, Beemer investigates on-the-ground difficulties experienced by medical and public health officials in implementing vital registration systems, changes in the structure of cause-of-death reporting, and evolving conceptions of disease, in two industrializing communities in Massachusetts. This community-level approach demonstrates that researchers must consider not only the quality of cause-of-death data but also the complex, contingent social, historical, and political interplay between lay communities, the state, and the medical community that created these sources.

Part 2: Epidemic Infectious Disease, Chronic Disease, and the Epidemiologic Transition

Interdisciplinary approaches can also shed light on declines in chronic and NCD mortality. Many aspects of modern, westernized environments are hazardous to human health: industrial pollution, sedentism, poor nutrition, obesity, and cigarette smoking, among others. However, since the risk of chronic and NCD mortality has declined over the 19th and 20th centuries (Preston, 1976; Gage, 1993, 1994, 2005), one or more yet unconfirmed aspects of modern environments must buffer humans from these risks. These include 20th-century lifestyle changes and medical advances, both too delayed to contribute much; the infectious origin of many degenerative conditions; direct interactions with infectious diseases; and indirect interactions with infectious disease mortality (Gage, 2005). Studies in this volume address the latter two.

In the 20th century, pandemics, particularly of respiratory diseases, such as the 1890 and 1918 influenza pandemics, were associated with excess chronic disease mortality (Lancaster, 1990; Azambuja and Duncan, 2002). The stress of these events may have contributed to chronic disease mortality, which then declined as infectious mortality declined. If so, the mechanisms are unknown (Gage, 2005). Indirect interactions may take several forms, including conditions wherein negative early life health experiences increase susceptibility to chronic and NCDs later in life (e.g., the Developmental Origins of Health and Disease Hypothesis (DOHaD)). Hallman and Gagnon (this volume) investigate the role of both direct and indirect effects on differential, age-dependent mortality in the 1918 pandemic using a biodemographic framework that draws from immunology, epidemiology, and human development. Their finding that 1918 mortality was a direct product of exposure to previous influenza pandemics not only potentially explains the dynamics of the 1918 pandemic, the last burst of crisis mortality, but also highlights the need to bring current scholarship on the DOHaD and fetal origins models into scholarship on the transition. Such models may be critical for understanding the segments of the population spared by epidemic infectious disease and therefore those who went on to experience continued declines in chronic and NCD mortality. Lastly, Orbann and colleagues (this volume) propose the use of agent-based modeling for studying epidemics in the context of the second transition and present a model designed for studying the dynamics of the 1918 pandemic in a small community. Agent-based modeling is widely employed in many different disciplines and is particularly useful for research questions that make use of incomplete and imperfect data but had yet to be applied to the second transition. The model enables analysis of how behavior patterns, particularly those related to social identity, affected disease spread. Future versions will be useful for addressing the interplay between overall health status (e.g., chronic disease presence) and modeling and potentially resolving many aspects of the transition: the effects of medical interventions and community-level health improvements on population structure and the relative importance of different factors in precipitating the transition.

Part 3: Regional and Temporal Variation in the Second Epidemiologic Transition

Researchers have roundly critiqued studies of the second transition and epidemiologic transition theory for their bias towards national-level data, Northern and Western European countries, and of large cities within these nations. This has led to questions about how well these models capture and explain global health trends and patterns of mortality and morbidity within different demographic units. It also leaves regional and temporal variation in the early phases of the second transition as a major source of debate (McKeown, 2009).

Several of the studies in this volume are the first to explore the epidemiologic transition—or aspects of the transition—within their respective areas or time periods. For instance, very little is known about the transition in Eastern Europe largely because of material issues. Therefore, Budnik (this volume) casts a wide net over a variety of evidence types to test the effects of local ecologies on mortality regimes in Poland, a late industrializer, and whether different regions experienced a transition similar to other parts of Europe or followed a unique model. Importantly, Budnik also uses various indices to test how environmental change associated with urbanization and industrialization affected forces of natural selection in Polish populations, demonstrating an avenue for assessing the successfulness of different adaptive strategies humans employed during the second transition.

Studies here also demonstrate routes for circumventing the scarcity of subnational-level data and exploring regional- and community-level heterogeneity in the transition. Kitson (this volume), for instance, employs family reconstitution, which is infrequently applied to urban, industrializing communities, to address a relatively neglected question: how, why, and when the transition played out outside of large urban centers. Kitson highlights the fact that the models' core assumptions about preindustrial society—that communities were small and wholly agrarian, or rooted in the manufacture and sale of agricultural products—are misleading in the face of evidence that preindustrial England instead featured numerous organizational paradigms, distributed differently across communities in time and space, with attendant heterogeneous mortality and fertility regimes. This accounts for one of the less commonly mentioned critiques of Omran's model: despite being created, in large part, to model the mortality patterns of 18th- to 19th-century England, it fails to map onto and thus explain many of these patterns in England. Kitson's unpacking of mortality patterns in relation to changing socioeconomic structures in a small English community draws attention to the fallacy of looking for a universal transition in epidemiologic regimes during the 18th and 19th centuries. Sattenspiel and Shattuck (this volume) also tackle problems involved in studying the demographic and epidemiologic transitions outside of large urban centers; smaller communities experienced major lags not only in sanitary improvements and therefore the transition but also in the initiation of vital registration systems. They propose a novel method for circumventing this evidentiary issue: headstone data, which are reliably available for many communities and time periods. Importantly, as this represents a highly accessible and valuable resource, the authors also discuss larger issues of whether and how local cemeteries can be used to reconstruct the transition in other regions and time periods.

These and other studies enrich our knowledge of variation in the transition and the relative importance of different factors in causing it in particular times and places. They also suggest a way forward for refining and expanding current models to more accurately reflect epidemiologic and demographic change both regionally and globally.

Part 4: Underrepresented Communities in the Second Epidemiologic Transition

Studies of the transition that rely on national-level data also mask variation in relation to sex, gender, race, class, and other aspects of social identity (Gaylin and Kates, 1997). However, even the original formulation posited that different experiences of mortality and longevity within populations, such as among children and reproductive-age women, would be key drivers of the transition and that different epidemiologic regimes would be found among whites and blacks (Omran, 1971; see Fleischer and McKeown, this volume). As women, children, and nonwhite ethnic groups are consistently underrepresented in the historical record, their biological experiences of the epidemiologic and demographic transitions remain poorly understood. Several studies in this volume demonstrate ways to circumvent these limitations, largely by embedding direct, biological evidence of health into textual evidence.

Several studies use skeletal evidence to do so. Importantly, skeletal remains can provide empirical evidence for the biological experiences of women and children that is largely free of the sex and age biases that affect documentary data (Grauer, 2003; Lewis, 2007). There is also a growing body of scholarship on reconstructing gender- and age-related identity in relation to health and disease (Hollimon, 2011; Lewis, 2007). However, few studies have used skeletal evidence to examine age-, sex-, and gender-based variation in experiences of the second transition. Here, DeWitte (this volume) explores sex-based patterns in mortality rates and frailty in preindustrial London. Koepke (this volume) uses stature data to test for evidence of gender discrimination in the net nutrition and biological standard of living in Europe. Guntupalli (this volume) also employs stature data, but from modern populations, specifically that of women experiencing the second transition in postcolonial India. Little is known about how the biological standard of living changed in India during the rapid population growth, shifting epidemiologic and demographic regimes, and slow economic growth of the postcolonial period. Guntupalli uses women's stature data to examine this process, but does not assume that gender is the most important variable in women's biological experiences. Instead, she takes an intersectional, multivariate approach, examining gender in relation to class, caste, religion, and other aspects of identity, and, by doing so, derives findings with substantial implications for economic policy.

Perry addresses children's experiences of the transition. As Lewis (2007) and others have emphasized, children come with a raft of interpretive issues that are very different from those of adults—and which are infrequently taken into consideration. Perry discusses these with an eye towards studies of the epidemiologic and demographic transition and an emphasis on how they can be accommodated and resolved. For instance, social definitions like "infant" carry cultural expectations, imply behavioral shifts, and therefore produce different mortality risks. Perry utilizes the Wellcome database to interrogate the validity of Omran's proposition that infant mortality and morbidity declined as part of the demographic and epidemiologic transitions. This study embodies the interdisciplinary approach favored in this volume; these skeletons represent the living populations upon whose vital statistics Omran's model is based, allowing an explicit, dynamic interrogation of these two lines of evidence.

While there are many ways that social inequality expressed during the transitions (see Schell, this volume), de la Cova (this volume) addresses the influence of race and class on populations undergoing the transition, specifically lower-class African-American and Euro-American communities. Using historical and well-documented skeletal evidence from anatomical collections, de la Cova investigates whether urbanization, industrialization, the Civil War, emancipation, reconstruction, and in-migration differentially affected the health of African-American and Euro-American males and whether traces of the second transition can indeed be found within these communities. This study demonstrates how scanty historical evidence can be integrated and interrogated against skeletal evidence to reconstruct the health experiences of otherwise invisible, highly underrepresented groups and how this evidence can be used to identify the intertwined effects of racial prejudice and tensions and poverty on experiences of the transition.

Part 5: The Environment and the Second Epidemiologic Transition

Studies of the historical relationship between environmental quality, industrial pollution, and the rise of chronic and NCDs can also greatly benefit from interdisciplinary approaches. Omran's model specified an increase in 'man-made' [sic] diseases, which anticipated the roles that pollution and other by-products of the industrial age currently play in the disease process (Caldwell, 2001a). As Schell (this volume) discusses, studies of modern populations suggest that exposure to industrial by-products, such as lead and mercury, has insidious effects on human health, particularly for lower-class, minority, and other marginalized communities, who tend to have higher rates of exposure. For instance, industrial water and air pollution has been linked to reduced life expectancy (Pope et al., 2009), allergies (Saxon and Diaz-Sanchez, 2005), and other conditions. Both Schell and Millard and colleagues (this volume) state that this dynamic likely existed during industrialization in the 19th and 20th centuries as well and that anthropogenic toxicant exposure should be evaluated as a defining characteristic of the second transition. However, few studies have assessed this issue.

One approach to this issue involves integrating geochemical evidence of toxicant exposure derived from human tissue with documentary evidence and data from environmental toxicology and human biology. Levels and sources of exposure as well as the biosocial consequences of exposure, such as impairment, can be reconstructed through skeletal and documentary data. This approach could, for instance, elucidate whether contemporary relationships between exposure, morbidity, mortality, and social inequality have time depth, linked to long-term processes of modernization, urbanization, and industrialization, or instead reflect purely current conditions. Here, Schell provides a historical framework for thinking about industrial toxicant exposure, highlighting morbidity issues among modern communities. Schell reminds us that social inequality—more so than direct proximity to polluted industrial areas—is one of the primary determinants of exposure in the present and likely in the past. Millard and colleagues present a case study demonstrating this interdisciplinary approach. They employ documentary data, trace element analysis, and lead isotopes to track the influence of rural versus urban environments, socioeconomic status, and gender on anthropogenic lead exposure in skeletons from 17th- to 18th-century London and investigate the source of lead exposure and evidence for chronic disease morbidity from toxicity. In addition to demonstrating the value—and limitations—of such an approach, this study and its findings complicate our understanding of which segments of urban, industrial populations suffered morbidity and mortality from exposure to industrial pollution.

In addition to chronic and NCDs like cardiovascular disease, diabetes, and stroke typically recognized in the second epidemiologic transition, some studies have also shown an increase in chronic inflammatory conditions such as allergic and autoimmune diseases in high-income nations since the mid-20th century (Aberg et al., 2005). As Zuckerman and Armelagos (this volume) highlight, many researchers have interpreted this through the hygiene hypothesis, which posits that sanitary improvements and the consequent lack of childhood exposure to parasitic and symbiotic microorganisms have deprived our immune systems of key modulatory stimuli and increased our inflammatory responses.

This hypothesis, embedded in evolutionary medicine, grants a historical trajectory to current diseases and therefore opens up the issue to investigation using archaeological and historical evidence. To encourage this, Zuckerman and Armelagos provide a set of best practice guidelines for conducting interdisciplinary research on levels of hygiene, sanitation, and exposure to symbiotic microorganisms and helminthic parasites in past and recent populations. The goal is to open dialogue between archaeologists, archaeoparasitologists, immunologists, and epidemiologists on this subject.

Reinhard and Pucu de Araújo (this volume) take up part of this challenge with an archaeoparasitological analysis of the first and second transitions in Europe and the Americas. This study fuses data and methods from parasitology with documentary and archaeological data to reconstruct household hygiene, sanitation levels, and changes in—and the effectiveness of—sanitation systems in various environments before and during the transitions. Archaeoparasitological evidence had yet to be applied to studies of the second epidemiologic transition but has the potential to not only provide direct evidence for our evolutionary relationships with parasites but also indirect insights into sanitation systems, water quality, and hygiene in areas and time periods backed by incomplete or absent documentary evidence.

Conclusion

This volume demonstrates how theoretical approaches, methods, evidence types, and data sets from diverse disciplines can be critically integrated to generate novel insights into the dynamics of the second epidemiologic and demographic transitions, particularly into issues that remain inscrutable in vital records, cause-of-death data, and census records. Health is most insightfully approached as a holistic phenomenon, with diverse psychological, social, cultural, biological, and ecological components, cross-cultural variation, and multiple levels of causation, and this can be best captured through diverse perspectives. Hopefully, other researchers will build upon and ultimately move past these studies to not only resolve lingering questions about the transitions but also generate new ones and devote this information to better understanding and predicting patterns of health and disease in contemporary populations.

References

Armelagos GJ, Barnes K. 1999. The evolution of human disease and the rise of allergy: epidemiological transitions. Med Anthropol 18:187–213.
Armelagos G, Brown P, Turner B. 2005. Evolutionary, historical and political economic perspectives on health and disease. Soc Sci Med 61:755–765.
Azambuja M, Duncan B. 2002. Similarities in mortality patterns from influenza in the first half of the 20th century and the rise and fall of ischemic heart disease in the United States: a new hypothesis concerning the coronary heart disease epidemic. Cad Saude Publica 18:557–577.
Barrett R, Kuzawa CW, McDade T, Armelagos GJ. 1998. Emerging and re-emerging infectious diseases: the third epidemiologic transition. Ann Rev Anthropol 27:247–271.
Buikstra J, O'Gorman J, Sutton C, editors. 2000. Never anything so solemn: An archeological, biological and historical investigation of the 19th century Grafton cemetery. Kampsville: Center for American Archeology.
Caldwell JC. 2001. Population health in transition. Bull World Health Org 79(2):159–160.
Colgrove J. 2002. The McKeown thesis: a historical controversy and its enduring influence. Am J Public Health 92:725–729.

Crimmins E, Hayward M, Saito Y. 1994. Changing mortality and morbidity rates and the health status and life expectancy of the older population. Demography 31:159–175.

Cutler DM, Miller G. 2005. The role of public health improvements in health advances: the twentieth-century United States. Demography 42:1–22.

Dobson MJ. 1997. Contours of death and disease in early modern England. New York: Cambridge University Press.

Gage TB. 1989. Biomathematical approaches to the study of mortality. Yrbk Phys Anthropol 32:185–214.

Gage TB. 1993. The decline of mortality in England and Wales 1861 to 1964: decomposition by cause of death and component of mortality. Pop Stud 47:47–66.

Gage TB. 1994. Population variation in cause of death: level, gender, and period effects. Demography 31:271–296.

Gage TB. 2005. Are modern environments really bad for us?: revisiting the demographic and epidemiologic transitions. Yrbk Phys Anthropol 48:96–117.

Gage T, DeWitte S. 2009. What do we know about the agricultural demographic transition? Curr Anthropol 50:649–655.

Gaylin DS, Kates J. 1997. Refocusing the lens: epidemiologic transition theory, mortality differentials, and the AIDS pandemic. Soc Sci Med 44:609–621.

Girard D. 2005. The cost of epidemiological transition: a study of a decrease in pertussis vaccination coverage. Health Policy 74:287–303.

Grauer A. 2003. Where were the women? In: Herring D, and Swedlund A, editors. Human biologists in the archives: Demography, health, nutrition, and genetics in historical populations. Cambridge: Cambridge University Press. p 266–288.

Harper KN, Armelagos GJ. 2010. The changing disease-scape in the third epidemiological transition. Int J Environ Res Public Health 7:675–697.

Hollimon S. 2011. Sex and gender in bioarchaeological research. In: Agarwal S, and Glencross B, editors. Social bioarchaeology. New York: Wiley Blackwell. p 149–182.

Huicho L, Trelles M, Gonzales F, Mendoza W, Miranda J. 2009. Mortality profiles in a country facing epidemiologic transition: an analysis of registered data. BMC Public Health 9:47.

Johansson SR. 1992. Measuring the cultural inflation of morbidity during the decline in mortality. Health Transit Rev 2:78–89.

Kirk D. 1996. Demographic transition theory. Popul Stud 50:361–387.

Lancaster H. 1990. Expectation of life. New York: Springer-Verlag.

Lewis ME. 2007. The bioarchaeology of children: perspectives from biological and forensic anthropology. Cambridge: Cambridge University Press.

Lindert P. 1994. Unequal living standards. In: Floud R, and McCloskey D, editors. The economic history of Britain since 1700. Cambridge: Cambridge University Press. p 357–386.

Marinho F, Soliz P, Gawryszewski V, Gerger A. 2013. Epidemiological transition in the Americas: changes and inequalities. Lancet 381:S89.

McKeown TH. 1976. The modern rise of population. New York: Academic Press.

McKeown RE. 2009. The epidemiologic transition: changing patterns of mortality and population dynamics. Am J Lifestyle Med 3:19–26.

McKeown TH, Record RG. 1962. Reasons for the decline in mortality in England and Wales during the nineteenth century. Pop Stud 16:94–122.

Morand O. 2004. Economic growth, longevity and the epidemiological transition. Eur J Health Econ 5:166–174.

Munoz-Tuduri M, Garcia-Moro C, Walker P. 2006. Time series analysis of the epidemiological transition in Minorca, 1634–1997. Hum Biol 78:619–634.

Omran AR. 1971. The epidemiologic transition: a theory of the epidemiology of population change. Milbank Mem Fund Q 49:509–538.

Omran AR. 1977. Epidemiological transition in the U.S. Popul Bull 32:1–41.

Perry MA. 2007. Is bioarchaeology a handmaiden to history? Developing a historical bioarchaeology. J Anthropol Archaeol 26:486–515.

Pope C, Ezzati M, Dockery D. 2009. Fine-particulate air pollution and life expectancy in the United States. N Engl J Med 360:376–386.

Preston SH. 1976. Mortality patterns in national populations. New York: Academic Press.
Riley J. 1989. Long-term morbidity and mortality trends: Inverse health transitions. In: Caldwell J, Findley S, Caldwell P, et al., editors. What we know about the health transition: The cultural, social and behavioral determinants of health. Canberra: Australian National University.
Riley J. 1992. From a high mortality regime to a high morbidity regime: is culture everything in sickness? Health Transit Rev 2:71–78.
Riley J, Alter G. 1989. The epidemiologic transition and morbidity. Ann Demogr Hist 199–213.
Roos N, Havens B, Black C. 1993. Living longer but doing worse: assessing health status in elderly persons at two points in time in Manitoba, Canada, 1971 and 1983. Soc Sci Med 36:273–282.
Saxon A, Diaz-Sanchez D. 2005. Air pollution and allergy: you are what you breathe. Nature 6(3):223–226.
Swedlund A, Herring D. 2003. Human biologists in the archives: demography, health, nutrition, and genetics in historical populations. In: Herring D, Swedlund A, editors. Human biologists in the archives: Demography, health, nutrition, and genetics in historical populations. Cambridge: Cambridge University Press. p 1–10.
Szreter S. 1988. The importance of social intervention in Britain's mortality decline, c. 1850–1914: a reinterpretation of the role of public health. Soc Hist Med 1:1–37.
Szreter S. 2002. Rethinking McKeown: the relationship between public health and social change. Am J Public Health 92(5):722–725.
Thompson WS. 1929. Population. Am J Soc 34(6): 959–975.
United Nations. 2012. World urbanization prospects, the 2011 revision: Highlights. New York: Department of Economic and Social Affairs, Population Division.
Usher BM. 2000. A multistate model of health and mortality for paleodemography: Tirup cemetery. Ph.D. Dissertation. University Park: Pennsylvania State University.
Vallin J. 1991. Mortality in Europe from 1720 to 1914: long-term trends and changes in patterns by age and sex. In: Schofield R, Reher DS, Bideau A, editors. The decline of mortality in Europe. Oxford: Clarendon Press. p 38–67.
WHO. 1948. Preamble to the constitution of the world health organization (definition of health). International Health Conference. New York: World Health Organization. p 100.
Wood JW, Milner GR, Harpending HC, Weiss KM. 1992. The osteological paradox. Curr Anthropol 33:343–370.
Woods R. 2000. The demography of Victorian England and Wales. Cambridge: Cambridge University Press.

Part 1
Causes of the Second Epidemiologic Transition

Chapter 2
Infectious Disease in Philadelphia, 1690–1807: An Ecological Perspective

Gilda M. Anroman
School of Pharmacy, Notre Dame of Maryland University, Baltimore, MD

Introduction

The epidemiologic transition model broadly describes the changing relationships between humans and their diseases and provides a means for understanding the evolution and spread of emerging diseases (Harper and Armelagos, 2010). The dramatic decline in mortality from infectious diseases (often referred to as the second epidemiologic transition) that became evident in the later years of the 19th century is often attributed to advancements in medical science and public health. Although these advancements likely contributed to this decline, they were by no means the only factors. The fundamental role of cultural, social, economic, and political conditions in modulating the ecological opportunities for infectious diseases is a dominant theme running through the ancient narrative of the interplay between human ecology and the microbial world. As early as 1940, Burnet (1940) argued that interactions between human beings and infectious agents are so complex that they can only be understood in the context of their mutual relationship to the global ecosystem. It was Dubos (1959), however, who conceived of the entire process as taking place within a *total environment*. He explained that "the process of living involves the interplay and integration of two ecological systems" (1959: 10). These systems include the community of interdependent cells, body fluids, and tissue structures that make up an organism's internal environment and all the living and inanimate things with which it comes into contact. Although he did not define it explicitly, the total environment evidently encompasses one's physical, biological, psychological, cultural, political, socioeconomic, and historical universe. As living things ordinarily achieve an unsteady and temporary *ecological* equilibrium sufficient for survival, any change in the *constellation of circumstances* under which the equilibrium evolved can upset the balance. Under these circumstances, it is possible for disease to swamp the host defenses if the change is too sudden for adaptive mechanisms (Dubos, 1959).

    This case study uses an ecological perspective to understand the patterns of infectious disease in Philadelphia between the years 1690 and 1807. This study interprets the historical record with conceptual guidance from the health sciences to set Philadelphia's health crises in a historical framework that shows people interacting *with*, rather than acting *within*, their *total environment* (Dubos, 1959; De Bevoise, 1995). Using an interdisciplinary model to incorporate thinking from the biological sciences, anthropology, and history, an ecological approach was fashioned here to understand the Philadelphia experience. Typically, three general forces can affect the burden of infectious diseases in a given population: change in abundance, virulence, or transmissibility of microbes; an increase in probability of exposure of individuals to microorganisms; and an increase in the vulnerability of hosts to infection and to the consequences of infection. A wide range of biological, behavioral, cultural, and social factors can influence one or more of these forces. Many are interrelated, and multiple synergies exist (Wilson, 1995). Consequently, an ecological approach allows us to see human health as an outcome of multiple, reciprocal, and continuing interactions between pathogens, hosts, and their pervasive environment.

The period between 1690 and 1807 was selected for this study for two reasons. First, it was a time during which the population of Philadelphia underwent a significant transition with respect to its disease environment. This period was characterized by events that considerably changed the risk factors for disease through alterations in population levels, pathogens, and human behavior. It was also characterized by intense change in the physical environment of the city. Short-term transitions are often distinguished by the introduction of new pathogens, new therapies, environmental modifications, or demographic changes brought about by contact with outside groups (Swedlund and Armelagos, 1990). Philadelphia experienced not one but many of these transition factors, and the extraordinarily high levels of morbidity and mortality in the city reflected this. As a result, this period in Philadelphia's history provides an excellent opportunity for scholars to study disease *emergence* as a dynamic feature in the interrelationships between people and their sociocultural and ecological environments. Second, the data for this type of study was readily available, thanks to the meticulous compilation and reconstruction of Philadelphia's vital rates by Klepp (1991), supplemented by church registers and other records.

Background

Historical inquiry can bring a valuable perspective to the understanding of disease emergence by focusing on "the consequences of human actions and the conditions that permit certain developments" (Morse, 1992: 38). The emergence and spread of microbial threats in 18th-century Philadelphia were driven by a complex set of factors, the convergence of which led to outcomes of disease much greater than any single factor might have suggested. Genetic and biological factors, for example, allow microbes to change and can make people more or less susceptible to infections. In addition, changes to the physical environment can impact the ecology of vectors and animal reservoirs, the transmissibility of microbes, and the activities of humans that expose them to certain threats (Smolinski et al., 2003). Human behavior, however, both individual and collective, is (and was) perhaps the most significant of all. The 18th century was a time of dynamic growth and change for Philadelphia. The size and mobility of the city's population increased the potential for pathogens to escape their prior geographic boundaries. High levels of immigration, coupled with high population densities, increased both interpersonal contact and contact between people and animals. Domesticated animals such as goats, sheep, cattle, pigs, and fowl provided novel reservoirs (Last, 2001) for zoonoses. Endemic diseases such as dysentery, malaria, and tuberculosis severely weakened their victims and increased their susceptibility to other infections. And industries such as tanning, sugar refining, and milling altered the physical environment and caused ecological disruptions. Infectious disease is a moving target, and as the climate and other sources of natural or anthropogenic change occur in a community, any disease that has an environmentally sensitive stage, reservoir, or vector will be affected (Wilcox and Colwell, 2005). It is quite possible that the sudden and abrupt nature of these changes was sufficient to disturb the delicate ecological equilibrium that existed in the city during the early years of settlement and paved the way for disease to take hold.

Disease patterns in the city were not constant, but changed in response to variations in both people and place. Seen from this perspective, disease in Philadelphia was less an expression of abnormality and pathology and more a mirror of the precarious balance that existed between human beings as biological organisms and the physical world they inhabited. Smallpox, for example, was the greatest killer of Philadelphians during the third quarter of the 18th century, accounting for most of the annual fluctuations in the death rate before the Revolutionary War (Smith, 1990). The decreasing virulence of smallpox after the 1770s most likely resulted from the disease becoming endemic in the city; density-dependent diseases like smallpox become endemic with increased population, meaning that they are common in childhood and therefore less lethal when they infect adolescents and young adults. Popular beliefs notwithstanding, changes in patterns of smallpox (as well as other diseases) at any given time may have had less to do with medical practice, or even conscious decisions by public health authorities, than on environmental factors. As the environment changed, the incidence of specific diseases changed, often radically (Grob, 1983, 2002).

The physical environment of Philadelphia after the first few decades of the 18th century set relatively high limits on the probability of contact with infection from outside reservoirs. The city was then situated entirely between the Delaware and Schuylkill Rivers, with little settlement beyond 7th Street (seven blocks from the shore of the Delaware River). The proximity of these rivers, together with an abundance of other natural resources, ensured that the region would be as much a magnet for European colonists as it had been for prehistoric people. William Penn founded the city in 1682, and, upon his arrival, there were already 10 houses and a tavern on the site and small settlements of Dutch and Swedish farmers nearby. With the arrival of Penn and the Quakers, the area rapidly became a major settlement. The first to follow Penn to Philadelphia were the British. Other groups included the Welsh, Scots, Scots-Irish, and enslaved people from the West Indies and Africa. Germans began arriving in large numbers starting in the 1730s and continuing until the beginning of the Revolutionary War, while the Irish came in two waves that peaked just before and after the War. Other ethnic groups were attracted to the city as well, and it became the political, administrative, and economic center of Pennsylvania. It also dominated the region of southern New Jersey and Delaware. By the middle of the 18th century, it was the largest city in British America. Additionally, Philadelphia hosted the Continental Congress, the Constitutional Convention, and was for 10 years the capital of the United States (Klepp, 1989a). As a result of its cosmopolitan character, the city housed many transients and visitors—people who could easily "pick up, process, carry and drop off" a wide variety of potentially infectious agents (Wilson, 2003).

Smallpox, Measles, and Yellow Fever: The Consequences of Dense Urban Living

Smallpox, measles, and yellow fever were serious threats to public health in 18th-century Philadelphia. Historians have focused much attention on both the vulnerability of Philadelphia's poor during epidemics of disease and the fundamental role of poverty in the emergence and spread of infection in the city (Smith, 1977, 1990; Klepp, 1989a, 1989b, 1994). Using poverty as a primary analytical tool, however, camouflages other relevant issues and obscures a more accurate understanding of how the convergence of any number of factors can create an environment in which infectious diseases emerge and become rooted in society. International trade and commerce, migration, public health policy, ecological changes, unsanitary conditions, poor hygiene, microbial adaptation, and disruptions due to War all played a role in disease emergence, but the increase in size and density of the population may have had the most dramatic effect on acute infectious disease. Demographically significant infectious diseases are generally categorized as either acute or chronic. The former have short latency and infectious periods and a short illness followed by either transient or permanent immunity. The latter have slow recovery rates, have long periods of infectiousness, and do not result in permanent immunity (Cvjetanović, 1982; Kunitz, 1984). Spectacular in their appearance, acute infectious diseases like smallpox, measles, and yellow fever each required large populations in order to become endemic, since smaller populations did not have a large enough annual input of susceptible hosts, and the disease would subsequently die out (Kunitz, 1984).

Since diseases are expressed biologically in individuals, it is often tempting to assume that the causes of disease and the solutions to their control and prevention are also biological and reside at the individual level. Many diseases, however, are caused only by the interaction of individuals within and between populations, and most are profoundly influenced by such interactions. Consequently, the causes of disease are often social and occur at the population level (Bhopal, 2002). For example, a high proportion of immune persons in a sparsely populated community generally precludes epidemic infectious diseases, because the capacity to transmit the pathogen from person to person was impaired. An increase in the number of susceptible individuals in a community, however, will create conditions conducive to the spread of a disease (Klepp, 1989b).

Smallpox

Smallpox was one of the greatest killers of Philadelphians during the 18th century, accounting for much of the annual fluctuations in the death rate prior to the start of the Revolutionary War (Klepp, 1997). Smallpox is caused by the variola virus, and two subgenera of smallpox exist: *variola major* (i.e., classical smallpox) and *variola minor*. It is likely that all smallpox cases in Philadelphia during the 18th century were *variola major*. There are five clinical types of *variola major*, classified according to the nature and development of the rash: hemorrhagic, flat, modified, noneruptive, and ordinary (Fenner et al., 1988). Generally, *variola major* has a mortality rate of approximately 30% in modern populations. Data do not exist on the distribution of cases among these types in the Philadelphia epidemics, nor is there information on how that distribution changed during periods of massive immigration into the city, concurrent epidemics with other diseases, War, or other stresses inflicted on the population. Modern research suggests that the most severe cases of smallpox seem to correlate with an inadequate immune response (Fenner et al., 1988), but what suppresses the response is largely unclear, as is the effect of the various nonspecific mechanisms that form part of the host's defenses. What *is* known, however, is that age is certainly a factor, since immune responses are less effective in the very young and the very old, and case fatality rates are accordingly the highest at the extremes of the age spectrum. It is also clear that malnutrition interferes with host defenses, but despite a few scattered indications that smallpox and malnourishment may be synergistic, the latter connection has never been proved (Fenner et al., 1988). Little is also known of the effects of multiple concurrent infections, as well as the effects of the psychological stress associated with wartime conditions (Fenner et al., 1988; De Bevoise, 1995). Notwithstanding the lack of empirical evidence, it is reasonable to assume that either susceptibility to or severity of smallpox increased as a consequence of the multitude of 18th-century assaults on the host defenses of the people of Philadelphia. Such assaults included a high level of urban density, facilitating high levels of person-to-person contact and therefore virus transmission, a continuous flow of sick and susceptible individuals into the community, chronic disease and debility, and haphazard inoculation procedures that further intensified the emergence and spread of the disease throughout the city.

Smallpox is mostly transmitted through a respiratory route, courtesy of saliva droplets. Modern investigators in Asia have established that propagation of smallpox often occurs as a result of close family contacts, particularly between those who sleep in the same room or bed. Transmission between persons living in the same house but who did not share sleeping quarters was the next most frequent method (Christie, 1974; Manson-Bahr and Apted, 1982; Fenner et al., 1988; De Bevoise, 1995). Importantly, sharing a bed with a family member was common practice in 18th-century Philadelphia. Unfortunately, prevention of smallpox transmission and epidemic spread requires strict isolation of the patient, effective vaccination, surveillance of contacts, proper burial, and disinfection of the premises, bedding, laundry, and personal items. Although quarantine was attempted in 18th-century Philadelphia, and inoculation practiced, other attempts to contain the spread of the disease were inadequate. Consequently, conditions that prevailed there did much to enhance smallpox's spread. Congested streets, crowded houses, and a continuous flow of susceptible individuals into the city were only parts of the problem, however. Nearly every city resident eventually contracted the disease either naturally, through the respiratory route, or through inoculation. In inoculation, matter taken from a pustule of someone who had caught the disease naturally was placed in several small cuts made by a needle or a lancet in the arm or leg muscles of the person receiving the inoculation. This method produced a milder, less fatal form of smallpox (Gehlbach, 2005), but the person was still contagious and could pass it on to others through contact (Rothenberg and Chapman, 2000). In the absence of both strict quarantine and a clear understanding of contagion, inoculation could easily start an epidemic, especially because inoculated individuals with mild infections often felt well enough to circulate in public. Abigail Adams, for example, who had expressed her own fear of the contagion, "attended publick worship constantly, except one day and a half" while she underwent inoculation in 1776 (Butterfield and Kline, 2002).

Philadelphia quickly became a center of inoculation for all of the British colonies in North America, attracting patients from as far away as the West Indies. Fenn (2001) has argued that the high incidence of smallpox in Philadelphia stemmed, at least in part, from the great frequency of this procedure and from the reluctance of the authorities to regulate it. While officials in other colonies were cautious or opposed it, those in Philadelphia may have been a bit indulgent and irresponsible with their policy. Quarantines were rare, and restrictions on the procedure were virtually nonexistent. As a result, the practice flourished (Fenn, 2001) in the city—and so did the disease.

No single group of people was more likely to carry smallpox away from Philadelphia than soldiers in the Continental army. Because of its central location on the Atlantic seaboard, the city was a logical stopping point for newly enlisted southern recruits on their way north to join Washington's army. Problematically, southerners were also among those most likely to be vulnerable to *variola* in the colonies. The southern colonies had neither the population density nor the transportation networks needed to sustain the ongoing, endemic presence of the virus, and as a result, years could pass between epidemics, allowing the number of susceptible individuals to greatly increase. Importantly, soldiers not only endangered their own lives but also the lives of those around them. This is because of smallpox's long incubation period, which extends for 12–14 days after initial infection. Consequently, a soldier could become infected in Philadelphia and then march for nearly 2 weeks before developing symptoms. Military authorities were aware of this, and Continental Congress President John

Hancock ordered regiments marching from Virginia to New Jersey to go around "Phil. On Acc. Of the Small Pox" (Fenn, 2001). But the problem persisted, creating epidemiologic issues for the army throughout the War and exasperating Washington. "I would wish to have the small Pox entirely out of Philadelphia," he wrote, but acknowledged that it simply could not be done (Washington, 1777). Only if the troops were immune could they pass through the Quaker city without risk.

Few people had more opportunity to observe the course of public health in Philadelphia than Elizabeth Sandwith Drinker (1735–1807). As a member of the upper class, her life was not fully representative of the 18th century, as this segment of society represented only a small percentage of the population of Philadelphia. However, the extraordinary span and sustained quality of her diary make it a very useful document for a number of historical endeavors. This is particularly true of the wealth of information contained in the dairy about the ailments in Drinker's family and the remedies that she and local physicians employed, which make it a particularly fertile source of information on the disease history of the city, as well as the history of drugs and pharmaceuticals in 18th-century America (Crane, 1991). The diary is sampled throughout the remainder of this chapter.

While many who suffered from smallpox eventually recovered, Elizabeth Drinker was correct in noting the differences in mortality between those who contracted it naturally, through respiratory transmission, and those who were inoculated. In an October 24, 1759, excerpt from her diary, Drinker wrote, "went this morning to Thos. Says, whose daughter Becky, lays ill, in the Small Pox, which she has taken the natural way; and to most that take it Naturally (at this time) it proves mortal" (Crane, 1991). If given a choice, inoculation was a better route than contracting the disease naturally. Accurate statistics do not exist for Philadelphia, but in the Boston epidemic of 1752, almost 10% of those who caught the disease naturally died, and in the 1764 epidemic, the figure was close to 18%. The comparable mortality figures for those inoculated were 1.4% and 0.9% (Hopkins, 1983; Crane, 1991).

Smallpox also continuously changed and evolved as its environment changed and evolved. In Europe, for instance, by the late 17th century, smallpox was assuming a more virulent form, possibly because a more lethal strain of *variola major* had been imported from Africa or the Orient, or because the virus mutated (Grob, 2002). Regardless of its origin, a far more malignant one superseded the relatively benign form of the disease that existed before 1630. In some areas, such as London, the more virulent strain combined with increasing population and population density to greatly exacerbate mortality from the disease; in London, where smallpox became endemic during this period, the average number of deaths per year between 1731 and 1765 was 2080 or 9% of total mortality (Duffy, 1953; Carmichael and Silverstein, 1987; Dobson, 1997; Grob, 2002). The pattern of smallpox in the colonies was quite different, however, because no colonial town approached the size of metropolitan London. In the British mainland colonies, an epidemic was succeeded by a period of years in which the disease was absent. During the interval, the number of susceptible individuals gradually increased, and the stage was set for another epidemic. However, this interval was particularly short in Philadelphia, where a steady flow of immigrants, soldiers, visitors, and delegates constantly refreshed the pool of susceptible individuals, meaning that Philadelphia suffered frequent epidemics.

The common treatment for smallpox was also quite dangerous and contributed to the general debility in Philadelphia. Mercury and mercury-containing compounds were widely used both to treat the disease and to prepare the body for inoculation (Wolman, 1974). Most of the preparations then in use—cinnabar, liquid mercury, mercurous chloride, and mercuric chloride—were quite toxic, and once in the body, they could accumulate in and damage the kidneys, brain, and other organ systems. Salivation and sore mouth, which some 18th-century physicians regarded as signs that the medications were active, are now known as early signs of mercury toxicity. Other early symptoms of poisoning include loss of appetite, loss of weight, nervous system irritability, and insomnia. More advanced exposure leads to damaged kidneys, muscle cramps, tremor, and changes in personality. This greatly exacerbated the morbidity associated with smallpox; Philadelphia physician Benjamin Rush noted that when used to treat smallpox or prepare for inoculation, mercury produced glandular swelling, loss of teeth, and, after infection had subsided, "weak habit of the body" (Wolman, 1974).

Smallpox became less virulent in Philadelphia in the latter portion of the 18th century. No one cause for this can be singled out from the changes in attitudes and practices that existed at the time. Public health measures, personal cleanliness, stricter enforcement of quarantine laws, and inoculation each may have played a role. More importantly, fewer migrants disembarking at the city's docks during the post-Revolutionary War period also resulted in less disease spread among the residents and, as a result, contributed fewer burials to the cemeteries (Smith, 1990). One of the most significant reasons for the decline in smallpox mortality in the city, however, was the reduction in the number of susceptible individuals. As more and more city residents developed immunity to the disease, the potential breeding ground for smallpox contracted steadily.

Measles

Measles is an acute viral infection, with characteristic symptoms including fever, spotted rash, and a cough. Measles is caused by a species of *Morbillivirus* and is a highly contagious disease with a case fatality rate of approximately 30% in modern populations. Like smallpox, it is primarily transmitted through a respiratory route through droplets, specifically from the nose, mouth, and throat secretions from infected persons. The incubation period from exposure to onset of the rash is approximately 14 days, and the disease is communicable for approximately 4 days after the start of the rash (Kim-Farley, 2003). Importantly, measles has no reservoir other than humans and needs a continuous chain of susceptible contacts in order to sustain transmission. It is generally regarded as a disease of childhood, if only because the loss of temporary immunity from the mother with weaning renders the infant susceptible at an early age. As with smallpox, epidemics often occur at intervals, when a sufficiently large pool of susceptibles accumulates. Although mortality rates varied, measles was not regarded as a particularly lethal disease in Europe during the 18th century (Grob, 2002). However, the disease's manifestations in Philadelphia and other colonial cities were quite different.

The first reference to measles in the Philadelphia records is in 1714, with subsequent epidemics throughout the 18th century. Overall, mortality was fairly high during the initial and subsequent epidemics, with some variation. The mortality rates in 1747, 1759, and 1772 were particularly devastating and contributed greatly to the overall death rates for these years. For instance, the Christ Church Bills of Mortality list 23 deaths from measles in 1747, 22 deaths in 1759, and 17 deaths in 1772 (Klepp, 1991). In a letter to his mother in October 1749, Benjamin Franklin rejoiced that his family had remained well despite much illness in town. The "measles and the flux," he reported, had "carried off many children," while a number of adults also had fallen victim to the disease (Sparks, 1839–1847; Duffy, 1953). Measles was often accompanied by conditions identified as *pleurisy*, which was used to describe respiratory disorders, and *flux* referring to diarrhea. For instance, the notation of 23 deaths from measles in 1747 was followed by 39 from pleurisy in 1748 in Christ Church Parish (Klepp, 1991). Measles is often accompanied by complications from secondary infections involving pneumonia and diarrhea, the latter of which is one of the most important causes of measles-associated mortality in developing countries today (Kim-Farley, 2003).

However, it is difficult to determine whether these conditions were independent but concurrent or whether pleurisy and the flux were complications of measles. Regardless, they do suggest that serious complications accompanied measles epidemics in Philadelphia, no doubt contributing to the disease's elevated mortality rates.

As was the case for smallpox, Philadelphia's ever-growing population provided a constant pool of susceptible individuals, creating the necessary environment for measles epidemics to occur. In her analysis of Philadelphia's bills of mortality, Klepp (1989b, 1991) notes that two of the most severe epidemics occurred in 1759 and 1772. In 1759, the population of Philadelphia was just over 18,000 people. That year, approximately 113 deaths were attributed to measles. By 1772, the population had grown to approximately 29,000; 180 deaths were recorded for measles. Although the reasons for such high mortality rates are not entirely clear, it is reasonable to assume that measles, which struck adults as well as children, in concert with other common diseases such as dysentery, influenza, diphtheria, and scarlet fever, caused large numbers of deaths. During the 1759 epidemic, for example, there were concurrent epidemics of measles, smallpox, and typhus. For 1759, the Christ Church Bills of Mortality alone list 106 deaths from smallpox, 22 from measles, and 9 from *nervous fever*—thought to be *typhus or typhoid* (Klepp, 1991). Concurrent infections certainly compromised the hosts and may have significantly increased overall mortality rates.

Malnutrition and overcrowding have long been implicated in measles mortality rates. For instance, Newberne and Williams (1970) have pointed out that until vaccines were developed, most children around the world became infected with the measles virus at some point, but death rates were up to 300 times higher in developing, low-income countries than developed, high-income ones. Newberne and Williams propose that the virus was not more virulent in these settings, nor that there were fewer medical services. Instead, higher levels of malnutrition and undernutrition in these contexts likely decreased host resistance to the virus. However, recent findings in genetic and epidemiologic research on measles have added a great deal to our understanding of the disease. In an evaluation of measles mortality rates, Aaby and colleagues (1984) found that host factors such as malnutrition and age of infection were not sufficient elements in mortality. Instead, mortality rates were much higher when the disease was contracted from a member of the same household or from a relative than when it was contracted from others in the community. In Guinea-Bissau, for example, the mortality rate for isolated cases and index cases was 8%, as compared with 23% for secondary cases in the household. Aaby (1988, 1991) conducted a study in Senegal that took 1.0 as the mortality rate for cases contracted in the community and found the rate would be 1.9 if contracted from a cousin, 2.3 from a half-sibling, and 3.8 from a full sibling. Other data further demonstrated the absence of any relationship between the severity of infection and the dose of the virus to which the individual was exposed. On the basis of these findings, Black (1992) has suggested that high death rates in the American colonies from European infectious diseases may be because a viral strain grown in one host becomes adapted to the immune response of that individual, and therefore, when introduced to a genetically similar host, the strain gains in virulence and produces higher mortality. Genetic homogeneity, in other words, may have enhanced viral virulence as it passed through ingrown communities and households. This may, for instance, explain the great lethality of the epidemics of measles experienced by Cotton Mather and his family during a 1714 epidemic of measles in Boston; four of his children and one member of his household staff perished from the disease (Caulfield, 1943).

Elevated mortality rates may also have stemmed, in part, from the failure of measles to become endemic in the city and American colonies. In England and on the Continent, densely populated urban areas always included large numbers of susceptible people. Consequently, many infectious diseases assumed an endemic character and became less lethal. Epidemics in relatively small 18th-century colonial communities, by contrast, reduced the number of susceptible persons to the point where no further epidemic was possible. Over a period of time, the number of susceptible individuals would increase, and the stage would be set for another epidemic of the disease. This cycle was certainly evident in Philadelphia, as epidemics occurred in 1747, 1759, 1772, 1778, 1783, 1789, 1795, and 1796 (Rush, 1815; Crane, 1991).

Sporadic epidemics of measles were common throughout the colonies, but the New England and Middle Atlantic regions were the hardest hit. The Chesapeake and southern colonies—areas where population density was lower—were somewhat less affected. In the areas where the disease was more prevalent, there was a steady decline in the intervals between epidemics. The interval in Boston declined from 30 to 11 years between its initial appearance in 1657 and 1772 (Caufield, 1943). A similar pattern was evident in Philadelphia, and after 1795, there was not much regularity to measles epidemics. The reason for this decline is that the disease had become endemic in Philadelphia, with a small number of deaths reported in Christ Church Parish nearly every year. There were two deaths reported in 1795, nine in 1796, one in 1797, one in 1799, and four in 1800 (Klepp, 1991).

Yellow Fever

Smallpox and measles were by no means the only imported diseases to affect Philadelphia. Yellow fever was another and, in many respects, a more terrifying disease. While its mortality rate (5–10% in modern populations) is lower than that of smallpox and measles, it was a relatively *new* disease to Europeans and colonials and consequently aroused greater fears than older but more familiar diseases, especially because of its unpredictable character. Its symptoms were dramatic; its appearance and disappearance were seemingly random; and it respected neither class, status, nor gender. However, unlike smallpox and measles, yellow fever is not spread through person-to-person contact, but through mosquitoes, especially *Aedes aegypti* (formerly *Stegomyia fasciata*). Yellow fever is an acute viral hemorrhagic disease. It remains endemic in tropical Africa and the Americas in a sylvan or jungle form, but historically, its greatest epidemiologic and demographic impacts have occurred through epidemics in urban environments. It presents symptoms ranging from mild to malignant, classically including fever, headache, jaundice, and gastrointestinal hemorrhage (black vomit). High mortality rates were recorded during epidemics (20–70%), but this elevated rate suggests that many cases were mild and undiagnosed (Cooper and Kiple, 2003).

It is now generally accepted that the cradle of yellow fever was West Africa. This means that both virus and vector had to be imported to the Americas, and the many challenges facing the pair in making a transatlantic voyage likely account for its relatively late debut there. The vector's migration most likely occurred within the last 400 years, and it is reasonable to assume that *A. aegypti* arrived in the Caribbean (e.g., West Indies) with early trading and slaving voyages. A ship in tropical waters with many individuals in a confined space and water casks close at hand likely provided the migrating insects with a nearly ideal habitat. The virus likely also made several early voyages, although not as easily or initially as successfully as its vector. Presumably, the virus would have boarded a ship in the bodies of nonimmune sailors or slaves (most likely the very young) during the first 3 to 5 days of illness when infected hosts are experiencing viremia but are still generally asymptomatic. To successfully make the link between virus and vector and transmit both to the Americas, a female *A. aegypti* mosquito onboard would also have to have bitten one of these hosts during this period, for hosts are no longer infective after viremia ceases and the disease becomes symptomatic. The infected mosquitoes would then have had to survive the virus's roughly 9- to 18-day incubation period before they would be able to transmit it to susceptible human host. After the incubation period, a mosquito generally remains infectious for the remainder of their lifespan, which is approximately 1 or 2 months but can occasionally exceed 180 days (Cooper and Kiple, 2003).

Philadelphia was the center of government and the center for mercantile trade and shipbuilding, making it both the economic and political heart of the colonies. These factors combined to draw large numbers of visitors, immigrants, and migrants to the city, dramatically increasing the population of the city as the century progressed while also making the city ripe for imported diseases, such as yellow fever. Immigrants and migrants converged on Philadelphia from three directions in the 1790s (Smith, 1990). European immigrants, mostly Irish and Scots-Irish, arrived after a long and often grueling sea voyage, which left many sick and malnourished. This influx, and their subsequent vulnerability, is evident in their steadily growing interments in the city's Catholic cemeteries during this time, as recorded in the bills of mortality. Refugees escaping the Haitian Revolution also arrived in the city. Unlike their European counterparts, who were usually poor to lower status, West Indian refugees were often merchants, doctors, lawyers, and planters (Powell, 1949). Arriving in large numbers in the summer of 1793 from areas of the Caribbean where yellow fever was endemic, this group most likely served as a viral reservoir for local *A. aegypti* mosquitoes, although mosquitoes accompanying them may also bear the blame. Regardless, their arrival was associated with an epidemic of yellow fever in Philadelphia, which occurred in 1793. Since viremia may last a week or more in human yellow fever infections, humans can serve as a reservoir in the urban form of the disease (Ryan and Ray, 2004). The final group to converge on Philadelphia consisted of rural migrants in search of improved economic prospects, who bore the burden of being immunologically naïve to most infectious diseases, and prime candidates for yellow fever (Smith, 1990).

Yellow Fever: The Vector

It is likely that the growth of the sugar industry in both the West Indies and the colonies assisted *A. aegypti* in becoming acclimated to its new surroundings. Goodyear (1978) argues that an important factor in the nurturance and travel patterns of these and other yellow fever vector mosquitoes was the paraphernalia of sugar syrup refining and transport. In addition to blood meals, most mosquitoes are attracted to and nourished by sweet fluids. Adult mosquitoes of both sexes in most species regularly feed on plant sugar throughout life, though only females feed on vertebrate blood (Mullen and Durden, 2002). The boiling of cane juices, the reboiling of molasses, and the piling of waste stalks around plantations certainly would have made a great deal of sucrose available to mosquitoes on a year-round basis. In addition, other features of sugar production would have contributed to mosquito proliferation, especially the use of artificial containers such as clay pots, which, when discarded or simply left outdoors to be filled with rainwater, became ideal breeding vessels (Kiple and Higgins, 1992).

Philadelphia was a major distribution center for imported sugars and molasses, and it is likely not a coincidence that the city's several visitations of yellow fever were all attributed to contact with the West Indies. The epidemic of 1762, in particular, was traced to a West Indies ship docked at Sugar House Wharf below South Street (LaRoche, 1855). By the end of the 18th century, Philadelphia had become an important sugar-refining center, with four operating refineries responsible for the "bulk of sugar consumed in the United States before 1789" (Vogt, 1908). Accordingly, this industry demanded large importations of raw sugar from the Caribbean. By providing exceptional feeding opportunities, imported sugars may have helped to temporarily increase mosquito population in an area like Philadelphia, which was otherwise not ecologically amenable to year-round mosquito infestation, and therefore drive epidemics of yellow fever in the city.

The habits of female *A. aegypti* also likely contributed to shaping the characteristics of these epidemics. *A. aegypti* is a domestic mosquito, which prefers to live close to and feed off of humans, and has become well adapted to breeding in anthropogenic water sources, such as pots, tires, or water storage jars. Its range is short, at most a few hundred yards, meaning that it requires a fairly dense human population to feed and therefore spread disease. Because standing water is key for its survival and reproduction, the mosquito is also adapted to environments with high levels of rainfall, meaning that yellow fever also occurs mostly in such areas. Lastly, *A. aegypti* also requires warm weather; it does not bite in temperatures below 62°F, hibernates in extended chilly weather, and is mostly active during daylight hours (Cooper and Kiple, 2003; Service, 2004). Yellow fever therefore flourished only during Philadelphia's warm summer months but was otherwise well suited to the densely populated urban area, especially given the city's orientation to maritime trade. Near the waterfront, many people were crowded into a complex maze of dirty, dark alleys and densely packed buildings. During the 1793 yellow fever epidemic, Samuel Jackson described the neighborhood near the waterfront in the following manner: "Those infected were scattered over the City in every quarter; many in Water Street, in various narrow & uncleanly alleys, & in small crowded & ill-ventilated dwellings" (Chervin, 1821). Mosquitoes aboard vessels docking in Philadelphia also found a hospitable environment near the waterfront, in the marshes west of the city, and in the vicinity of Dock Creek, a stream that "barely oozed from the swamps through the city's heart to the Delaware River" (Klepp, 1995).

Yellow Fever: The Virus

It is evident that much was needed for the mosquito to survive and thrive in Philadelphia, but the yellow fever virus also had some distinctive requirements for transmission—a process in which humans are best thought of as the site where the virus changes mosquitoes. As detailed previously, this exchange can only take place during viremia, and infected mosquitoes can only transmit the virus to new, susceptible human hosts after their own incubation stage. These hosts can also act as a source of virus spread and transmission, since noninfected mosquitoes can feed on them and become infected, thereby increasing the frequency and scope of transmission (Ryan and Ray, 2004). The many factors associated with the pathogenesis of yellow fever highlight the complex relationships that existed in Philadelphia between pathogens, vectors, human hosts, and their environment. The virus, for example, must establish a cycle that allows indefinite transfer from mosquito to host to mosquito. This also requires a large number of mosquitoes. Without them, the virus cannot move from person to person rapidly enough; humans are infected for only 7–10 days and are in viremia for only 3–6 days. The cycle also requires a favorable ratio of susceptible to infected hosts available for mosquitoes to feed upon as these vectors have fairly short lifespans, necessitating a continuous, critical mass of both hosts and vectors; the cycle breaks only when mosquitoes transmit the virus only into immune, nonsusceptible hosts (McNeill, 1999). Philadelphia's densely populated city streets, along with standing water and a constant influx of susceptible individuals and disease-carrying mosquitoes, accommodated this cycle very well throughout the 18th century.

Yellow Fever: The Disease

Yellow fever epidemics arose where certain conditions prevailed: the presence of the virus, the insect vector, and a sufficiently large number of both infected and susceptible hosts. As extensive commercial links were established between Philadelphia and regions of the world where yellow fever was endemic, and the susceptibility of Philadelphia's population increased due to fertility from residents but much more so constant waves of immigration and migration, the possibility of an epidemic was always present. The disease had a long history in Philadelphia, with the first epidemic in 1699 perhaps being the most lethal. The population of the city at the time was just over 2200, and more than 300 deaths were attributed to the disease (Klepp, 1989b). While many fled the city during the epidemic, all of those who remained were immunologically naïve to the new disease and presumably extremely susceptible to infection. Most of what is known about this epidemic comes from letters written by residents. According to one witness, the disease was introduced into Philadelphia by "a ship from Barbados whose cargo consisted of cotton in bags which were landed at a wharf between Market Street and the draw-bridge and there stored for sale" (College of Physicians of Philadelphia, 1806; Wolman, 1974). The disease emerged first in this neighborhood and then spread gradually through the city with *great mortality*. Another spectator ascribed its origin to the stench of the pits from the two tanyards on Dock Street fronting the river. Perhaps supporting this, an owner of one of these yards died within 2 days from a violent attack of yellow fever, and, soon afterward, many families in the area manifested symptoms (Hazard, 1828–1835). As with later epidemics, the epidemic intensified in late August and early September and did not subside until after cooler weather prevailed.

Yellow fever was then absent from the city for 42 years and did not reappear until 1741. This epidemic, while not as lethal as the one in 1699, lasted from early June through early October. It followed a characteristic pattern of infecting those new to the city, who had not been exposed to previous epidemics. Because of this pattern, yellow fever was often known as *the palatine fever* since it prevailed among recently arrived German immigrants who had no prior exposure. Newly arrived Irish immigrants also were highly susceptible and were promptly blamed as the source of the disease in several accounts. For example, one physician, Dr. Phineas Bond, claimed the disease originated among "a number of convicts from the Dublin jail" (Currie, 1800; LaRoche, 1855). However, once in the city, reports suggest that the disease progressed along the waterfront, attacking both individuals who had been born in the interval since the last epidemic and new arrivals in these neighborhoods.

Yellow fever next visited Philadelphia in 1747, arriving in late June and persisting through the beginning of October. The CDR reached a high of 57 per 1000 that year (Klepp, 1989b), with epidemics of yellow fever, measles, malaria, and influenza all simultaneously raging in the city. The Christ Church Bills of Mortality alone list 23 deaths from measles and 42 deaths from *fever* in 1747, and in 1748, 39 deaths from pleurisy and 33 from *fever* (Klepp, 1991). The epidemic followed the previous pattern, appearing first along the waterfront, this time the wharves on the Delaware River and confining itself to the southern part of town. Records suggest that the epidemic was particularly lethal in the neighborhood near the Dock, which, in 1747, was a muddy stream that crossed three major city streets. The Dock was used widely as a garbage dump and was the focus of several legislative attempts to clean up the city. During the 1790s, the city finally allocated funds to cover portions of the creek. Although this was a significant measure to combat disease, those living in the vicinity of Dock Street still experienced significant mortality during subsequent epidemics of yellow fever (Smith, 1997).

Yellow fever appeared again in Philadelphia in August 1762 and did not subside until October of that year. This was the first epidemic to be mentioned in Elizabeth Drinker's diary, but little information is provided, except for the brief entry in mid-September wherein Drinker notes, "A Sickley time in Phila. many Persons are taken down, with Something very like the Yallow-Feaver" (Crane, 1991). The Drinkers were out of town at their summer home in Frankfort during most of the epidemic, although Elizabeth's husband, Henry Drinker, and her sister, Mary, made frequent visits to the city to conduct business. Yellow fever was only one of three epidemics to visit Philadelphia that year: smallpox and typhoid fever also were present. Given this, it is perhaps unsurprising that the CDR reached a high of 59 deaths per 1000 that year (Klepp, 1989b). The Christ Church Bills of Mortality for 1762 list 13 deaths from *consumption* and 26 from *decay* (i.e., tuberculosis), 20 from *fits*, 16 from *fever*, 9 from *nervous fever* (i.e., typhus or typhoid), 22 from *purging and vomiting*, 30 from smallpox, and 16 from yellow fever (Klepp, 1991). Chronic diseases like tuberculosis, malaria, and dysentery likely interacted synergistically with yellow fever during this epidemic and added substantially to the mortality rate.

Yellow fever returned again to Philadelphia during the summer of 1793, after a 30-year break, and followed by a cluster of subsequent epidemics over the next two decades. It is possible that the lull from 1762 to 1793 was directly linked to the absence of yellow fever in the Lesser Antilles during this period (Blake, 1968), while the 1793 epidemic and the tight cluster of epidemics in the colonies that followed it were likely directly attributable to the dramatic upsurge of yellow fever cases in the West Indies following the lull. This escalation can be linked to a variety of factors, including the Anglo-French wars, the Haitian revolt, and the large-scale introduction of susceptible European troops into the region (Geggus, 1979). The 1793 epidemic was particularly devastating for the city, largely because of its dramatically high mortality—the CDR was triple that of preceding epidemics; according to Klepp (1997), death rates among Philadelphia's inhabitants averaged between 64 and 98 per 1000 of those infected during the epidemic (Klepp, 1997). Bills of mortality for the year suggest high rates of several endemic infections, such as dysentery and tuberculosis, which likely synergistically exacerbated the vulnerability of many residents to yellow fever. The mortality rate was also considerably higher among those who did not flee the city: as many as one in five died. In this epidemic, the causal virus was likely imported from Santo Domingo, which was then experiencing a slave rebellion. Records suggest that approximately 2000 refugees arrived in Philadelphia from the islands, some of whom were ill with yellow fever or who were accompanied by infected mosquitoes (or perhaps both). A hot and humid summer in Philadelphia provided ideal conditions for the proliferation of the vectors, and by August, the city faced an epidemic of catastrophic proportions. "It was indeed melancholy," one resident subsequently recalled, "to walk the streets, which were completely deserted, except by carts having bells attached to the horses' heads, on hearing which the dead bodies were put outside on the pavements and placed in the carts by the negroes, who conveyed their charge to the first grave yard, when they returned for another load" (Simpson, 1788–1807, 1949).

Following this catastrophic epidemic, epidemics of yellow fever repetitively occurred in Philadelphia in 8 out of the 12 subsequent years, killing some 10,000 citizens over that period. One of these epidemics, occurring in 1798, may have rivaled the devastation and mortality of the 1793 catastrophe (Pascalis-Ouvière, 1798; Middleton, 1964; DeClue and Smith, 1998). Records suggest that the CDR for 1798 was particularly high, at 68 deaths per 1000 (Klepp, 1989b), and a total of 3500 deaths reported for a 4-month segment of the epidemic, though not all of these can be directly and unambiguously attributed to yellow fever. However, while the overall mortality was not greater than in 1793, records do suggest that it was greater in proportion to the number of individuals perceived to be infected or those who "continued [to be] exposed to the infection" (LaRoche, 1855, Vol. 1). Intriguingly, the high mortality persisted in the face of attempts by city officials to quarantine incoming ships between May and October (Condie and Folwell, 1798) and an early and extensive evacuation of the city, which greatly exceeded the attempts at evacuation in all previous epidemics (Shannon and Cromley, 1982).

In part, the high mortality during this epidemic may have been due to sanitary conditions within the city, which had not been substantially improved since 1793. Eyewitness accounts of the 1798 epidemic made particular note of numerous insects in the city, with one observer remarking that "many tribes of insects were uncommonly numerous; as mosquitoes, ants, crickets, cockroaches…" (Condie and Folwell, 1798: 35). In addition, many environmental factors in the city that did not directly influence the number of yellow fever vectors may have also contributed to diseases among the populace that synergistically weakened host resistance to yellow fever, such as malaria, dysentery, and tuberculosis. For instance, Philadelphia lacked an efficient sewage system, and outhouses and industrial wastes persistently polluted local water sources. Animal carcasses lay rotting on the banks of the Delaware River and in public streets, particularly in the market area along High Street (Foster et al., 1998). A severe epidemic of dysentery in July further added to the generally unsanitary and unhealthy state of the city. Young children were particularly vulnerable to dysentery, and surviving individuals were greatly weakened by the infection, and this was exacerbated by a resurgence of yellow fever in August (Shannon and Cromley, 1982). All of these factors likely magnified the 1798 epidemic's mortality.

Although sporadic epidemics of yellow fever occurred in Philadelphia during the early 19th century, the CDR steadily dropped, presumably as the number of susceptible individuals in the population steadily decreased. After 1798, the CDR exceeded 30 deaths per 1000 only twice, in 1802 and in 1805 (Klepp, 1989b). In the 1802 epidemic at least, synergy with at least one other infectious disease may be to blame for the elevated CDR; both measles and scarlet fever may have been present in minor epidemic form in the city (Mansfield, 1949). The Christ Church Bills of Mortality for 1802 list 11 deaths from measles, 9 from scarlet fever, 11 from consumption, and 24 from decay (Klepp, 1991). This epidemic receded over the following year, but not before affecting Elizabeth Drinker's family. Her diary records that Drinker's last intimate contact with yellow fever happened in 1803 when Sally Dawson, a favorite servant, died from the illness. This was despite adequate medical treatment; as members of the family, the Drinker's servants received much the same medical treatment as everyone else in the household (Crane, 1991).

The history of yellow fever in Philadelphia suggests that public fears and apprehensions bear little or no relationship to the actual impact or demographic significance of a specific disease. When mortality from endemic, common diseases, such as tuberculosis, was regular and predictable, there was relatively little concern; death was accepted as a part of life. However, when uncommon diseases, like yellow fever, arrived as epidemics, appeared at irregular intervals, and resulted in mortality spikes in an already unhealthy community, public fears intensified. Disease was commonplace in Philadelphia, but the spectacular nature of yellow fever epidemics disrupted community life on a grand scale. This is at odds with the demographic data, which suggest that the high mortality associated with these epidemics had a smaller total impact on population size in Philadelphia throughout the 18th century than did mortality from the conditions that were endemic in the city.

Conclusion

The emergence and spread of acute infectious diseases in Philadelphia were driven by a complex set of ecological, social, economic, and demographic factors, all of which constituted the city and its inhabitants' *total environment*. The urban environment of Philadelphia contained the epidemiologic factors necessary for the growth and propagation of a wide variety of infectious agents, while the social, demographic, and behavioral characteristics of the people of the city provided the opportunity for novel diseases to emerge and spread. What emerges from this study is a complex picture of a city undergoing rapid demographic, ecological, social, and epidemiologic changes. Large-scale immigration supplied a susceptible population group, as international trade, densely packed city streets, unsanitary living conditions, a contaminated water supply, and a hot and humid summer climate combined to create ideal circumstances for the proliferation of both pathogens and, in the case of yellow fever, vectors. These factors combined to set the stage for the many public health crises that plagued Philadelphia throughout the 17th and 18th centuries.

The people of Philadelphia were fearful of such diseases as smallpox, measles, and yellow fever if only because of their relative novelty, visibility, the dramatic nature of their symptoms, and their unpredictably intermittent appearance. Despite their high tolls in lives, however, these diseases were by no means the most significant determinants of morbidity and mortality in the city. Certain endemic diseases, notably dysentery, tuberculosis, and malaria, took a far higher demographic toll, even though their omnipresent nature tended to reduce public fear. Although these diseases did not often kill their victims directly, they did weaken them and reduce their immunological resistance to other infections. Dysentery, for instance, was widespread in Philadelphia throughout the 18th century and consisted of a number of causal agents including shigella and salmonella, as well as certain disease-causing amoebas; each posed a potentially serious threat to health, especially for weaned infants and children (Patterson, 2003). What is important about these more common, endemic conditions is that they do not evoke lifelong immunity, and as a result, individuals were susceptible throughout their lifetimes. This meant that they could become endemic in smaller populations than were required to support measles and smallpox, which only became endemic childhood diseases at the end of the colonial period (Duffy, 1953; Kunitz, 1984). By the end of the 18th century, a new equilibrium had been established in Philadelphia. The density-dependent epidemic diseases declined, leaving behind a relatively stable group of endemic diseases coupled with a lower mortality rate. If disease and environment are mirror images of each other, it stands to reason that a change in one will merely reflect a change in the other. As the physical, demographic, social, and economic environments of the city changed in the course of the century, so, too, did its disease environments.

Future Initiatives

No issue could be a more fundamental measure of sustainability than public health, and the increasing emergence and reemergence of infectious diseases globally is possibly the world's most challenging public health problem today. However, this problem is incomprehensible without a vastly broadened research perspective, if not an entirely new paradigm (Wilcox and Colwell, 2005). Public health and epidemiology have been tracing acute infectious diseases back to their point of origin for over a century, but the scale of the approach is widening, and a series of novel strategies for studying complex disease dynamics are being adopted. The challenge to researchers here is to break down disciplinary divides between, for example, medicine and ecology, virology and wildlife biology, and sociology and epidemiology to better understand the combined ecological and social dynamics at play behind the emergence and reemergence of infectious diseases (Daszak, 2005).

This ambitious goal will not be reached easily and will require science and education initiatives that cut across disciplinary as well as institutional, societal, and cultural boundaries (Kaneshiro et al., 2005). Historians can contribute to this endeavor by studying disease in past populations where the *global* scope was smaller, the rhythm of life was slower, and the variables influencing the emergence of disease were fewer. As demonstrated here, epidemic and endemic diseases on the scale experienced by Philadelphia reveal a population out of harmony with itself and its total environment. Clearly, by appreciating the complex dynamic between social, cultural, and ecological processes in the emergence of disease in this 17th- and 18th-century city, we could potentially gain insights into the underlying causes of the recent upsurge in emerging infectious diseases.

If a practical lesson emerges from the Philadelphia experience for policymakers today, it is that disease will result from actions and initiatives that disturb equilibrium in the total environment. Changes in human culture, technology, and environmental incursions nearly always have consequences for human health and patterns of disease. The continuing interplay between human culture and the world of pathogens is growing increasingly complex. Antibiotic overuse, increased human mobility, long-distance trade, urbanization, expanding numbers of refugees, and the exacerbation of poverty in inner-city and peri-urban slum communities all have great consequences for infectious diseases. While those who wanted to clear the forests to expand Philadelphia can be excused for their unawareness of ecological principles and probable results, we have no such excuse today. A keen awareness and understanding of the processes by which interventions yield epidemiologic consequences should guide our policymakers to proceed with increased caution, insight, and wisdom.

References

Aaby P. 1988. Malnutrition and overcrowding/intensive exposure in severe measles infection: review of community studies. Rev Infect Dis 10:478–491.
Aaby P. 1991. Determinants of measles mortality: Host or transmission factors? In: De la Maza LM, Peterson EM, editors. Medical virology 10: Proceedings of the 1990 symposium on medical virology. New York: Plenum. p 83–116.
Aaby P, Bukh J, Lisse IM, Smits AJ. 1984. Overcrowding and intensive exposure as determinants of measles mortality. Am J Epidemiol 120:49–63.
Bhopal R. 2002. Concepts of epidemiology: An integrated introduction to the ideas, theories, principles and methods of epidemiology. Oxford: Oxford University Press.
Black FL. 1992. Why did they die? Science 258:1739–1740.
Blake JB. 1968. Yellow fever in eighteenth-century America. Bull N Y Acad Med 44:673–686.
Burnet FM. 1940. The biological aspects of infectious disease. Cambridge: Cambridge University Press.
Butterfield LH, Kline MJ, editors. 2002. The book of Abigail and John: Selected letters of the Adams family, 1762–1784. Boston: Northeastern University Press.
Carmichael AG, Silverstein AM. 1987. Smallpox in Europe before the seventeenth century: Virulent killer or benign disaster. J Hist Med Allied Sci 42:147–68.
Caufield E. 1943. Early measles epidemics in America. Yale J Biol Med 15:531–556.
Chervin N. 1821. Nicolas chervin collection. Philadelphia: College of Physicians of Philadelphia.
Christie AB. 1974. Infectious diseases: Epidemiology and clinical practice. Edinburgh: Churchill Livingstone.
College of Physicians of Philadelphia. 1806. Additional facts and observations relative to the nature and origins of the pestilential fever. Philadelphia: College of Physicians of Philadelphia.
Condie T, Folwell R. 1798. History of the pestilence, commonly called yellow fever, which almost desolated Philadelphia, in the months of August, September, & October, 1798. Philadelphia: R. Folwell.
Cooper DB, Kiple KF. 2003. Yellow fever. In: Kiple KF, editor. The Cambridge historical dictionary of disease. Cambridge: Cambridge University Press. p 365–369.
Crane EF, editor. 1991. The diary of Elizabeth drinker. Boston: Northeastern University Press.
Currie W. 1800. A sketch of the rise and progress of yellow fever, and the proceedings of the board of health, in Philadelphia, in the year 1799: To which is added a collection of facts and observations respecting the origin of the yellow fever in this country, and a review of the different modes of treating it. Philadelphia: Budd & Bartram.
Cvjetanović B. 1982. The dynamics of bacterial infections. In: Anderson RM, editor. Population dynamics of infectious diseases. London: Chapman and Hall, Ltd. p 209–241.
Daszak P. 2005. Emerging infectious diseases and the socio-ecological dimension. EcoHealth 2:239–40.
De Bevoise K. 1995. Agents of apocalypse: Epidemics in the colonial Philippines. Princeton: Princeton University Press.

DeClue A, Smith BG. 1998. "Wrestling the Pale Faced Messenger": The diary of Edward Garrigues during the 1798 Philadelphia yellow fever epidemic. Pa Hist 65:243–68.

Dobson MJ. 1997. Contours of death and disease in early modern England. New York: Cambridge University Press.

Dubos R. 1959. Mirage of health: Utopias, progress, and biological change. New York: Harper & Row.

Duffy J. 1953. Epidemics in colonial America. Baton Rouge: Louisiana State University Press.

Fenn EA. 2001. Pox Americana: The great smallpox epidemic of 1775–82. New York: Hill and Wang.

Fenner F, Henderson A, Arita I, Ježek Z, Ladnyi ID. 1988. Smallpox and its eradication. Geneva: World Health Organization.

Foster KR, Jenkins MF, Toogood AC. 1998. The Philadelphia yellow fever epidemic of 1793. Sci Am 278:88–93.

Geggus D. 1979. Yellow fever in the 1790s: The British army in occupied Saint Domingue. Med Hist 23:38–58.

Gehlbach SH. 2005. American plagues: Lessons from our battles with disease. New York: McGraw-Hill.

Goodyear JD. 1978. The sugar connection: A new perspective on the history of yellow fever. Bull Hist Med 52:5–21.

Grob GN. 1983. Disease and environment in American history. In: Mechanic D, editor. Handbook of health, healthcare, and the health profession. New York: The Free Press.

Grob G. 2002. The deadly truth: A history of disease in America. Cambridge: Harvard University Press.

Harper K, Armelagos G. 2010. The changing disease-scape in the third epidemiological transition. Int J Environ Res Public Health 7:675–597.

Hazard S, editor. 1828–1835. Register of Pennsylvania. 16 vols. Vol. 9. Philadelphia.

Hopkins DR. 1983. Princes and peasants: Smallpox in history. Chicago: University of Chicago Press.

Kaneshiro KY, Chinn P, Duin K, Hood AP, Maly K, Wilcox BA. 2005. Hawaii's mountain-to-sea ecosystems: Social-ecological microcosms for sustainability science and practice. EcoHealth 2:349–360.

Kim-Farley RJ. 2003. Measles. In: Kiple KF, editor. The Cambridge historical dictionary of disease. Cambridge: Cambridge University Press. p 211–214.

Kiple KF, Higgins BT. 1992. Yellow fever and the Africanization of the Caribbean. In: Verano JW, Ubelaker DH, editors. Disease and demography in the Americas. Washington, DC: Smithsonian Institution Press. p 237–248.

Klepp SE. 1989a. Philadelphia in transition: A demographic history of the city and its occupational groups, 1720–1830. New York: Garland Publishing, Inc.

Klepp SE. 1989b. Demography in early Philadelphia, 1690–1860. Proc Am Philos Soc 133:85–111.

Klepp SE. 1991. "The Swift Progress of Population": A documentary and bibliographic study of Philadelphia's growth, 1642–1859. Philadelphia: American Philosophical Society.

Klepp SE. 1994. Seasoning and society: Racial differences in mortality in eighteenth-century Philadelphia. WMQ 51:473–506.

Klepp SE. 1995. Zachariah Poulson's bills of mortality, 1788–1801. In: Smith BG, editor. Life in early Philadelphia: Documents from the revolutionary and early national periods. University Park: The Pennsylvania State University Press. p 219–242.

Klepp SE. 1997. Appendix I: "How Many Precious Souls Are Fled"? The magnitude of the 1793 yellow fever epidemic. In: Estes JW, Smith BG, editors. A Melancholy scene of devastation: The public response to the 1793 yellow fever epidemic. Philadelphia: Science History Publications. p 163–182.

Kunitz SJ. 1984. Mortality change in America, 1620–1920. Hum Biol 56:559–582.

LaRoche R. 1855. Yellow fever, considered in its historical, pathological, etiological and therapeutical relations. Vol. 1. Philadelphia: Blanchard and Lea.

Last JM. 2001. A dictionary of epidemiology. Oxford: Oxford University Press.

Mansfield MAF. 1949. Yellow fever epidemics in Philadelphia, 1699–1805. M.A. thesis. Pittsburgh: University of Pittsburgh.

Manson-Bahr PEC, Apted FIC. 1982. Manson's tropical diseases. 18th ed. London: Saunders.
McNeill JR. 1999. Ecology, epidemics and empires: Environmental change and the geopolitics of tropical america, 1600–1825. Environ Hist 5:175–184.
Middleton WS. 1964. Felix Pascalis-Ouviere and the yellow fever epidemic of 1797. Bull Hist Med 38:497–515.
Morse SS. 1992. Aids and beyond: Defining the rules for viral traffic. In: Fee E, Fox DM, editors. Aids: The making of a chronic disease. Berkeley: University of California Press. p 38.
Mullen G, Durden L, editors. 2002. Medical and veterinary entomology. San Diego: Academic Press.
Newberne PM, Williams G. 1970. Nutritional influences on the course of infections. In: Dunlop RH, Moon HW, editors. Resistance to infectious disease. Saskatoon: Saskatoon Modern Press. p 93.
Pascalis-Ouviere F. 1798. An account of the contagious epidemic yellow fever which prevailed in Philadelphia in the summer and autumn of 1797. Philadelphia: Snowden and McCorkle.
Patterson DK. 2003. Bacillary dysentery. In: Kiple KF, editor. The cambridge historical dictionary of disease. Cambridge: Cambridge University Press. p 43–4.
Powell JM. 1949. Bring out your dead: The great plague of yellow fever in Philadelphia in 1793. Philadelphia: University of Pennsylvania Press.
Rothenberg MA, Chapman CF. 2000. Dictionary of medical terms. Hauppauge: Barron's Educational Series, Inc.
Rush B. 1815. An account of the measles, as they appeared in Philadelphia in the spring of 1789. In: Rush B, editor. Medical inquiries and observations. Philadelphia: Griggs & Dickinson. Vol. 2, p 255–261.
Ryan KJ, Ray CG, editors. 2004. Sherris medical microbiology: An introduction to infectious disease. New York: McGraw-Hill Companies, Inc.
Service MW. 2004. Medical entomology for students. 3rd ed. Cambridge: Cambridge University Press.
Shannon GW, Cromley RG. 1982. Philadelphia and the yellow fever epidemic of 1798. Urban Geogr 3:355–370.
Simpson R. 1788–1807. Robert simpson letterbook, 1788–1807. Philadelphia: Historical Society of Pennsylvania.
Simpson R. 1949. Narrative of a Scottish adventurer. J Presbyt Hist Soc 27:50.
Smith BG. 1977. Death and life in a colonial immigrant city: A demographic analysis of Philadelphia. J Econ Hist 37:863–89.
Smith BG. 1990. The "Lower Sort": Philadelphia's laboring people, 1750–1800. Ithaca: Cornell University Press.
Smith BG. 1997. Comment: Disease and community. In: Estes JW, Smith BG, editors. A melancholy scene of devastation: The public response to the 1793 Philadelphia yellow fever epidemic. Philadelphia: Science History Publications. p 150–151.
Smolinski MS, Hamburg MA, Lederberg J, editors. 2003. Microbial threats to health: Emergence, detection, and response. Washington, DC: The National Academies Press.
Sparks J, editor. 1839–1847. The works of Benjamin Franklin. 10 vols. Vol. 7. Boston: Charles Tappan.
Swedlund AC, Armelagos GJ, editors. 1990. Disease in populations in transition: anthropological and epidemiological perspectives. New York: Bergin and Garvey.
Vogt PL. 1908. The sugar refining industry in the United States: Its development and present condition (publications of the university of Pennsylvania series in political economy and public law. no. 21). Philadelphia: University of Pennsylvania Press.
Washington G. 1777. Writings of George Washington. 12 vols. Vol. 7. Boston: American Stationer's Company.
Wilcox BA, Colwell RR. 2005. Emerging and reemerging infectious diseases: Biocomplexity as an interdisciplinary paradigm. EcoHealth 2:244–57.
Wilson ME. 1995. Infectious diseases: An ecological perspective. BMJ 311:1681–1684.
Wilson ME. 2003. The traveller and emerging infections: Sentinel, courier, transmitter. J Appl Microbiol 94:1S–11S.
Wolman RS. 1974. Some aspects of community health in colonial Philadelphia. Ph.D. dissertation. Philadelphia: University of Pennsylvania.

Chapter 3
Modeling the Second Epidemiologic Transition in London: Patterns of Mortality and Frailty during Industrialization

Sharon N. DeWitte
Department of Anthropology, University of South Carolina, Columbia, SC

Introduction

The second epidemiologic transition is a model, originally detailed by Omran (1971), that describes changes in the cause-of-death structure in human populations. The general pattern of the transition, which has been observed in several populations, is an overall decline in the numbers of deaths resulting from infectious diseases and a rise in chronic or degenerative diseases as leading causes of death. The *transition* refers to the point at which chronic/degenerative diseases surpass infectious diseases as predominant causes of death. The transition is a model, and as such, it oversimplifies a process that a number of studies suggest varies with respect to sex, socioeconomic status, and other factors (e.g., Gage, 1994; Barrett et al., 1998). Furthermore, recent research has revealed the infectious etiology of numerous diseases that were once considered to be chronic (e.g., chronic liver disease resulting from infection with hepatitis B and coronary heart disease resulting from infection with *Helicobacter pylori*), which indicates that a simple dichotomous view of disease is itself also an oversimplification (Pasceri et al., 1998; Sumi et al., 2003). This transition is associated with (and is often conflated with) the demographic transition, in particular the decreased levels of mortality and increases in life expectancy at birth that occur at later stages of the demographic transition. Various researchers have suggested several ultimate causes of the second epidemiologic and demographic transitions, including improvements in sanitation, the development of modern medicine (and improved access thereof), improvements in nutrition, and evolution of pathogens or their human hosts (Omran, 1971; McKeown, 1976, 2009); however, the exact causes or combinations of causes responsible for the transitions are still a matter of debate.

Most of the research on the demographic and second epidemiologic transitions that has been done over the last several decades has been based upon demographic, epidemiologic, or census data (see Gage, 2005; Coste et al., 2006; Hill et al., 2007; Malina et al., 2008; Stevens et al., 2008; Ahsan Karar et al., 2009; Huicho et al., 2009; Risquez et al., 2010; Carter et al., 2011). For example, there exist relatively good historical records from the time of the second transition in England. From these records, we know that infectious disease mortality declined there during the mid-1800s and that this was accompanied by general declines in mortality (Gage, 2005). This was followed by the emergence of chronic diseases as leading causes of death. More recently, in Chile, data from census and health records reveal that in the latter half of the 20th century, degenerative disease mortality increased, ultimately causing nearly 75% of all deaths, and life expectancies have increased by as much as 9 years (Albala and Vio, 1995).

There are many populations throughout the world, and throughout history, for which historical documents from the time of their particular transitions do not exist. Alternatively, for some populations, records do exist, but they are not sufficiently complete to allow for a full reconstruction of the transition. Furthermore, for those populations for which we actually do have relatively good historical data on age at death and cause of death during the transition, we generally lack comparatively good historical data on the population-level patterns of health and mortality *before* the transition. In England, for example, there are very few existing data on cause of death before the mid-1800s (Gage, 2005). The London Bills of Mortality do record causes of death beginning in the 17th century; however, such early determinations of cause of death were generally made by visual assessment of the dead and by asking the opinions of people who were present near the time of death (Landers, 1993), and thus, these diagnoses lack the precision possible with more recent data. Even if we had complete, accurate data on cause of death going back long before the transition in England and other populations, just knowing the numbers of people who died from various causes does not provide a clear picture of the general health of a population, as health includes the disease experiences of *living* people, not just how they ultimately died.

Given the limitations of available data, even when we know the timing and the pattern of the second transition for a particular population, it is also not always clear why the population made the transition at the time that it did. Yet, it is the reasons underlying the transition—why people start dying from some causes but not others and become capable of surviving to older adult ages—that are ultimately of greatest interest to many researchers, particularly those in public health and epidemiology. Information about the particular circumstances within populations that existed prior to their transitions would allow for a better understanding of the variation that characterizes the second transition and for an evaluation of its proposed ultimate causes.

Skeletal samples from archaeological sites potentially provide empirical data on health conditions and mortality patterns just before and during the second transition that are not available from existing historical documents. Although methods of skeletal age estimation, sex determination, and pathological lesion identification are not without their own problems, skeletal samples might in some cases provide the only surviving health and demographic data for a past population. Skeletal data, therefore, can be used to investigate the second transition in populations without relevant historical data or to provide a richer understanding of the transition independent of that provided by existing (and potentially incomplete or biased) historical documents. Bioarchaeological and paleodemographic approaches, such as those presented here, will potentially allow us to increase the number of populations in which we can examine the transition and thereby expand the geographic and temporal scope of our investigations. Data from a large number of populations across space and throughout history will allow for a more thorough understanding of the variation among populations in the timing of the transition, of the differential experiences of the sexes and various age groups during the transition, and of the factors affecting the progression (or lack thereof) of the transition.

One possible approach to investigating the second transition, which can make efficient use of relatively limited bioarchaeological data from past populations, is to examine the patterns of frailty within a population. Frailty refers to an individual's relative risk of death compared to other members of the population (Vaupel et al., 1979) and can be affected by, among other things, biological sex, genetics, nutritional status, and environment. A focus on intrapopulation variation in frailty can address such questions as: What were the age and sex patterns of disease and mortality before and during the transition? How did morbidity and mortality patterns at earlier (e.g., childhood) ages affect frailty at later ages? Were people succumbing more easily to or more strongly resisting death following exposure to various physiological stressors (e.g., malnutrition or infectious disease)? Were people surviving longer with diseases? Did any of these patterns change over time in systematic ways and how did they affect the cause-of-death structure? Such an approach acknowledges and allows for variation in the health and mortality experiences of individuals within populations making the transition and has the potential to uncover factors responsible for the second transition.

Before using skeletal data to investigate the second transition in past populations that lack sufficient, relevant historical data, it would be useful to determine whether skeletal samples can, in fact, yield patterns comparable to those discernable from historical documents. The current study represents a step toward achieving that goal. This study examines the skeletal remains of people who died before the transition in London to determine the patterns of health and mortality that existed within the population at the eve of the transition. These analyses will contribute to an understanding of the circumstances that contributed to the subsequent transition within the population of London in particular. More generally, however, it is hoped that the results of skeletal analyses, from a population for which we know the timing of the transition fairly well, will provide *paleodemographic signatures* that can be used as baseline expectations when examining skeletal samples from other populations for which there are no comparable existing historical data.

This study addresses two primary questions. First, how did age patterns of mortality in London change, if at all, in the period leading up to the second transition? That is, what were mortality patterns like in London right *before* the population experienced decreasing numbers of deaths from infectious diseases and dramatic changes in overall mortality levels, both of which are observed in historical documents? Second, were there changes in patterns of frailty in the population right before the transition? This study specifically addresses whether there were changes in the way that previous exposure to physiological stress (i.e., health history) affected risks of death in the years leading up to the transition that might have contributed to the subsequent changes in cause-of-death structure and decreases in mortality that occurred in mid-19th-century England.

Materials and Methods

Skeletal Samples

The data for this study come from four London cemeteries, all of which were used for burials of lower socioeconomic status individuals. Two of the cemeteries, Broadgate (c.1540–1714, $n=150$) and St. Thomas' Hospital (c.1540–1714, $n=193$), predate the Industrial Revolution and comprise the *preindustrial* sample for the analyses described later. The remaining two cemeteries, Cross Bones (c.1800–1853, $n=148$) and St. Bride's Lower Churchyard (c.1770–1849, $n=544$), date to the time of the Industrial Revolution and represent mortality patterns at the eve of the second transition in London; together, these two cemeteries comprise the *industrial* sample for this study.

All of the data from these cemeteries used in this study were collected by researchers at the Museum of London Centre for Human Bioarchaeology and were obtained from the Centre's online database.[1]

The Preindustrial Sample

BROADGATE CEMETERY

Broadgate Cemetery was established in 1559 by the City of London as the *New Graveyard* in east central London in an effort to relieve the burden on parish cemeteries in the city, and it was used until 1714 (Ogden et al., 2007). For most of its existence as an active burial ground, Broadgate was used primarily for the internment of poor individuals, and the majority of people were buried without coffins. The Museum of London Department of Urban Archaeology excavated the cemetery between 1984 and 1987, revealing several hundred individuals. Approximately 400 were excavated from the site and are curated at the Museum of London and available for analysis; data on age, sex, and the presence of pathological conditions are currently available for 150 individuals from the Museum's online database.

ST. THOMAS' HOSPITAL CEMETERY

The St. Thomas' Hospital in South London had an associated cemetery that was excavated by the Museum of London in 1991. Excavation of the site revealed mass burial trenches, which have been interpreted as either epidemic burials or pauper graves. The individuals used in this study were all recovered as articulated individuals found beneath a large number of disarticulated remains that were likely part of a charnel pit. The presence of coffin nails associated with the articulated remains indicates that many were buried in coffins. Of the 227 articulated skeletons excavated from the site, 193 individuals were analyzed for inclusion in the Museum of London database.

The Industrial Sample

ST. BRIDE'S LOWER CHURCHYARD

The St. Bride's Lower Churchyard, located in the eastern part of central London, was established to deal with overcrowding of the churchyard at St. Bride's Church. St. Bride's Lower Churchyard was used for burials of people from the nearby Bridewell workhouse and Fleet prison and other individuals of low socioeconomic status. Most of the individuals excavated from St. Bride's were buried in coffins, and several coffins (up to 10) were stacked on top of each other in large pits. These large burial pits were kept open until filled, and burials at the top of the pits tended to be reserved for infants and children (Brickley and Miles, 1999). Excavation of the cemetery by the Museum of London yielded 606 individuals, and data from 544 individuals, representing an estimated 50% of the original burial population, are available from the Museum database.

CROSS BONES CEMETERY

The Cross Bones burial ground in South London was established as early as the 17th century and is believed to have originally been a burial ground for prostitutes (Brickley and Miles, 1999). By the mid-18th century, and continuing until its closure in 1853, Cross Bones was used as a pauper's cemetery. The cemetery was excavated by the Museum of London between 1991 and 1998, and 148 individuals were recovered. Data on all 148, all of whom have been dated to 1800 to 1853 based on historical evidence, are available from the Museum of London database. As in the contemporaneous St. Bride's cemetery, burials

---

[1] http://www.museumoflondon.org.uk/Collections-Research/LAARC/Centre-for-Human-Bioarchaeology/

in Cross Bones consisted of large burial pits with stacked coffins. None of the burial pits in Cross Bones were completely excavated, however, so it is not clear whether burials at the tops of the pits contained more infants and children than did lower levels; however, given patterns of burials observed at contemporaneous sites in London, it is possible that the excavated sample has a higher proportion of infants than was true of the original burial population (Brickley and Miles, 1999). The possible effect of infant overenumeration in the Cross Bones cemetery will be discussed later.

Age and Sex Estimation

Age Estimation

Age estimates for subadults (i.e., individuals <18 years old) are based on the diaphyseal lengths of major long bones, epiphyseal fusion, and dental development and eruption (Moorees et al., 1963; Gustafson and Koch, 1974; Scheuer et al., 1980; Smith, 1991; Scheuer and Black, 2000). Estimates of adult ages (i.e., individuals 18 years or older) are based on tooth wear (Brothwell, 1981) and age-related changes of the pubic symphysis (Brooks and Suchey, 1990), iliac auricular surface (Lovejoy et al., 1985; Buckberry and Chamberlain, 2002), and sternal rib ends (Iscan et al., 1984, 1985). These age-estimation methods yield interval estimates as shown in Table 3.1. Only those individuals for whom at least one of the methods listed earlier could be used are included in the analyses presented here; that is, individuals who were too poorly preserved or too highly fragmented to be aged using one of these methods were excluded from the study.

Sex Estimation

Sex was assessed only in adult skeletal remains (i.e., in those individuals in which all epiphyses had completely fused) because of the current lack of reliable methods for determining sex macroscopically from juvenile skeletal remains. Sex was determined from sexually dimorphic features of the skull and pelvis using the standards described in Buikstra and Ubelaker (1994). The following features of the skull were scored: glabella/supraorbital ridge, supraorbital margin, mastoid process, external occipital protuberance/nuchal crest, and the mental eminence. The following features of the pelvis were also evaluated for sex determination: the ventral arc of the pubis, subpubic concavity, ischiopubic ramus ridge, and the greater sciatic notch. Numerous studies have shown that the sex determination based on these individual traits, or combinations thereof, has accuracy ranges from 68 to

Table 3.1. Age-at-death distributions for the preindustrial and industrial samples.

Age	Preindustrial n (%)	Industrial n (%)	Preindustrial n (%)	Industrial n (%)
<0	32 (0.126)	126 (0.205)		
1–4.99	20 (0.079)	116 (0.189)		
5–9.99	23 (0.091)	18 (0.029)		
10–14.99	26 (0.103)	11 (0.018)		
15–29.99	46 (0.182)	13 (0.021)	46 (0.303)	13 (0.038)
30–39.99	24 (0.095)	48 (0.078)	24 (0.158)	48 (0.140)
40–49.99	48 (0.190)	106 (0.173)	48 (0.316)	106 (0.309)
50+	34 (0.134)	176 (0.287)	34 (0.224)	176 (0.513)
Total	253	614	152	343

over 96% (Sutherland and Suchey, 1991; Rogers, 2005; Williams and Rogers, 2006). Multiple skeletal indicators of sex are used for this study given that including more than one indicator improves the accuracy of sex determination (Meindl et al., 1985; Rogers, 2005; Williams and Rogers, 2006; Walker, 2008).

Osteological Stress Markers

To examine the relationships among physiological stress, frailty, and mortality, the associations between certain pathological conditions evident on the skeleton and risk of mortality were assessed in the industrial and preindustrial samples. Two pathological conditions, periosteal new bone growth (i.e., periosteal lesions) and cribra orbitalia, were selected because they have been shown by previous research to be associated with frailty such that individuals with these skeletal lesions face elevated risks of mortality compared to individuals without them (Usher, 2000; DeWitte and Wood, 2008; DeWitte, 2010b).

Periosteal Lesions

Periosteal new bone formation occurs in response to stimuli that tear, stretch, or otherwise traumatize the periosteum. They can also result from local or systemic infection or inflammation associated with a variety of factors (Larsen, 1997; Ortner, 2003; Weston, 2008). Periosteal lesions are often used by bioarchaeologists as nonspecific indicators of physiological stress (Larsen, 1997; Weston, 2008), and that is how they are interpreted for this study (i.e., no attempt was made to diagnose specific pathologies).

Periosteal lesions were scored on the tibia for this study because studies have demonstrated that the tibia is commonly affected by such lesions (Eisenberg, 1991; Milner, 1991; Larsen, 1997; Roberts and Manchester, 2005) and, because it is robust, the tibia is often well preserved in skeletal samples. Only the surface of the tibial diaphysis was assessed for periosteal lesions; the metaphyses were avoided as these surfaces have muscle markings that can interfere with the identification of periosteal lesions. Periosteal lesions were identified macroscopically and scored as present if there was at least one distinct patch of woven or sclerotic bone (or a combination of the two) laid down on the surface of the diaphysis. Only tibiae with diaphyseal surfaces that were free of both periosteal new bone growth and postmortem damage were scored as lacking periosteal lesions; tibiae with no visible lesions but with postmortem damage that prevented visual assessment of the anterior surface were given a score of *unobservable* with respect to periosteal lesions and thus excluded from analysis.

Cribra Orbitalia

Cribra orbitalia is a lesion that forms on orbital roofs and is characterized by a porous appearance of the outer table of the affected bone, which is often associated with expansion of the underlying diplöic bone (Mensforth et al., 1978; Ortner, 2003). Cribra orbitalia is often attributed to anemia or other causes during childhood that result in an expansion of the bone marrow and thus an expansion of the surrounding diplöic bone. Though it typically forms during childhood, once formed, cribra orbitalia can be retained into adulthood (Walker et al., 2009). For the current study, the roofs of both orbits were scored for the lesion, and lesions were identified macroscopically. Cribra orbitalia was scored according to the categories described by Stuart-Macadam (1985), and lesions accordingly ranged from scattered small foramina to outgrowths of the underlying diploe that altered the normal contour of the bone. Scores for various levels of severity were collapsed into presence/absence scores for the purposes of the analyses described later.

Models

Age-at-Death Distributions

The age-at-death distributions from the preindustrial and industrial samples are compared to each other using a Kolmogorov–Smirnov test. This represents a preliminary step in the analysis of age patterns of mortality and allows for the identification of broadscale differences in the proportions of each sample dying within each age interval.

Comparison of the raw age-at-death distributions alone, however, is insufficient for a complete understanding of mortality patterns in past populations. It is possible that age-at-death distributions estimated from skeletal samples *might* directly reflect the age structure and mortality patterns of the associated once-living population and any observed differences between cemeteries *might* therefore directly reflect differences in living age structures or age-specific mortality rates. However, skeletal samples are subject to various processes of selection that can, individually and combined, create a sample that is biased and thus not a perfect representation of the age structure or mortality patterns of a past population. Soil pH, moisture, and other factors can affect the preservation of bone following burial, and these factors can vary within and between cemeteries. Larger bones with thicker cortices are more likely to preserve over centuries of burial than are smaller, thinner bones; thus, in certain burial contexts, larger adults are more likely to eventually be recovered by excavators than are small individuals (e.g., infants and children) or adults with pathologies that reduce bone mineral density (Lewis, 2007; Milner et al., 2008). If the small, relatively fragile bones of infants and children are less well preserved than those of adults in a cemetery, this can result in infant and child underenumeration and consequently underestimation of the risk of childhood mortality for that population. Additionally, not all individuals who die in a population are buried in the same area within a cemetery, and in some cultures, certain individuals are excluded from the primary village, town, or parish cemetery (Milner et al., 2008). Some individuals may also perish while outside of their community and thus not be represented in the sample. Systematic variation in burial location, coupled with incomplete excavation of a site (which is often the case as a result of time and financial limitations), can result in samples that are missing certain segments of the population and thus underestimation of mortality risks for those subpopulations.

For these reasons, the age patterns of mortality in the preindustrial and industrial samples are also assessed using a competing hazards model. One advantage of using hazards models, like the Siler and Gompertz–Makeham models used in this study and described later, is that they can be applied to small samples. These models smooth the random variation in mortality data that is often an artifact of small sample size (i.e., as typical of cemetery samples), without imposing any particular age pattern on the data (Gage, 1988).

Age Patterns of Mortality

The age patterns of mortality in the preindustrial and industrial samples are assessed using the Siler model of mortality, which has been shown to fit a variety of human mortality patterns. The Siler model describes a general pattern of high risks of mortality immediately after birth, relatively low risks of mortality during later childhood and continuing through early adulthood, and relatively high risks for elderly adults (Gage, 1991; Wood et al., 2002):

$$h(a_i) = \alpha_1 e^{-\beta_1 a} + \alpha_2 + \alpha_3 e^{\beta_3 a} \tag{3.1}$$

In this model, a_i is the age of the *i*th skeleton in years. The first component of the Siler model, $\alpha_1 e^{-\beta_1 a}$, represents the juvenile risk, which is high at birth and decreases rapidly thereafter; α_1 is the risk of death at birth associated with immaturity, and β_1 is the rate at which this risk decreases exponentially with age a. The second component of the model, α_2, is the constant age-independent risk that everyone within the population faces (i.e., the chance of dying from causes that are unrelated to aging, including accidental causes). The third component of the model, $\alpha_3 e^{\beta_3 a}$, is the senescent risk, which increases exponentially with age; α_3 is the risk of death associated with senescence at birth, and β_3 is the rate at which this risk increases with age (Gage, 1988). The three components of the Siler model are independent, so surviving one component of mortality does not affect an individual's risk of death during the others (Wood et al., 2002). For example, individuals who survive infancy and childhood would, according to this model, face the same risks as everyone else of dying from senescent causes or age-independent causes; survival through childhood does not raise or lower risks of dying at other ages.

For this analysis, a pooled sample that includes all individuals from the preindustrial and industrial cemeteries is used to estimate the baseline Siler hazard. To evaluate changes in age patterns of mortality over time, time period (preindustrial vs. industrial) is modeled as a covariate affecting the baseline Siler model using a proportional hazards specification:

$$h(t|x\rho) = h(t)e^{(x\rho)} \tag{3.2}$$

where t is age, x is time-period covariate (0 or 1), and ρ is the effect of the time-period covariate. Individuals in the two preindustrial cemeteries were assigned a covariate score of *0*, and individuals in the industrial cemeteries were assigned a score of *1*. Using these specifications, a positive estimated value of the parameter representing the effect of the time-period covariate would indicate an elevated risk of mortality in the industrial sample compared to the preindustrial sample. Conversely, a negative estimated value of the parameter representing the effect of the time-period covariate would indicate that the risk of mortality was lower in the industrial sample compared to the preindustrial sample.

By using the proportional hazards specification, this model does not allow for variation across ages in the effect of time period on risk of death; this specification might not necessarily reflect reality, but it has the benefit of requiring the estimation of a relatively small number of parameters. Such a consideration is necessary when working with the small samples typical of paleodemography, as increasing the number of model parameters requires ever-increasing sample sizes in order to recover good estimates of those parameters. To allow for at least some variation with age in the effect of the time-period covariate on risk of mortality, in addition to modeling the covariate on the Siler model as a whole (and, thus, on all ages simultaneously), this study also models time period as a covariate affecting the juvenile and senescent components of the Siler model separately (i.e., affecting $\alpha_1 e^{-\beta_1 a}$ and $\alpha_3 e^{\beta_3 a}$ parts of the Siler hazard individually).

Patterns of Frailty

To examine the temporal variation in the effect of physiological stress on risk of death in London, the excess mortality associated with skeletal pathological conditions—cribra orbitalia and tibial periosteal lesions—is compared between the preindustrial and industrial samples using the Usher model (Usher, 2000), shown in Figure 3.1. This multistate model of health and mortality has three nonoverlapping states, two of which (States 1 and 2) are living states and the third of which (State 3) represents death. Everyone in the cemetery samples is, of course, observed in State 3, so the two living states represent the states that individuals could have been in before they died. State 1 includes those people who have no visible periosteal lesions or cribra orbitalia, and State 2 includes individuals with one or both of these pathological conditions. Transitions between each of the living states and death are determined by age-specific hazard rates. For this study, the baseline risk of death

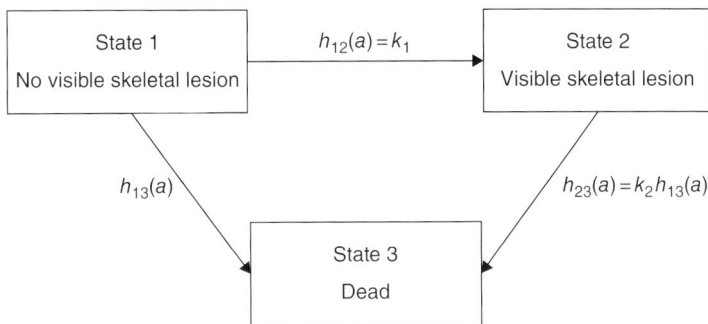

Figure 3.1. Multistate model of health and mortality; modified from Usher (2000).

(i.e., the transition from State 1 to death), $h_{13}(a)$, is modeled variously as a Siler model, a Gompertz–Makeham model, and a negative Gompertz model, as described further later. In the model, the hazard of developing lesions, $h_{12}(a)$, is estimated as a constant k_1, as the age at which an individual experienced a physiological stressor sufficient to result in a skeletal lesion often cannot be determined from bioarchaeological data. For simplicity, in this study, the age of onset of skeletal lesions is specified as an exponential random variable. The Usher model allows for the estimation of the difference in risk of death associated with the two living states, and this difference in risk is indicated by the k_2 parameter in the hazard for dying from State 2:

$$h_{23}(a) = k_2 h_{13}(a) \qquad (3.3)$$

The k_2 parameter indicates the difference in risk of death between individuals with and without skeletal lesions. The value of the parameter can be greater than 1 (indicating that individuals with lesions faced higher risks of dying than their age peers without them), less than 1 (indicating that individuals with lesions were at reduced risks of dying), or equal to 1 (indicating no significant difference in risk between people with and without skeletal lesions). The estimated value of k_2 provides a measure of the excess mortality (or lack thereof when k_2 is equal to 1) associated with skeletal lesions.

To estimate the differences between the preindustrial and industrial samples in the excess mortality associated with skeletal lesions, time period is modeled as a covariate affecting the k_2 parameter in the Usher model. For this analysis, a pooled sample including all individuals from all four cemeteries is used to estimate the baseline hazard, and covariate scores were assigned as described earlier (preindustrial = 0, industrial = 1). A positive estimated value of the parameter representing the effect of time-period covariate in this analysis would indicate greater excess mortality associated with skeletal lesions in the industrial sample compared to the preindustrial sample. A negative estimated covariate effect would indicate that excess mortality associated with lesions was lower in the industrial sample. The Usher model, with the time-period covariate, is fit to data from all individuals in the samples using the Siler model to assess the effect of physiological stress on risk of mortality across all ages simultaneously. Additionally, to allow for some variation with age in the effect of physiological stress on risk of mortality, the Usher model is fit separately to data from juveniles and adults. To examine the pattern for juveniles, the baseline hazard is modeled as a negative Gompertz model,

$$h(a) = \alpha_1 e^{-\beta_1 a} \qquad (3.4)$$

(i.e., the first, exponentially decreasing, component of the Siler model). To examine the pattern for adults, the baseline hazard is modeled as a Gompertz–Makeham model,

$$h(a_i) = \alpha_1 + \alpha_2 e^{\beta a} \quad (3.5)$$

(i.e., the last two components of the Siler model) (Gage, 1988; Wood et al., 2002).

Sex Differences in Mortality

To determine whether the sex patterns of mortality differed (i.e., whether males and females faced similar risks of dying) between the preindustrial and industrial periods, sex is modeled as a covariate affecting adult mortality within the two samples. Sex differences in mortality among children and adolescents are not assessed given the difficulties associated with determining sex in juvenile remains.

Adult mortality is modeled using the Gompertz–Makeham mortality model. To evaluate whether males and females differed in their risks of mortality within each of the samples, sex is modeled as a covariate acting upon the parameters of the Gompertz–Makeham model. For this analysis, females were coded as 0 and males as 1. A significant positive estimate for the parameter representing the effect of the covariate on the hazard would suggest males were at an increased risk of death compared to females. Conversely, a significant negative estimate would indicate the risk for males was lower than that for females. The model is fit separately to the data from preindustrial and industrial samples.

For all the analyses in this study, the model parameters are estimated using maximum likelihood analysis with the program *mle* (Holman, 2005).

Results

Age-at-Death Distributions

The age-at-death distributions from the preindustrial and industrial samples are shown in Table 3.1. Given the possibility of infant overenumeration in the Cross Bones cemetery, the age-at-death distributions of just the adults from each sample are also compared. The results of the Kolmogorov–Smirnov tests indicate that the preindustrial and industrial age-at-death distributions are significantly different (for both comparisons of age-at-death distributions, $p < 0.05$). As seen in Table 3.1, the industrial sample has more individuals between the ages of 0 and 5 years and above the age of 50 but fewer individuals between the ages of 5 and 30 than in the preindustrial sample. The two samples have a similar proportion of people in the 30- to 50-year category. The abundance of infants in the industrial sample might suggest higher infant mortality or increased fertility on the eve of the second transition, and the high proportion of older adults might indicate increased longevity in the industrial sample. However, such conclusions are premature given the possibility of differential preservation, variation in burial location, and incomplete excavation processes within and between the cemeteries used in this study.

Age Patterns of Mortality

The estimated values of the parameter representing the effect of the time-period covariate on age patterns of mortality are shown in Table 3.2. When the time-period covariate is modeled on the entire Siler model (i.e., all ages are assessed simultaneously, with no variation in covariate effect across age), the results suggest that mortality risks in general were

Table 3.2. Estimated values of the parameter representing the effect of time period on age patterns of mortality.

	Siler all ages	Juvenile mortality	Senescent mortality
Time period = industrial	−0.23 (0.09)	0.71 (0.16)	−0.74 (0.09)

Standard errors are shown in parentheses.

Table 3.3. Estimated values of the excess mortality associated pathological conditions and the effect of the time-period covariate on excess mortality.

		Effect of industrial time period		
	Excess mortality (k_2)	All ages	Juveniles	Adults
Tibial periostitis	2.18 (0.3)	0.027 (0.2)	0.54 (0.4)	−0.72 (0.2)
Cribra orbitalia	7.23 (2)	0.098 (0.3)	0.94 (0.3)	−1.25 (0.2)

Standard errors are shown in parentheses.

lower in the industrial sample. However, when time period is modeled as a covariate separately on the juvenile and senescent components of the Siler model, the results indicate a higher risk for children in the industrial sample compared to the preindustrial sample, but a lower risk for adults in the industrial sample compared to the preindustrial sample.

Patterns of Frailty

The estimated values of the k_2 parameters (i.e., excess mortality) associated with tibial periosteal lesions and cribra orbitalia and the parameter representing the effect of the time-period covariate (industrial = 1) on the excess mortality associated with each lesion are shown in Table 3.3. The estimated values of the k_2 parameters for both pathological conditions indicate that they were both associated with elevated risks of mortality. For both lesions, the estimated effects of the time-period covariate are not significantly different from 0 when all ages are assessed simultaneously (i.e., with the Siler model as the baseline hazard). However, when only adults are included in the analyses (with the Gompertz–Makeham model as baseline hazard), the estimated values of the time-period covariate for both lesions are significantly less than 0, indicating that there was lower excess mortality associated with these lesions among adults in the industrial period compared to preindustrial period. Further, when only children are included in the analyses (with the negative Gompertz model as the baseline hazard), the estimated values of the time-period covariate for both lesions are significantly greater than 0. This indicates that there was greater excess mortality associated with skeletal lesions among children in the industrial period compared to the preindustrial period. These results suggest that there were differential changes in the patterns of frailty among children and adults at the eve of the second epidemiologic transition.

Sex Differences in Mortality

The estimated values of the parameter representing the effect of the sex covariate on adult mortality within the preindustrial and industrial samples are shown in Table 3.4. The estimated value of the sex-covariate effect in the preindustrial sample is not significantly different from 0, but it is greater than 0 in the industrial sample. These results indicate that

Table 3.4. Estimated values of the parameter representing the effect of the sex covariate on adult mortality.

	Preindustrial	Industrial
Effect of being male	0.12 (0.19)	0.24 (0.1)

Standard errors are shown in parentheses.

in the preindustrial population, there was no significant difference between males and females in the overall risk of mortality but that in the industrial sample, the risk of mortality was lower for females compared to males.

Discussion

Age Patterns of Mortality

The estimated parameters of the hazards models indicate that overall mortality and adult mortality in particular were declining prior to the second epidemiologic transition but that childhood mortality was simultaneously increasing. These results are also suggested, in part, by the age-at-death distributions, given the higher proportions of young children and older adults in the industrial sample. As mentioned previously, however, it is possible that the Cross Bones cemetery has an overrepresentation of infants and children, given patterns of burial at contemporaneous sites (such as St. Bride's lower) and the incomplete excavation of burial pits in Cross Bones. The patterns for children therefore might be an artifact of biases in the sample, rather than truly reflecting age-specific mortality patterns in the once-living industrial population. However, the excess of older adults in the industrial sample compared to the preindustrial sample is independent of the childhood patterns, so conclusions about positive changes in adult mortality and overall lower mortality risks in the industrial sample are unaffected by any possible infant/child overenumeration. Further, possible infant/child overenumeration is not likely to affect the results of the analyses of frailty, as the Usher model compares the risks of death between those with and without skeletal lesions within each age group (i.e., it compares differences in risk between age peers). A difference between the two sites in the numbers of children is unlikely to affect the overall patterns that emerge from the analysis of the effect of physiological stress on risk of mortality.

The apparent increases in childhood mortality might reflect shifts from epidemic diseases to endemic diseases that occurred during industrialization, which would have more severely affected young children with naïve immune systems (Fenner, 1970; McNeill, 1980; Gage, 2005). Increases in childhood mortality might have influenced the decreases in adult mortality, as high infant/juvenile mortality would have weeded out the frailest children, resulting in a surviving adult cohort with immunity to the endemic diseases and with lower average frailty.

Patterns of Frailty

The estimates of the k_2 parameters for cribra orbitalia and tibial periosteal lesions indicate that both pathological conditions were associated with elevated risks of mortality in the preindustrial and industrial populations of London. These results are consistent with previous studies from other populations, which have also shown that individuals with these lesions were at higher risks of mortality than their peers without them (Usher, 2000; DeWitte and Wood, 2008; DeWitte, 2009; Redfern and DeWitte, 2011). As shown in Table 3.1, for both pathological conditions, the estimated value of the effect of the time-period covariate is not significantly different from 0 when all ages are assessed simultaneously. However, when only adults are included in the analyses, the results indicate that there was lower excess mortality associated with both lesions in the industrial period compared to the preindustrial. When only children are assessed, the results indicate that there was greater excess mortality associated with the lesions in the industrial period compared to the preindustrial. Together, these results indicate that compared to the earlier period, at the eve of the second transition, adults exposed to physiological stress were better able to resist dying (i.e., previous physiological stress was less strongly associated with risks of mortality), but physiologically stressed children were at higher risks of dying. It appears that adults were experiencing positive changes in health but that childhood health was relatively compromised right before the transition in London.

The analyses of the age patterns of mortality and the patterns of frailty both indicate that there were positive changes in adult health and mortality right before the second transition in London but that children were experiencing negative changes in health and mortality at that same time. The results of this study suggest that the effects of industrialization—including increased size of urban centers, lack of efficient public sanitation, increased infectious diseases, inadequate or contaminated food in urban centers (particularly for lower socioeconomic status people), and increased environmental pollutants (Lewis, 2002; Lewis and Gowland, 2007; Mays et al., 2008)—more negatively affected children than adults. Children appear to have been less well buffered against or more vulnerable to the stressors associated with urbanization and industrialization. As mentioned previously, there might have been a relationship between childhood and adult frailty in the industrial population of London, that is, factors that disproportionately affected the health of children resulted in strong selective mortality among children, which, in turn, resulted in cohorts of adults who were generally less frail compared to those in the preindustrial population.

The results of this study are generally consistent with those obtained from other bioarchaeological studies. For example, Lewis (2002) found that there was a higher frequency of cribra orbitalia in children younger than 6 months of age and a higher frequency of *in utero* enamel hypoplasia among children in a sample from industrial London compared to medieval samples from both urban and rural locations. They also found a higher frequency of skeletal lesions associated with metabolic diseases such as rickets and scurvy in the industrial London sample compared to the medieval samples, and children in industrial London were shorter than those in the earlier cemeteries. Together, these results indicate the negative effects of industrialization on the health of children in London. A comparison of skeletal growth profiles from medieval and early 20th-century Portugal (Cardoso and Garcia, 2009) found no significant differences in femoral length in children between the two samples, but significantly longer femora in medieval adults. These results suggest that child-labor practices and greater risks of undernutrition in industrial Portugal have negatively affected the potential for catch-up growth in the more recent sample.

The patterns from the skeletal data are also consistent with those obtained by some investigations of historical data. For example, in Britain in general, there were declines in overall mortality rates during the 19th century (Wrigley and Schofield, 1981; Floud and Harris, 1997), but declines in infant mortality did not occur until the turn of the 20th century, nearly 40 years later (Millward and Bell, 2001). Further, in industrialized towns in northern England, infant mortality actually rose during the first half of the 19th century (Huck, 1995). According to Floud and Harris (1997), there were declines in the average height of British men born between 1820 and 1860, which indicates compromised health in children during the mid-19th century. However, not all such studies have revealed a general pattern of relatively poor health for children at the beginning of the 19th century. For example, data from the London Bills of Mortality indicate that infant mortality in the city dropped by nearly half between the early 18th century and the mid-19th century (Landers, 1993). Overall mortality rates in London, estimated from both the Bills and Quaker records, were higher in the 17th century than elsewhere in England, but they declined rapidly from the latter quarter of the 18th century onward, and Landers attributes this decline primarily to decreases in neonatal, childhood, and early adulthood mortality. These differences in estimated health and mortality patterns might reflect variation in sample composition (e.g., level of aggregation, how representative they are of different socioeconomic statuses) or real variation in the pattern and timing of changes within and between populations (e.g., Landers (1993) highlights the variation in mortality and epidemiologic *regimes* within London alone).

Sex Differences in Mortality

The results of the analysis of the effect of sex on the risk of mortality among adults indicate that in the preindustrial sample, males and females faced approximately equal risks of dying. However, in the industrial sample, the risk of mortality was slightly lower for females compared to males. According to Gage (1994), declines in mortality and the second transition are associated with increased sex differentials in mortality that favor women. In most living populations, females are generally less frail than males, and both life expectancies are longer and age-specific mortality rates are lower for females in comparison to males (Heligman, 1983; Coale, 1991; Hill and Upchurch, 1995). These differentials occur because females are more resistant to many diseases and generally more highly buffered against environmental stressors than males are (Stinson, 1985). Males suffer more severe symptoms or face elevated risks of mortality from a variety of parasitic, infectious diseases and chronic or degenerative diseases (e.g., Hoff et al., 1979; Brabin and Brabin, 1992; Acuna-Soto et al., 2000; Klein, 2000; Blessmann et al., 2002; Laupland et al., 2003; Jansen et al., 2007; Falagas et al., 2008; Noymer, 2009).

These sex differentials in morbidity and mortality among living populations have both behavioral and biological causes. Sex hormones are strongly implicated as causal in these sex differentials because, generally, estrogens enhance immunocompetence and androgens reduce it (Waldron, 1984). Sex hormones also affect behaviors, such as aggression or ranging patterns, which can influence an individual's risk of exposure to disease (Grossman, 1985; Ansar Ahmed et al., 1999; Klein, 2000; Roberts et al., 2001; Janele et al., 2006). Sex hormones may also play a role in the development of some degenerative diseases, for example, by affecting atherogenesis (plaque formation within arteries) and thereby influencing risks of cardiovascular disease (Klein, 2000). Behavioral differences between men and women can also strongly affect their respective risks of degenerative diseases; for example, men smoke cigarettes and consume alcohol to a greater extent than women in most populations, and these behaviors are associated with excess mortality from coronary heart disease, certain cancers, and cirrhosis of the liver (Choi and McLaughlin, 2007).

The results of the analysis done here, which show a higher risk of mortality for males in the industrial sample, indicate that sex differentials favoring females predate the second transition. Similar excess mortality among males during industrialization has also been observed in some historical data, and some researchers have attributed the differentials to social factors, including accidents at work (Vallin, 1991). The difference between the preindustrial and industrial samples (i.e., a difference in mortality risk between males and females in the latter, but not in the former) in this study might reflect the existence of a higher infectious disease burden within London at the time of the Industrial Revolution compared to the previous period. That is, males might have been generally more frail than females in both samples, because of innate biological differences that favor females, but it was only in the context of an environment with more infectious diseases (or higher rates of transmission) in the industrial sample, which could have more strongly affected the health and mortality risks of males than those of females, that a discernable difference in mortality risk between the sexes emerged.

Conclusion

The results of this study reveal discernable changes in the patterns of frailty and mortality, estimated from skeletal samples, that occurred just before the second epidemiologic transition itself took place in London. In general, risks of adult mortality declined from the preindustrial to the industrial period, and physiological stress was less strongly associated with risks of adult mortality in the later period. Females appear to have benefited more than males did from positive changes in health (or, alternatively, were less strongly impacted than males by negative changes) associated with living in industrialized London. However, the period right before the transition was not similarly favorable for children in London, as they appeared to face increased risks of mortality and greater excess mortality as a result of physiological stress compared to adults.

Though this study reveals new information about variation in health in the population of London prior to the second transition, it is not without limitations. One limitation is the lack of detailed information about mortality patterns at later adult ages resulting from the use of traditional methods to estimate adult ages in these cemeteries. The traditional age-estimation methods used by researchers at the Museum of London, which are widely used by bioarchaeologists, yield interval estimates with a terminal age category of 50+ years. The lack of more precise estimates for older adults makes it difficult to examine the changes in longevity (e.g., life expectancy at birth) that might have occurred before and during the transition. Evidence of increases in longevity in the industrial sample would bolster the results of this study that indicate decreases in frailty among adults before the transition. In the future, analyses of these cemeteries would benefit from the application of age-estimation methods that yield point estimates of age, particularly for older adults. One such method is transition analysis (Boldsen et al., 2002). In addition to allowing for the estimation of point estimates of age at all possible adult ages, this method is preferable to other age-estimation methods because it uses statistical methods to avoid the problem of age mimicry associated with traditional methods (age mimicry refers to estimated ages that are biased toward the age distribution of a known-age reference sample) (Bocquet-Appel and Masset, 1982; Boldsen et al., 2002). Transition analysis has been informatively applied to several bioarchaeological studies (Boldsen, 2005; DeWitte, 2009; DeWitte, 2010a, ,2010b), and there is dedicated software to apply the method to age-related changes on the pubic symphysis, iliac auricular surface, and cranial sutures (Boldsen et al., 2002).

Another limitation of the current study is its focus on lower socioeconomic status people. This narrow focus raises questions about what was happening among higher-status individuals and whether there were differences in the health and mortality patterns among different status groups in London before the transition. Fortunately, the Museum of London Centre for Human Bioarchaeology collection includes the Chelsea Old Church cemetery, which was used primarily for the burial of high-status people during the 18th to 19th centuries, so such a comparison is possible in the future. A comparison of high- and low-status cemeteries would allow for an examination of the assumption that high status acted as a buffer against the stressors associated with industrialization.

Acknowledgments

I am very grateful to the Museum of London Centre for Human Bioarchaeology, and in particular Jelena Bekvalac and Rebecca Redfern, for providing access to data from the cemeteries used in this study. I would also like to thank Molly Zuckerman for organizing and inviting me to participate in the 2011 SCIAA Post-Doctoral Fellows Conference.

References

Acuna-Soto R, Maguire JH, Wirth DF. 2000. Gender distribution in asymptomatic and invasive amebiasis. Am J Gastroenterol 95:1277–1283.

Ahsan Karar Z, Alam N, Kim Streatfield P. 2009. Epidemiologic transition in rural Bangladesh, 1986–2006. Glob Health Action 2. Available at: http://www.globalhealthaction.net/index.php/gha/article/view/1904/2301 (accessed November 4, 2013).

Albala C and Vio F. 1995. Epidemiologic transition in Latin America: the case of Chile. Public Health 109:431–442.

Ansar Ahmed SA, Hissong BD, Verthelyi D, Donner K, Becker K, Karpuzoglu-Sahin E. 1999. Gender and risk of autoimmune diseases: possible role of estrogenic compounds. Environ Health Perspect 107(S5):681–686.

Barrett R, Kuzawa CW, McDade T, Armelagos GJ. 1998. Emerging and re-emerging infectious diseases: the third epidemiologic transition. Annu Rev Anthropol 27:247–271.

Blessmann J, Van Linh P, Nu PA, et al. 2002. Epidemiology of amebiasis in a region of high incidence of amebic liver abscess in central Vietnam. Am J Trop Med Hyg 66:578–583.

Bocquet-Appel JP and Masset C. 1982. Farewell to paleodemography. J Hum Evol 11:321–333.

Boldsen JL. 2005. Analysis of dental attrition and mortality in the Medieval village of Tirup, Denmark. Am J Phys Anthropol 126:169–176.

Boldsen JL, Milner GR, Konigsberg LW, Wood JW. 2002. Transition analysis: A new method for estimating age from skeletons. In: Hoppa RD and Vaupel JW, editors. Paleodemography: Age distributions from skeletal samples. Cambridge: Cambridge University Press. p 73–106.

Brabin L, Brabin BJ. 1992. Parasitic infections in women and their consequences. Adv Parasitol 31:1–60.

Brickley M, Miles A. 1999. The cross bones burial ground, redcross way, Southwark, London: Archaeological excavations (1991–1998) for the London underground limited jubilee line extension project (MoLAS monograph 3). London: Museum of London.

Brooks S, Suchey J. 1990. Skeletal age determination based on the os pubis: a comparison of the Ascadi-Nemeskéri and Suchey-Brooks methods. Hum Evol 5:227–238.

Brothwell D. 1981. Digging up bones: The excavation, treatment and study of human skeletal remains. London: British Museum of Natural History.

Buckberry J, Chamberlain A. 2002. Age estimation from the auricular surface of the ilium: a revised method. Am J Phys Anthropol 119:231–239.

Buikstra JE, Ubelaker DH, editors. 1994. Standards for data collection from human skeletal remains: Proceedings of a seminar at the field museum of natural history (Arkansas archaeology research series 44). Fayetteville: Arkansas Archeological Survey Press.

Cardoso HFV, Garcia S. 2009. The Not-so-Dark Ages: ecology for human growth in medieval and early twentieth century Portugal as inferred from skeletal growth profiles. Am J Phys Anthropol 138:136–147.

Carter K, Soakai TS, Taylor R, et al. 2011. Mortality trends and the epidemiologic transition in Nauru. Asia Pac J Public Health 23:10–23.

Choi BG, McLaughlin MA. 2007. Why men's hearts break: cardiovascular effects of sex steroids. Endocrinol Metab Clin North Am 36:365–377.

Coale AJ. 1991. Excess female mortality and the balance of the sexes in the population: an estimate of the number of "missing females." Popul Dev Rev 17:517–523.

Coste J, Bernardin E, Jougla E. 2006. Patterns of mortality and their changes in France (1968–99): insights into the structure of diseases leading to death and epidemiologic transition in an industrialised country. J Epidemiol Community Health 60:945–955.

DeWitte SN. 2009. The effect of sex on risk of mortality during the Black Death in London, A.D. 1349–1350. Am J Phys Anthropol 139:222–234.

DeWitte SN. 2010a. Age patterns of mortality during the Black Death in London, A.D. 1349–1350. J Archaeol Sci 37:3394–3400.

DeWitte SN. 2010b. Sex differentials in frailty in medieval England. Am J Phys Anthropol 143:285–297.

DeWitte SN, Wood JW. 2008. Selectivity of the Black Death with respect to preexisting health. Proc Natl Acad Sci U S A 105:1436–1441.

Eisenberg LE. 1991. Mississippian cultural terminations in Middle Tennessee: What the bioarcheological evidence can tell us. In: Powell ML, Bridges PS, Mires AM, editors. What mean these bones. Tuscaloosa: University of Alabama Press. p 70–88.

Falagas ME, Vardakas KZ, Mourtzoukou EG. 2008. Sex differences in the incidence and severity of respiratory tract infections. Respir Med 102:627.

Fenner F. 1970. The effects of changing social organization on the infectious diseases of man. In: Boyden S, editor. The impact of civilization on the biology of man. Canberra: Australian National University Press. p 48–76.

Floud R, Harris B. 1997. Health, height, and welfare: Britain, 1700–1980. In: Steckel RH and Floud R, editors. Health and welfare during industrialization. Chicago: University of Chicago Press. p 91–126.

Gage TB. 1988. Mathematical hazard models of mortality: an alternative to model life tables. Am J Phys Anthropol 76:429–441.

Gage TB. 1991. Causes of death and the components of mortality: testing the biological interpretations of a competing hazards model. Hum Biol 3:289–300.

Gage TB. 1994. Population variation in cause of death: level, gender, and period effects. Demography 31:271–296.

Gage TB. 2005. Are modern environments really bad for us? Revisiting the demographic and epidemiologic transitions. Yearb Phys Anthropol 48:96–117.

Grossman CJ. 1985. Interactions between the gonadal steroids and the immune system. Science 227:257–261.

Gustafson G, Koch G. 1974. Age estimation up to 16 years of age based on dental development. Odontol Revy 25:297–306.

Heligman L. 1983. Patterns of sex differentials in mortality in less developed countries. In: Lopez AD and Ruzicka LT, editors. Sex differentials in mortality: Trends, determinants, and consequences. Canberra: Department of Demography, Australian National University. p 7–32.

Hill K, Upchurch DM. 1995. Gender differences in child health: evidence from the demographic and health surveys. Popul Dev Rev 21:127–151.

Hill K, Vapattanawong P, Prasartkul P, Porapakkham Y, Lim SS, Lopez AD. 2007. Epidemiologic transition interrupted: a reassessment of mortality trends in Thailand, 1980–2000. Int J Epidemiol 36:374–384.

Hoff R, Mott KE, Silva JF, et al. 1979. Prevalence of parasitaemia and seroreactivity to *Trypanosoma cruzi* in a rural population of northeast Brazil. Am J Trop Med Hyg 28:461–466.

Holman DJ. 2005. mle: A programming language for building likelihood models. Version 2.1 ed. Seattle.

Huck P. 1995. Infant mortality and living standards of English workers during the industrial revolution. J Econ Hist 55:528–550.

Huicho L, Trelles M, Gonzales F, Mendoza W, Miranda J. 2009. Mortality profiles in a country facing epidemiologic transition: an analysis of registered data. BMC Public Health 9:47.

Iscan MY, Loth SR, Wright RK. 1984. Age estimation from the rib by phase analysis: white males. J Forensic Sci 29:1094–1104.

Iscan MY, Loth SR, Wright RK. 1985. Age estimation from the rib by phase analysis: white females. J Forensic Sci 30:853–863.

Janele D, Lang T, Capellino S, Cutolo M, Da Silva JA, Straub RH. 2006. Effects of testosterone, 17beta-estradiol, and downstream estrogens on cytokine secretion from human leukocytes in the presence and absence of cortisol. Ann N Y Acad Sci 1069:168–182.

Jansen A, Stark K, Schneider T, Schoneberg I. 2007. Sex differences in clinical leptospirosis in Germany: 1997–2005. Clin Infect Dis 44:e69–e72.

Klein SL. 2000. The effects of hormones on sex differences in infection: from genes to behavior. Neurosci Biobehav Rev 24:627–638.

Landers J. 1993. Death and the metropolis: Studies in the demographic history of London, 1670–1830. Cambridge: Cambridge University Press.

Larsen, C. 1997. Bioarchaeology: Interpreting behavior from the human skeleton. Cambridge: Cambridge University Press.

Laupland KB, Church DL, Mucenski M, Sutherland LR, Davies HD. 2003. Population-based study of the epidemiology of and the risk factors for invasive *Staphylococcus aureus* infections. J Infect Dis 187:1452–1459.

Lewis ME. 2002. Impact of industrialization: comparative study of child health in four sites from medieval and postmedieval England (A.D. 850–1859). Am J Phys Anthropol 119:211–223.

Lewis ME. 2007. The bioarchaeology of children: Perspectives from biological and forensic anthropology. Cambridge: Cambridge University Press.

Lewis ME, Gowland R. 2007. Brief and precarious lives: infant mortality in contrasting sites from medieval and post-medieval England (AD 850–1859). Am J Phys Anthropol 134:117–129.

Lovejoy C, Meindl R, Pryzbeck T, Mensforth R. 1985. Chronological metamorphosis of the auricular surface of the ilium: a new method for determining adult age at death. Am J Phys Anthropol 68:15–28.

Malina RM, Peña Reyes ME, Little BB. 2008. Epidemiologic transition in an isolated indigenous community in the Valley of Oaxaca, Mexico. Am J Phys Anthropol 137:69–81.

Mays S, Brickley M, Ives R. 2008. Growth in an English population from the industrial revolution. Am J Phys Anthropol 136:85–92.

McKeown T. 1976. The rise of modern populations. New York: Academic Press.

McKeown RE. 2009. The epidemiologic transition: changing patterns of mortality and population dynamics. Am J Lifestyle Med 3:19S–26S.

McNeill W. 1980. Migration patterns and infection in traditional societies. In: Stanley N and Joske R, editors. Changing disease patterns and human behaviour. London: Academic Press. p 28–36.

Meindl RS, Lovejoy CO, Mensforth RP, Carlos LD. 1985. Accuracy and direction of error in the sexing of the skeleton: implications for paleodemography. Am J Phys Anthropol 68:79–85.

Mensforth RP, Lovejoy CO, Lallo JW, Armelagos GJ. 1978. The role of constitutional factors, diet, and infectious disease in the etiology of porotic hyperostosis and periosteal reactions in prehistoric infants and children. Med Anthropol 2:1–58.

Millward R and Bell F. 2001. Infant mortality in Victorian Britain: the mother as medium. Econ Hist Rev 54:699–733.

Milner GR. 1991. Warfare in late prehistoric west-central Illinois. Am Antiq 56:581–603.

Milner GR, Wood JW, Boldsen JL. 2008. Paleodemography. In: Katzenberg M and Saunders S, editors. Biological anthropology of the human skeleton. New York: Wiley-Liss. p 561–600.

Moorees CFA, Fanning EA, Hunt EE. 1963. Formation and resorption of three deciduous teeth in children. Am J Phys Anthropol 21:205–213.

Noymer A. 2009. Testing the influenza-tuberculosis selective mortality hypothesis with Union Army data. Soc Sci Med 68:1599–1608.

Ogden AR, Pinhasi R, White WJ. 2007. Gross enamel hypoplasia in molars from subadults in a 16th-18th century London graveyard. Am J Phys Anthropol 133:957–966.

Omran AR. 1971. The epidemiologic transition. A theory of the epidemiology of population change. Milbank Mem Fund Q 49:509–538.

Ortner DJ. 2003. Identification of pathological conditions in human skeletal remains. Amsterdam: Academic Press.

Pasceri V, Cammarota G, Patti G, et al. 1998. Association of virulent *Helicobacter pylori* strains with ischemic heart disease. Circulation 97(17):1675–1679.

Redfern RC, DeWitte SN. 2011. A new approach to the study of Romanization in Britain: a regional perspective of cultural change in late iron age and roman dorset using the siler and gompertz-makeham models of mortality. Am J Phys Anthropol 144:269–285.

Risquez A, Echezuria L, Rodriguez-Morales AJ. 2010. Epidemiologic transition in Venezuela: relationships between infectious diarrheas, ischemic heart diseases and motor vehicles accidents mortalities and the Human Development Index (HDI) in Venezuela, 2005–2007. J Infect Public Health 3:95–97.

Roberts CA, Manchester K. 2005. The archaeology of disease. Ithaca: Cornell University Press.

Roberts CW, Walker W, Alexander J. 2001. Sex-associated hormones and immunity to protozoan parasites. Clin Microbiol Rev 14:476–488.

Rogers TL. 2005. Determining the sex of human remains through cranial morphology. J Forensic Sci 50:493–500.

Scheuer L, Black S. 2000. Developmental juvenile osteology. New York: Academic Press.

Scheuer JL, Musgrave JH, Evans SP. 1980. The estimation of late fetal and perinatal age from limb bone length by linear and logarithmic regression. Ann Hum Biol 7:257–265.

Smith BH. 1991. Standards of human tooth formation and dental age assessment. In: Kelley M and Larsen CS, editors. Advances in dental anthropology. New York: Wiley-Liss. p 143–168.

Stevens G, Dias RH, Thomas KJA, et al. 2008. Characterizing the epidemiologic transition in Mexico: national and subnational burden of diseases, injuries, and risk factors. PLoS Med 5:e125.

Stinson S. 1985. Sex differences in environmental sensitivity during growth and development. Am J Phys Anthropol 28:123–147.

Stuart-Macadam P. 1985. Porotic hyperostosis: representative of a childhood condition. Am J Phys Anthropol 66:391–398.

Sumi H, Yokosuka O, Seki N, et al. 2003. Influence of hepatitis B virus genotypes on the progression of chronic type B liver disease. Hepatology 37(1):19–26.

Sutherland LD, Suchey JM. 1991. Use of the ventral arc in pubic sex determination. J Forensic Sci 36:501–511.

Usher BM. 2000. A multistate model of health and mortality for paleodemography: Tirup cemetery. Ph.D. Dissertation. University Park: Pennsylvania State University.

Vallin J. 1991. Mortality in Europe from 1720 to 1914: Long-term trends and changes in patterns by age and sex. In: Schofield R, Reher DS, Bideau A, editors. The decline of mortality in Europe. Oxford: Clarendon Press. p 38–67.

Vaupel JW, Manton KG, Stallard E. 1979. The impact of heterogeneity in individual frailty on the dynamics of mortality. Demography 16:439–454.

Waldron I. 1984. The role of genetic and biological factors in sex differences in mortality. In: Lopez AD and Ruzicka LT, editors. Sex differentials in mortality: Trends, determinants, and consequences. Canberra: Department of Demography, Australian National University. p 141–164.

Walker PL. 2008. Sexing skulls using discriminant function analysis of visually assessed traits. Am J Phys Anthropol 136:39–50.

Walker PL, Bathurst RR, Richman R, Gjerdrum T, Andrushko VA. 2009. The causes of porotic hyperostosis and cribra orbitalia: a reappraisal of the iron-deficiency-anemia hypothesis. Am J Phys Anthropol 139:109–125.

Weston D. 2008. Investigating the specificity of periosteal reactions in pathology museum specimens. Am J Phys Anthropol 137(1):48–59.

Williams BA, Rogers TL. 2006. Evaluating the accuracy and precision of cranial morphological traits for sex determination. J Forensic Sci 51:729–735.

Wood JW, Holman DJ, O'Connor KA, Ferrell RJ. 2002. Mortality models for paleodemography. In: Hoppa RD and Vaupel JW, editors. Paleodemography: Age distributions from skeletal samples. Cambridge: Cambridge University Press. p 129–168.

Wrigley EA, Schofield RS. 1981. The population history of England, 1541–1871: A reconstruction. London: Edward Arnold.

Chapter 4
The Wider Background of the Second Transition in Europe: Information from Skeletal Material

Nikola Koepke
Departament d'Història i Institucions Econòmiques, Universitat de Barcelona, Barcelona, Spain

Introduction

The central topics at hand here are the potential causes and consequences of the second epidemiologic transition. But what does a transition mean in this context? Here, the concept does not—as is sometimes the case—relate to an advanced step in economic and political changes; instead, this study addresses aspects of the demographic and *classic* epidemiologic transitions. The demographic transition is a multistage, descriptive model of secular declines in mortality, fertility, longevity, and consequent population growth. The epidemiologic transition, first described by Omran (1971), models the trends in cause-specific mortality that comprised the demographic transition, specifically secular declines in mortality from acute, epidemic infectious disease with increased mortality from degenerative disease (see Gage, 2005). Having been placed into a larger evolutionary context that recognizes multiple transitions throughout time—namely, the first, a rise in mortality from acute, epidemic infectious disease coincident with the Neolithic transition, and the third, which is ongoing and consists of increased mortality from emerging, reemerging, and antibiotic-resistant infections—the classic epidemiologic transition is now known as the second transition (Barrett et al., 1998). The demographic and second epidemiologic transitions coincided with the industrial revolution in Western Europe and the United States; however, it is not yet possible to completely unfold the relationships between the transitions and the large-scale social, economic, and ecological changes that accompanied industrialization. So far, research indicates that industrialization started first, followed by the transitions, but more research is required to understand how these two phenomena fit into the gradual, continuous process of industrialization, urbanization, and the creation of modern environments.

However, before starting to disentangle the more proximate causes of the second transition, it is important to address more ultimate factors and examine the foundations of the transition, namely, the standard of living in the centuries preceding the transition. For Europe, it is often assumed that the biological standard of living stagnated between the first epidemiologic transition, associated with the emergence of agriculture and sedentism, and the second. This picture, however, is based on evidence generated from single sites. Is this conclusion correct, overall, or does the archaeological record yield evidence of changes in biological well-being that predate the second transition?

This paper has two aims. The first is to present evidence on the standard of living across Europe for the centuries preceding the second transition and test whether conditions were indeed stagnant. The second is to assess whether regional differences in the standard of living existed, which in turn might be responsible for regional and national-level differences in the trajectory of the transition throughout the continent. As the centuries under investigation are primarily represented by archaeological rather than historical evidence, nutritional status, represented by adult stature estimated from skeletal remains, is used as a proxy for the biological standard of living and biological well-being. The nutritional status, or net nutrition, of a population is a sensitive indicator of the many environmental conditions that influence human health and well-being. It is comprised of several direct factors, namely, dietary quality and quantity, disease burden, and physical stress (workload). These are both directly (Curtis et al., 2005) and indirectly influenced by a given populations' environment (e.g., Al-Dabbagh and Ebrahim, 1984; Scott and Duncan, 2002; Ulijaszek, 2006).

This study quantitatively assesses the net nutrition of Europeans from the 8th century BC to the 18th century AD, differentiating between three major regions: the Mediterranean Central-Western and North-Eastern Europe. In addition to the temporal and regional comparison, this study also discusses potential determinants of patterns and differences detected and tests their effects by applying quantitative methods. This type of broad-sweeping, long-term, and multiregional analysis is rare in the literature, as in general only data series on single regions or shorter time frames are available (e.g., Angel, 1984; Steckel, 2004; Koepke and Baten, 2005; Cardoso and Gomez, 2009; Mummert et al., 2011). The data on stature used here comes from archaeologically derived skeletons recovered from cemeteries throughout Europe. Similarly, the data used to assess dietary quantity and quality, disease load, and other influences on stature were mostly derived from archaeological evidence. The quantitative methods employed, however, are those of applied econometric historic research. Thus, this study follows an interdisciplinary approach.

The analysis addresses the following questions: What was the overall trajectory of the nutritional status of premodern Europeans? Which periods involved improved environmental conditions, as reflected by nutritional status? How did environmental conditions alter in various regions from the early Iron Age onwards? What are the underlying causes of any temporal changes and regional differences in nutritional status detected, with an understanding that these constitute the biological background conditions of the second transition?

Background

Mean Stature as an Indicator of Nutritional Status

Longitudinal growth in humans is not only influenced by endogenous factors like genetics and hormones, but also by exogenous factors such as nutrition and the disease-scape (Henry and Ulijaszek, 1996; Cameron, 2002). Using mean adult stature as an indicator of a population's well-being is based on the effect of nutritional status on growth and development during the prenatal period, childhood, and adolescence. UNICEF (1990, 1998) recognizes three interrelated, hierarchical levels of determinants affecting child growth and development in relation to nutritional status. These include basic determinants, such as potential resources and cultural, social, economic, and political structures in the larger society, family, and community resources, information, and education; underlying determinants, such as access to health services and the healthfulness of the environment, household food security, psychosocial care, hygienic practices, breastfeeding, and care for women; and finally intermediate determinants, namely, the health of the child and the adequacy of their dietary intake. The child's survival, growth, and development, as well as whether they attain their genetic stature potential (see WHO standards, 2007), represent the outcome of these determinants. Therefore, temporal or regional variation in the mean stature of a genetically homogenous population is driven by differences in a diverse set of interrelated environmental factors. Focusing on mean stature of a population largely levels out the influence of genetics, allowing direct comparisons across populations (see van Wieringen, 1972; Mascie-Taylor and Bogin, 1995; Bogin, 1999; Silventoinen, 2003; Curtis et al., 2005; Floud et al., 2011).

During growth and development, chronic undernutrition can lead to reduced growth as the body allocates resources away from developmental processes and towards basic bodily processes. Depending on the degree of undernutrition, this can result in growth retardation or cessation for a period of time. In the case of a prolonged deficiency, somatic growth can even be permanently impaired, which can result in stunted final stature (Waterlow, 1972; Eveleth and Tanner, 1976; Floud et al. 1990; Rey and Bresson, 1997; Wahl and Kokabi, 1999; Stinson, et al. 2012). More temporary insults can be compensated for by catch-up growth (i.e., stature velocity above the limits of normal for age); however, this requires a marked improvement in nutritional status in order to occur (Whitehead, 1977). In general, nutritional status during infancy and weaning have the greatest impact on attainment of stature potential because growth velocity is the highest during this period. Additionally, weaning also exposes toddlers to substantial health threats (Poskitt, 1999; Scott and Duncan, 2002).

Risk factors for impaired growth often occur in concert, amplifying their detrimental impact (Walker et al., 2007). For instance, inadequate housing and sanitation are frequently found alongside with qualitative and quantitative dietary inadequacy. As a result, populations who have enjoyed greater buffering from risk factors, such as those of higher social status, often have a comparably taller mean stature than do those who have been insufficiently buffered or subject to increased stressors, such as those of lower status (see Huber, 1967; Larsen, 1997).

Therefore, final adult stature represents the "cumulative record of the nutritional and health history of a person or population" (Bogin and Keep, 1999: 333). In turn, mean stature is a reliable proxy for the nutritional status of a population and its biological standard of living, both in the present (WHO, 1995) and the past (Steckel, 2009). The fact is of special interest for periods for which any reasonable statistically quantifiable data on welfare measures are scarce or unavailable.

The Skeleton as a Source of Information on Health and Well-Being

The skeleton records numerous indicators of overall health and exposure to stressors. However, most skeletal stress indicators are nonspecific, such as linear enamel hypoplasias, meaning that they can have numerous, often indistinguishable causes (Larsen, 1997). These indicators most importantly generally record single episodes of stress, rather than cumulative exposure to stressors. In contrast, final adult stature bears the advantage of measuring outcomes, not inputs (Steckel, 1995), and represents cumulative living conditions over the complete course of growth and bodily development, including possible catch-up growth.

Adult stature is estimated from long bone length, specifically that of the femur, due to the tight correlation between these metrics (Verhoff et al., 2006). Femora also tend to preserve well in the archaeological record, meaning that femoral stature data is available for many time periods and regions that are underrepresented—or unrepresented—by other types of data on health and well-being. Importantly, their availability enables a long-term, multiregional, quantitative assessment of the effect of various potential determinants on stature across populations.

Potential Determinants of Stature Development

As discussed earlier, numerous interrelated determinants influence nutritional status, growth and development, and the attainment of final adult stature. Given the fluid dynamics between these factors, it can be analytically difficult to parse them out into different levels, particularly when dealing with historical and archaeological evidence, as in the present study. Additionally, while some of the determinants likely remained stable over substantial time periods in the period under study, others shifted markedly over time. Little substantial progress was made in health care or sanitation prior to the 18th century AD in Europe (Scheidel, 2010). Therefore, we can assume that on average disease burden and parasite exposure did not lessen substantially over time (e.g., malaria was widely common) (see Reinhard and Pucu de Araújo, this volume). Similarly, child labor (determining the physical stress exposure on growth) seem not to have changed considerably (Hindman, 2009). In contrast, dietary quality and quantity are expected to have changed markedly over time and vary greatly by region and cultural period. Importantly, these two factors are multilayered: there are many indirect as well as direct causes of nutritional deprivation (Sen, 1984), ranging from food shortages related to agricultural practices or climate change to a lack of cultural knowledge about dietary adequacy because of educational shortfalls. However, due to the constrained availability of relevant archaeological data, not all possible factors could be assessed in the present study. The following surveys the direct and indirect influences on net nutrition that could be assessed and included.

Dairy and Beef Consumption

The consumption of milk, a high quality protein, entails ingesting a range of vitamins and microfibers and has been found to be of particular importance to overall health and attainment of adult status in more recent centuries (Scott and Duncan, 2002; Haines, 1998; Baten, 1999; Steckel and Prince, 2001; Moradi and Baten, 2005; Dror and Allen, 2011). For instance, scholars have found that in the centuries preceding refrigeration, a direct relationship existed between mean stature and the geographical proximity and supply of milk (Komlos, 1989; Woolgar, 2006). Here, it is assumed that there is a direct relationship between the geographical proximity and supply of dairy and beef products their consumption and thus the mean stature in a region (see Koepke and Baten, 2008) and mean stature in the past.

Land per capita

This factor represents the maximum space available for food production in relation to population density, and is a key indicator of the environmental health and environmental quality (Cohen and Armelagos, 1984; Armelagos, 1990; Livi-Bacci, 1991; Steckel and Rose, 2002). Overall, it is predicted that nutritional status and land *per capita* were positively correlated in the past.

Climate Change

Climatic conditions can exert a great effect upon agricultural productivity, the quality and quantity of food supply, and general living conditions. For instance, in recent centuries, particularly in Central-Western and North-Eastern Europe, a negative effect has been detected between agricultural production (harvest and husbandry) and cooler periods, whereas warmer periods were associated with greater productivity (Pfister, 1988; Baten, 2002; Grove, 2002). Therefore, a positive correlation is expected between temperature and nutritional status since in the pre-industrial period effective technological adaptations to climate did not exist.

Cultivation Methods

A change in cultivation methods can exert a substantial positive influence on agricultural productivity (Wiese and Zils, 1987; Grupe, 2003), reducing risk factors and increasing outputs. Such innovations include iron plows and the replacement of two-field rotation with three-field rotation during the medieval period. In two-field rotation, land was divided into two sections: one was planted, while the other lay fallow 1 year, and the pattern was reversed the next year. In three-field, land was divided into three parts: the first was planted in the autumn with winter wheat or rye, the second in the spring with legumes, and the third kept fallow; they were rotated so that every 3 years each section was fallow. This increased both productivity and soil fertility—courtesy of the legumes—and improved overall nutrition in the areas in which it was employed (Küster, 2006).

Urban Share

In the centuries preceding the industrial revolution and the second transition, this factor, defined as the percentage of urban-like settlements within the total of settlements, encompasses all of the conditions potentially involved in settlements with high population density: inadequate sanitary conditions, agglomerations of refuse disposal, and cramped housing (e.g., parasitism, interpersonal violence). Here, in the 8th century BC to the 18th century AD, the urban share is expected to be linked to a higher risk of infectius disease, with chronic, nonepidemic conditions most likely preceding the first transition and acute, epidemic conditions following it (see Reinhard and Pucu de Araújo, this volume). Moreover, urban settings may be linked to reduced dietary quality and quantity, as urbanites are more separated from *de facto* nontradable goods and have less access to subsistence possibilities than do rural residents. Overall, for the centuries under consideration, it is expected that an increased urban share will exert a negative effect upon nutritional status.

War and Conflict

War and conflict have a decidedly negative impact on the health and well-being of those directly involved in the conflict but can also detrimentally affect involved civilians. Consequences for civilians can range from psychosocial stress, interrupted food supplies, and a greater incidence of acute epidemic infectious disease due to troop movement or social unrest to decreased agricultural productivity from fields left fallow or scorched earth tactics. It is predicted that periods of war and conflict will be related to worsened nutritional status.

The Disease-Scape

Between the 8th cent. BC and 18th century AD, various characteristics of the disease-scape are expected to have complex effects upon population nutritional status and, consequently, mean stature. If a disease event, such as an epidemic, was long lasting and occurred on a supraregional level, nutritional status would likely worsen. However, a positive impact is also possible if, as with the 14th-century Black Death, the demographic impact was sufficiently large to improve living conditions for the survivors.

The "Roman Impact": Roman Impact Integration to the Roman Empire

Romanization, the "Roman Impact" or inclusion in the Roman Empire, is another factor under consideration. Romanization is frequently regarded as having an ultimately positive effect on health and well-being, due to such features as a secured basic food supply as part of the system of organized provisions for Roman citizens (see Ward-Perkins, 2005). However, it presumably could also exert a negative effect upon health, due to factors such as the great increase in trade and troop movements facilitating disease epidemics and increased exposure to lead through contact with water pipes, toys, etc., which can negatively affect growth (Stinson, 2000). Additionally, weaning age was low in Roman society in comparison with contemporaries (Dittmann and Grupe, 2000), milk consumption was likely low in Romanized communities, as it was seen as barbaric (see Poseidonius, Tacitus, and Caesar in Thüry, 2007), and the introduction of sanitary systems may have often increased the risk of waterborne diseases (Thüry, 2001). The effects of this, as evaluated here, may have been complex, varying between core regions of the empire and Romanized provinces. Overall, there may be no clear effect, as the various subsumed aspects may have had equalized each other.

Gender Inequality

Historically, Europe seems to have consisted mainly of patriarchal societies, which raises the question of whether this, and associated degrees of gender inequality, may have affected health and well-being. While females generally have a smaller mean stature than males due to biological factors, stature dimorphism can also vary due to gender-related inequalities in environmental and socioeconomic conditions (Klasen, 2002; George, 2006; Olds, 2006). There are two important interpretive considerations involved in analyzing for gender inequality-produced stature dimorphism. First, that the relative stature difference does not increase with mean stature (Gustafsson and Lindenfors, 2004), and second, that the biological buffering that females enjoy over males, likely courtesy of the immunomodulatory effects of sex steroid hormones (see Fish, 2008), does not seem to exert a substantial effect on sexual dimorphism in growth, development, and stature (Stinson, 2000). Instead, gender-biased parental investment, such as female discrimination and neglect, has been found to have a greater influence on stature (e.g., Bogin, 1999; Guntupalli and Moradi, 2009; Harris et al., 2009). While unfavorable nutritional status among girls can affect the statures of both genders through intergenerational effects—because adverse conditions *in utero* hurt both male and female fetuses (Osmani and Sen, 2003)—in the postnatal period, girls are additionally disadvantaged through patterns of neglect and discrimination in strongly patriarchal societies, such as in India (Sen, 1984; Perkins et al., 2011; Guntupalli, this volume). Overall, females are predicted to have experienced comparably worse nutritional status than males in the period under study, manifesting as reduced mean stature.

Methods and Materials

Data on Stature

Data was compiled from 18,502 adult (≥18 years) individuals derived from 484 archaeological sites distributed throughout three major regions in Europe (for data sources, see Koepke, 2008). Here, the Mediterranean includes Italy, Southern France, Spain, Portugal, and the Balkans. Central-Western Europe refers to the Benelux, Northern and Eastern France, the United Kingdom, Western and Southern Germany, Switzerland, and Austria. Northern and Eastern Europe includes Scandinavia, Poland, Northern and Eastern Germany, and Hungary. These sites span the 8th century BC to the 18th century AD, but in terms of a demographic analysis, the crude temporal resolution of much of the archaeological data allows for birth cohorts of 100-year duration rather than a narrower range. Only individuals recovered from cemeteries associated with attritional mortality and non-elite communities were included in an attempt to span the full spectrum of social strata and communities. When possible, socioeconomic status, reconstructed from mortuary context, was controlled for in order to evaluate social composition and potential social selectivity in relation to stature trends. However, a previous study has shown that for the period—and populations—under study, high- to middle-status individuals were significantly but only marginally taller than those of lower status for the whole of Europe and insignificantly taller when parsed out into the three regions studied here (Koepke, 2008).

Lastly, genetic growth potential and migration must be controlled for to guarantee reliable results in relation to trends in stature. Various lines of evidence suggest that the growth potential for stature across Europe is highly uniform, both in the present and past (Roth, 1984; Quiroga Valle, 1998; Komlos et al., 2003; Lao et al., 2008), and that the genetic landscape of Europe is very homogenous (Bertranpetit, 2012). Migration and geographical mobility are critical because of the importance of environmental conditions during the first years of life for adult stature. For the period under study, the most important migratory trend involved population movement, mainly by the Roman army, from the Mediterranean into Central and Western Europe between the 1st and 3rd centuries AD and migration by the North-Eastern European tribes into Central and Southern Europe between the 4th and 6th centuries. However, Koepke (2008) found that in both cases, migrants (identified by mortuary context) were not significantly different in stature from nonmigrants. Lastly, the populations under study must have an equivalent growth potential.

Stature Reconstruction and Conversion of Stature Reconstructions

Data on stature was derived from archival data as well as published analyses and reports on skeletons recovered from archaeological sites (see Koepke, 2008 for sources). These contained stature estimates rather than raw data—femoral metrics—which had been arrived at using a number of different regression equations. Problematically, however, different formulae tend to generate slightly different stature reconstructions, threatening comparability across studies and complicating their use in this study. To overcome this issue, conversion equations for recovering the original femoral metric data, in this case maximum femoral length, were generated and used to calculate standardized normative stature estimates from the recovered data.

The most commonly used regression equations for stature reconstruction of European inhumations—and those employed in the archival materials and published studies used here—are Breitinger (1937) for males, in combination with Bach (1965) for females; Trotter and Gleser (1952) for whites; and Pearson (1899), WolaDski (1953), and Manouvrier (1893). Conversion equations were generated for each of these formulae. The basic regression equations used to generate the standardized normative stature estimates, known as B & B, is are the formulae by Breitinger (1937) and Bach (1965). Breitinger and Bach's formulae were used as they are regarded as being more accurate than Trotter and Gleser (Formicola, 1993) for the relevant stature range. Also, Trotter and Gleser's formula tends to result in higher, and Pearson in lower, stature estimates, whereas B & B estimates fall between them; therefore, B & B estimates constitute a compromise estimate for these three most commonly used methods. Lastly, Breitinger (1937) and Bach (1965) were the most commonly employed formulae in the archival and original published stature reconstructions used in this study.

Equations

As mentioned earlier, Breitinger, Bach, Pearson, and Trotter and Gleser use maximum femur length. This metric, represented by $f1$, includes the proximal extremity and is measured with the femur in anatomical position. The conversion equations used to derive the original femoral metric data invert the stature formula used by the original authors:

$$H_X = b_X * f + a_X \qquad (4.1)$$

Here, H_X is the known stature estimate reconstructed from the femur length, f, using method X; b_X is the constant and a_X the intercept of the specific regression formula in use. Thus, the femur length, f, is given as

$$f = \frac{(H_X - a_X)}{b_X} \qquad (4.2)$$

In the second step, to generate a standardized, normative stature, the recovered femur length, f, is inserted into the stature equations from Breitinger (for males) and Bach (for females):

$$H_B = b_B * f + a_B \qquad (4.3)$$

By combining equation 4.2 with 4.3 H_X can be converted to H_B:

$$H_B = H_X \frac{b_B}{b_X} + a_B - a_X * \frac{b_B}{b_X} \qquad (4.4)$$

When femur length was actually provided in the original archival or published material, it was directly incorporated into the study, using Equation 4.5 or 4.13, depending on the sex. Sometimes, a slight asymmetry occurs between the lengths of the left and right femora in a single individual. This is insignificant for stature reconstruction (Byers, 2002), and in the cases when two separate, differing measurements for a pair of femora were provided, the mean of the two values was used. Reconstructed stature values are approximate, and estimation error might be relatively high for an individual. However, the use of mean stature makes this issue negligible.

STATURE ESTIMATIONS BY SEX
Since these formulae are different for males and females, separate conversion equations had to be generated for each sex. Importantly, when male and female values were mixed in the original archival or published materials, they were not used here. Equations for estimating male stature Breitinger's basic equation for males is

$$H_{Bmale} = 1.645f + 94.31 \qquad (4.5)$$

This forms the basis of the standardizing formula for estimating male stature.
   Pearson's equation, used for males, is

$$H_P = 1.88f + 80.0 \qquad (4.6)$$

Combining Pearson with Breitinger gives a conversion equation of

$$H_{Bmale} = 0.875 * H_P + 24.308 \qquad (4.7)$$

Pearson (1899) and Wolańksi's (1953) formulae are equivalent and were subject to the same conversion formulae.
   Trotter and Gleser's formula, for use on white males, is

$$H_{TG} = 2.58f + 52.5 \qquad (4.8)$$

Combining Trotter and Gleser (1952) with Breitinger gives a conversion equation of

$$H_{Bmale} = 0.638 * H_{TG} + 60.836 \qquad (4.9)$$

Manouvrier (1893) is the most commonly used equation for estimating the stature of skeletons recovered from France and Italy. Unlike the other formulae, however, Manouvrier is based on a nomogram, rather than regression, and employs total femoral length ($f2$), which includes the distal portion of the bone. To access the original femoral metric data used in published and archival stature reconstructions that employ Manouvrier, a separate conversion formula was generated. Additionally, two formulae were needed for this process because $f2$ must be translated into $f1$ to be used in Breitinger's formula for males. The first formula, with an adaptation factor, is

$$\Delta f = \frac{(f1 - f2)}{f2} \qquad (4.10)$$

For the purposes of this equation, $f1$ is identified here as f_M. The second formula is

$$f_M = (1 + \Delta f) * f2 \qquad (4.11)$$

Δf values were calculated for different stature groups on the basis of Manouvrier's nomograms, which give $f2$ as a ratio between H_M and stature-dependent factors. Using samples of individuals for which both $f1$ and $f2$ values were given, Δf was calculated. The plotting is based on 204 observations from the data compiled for this study. Since $f2$ is not based on a regression, the normative, derived stature following Breitinger is calculated directly from Equation 4.5, using f_M:

$$H_{Bmale} = 1.645 f_M + 94.31 \qquad (4.12)$$

EQUATIONS FOR ESTIMATING FEMALE STATURE
The basic equation for reconstructing female stature from Bach (1965) is

$$H_B = 1.313 f + 106.69 \qquad (4.13)$$

This forms the basis of the standardizing formula for estimating female stature.
 Pearson's equation for females is

$$H_P = 1.95 f + 71.66 \qquad (4.14)$$

Therefore, the conversion equation, for females from Pearson (1899) to Bach (1965), is

$$H_{Bfemale} = 0.673 * H_P + 58.49 \qquad (4.15)$$

Trotter and Gleser's equation, developed for white females, is

$$H_{TG} = 2.47 f + 54.1 \qquad (4.16)$$

Trotter and Gleser's equation (1952) results in the conversion equation

$$H_{Bfemale} = 0.532 * H_{TG} + 77.932 \qquad (4.17)$$

The conversion formula for Manouvrier's (1893) equation for females was calculated in the same manner as the male formula, using Equations 4.10 and 4.11. The plotting of Δf is based on 170 observations from the collected data. Applying Bach, and using f_M, the conversion equation is

$$H_{Bfemale} = 1.313 f_M + 106.69 \tag{4.18}$$

Stature Data from Cremains

The epiphyseal diameters of long bones, particularly the femur, are also used for stature reconstruction, and epiphyseal metric data, specifically that derived from cremated remains (cremains), was also used to generate mean stature. There is a tight correlation between epiphyseal diameter and long bone length (Rösing, 1977; Herrmann et al., 1990; May, 1997; Gehring and Graw, 2001), and the caput femoris ($f18$), or vertical diameter of the femoral head, is a commonly used metric for stature reconstruction (Herrmann et al., 1990). Numerous regression models exist for estimating stature from these metrics (see Wahl, 1982; Heußner, 1987); Rösing (1988) is one of the most commonly used and was employed here. Importantly, results from this model can be directly compared with those from Breitinger and Bach. Although, the standard error for stature reconstructions from cremains, including the caput femoris, is higher than that from inhumations, the R2 for Rösing's formula is approximately 0.7 (2002, personal communication, University of Ulm and Praxis Forensische Anthroplogie/Praxis Forensic Anthropology). Therefore, this study utilized stature reconstructions provided by Rösing (1977).

Information on Stature Determinants

Dairy Consumption

Values for dairy and beef supply (and consumption, by proxy) were generated from the only available source: cattle bones from archaeological faunal assemblages. While taphonomic biases made it impossible to determine the absolute number of animals available *per capita*, it was possible to determine the relative proportion of cattle bones to that of other large, domesticated animals present at a given site and therefore—roughly—their availability to a given population. While cattle husbandry was always to some extent multipurpose (see Luff, 1993; Crabtree, 1996; Bartosiewicz et al., 1997; Seetah, 2005), various archaeological and historical sources suggest that dairy was the central focus as it has a higher efficiency than meat as a protein source (Foley et al., 1972; Sherratt, 1981; Davis, 1987). The cattle share proxy was generated using faunal data ($N = \sim 2,000,000$ large animal bones) from 415 habitation sites in the three regions under study, dated to the 10th century BC until the 17th century AD, published in King (1984, 1999a, 1999b), Benecke (1986), and others (see Koepke, 2008). Figure 4.1 demonstrates that the cattle share was lowest in the Mediterranean, highest in North-Eastern Europe, and intermediary in Central-Western Europe.

Land per capita

Land *per capita* values were generated from estimates of population density compiled by McEvedy and Jones (1980), Allen (2003), Dupâquier (1988), and Wrigley and Schofield (1981). Figure 4.2 shows that population density increased gradually in all three regions, with a more marked increase in the Mediterranean and Central-Western Europe during the Roman Empire. Population density was highest for the Mediterranean during this period but peaked in Central-Western Europe towards the 17th century. From the 10th century

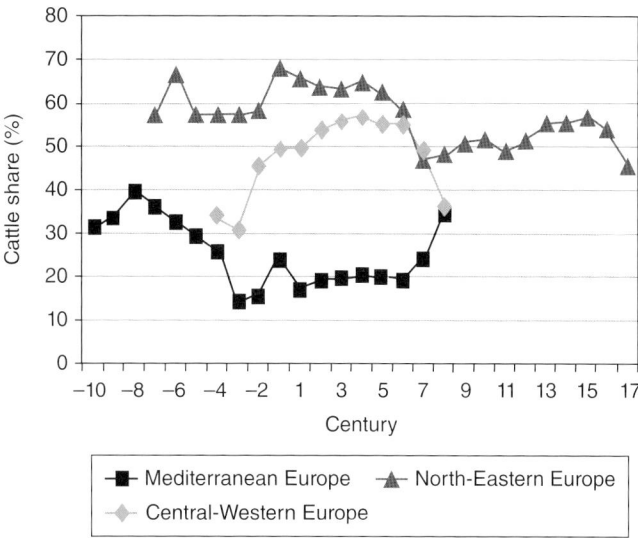

Figure 4.1. Cattle shares over time (10th century BC–17th century AD) in the Mediterranean, North-Eastern, and Central-Western Europe.

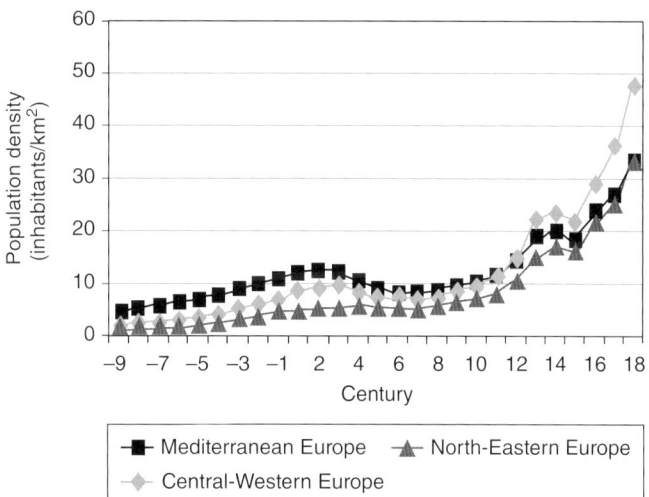

Figure 4.2. Changes in land *per capita* over time in the Mediterranean, North-Eastern, and Central-Western Europe.

onwards, there was a demographic surge, followed by stagnation linked to the Great Famine, the Black Death, and the Hundred Years' War. In the quantitative analysis, land *per capita* was calculated as km² per inhabitant in logarithmic form to account for decreasing marginal product.

Climate Change

Evidence on climate change consisted of a data series of temperature reconstructions from Mann and Jones (2003), which provide the longest time frame available. Problematically, data

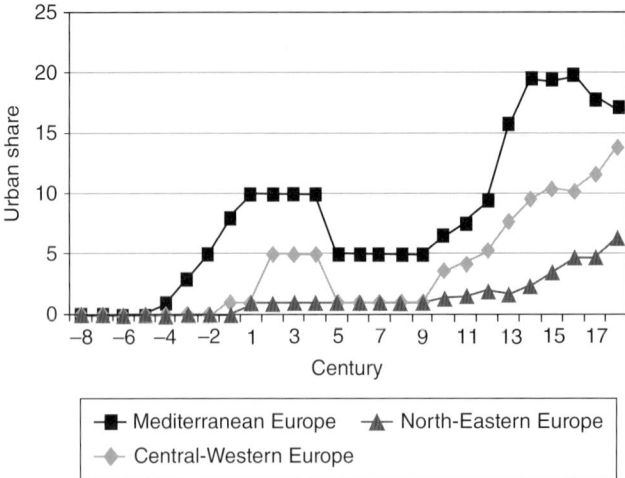

Figure 4.3. The urban share over time in the Mediterranean, North-Eastern, and Central-Western Europe.

is only available for the Northern Hemisphere as one unit, meaning that regional effects could not be investigated without using regional dummies. Although there is a large degree of variability between the different published series (IPCC et al., 2007), in general, they show warming during the medieval period and cooling during the Little Ice Age, which spanned the 16th to 19th centuries AD. Overall, temperature changes between the 8th century BC to the 18th century AD—prior to a pronounced anthropogenic influence—are not very strong.

Cultivation Methods

During the time period of interest, the introduction of three-field rotation was the solitary major, testable, widespread change in cultivation method that occurred (Küster, 2006); the iron plow had already been invented, and limited information exists on fertilizer use. A dummy variable, three-field rotation, was generated to control for this technology's potential influence: *0* represents unaffected centuries and *1* the 11th century AD onwards, when three-field rotation was in use.

Urban Share

Data on the urban share was compiled from Allen (2003), Bairoch (1976), and Federico and Malanima (2004). Measured as percent per overall number of settlements, the urban share is shown in Figure 4.3. Trends in the urban share are similar to those of land *per capita*; during the Roman Empire, the Mediterranean was highly urbanized, Central-Western Europe less so, and the North-East not at all. The urban share greatly decreased after the end of the Empire in the Mediterranean, but from the 7th to 8th century AD onwards, it increased throughout Europe.

War and Conflict and the Disease-Scape

To control for the potential effects of war and conflict and aspects of the disease-scape, dummy variables—war and conflict and plague—were generated. The centuries during which large-scale episodes of war, conflict, and disease epidemics occurred were coded as *1*, while centuries lacking these phenomena were coded as *0*.

Integration into the Roman Empire

The various impacts of inclusion in the Roman Empire cannot be tested separately. To generally assess its potential overall effect, a dummy variable, Roman Empire, was generated. It was coded as *1* for centuries and regions belonging to the Roman Empire and *0* for unaffected parts.

Gender Inequality

Gender inequality is difficult to assess using archaeological and historical sources for the period and regions of interest as they do not consider or are not informative about all social strata and only cover small time intervals. Therefore, relative stature differences from the data set were used as a function of regions and centuries.

Results

Temporal Trends in Mean Stature

Changes in mean stature in the Mediterranean and Central-Western and North-Eastern Europe from the 8th century BC until the 18th century AD are depicted in Figure 4.4. Estimations of mean stature are based on regression equations reconstructing stature on the individual level, with females and males pooled. This study is based on centuries with more than 35 available observations (see Koepke, 2008). Over the long run, there is a slight tendency towards increased mean stature in all of the regions; aggregating all of the observations, it amounts to an average of 0.5 cm every 1000 years. However, remarkable variation is evident

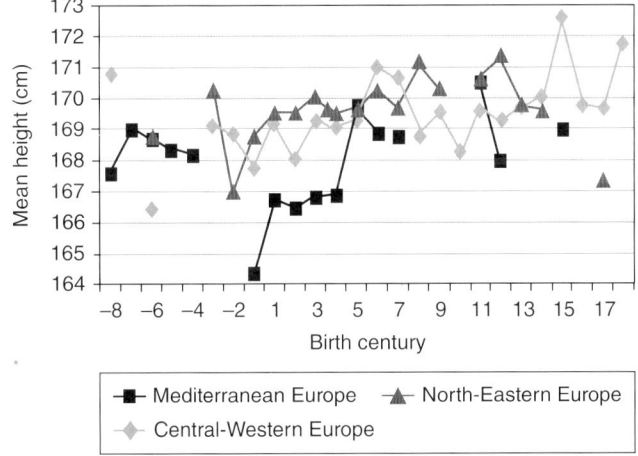

Figure 4.4. Mean stature in the Mediterranean, Central-Western, and North-Eastern Europe between the 8th century BC and the 18th century AD. [Considered data (*read as*: century: number of observations). **Mediterranean Europe** (total number of observations = 3572): −8:172; −7:140; −6:236; −5:404; −4:158; −1:114; 1:211; 2:445; 3:124; 4:566; 5:468; 6:150; 7:56; 11:51; 12:130; 15:61. **Central-Western Europe** (total number of observations = 8464): −8:79; −6:46; −3:39; −2:82; −1:41; 1:88; 2:1146; 3:407; 4:1225; 5:222; 6:1387; 7:1477; 8:266; 9:327; 10:153; 11:136; 12:216; 13:189; 14:242; 15:55; 16:455; 17:66; 18:103. **North-Eastern Europe** (total number of observations = 6466): −6:96; −3:117; −2:81; −1:37; 1:217; 2:174; 3:181; 4:318; 5:125; 6:198; 7:279; 8:787; 9:533; 10:287; 11:1423; 12:462; 13:358; 14:554; 17:80.]

between the three regions and over time. Overall, this suggests that there was no straightforward, continuous improvement in nutritional status.

A closer look at the temporal trajectory reveals that more beneficial and detrimental centuries alternate, with statistically significant changes present (see Koepke, 2008 for the background regressions), and that regional differences are evident. Overall, populations in the Mediterranean fared poorly in terms of nutritional status: on average, the stature series from this region has the smallest values. This is roughly consistent with historical and archaeological evidence suggesting that net nutrition was negatively impacted by various factors operating in the Mediterranean, such as high rates of urbanization; high population density, land *per capita*; low dairy and beef consumption; and being a part of the Roman Empire.

On average, mean stature in North-Eastern Europe is greater than that of the other two regions. This is consistent with historical and archaeological evidence on environmental and epidemiologic conditions in the region: low population density, a low degree of urbanization, and a diet rich in animal protein from dairy products. However, the stature series indicates that North-Eastern Europe lost this advantage in the 17th century AD. The decline may be due to the Little Ice Age and associated declines in agricultural productivity and increased food insecurity—the cooling period had a particularly strong impact on North-Eastern Europe.

The stature series from Central-Western Europe is intermediate to the other two regions. Intriguingly, this regional pattern corresponds to that of the cattle trend; the lowest cattle share was found in the Mediterranean and the highest in the North-East, with the Central-Western in between. It also corresponds with the degree of urbanization, with the Central-Western region bearing a moderately high urban share. However, with changes in population density, the pattern is more ambiguous over the centuries.

There are also distinct temporal trends. Mean stature trends suggest that the Roman Empire had a particularly negative effect on net nutrition in affected areas; the Mediterranean series shows extreme deterioration in the 1st century BC, but is more or less constant in the 8th to 3rd centuries BC. The decline could be explained by changes in the populations' access to animal protein; integration in the Roman Empire was associated with a shift towards a grain-/cereal-intensive agricultural system (Garnsey, 1999). It might also be linked to the first major wave of Roman expansion, which likely precipitated a diversion of resources away from the populace and towards the demands of the military.

Supporting explanatory evidence for North-Eastern and Central-Western Europe during the 8th to 1st centuries BC is scarce in comparison to that for the Mediterranean, making interpretation difficult. The likely conclusion is that environmental conditions were largely stagnant. In Central-Western Europe, the decline in mean stature in the 1st century BC may be partly attributable to negative impacts from military infringement associated with the Roman Empire. This entails an influx of migrants to a given region associated with the mobility of the Roman administration, such as traders and troops. Regional climatic conditions may also have worsened during this period (Schönwiese, 1995), precipitating worsened environmental conditions.

In the 1st century, mean stature increased in the Mediterranean and Central-Western Europe, likely due to political factors: an economic and political stabilization associated with the *pax Augustae* under Emperor Augustus, which followed a long period of conflict and unrest. However, the series indicate that this positive trend did not persist into subsequent centuries; mean stature stagnated or decreased in the 2nd century AD throughout the three regions. As this century is regarded as the heyday of Roman rule, accompanied by expansion of the empire and an economic upswing (Bowman and Wilson, 2009), the decrease is striking and unexpected at the first glance. This trend seems to represent an economic growth puzzle, much like the antebellum growth puzzle documented in the early period in the USA and throughout Europe (e.g., Komlos, 1998; Haines et al., 2003), wherein stature paradoxically declined in the face of economic prosperity. In Central-Western Europe, this puzzling decline may be attributable to factors associated with the economic boom, such as market integration resulting in reduced access to vital goods that were previously accessible in indigenous peripheral areas (Komlos, 1989), or worsened environmental conditions and increased infectious disease to expanded trade and therefore person-to-person contacts. Confirmation for the malign impacts of the Roman Empire comes from the absence of a decisive decline in stature in North-Eastern Europe, which was not immediately affected by the empire during this time.

Political and epidemiologic issues associated with the Roman Empire continued net nutrition and therefore stature into the 4th century. Stature slightly increased in the 3rd century, possibly due to reduced population pressure brought about by the demographic costs of the Antonine Plague between 165 and 180 AD. Declines in the 4th century are likely linked to worsened living conditions associated with the decline of the Roman Empire, with political conflict and associated degradation so widespread as to affect North-Eastern Europeans now, which corresponds with archaeological evidence suggesting that there was now a Roman presence in the region (Menghin and Planck, 2005).

The fall of the Empire in the 5th century AD (in 476 AD) had a pronounced positive effect on stature in the Mediterranean. The fall of the Empire precipitated increased migration, known as the Migration Period, which redistributed populations throughout Europe, resulting in Mediterraneans' stature adjusting to that found in other regions. This suggests that the changes in living conditions associated with the Migration Period resulted in a similar nutritional status across Europe, with greater uniformity in living conditions. The trend continues into the 6th century; increased stature is likely attributable to increased proximity to protein-rich meat and dairy sources. Pastoralism became more widespread with the fall of the Empire, reduced urbanization with the total disintegration of the Roman world in the west in 476 AD and presumably reduced population pressure from the Justinian Plague in the 6th to 8th centuries AD (Harbeck et al., 2013).

During the 11th and 12th centuries, mean stature increased in Central-Western and especially North-Eastern Europe, but not in the Mediterranean. This might be attributable to the medieval warming period, or climatic optimum, c. AD 950–1250. Increased temperature may have had a greater, more beneficial effect on agricultural productivity and environmental conditions in typically cooler northern regions, while having an opposite effect on already warm regions, such as the Mediterranean.

Interestingly, no declines occurred in the 14th century, despite the Hundred Years' War (AD 1337–1453); the Great Famine (AD 1315–1317), a period of crop failures, mass death, and disease epidemics across Europe, which occurred in the midst of a hundred year's worth of smaller, repetitive famines; and the Black Death (c.1348–AD 1350). This puzzling result may be due to inadequacies in the temporal resolution of the data set. However, it is also possible that the great demographic impact of these events relieved population pressures sufficiently to counteract their biological costs; this has already been documented for the Black Death, which left increased agricultural productivity and surpluses, decreased population density and urbanization, and improved living standards across Europe in its wake (Kelly, 2006).

The 15th to 18th centuries are represented by very little data, making interpretation difficult. Nevertheless, the 17th century decline in North-Eastern Europe could well be explained by the coldest phase of the Little Ice Age (Jordan, 1996), a cooling phase starting in the 16th century, which had a greater negative impact on environmental conditions, food security, and agricultural productivity in northern than southern regions. It is likely that nutritional status was also negatively affected by the Thirty Years War (AD 1618–1648), a series of conflicts in Central Europe that was one of the longest and most destructive conflicts in European history.

Overall, the data series suggest that despite variation between the regions, Europeans did not experience stagnating environmental conditions and biological well-being from the Iron Age onwards. Instead, they experienced a general increase in their standard of living prior to industrialization and the second epidemiologic transition.

Determinants Influencing Stature

In order to test which of the factors exerted an influence on biological well-being and therefore stature, panel data analysis was applied on a regional level. Data from Central-Western Europeans in the early Middle Ages was used as a reference category (captured by the constant in Table 4.1), as they are the best represented group in the available data.

Table 4.1 contains results of several of the regressions conducted (with the coefficients and p-values of each model displayed in two adjacent columns). The regressions are weighted least square estimates (the square root of the sample size was applied to adjust for nonuniform data distribution) with regional dummies (equivalent to fixed effects) and period dummies (to control for intertemporal heterogeneity).

The first model (Table 4.1 columns 2 and 3; adj. R^2 of 0.39) includes the variable that controls for the effect of climate change. The model is based on data from the 1st to 18th centuries AD only, as there is insufficient data from preceding centuries. Results suggest that higher temperatures have no significant effect on mean stature in all three regions.

Modelling the effect of all of the other testable determinants—excluding climate change—demonstrates that it has an explanatory power of 49% (columns 4 and 5). In this model most of the variables have no statistically significant impact on the very-long-term development of the net nutrition. The exceptions are the period dummies that proxy late prehistory and antiquity; both are statistically significant at 5% and 1% levels. In both periods, populations were on average smaller in stature than during the early medieval period. In terms of economic significance, higher cattle share resulted in a 0.59 cm increase in stature, whereas land per capita is also economically insignificant.

Not controlling for cattle share and land *per capita*, the adjusted R^2 is 0.43, and the regional dummy denominating the Mediterranean becomes statistically significant at the 5% level (columns 6 and 7). Overall, Mediterranean Europeans were smaller in stature than those in the other two regions. These remarkable regional differences in mean stature indicate regional disparities in net nutrition, which is specified by the missing impact of the two excluded variables—cattle share and land *per capita*. These are proxies for access to high-quality protein, specifically dairy and beef, and population density and therefore environmental quality, respectively. The data suggest that these had a great impact on health and net nutritional status in the Mediterranean. This is consistent with archaeological and historical evidence discussed, suggesting reduced cattle husbandry in southern Europe, which would have produced reduced access to protein from dairy and beef, in comparison to other regions of Europe. Similarly, the agricultural emphasis on cattle husbandry may have particularly affected local nutrition if land *per capita* was high. This result suggests that regional variation in European stature is environmentally driven, confirming earlier studies such as Quiroga Valle (1998) or Komlos et al. (2003).

A fifth model (columns 8 and 9) tests the specific impact of these two variables, cattle share and land *per capita*, alone. In this model (adj. R^2 of 0.30), cattle share is significant at the 1% level; in terms of economic significance, a one standard deviation higher cattle share would imply 0.72 additional centimeters in stature (see Table 4.2). But even in this reduced model, land *per capita* is still insignificant statistically and economically as a determinant of

Table 4.1. Regressions and decriptives: determinants affecting mean stature between the 8th century BC and 18th century AD.

1	2	3	4	5	6	7	8	9	10	11	12	13
Constant	167.89	0.00	167.51	0.00	169.63	0.00	165.95	0.00	168.57	0.00	167.41	0.00
Mediterranean Europe	0.09	0.96	0.34	0.79	−1.10	0.02			0.26	0.82		
North-Eastern Europe	0.24	0.89	−0.12	0.86	0.44	0.25			0.02	0.96		
Early prehistory			−1.61	0.23	−0.89	0.18			−1.85	0.06		
Late prehistory			−1.36	0.08	−1.07	0.07			−1.56	0.01		
Antiquity	1.13	0.14	−1.72	0.01	−1.39	0.01			−0.75	0.21		
High medieval period	0.46	0.72	−0.14	0.91	−0.81	0.37			−0.05	0.96		
Late medieval period	0.35	0.90	−1.00	0.60	−1.05	0.38			−1.19	0.38		
Modern	2.54	0.52	−3.17	0.19	−1.70	0.19			−2.92	0.11		
Cattle share (dairy consumption)	0.03	0.54	0.04	0.22			0.05	0.01	0.01	0.66	0.04	0.01
Three-field rotation (cultivation methods)			0.79	0.55	0.72	0.45			0.82	0.46		
Land per capita (log)	0.29	0.91	0.31	0.82			0.20	0.70			−0.22	0.65
Urban share	0.96	0.92							−0.29	0.01		
Climate change			0.05	0.90	−0.04	0.92			0.10	0.80		
War and conflict			−0.08	0.88	−0.09	0.86			0.27	0.59		
Plague (the disease-scape)			1.22	0.30	0.53	0.59	1.67	0.15	1.20	0.24		
Gender inequality	1.04	0.69										
Roman Empire											−1.43	0.00
Adj. R^2	0.39		0.49		0.43		0.30		0.61		0.43	
N	24		42		52		42		42		43	

Weighted least squares regression: number of cases adjusted for aggregated observations using square root. Constant refers to stature estimation for the reference group "Early Middle Ages—Central-Western Europe." p-values in columns 3, 5, 7, and 9 in italics. Statistically significant coefficients highlighted grey.

Table 4.2. Descriptives.

1	2	3	4
	N	Mean	Std. deviation
Cattle share (dairy consumption)	53	45.04	15.79
Land *per capita* (log)	78	−2.15	0.77
Urban share	78	4.49	5.33
Climate change	24	9.32	0.06
Gender inequality	56	0.95	0.20

stature over the entire time period under study. Hence, the potential influence of access to high-quality protein on stature seems to be particularly decisive and much more important than population density in the long term. The negligible influence of land *per capita* on net nutrition may be due to low population numbers during the majority of the centuries under study, resulting in a limit to the influence that density could exert on health—overall, sufficient space existed for contemporary populations. Correspondingly, Koepke and Baten (2005) found that increased population density was only significant in having a negative effect on stature for the 9th century onwards.

To further test whether poor living conditions affected stature in the period under study, an additional model controls for the urban share (excluding land *per capita* to prevent multicollinearity) (columns 10 and 11). This model explains 61% of the total variation in mean stature; here, a higher urban share has a significant—namely, the expected negative—effect on mean stature at the 1% level. However, none of the other variables are significant except for the time dummies comprising both prehistoric periods (on the 5% and 1% levels). Both regional dummies become insignificant if the urban share is not controlled for. Mediterranean populations may have been biologically harmed not only by their limited access to high-quality animal protein but also by urbanization and its well-established detrimental effects on health in the past.

The last model (columns 12 and 13) controls for the effects of inclusion in the Roman Empire (instead of the time dummies) as well as cattle share and land *per capita* (adj. R^2 is 0.43). The regression reveals that the gross effect of the Roman Empire was detrimental for health; the Roman Empire dummy has a significant negative impact on mean stature of −1.43 cm (on a 1% level). The cattle share is positively significant, also at 1%. In terms of economic significance, the positive impact of a higher cattle share exceeds the negative ancient Roman impact: +0.6 cm versus −0.5 cm. Inclusion in the Roman Empire also has a significant effect when evaluated in a model that includes all of the other determinants (not shown). Again, none of the other determinants have a significant effect, but remarkably, the insignificance of the Mediterranean Europe dummy disappears when cattle share and land *per capita* are excluded. The Roman Empire, as evaluated here, also has a considerable, detrimental effect outside of the Italic heartland; overall, it is negatively significant everywhere. However, testing the effect of inclusion in the Empire for each of the three regions alone reveals that the impact was much greater for the Mediterranean than for Central-Western Europe (and, of course, no effect for the North-East).

Why do so many of these determinants have insignificant effects on stature and biological well-being? First, much of this may be due to inadequate data, whether insufficiencies in the data series or oversimplified dummies. For instance, more accurately assessing the effects of climate change would require consideration of its different potential effects in different climatic zones, and inclusion of data on precipitation, which, in the form required for this analysis, does not yet exist. Second, the insignificance may be due to 'equalizing effects' or the balancing out of positive and negative effects. For instance, this phenomenon may be responsible for the nonexistent impact of war and conflict and the disease-scape and cultivation methods on stature, as well as land *per capita* in the long term. Lastly, the reason could also be that the data depicts a determinant correctly, and that in fact these determinants did not substantially impact biological well-being captured by mean stature.

Conclusion

What Drove the Second Epidemiologic Transition?

This study operates off of the premise that a better understanding of the changes that occurred during the second epidemiologic and demographic transitions can be gained from uncovering the environmental, dietary, and epidemiologic conditions that prevailed in Europe in the preceding centuries. In particular, it is assumed that understanding these background conditions can shed key insights into whether biological well-being stagnated during these centuries or was subject to other trends, what initiated the transition, and why it played out differently in timing, intensity, and scope in different regions of Europe.

The findings suggest that well-being, represented by net nutrition and therefore mean stature, did not stagnate prior to the second transition, but instead experienced substantial variation by region and over the centuries. However, overall, the data show only a very small positive trend in mean stature between the 8th century BC and the 18th century AD of approximately 0.5 cm per 1000 years. Moreover, the results suggest that increased rates of urbanization throughout time and the declines in environmental quality that they represent have presented a major impediment to improvements in health over time. In contrast, access to and presumably consumption of high-quality protein, specifically dairy and beef (i.e., cattle share), acted as a key determinant of health in the centuries preceding the second epidemiologic and demographic transitions.

What can these findings on background health tell us about the second epidemiologic transition and the demographic transition in Europe? Most discussions of the causes of the second transition, especially by medical historians, have taken a black-and-white perspective, concentrating solely on diet and nutritional adequacy or medical advances and health care as key drivers (e.g., McKeown, 1983). However, more recent, more nuanced studies suggest that multiple, mixed causes are more likely. The important consideration for Europe is that the second transition was staggered across the continent and included multiple phases (see Arora, 2012). Therefore, it seems unlikely that a singular common cause can be identified.

As complex interactions exist between fertility, longevity, economic improvements, nutrition, health, and education, the second transition presumably occurred in response to a multifaceted, mutual interaction and interdependence of various factors (e.g., Birchenall, 2007; Dalgaard and Strulik, 2009; Mokyr and Voth, 2009). Studies on dietary quality and quantity even indicate that these factors are important enough to be considered separately (Lunn, 1991; Birchenall, 2007) and suggest that diet may have played an especially key role in the transition. Malnutrition, particularly undernutrition, has a close interrelationship with susceptibility to infectious disease (e.g., Stephensen, 1999; Calder and Jackson, 2000; Rodríguez et al., 2011). Moreover, the interaction of diet and human capital is essential to forming the foundation for improved living conditions and consequent improvements in health. In regards to the 19th and 20th centuries and the second transition, health-related human capital was in turn affected by technophysio evolution, which seems to have been a synergistic process between rapid technological change and improvement in human physiology. The findings of this and other studies suggest that this beneficial process initiated previous to modern medical advances (Fogel and Costa, 1997; Floud et al., 2011), with agriculture—specifically animal husbandry and improved access to high-quality protein—proving to be particularly vital. In this form of evolution, better nutrition results in reduced organ damage during bodily growth, also enabling improved cognitive abilities and leading to the innovations associated with industrialization.

Another key aspect to consider within this dynamic is that bodily growth does not start with parturition but instead commences with maturation of the oocyte and is therefore affected by the living conditions, including the diet, of the mother. Therefore, significant improvements in the dietary status of females influence the overall biological well-being of the entire population (Osmani and Sen, 2003). This undoubtedly complements the contributions made by hygienic improvements and, later, medical advances, in improving maternal health and reducing infant mortality and morbidity during the course of the demographic and second epidemiologic transitions.

Overall, diet seems to have played a key role in the various stages of the epidemiologic and demographic transitions. In the first step, declining mortality due to infectious disease may have been triggered by improvements in dietary quality and quantity (e.g., Harris, 2004). In a second step, medical advances and improved health care may have accelerated the decline (Birchenall, 2007)—the 19th and early 20th centuries were characterized by marked improvements in medical knowledge and practice. Additionally, factors such as improved sanitation, ranging from sanitized water sources to sewage systems to improved hygiene, from awareness to products, exerted both a direct, contagion prevention, and indirect, strengthened body composition due to improved overall nutrition, influence on the spread of acute epidemic infectious diseases.

In relation to the demographic transition, recent findings indicate that a mutual causality between education and fertility (a quality–quantity trade-off to human capital) played a substantial role while also affecting the shift from Malthusian stagnation to sustained economic growth (Becker et al., 2010, 2012). Another key, but often neglected, aspect is the introduction of child-labor regulations (Doepke, 2003). Importantly, "once households start investing in quality rather than the mere quantity of children, the economy ventures onto a trajectory characterized by rising living standards" (Dalgaard and Strulik, 2009: 2). A spiral of positive developments sets in, resulting in further improvements in health care, education, and diet, which in turn manifest as a population's nutritional status (and therefore stature). Overall, the second epidemiologic and demographic transitions can be understood as results of a positive, reinforcing feedback loop between nutrition, health, and education.

In general, one can assume that a population with a high nutritional status likely possessed a better biological foundation for setting off into the second epidemiologic and demographic transitions. A second assumption, based on the results of this study, is that this condition existed to a greater extent in regions with environmental conditions that produced high nutritional status, such as a lower urban share and greater access to high-quality protein, over the course of several centuries. Correspondingly, the historical trajectories of the determinants assessed in this study, with their variation by region, indicate that these might have made for different epidemiologic and demographic trajectories in different parts of Europe. However, a legacy will not hold over the centuries if the reasons for the conditions change. For example, results from this study demonstrate that the detrimental impact of integration in the Roman Empire does not result in a poor biological legacy for the Mediterranean; health, measured by stature, equalizes across all three regions almost instantly after the decline and fall of the Roman Empire.

The long-term, average development of human health and biology, represented by stature, cannot be used to generate insights into the immediate causes of the second epidemiologic and demographic transitions. However, evidence on living conditions in the past reveals the wide variability of factors that influence well-being and therefore also the starting conditions for the second transition. Unfortunately, the archaeological data employed here does not necessarily grant the fine temporal resolution needed to determine the immediate, short-term triggers of the second transition. For this, a better temporal and regional resolution would be necessary. Additionally, before this endeavor can be broached, it is also necessary to come to a consensus on defining research strategies for studying the timing and process of the second transition and engaging in more detail with issues of simultaneity and endogeneity. Instead, this study reasserts that it is not only essential to consider a wide variety of factors in order to best conceptualize the causes of the second epidemiologic and demographic transitions. It is also critical to avoid pursuit of one specific cause for the transitions; a safe assumption to guide future research is that mutual reinforcement between numerous factors caused the second epidemiologic transition.

References

Al-Dabbagh A, Ebrahim G. 1984. The preventable antecedents of malnutrition. J Trop Pediatr 30:50–52.
Allen R. 2003. Progress and poverty in early modern Europe. Econ Hist Rev 56(3):403–443.
Angel J. 1984. Health as a crucial factor in the changes from hunting to developed farming. In: Cohen M, Armelagos G, editors. Palaepathology at the origins of agriculture. Orlando: Academic Press. p 51–74.
Armelagos G. 1990. Health and disease in prehistoric populations in transition. In: Swedlund A, Armelagos G, editors. Disease in populations in transition. New York: Bergin and Garvey. p 127–144.
Arora S. 2012. Understanding aging during the epidemiologic transition. Working Papers 12–07, Association Française de Cliométrie.
Bach A. 1965. Zur Berechnung der Körperhöhe aus den langen Gliedmaßenknochen weiblicher Skelette. Anthropol Anz 29:12–21.
Bairoch P. 1976. Population urbaine et taille des villes en Europe de 1600 á 1070. Econ Hist Rev 53:304–335.
Barrett R, Kuzawa C, McDade T, Armelagos G. 1998. Emerging and re-emerging infectious diseases: the third epidemiologic transition. Annu Rev Anthropol 27:247–271.
Bartosiewicz K, Van Neer W, Lentacker A. 1997. Draught cattle: their osteological identification and history. *Annalen zoölogische Wetenschappen 281*. Tervuren: Koniklijk Museum voor Midden-Afrika.
Baten J. 1999. Ernährung und wirtschaftliche Entwicklung in Bayern, 1730–1880. Stuttgart: Franz Steiner Verlag.
Baten J. 2002. Climate, grain production and nutritional status in 18th century southern Germany. Journal of European Economic History 30(1): 9–47.
Becker S, Cinnirella F, Woessmann L. 2010. The trade-off between fertility and education: evidence from before the demographic transition. J Econ Growth 15(3):177–204.
Becker S, Cinnirella F, Woessmann L. 2012. The effect of investment in children's education on fertility in 1816 Prussia. Cliometrica 6(1):29–44.
Benecke N. 1986. Die Entwicklung der Haustierhaltung im südlichen Ostseeraum. Weimarer Monographien zur Ur- und Frühgeschichte, 19. Weimar: Beiträge zur Archäozoologie V. Volkswacht Gera.
Bertranpetit J. 2012. Tracing the origins of European populations through the analysis of genomes. I Jornades UPF d'Arqueologia, un fòrum de debat sobre el fenomen migratori a la prehistòria d'Europa. Migrations in the Prehistory of Europe: Archaeological, Linguistic and Genetic Evidence. Barcelona.
Birchenall J. 2007. Economic development and the escape from high mortality. World Dev 35(4): 543–568.

Bogin B. 1999. Patterns of human growth. Cambridge: Cambridge University Press.

Bogin B, Keep R. 1999. Eight thousand years of economic and political history in Latin America revealed by anthropometry. Ann Hum Biol 26(4):333–351.

Bowman A, Wilson A, editors. 2009. Quantifying the Roman economy. Oxford: Oxford University Press.

Breitinger E. 1937. Zur Berechnung der Körperhöhe aus den langen Gliedmaßenknochen. Anthropol Anz 14:249–274.

Byers S. 2002. Introduction to forensic anthropology. Boston: Allyn and Bacon.

Calder P, Jackson A. 2000. Undernutrition, infection and immune function. Nutr Res Rev 13(1):3–29.

Cameron N. 2002. Human growth and development. San Diego: Academic Press.

Cardoso H, Gomez J. 2009. Trends in adult stature of peoples who inhabited the modern Portuguese territory from the Mesolithic to the late 20th century. Int J Osteoarchaeol 19:711–725.

Cohen M, Armelagos G. 1984. Palaeopathology at the origins of agriculture. Orlando: Academic Press.

Crabtree P. 1996. Production and consumption in an early complex society: animal use in Middle Saxon East Anglia. World Archaeol 28:58–75.

Curtis T, Kvernmo S, Bjerregaard P. 2005. Changing living conditions, life style and health. Int J Circumpolar Health 64(5):442–450.

Dalgaard C-J, Strulik H. 2009. A bioeconomic foundation of the Malthusian equilibrium: body size and population size in the long-run. Diskussionspapiere der Wirtschaftswissenschaftlichen Fakultät der Leibniz Universität Hannover. dp-373.

Davis S. 1987. The archaeology of animals. New Haven: Yale University Press.

Dittmann K, Grupe G. 2000. Biochemical and palaeopathological investigations on weaning and infant mortality in the early Middle Ages. Anthropol Anz 58:345–355.

Doepke M. 2003. Accounting for fertility decline during the transition to growth. UCLA Department of Economics Working Paper No. 804, Los Angeles.

Dror DK, Allen LH. 2011. The importance of milk and other animal-source foods for children in low-income countries. Food Nutr Bull 32(3):227–243.

Dupâquier J. 1988. Histoire de la population Française. Paris: Presses University de France.

Eveleth P, Tanner J. 1976. Worldwide variation in human growth. Cambridge: Cambridge University Press.

Federico G, Malanima P. 2004. Labour productivity in Italian agriculture 1000–2000. Econ Hist Rev 57(3):437–474.

Fish E. 2008. The X-files in immunity: sex-based differences predispose immune responses. Nat Rev Immunol 8:737–744.

Floud R, Watcher K, Gregory A. 1990. Height, health and history. Nutritional status in the United Kingdom, 1750–1980. Cambridge: Cambridge University Press.

Floud R, Fogel R, Harris B, Hong S. 2011. The changing body. Health, nutrition, and human development in the western world since 1700. Cambridge: Cambridge University Press.

Fogel R, Costa D. 1997. A theory of technophysio evolution, with some implications for forecasting population, health care costs, and pension costs. Demography 34(1):49–66.

Foley R, Bath D, Dickinson F, Tucker H. 1972. Dairy cattle: principles, practices, problems, profits. Philadelphia: Lea and Febiger.

Formicola V. 1993. Stature reconstruction from long bones in ancient population samples: an approach to the problem of its reliability. Am J Phys Anthropol 90(3):351–358.

Gage TB. 2005. Are modern environments really bad for us? Revisiting the demographic and epidemiologic transitions. Yearb Phys Anthropol 48:96–117.

Garnsey P. 1999. Food and society in classical antiquity. Cambridge: Cambridge University Press.

Gehring KD, Graw M. 2001. Körperhöhenbestimmung anhand des Femurs und von Femurfragmenten. Arch Kriminol 207:170–180.

George S. 2006. Millions of missing girls: from fetal sexing to high technology sex selection in India. Prenat Diagn 26(7):604–609.

Grove J. 2002. Climatic change in northern Europe over the last two thousand years and its possible influence on human activity. In: Wefer G, Berger W, Behre KE, Jansen E, editors. Climate development and history of the north Atlantic realm, Hanse conference on climate history. Berlin/Heidelberg: Springer. p 313–326.

Grupe G. 2003. Stable nitrogen isotope analysis of archaeological human bone from medieval times. In: Noël R, Paquay I, Sosson J-P, editors. Au-Delà de l'Ècrit. Les hommes et leurs vécus matériels au Moyen Âge à la lumière des sciences et des techniques. Nouvelles Perspectives. Actes du Colloque international de Marche-en-Famenne oct. 2002. Court-Saint-Étienne: Brepols. p 281–294.

Guntupalli AM, Moradi A. 2009. What does gender dimorphism in stature tell us about discrimination in rural India, 1930–1975? In: Pal M, Bharati P, Vasulu T, editors. Gender and discrimination: Health, nutritional status, and role of women in India. New Delhi: Oxford University Press. p 258–277.

Gustafsson A, Lindenfors P. 2004. Human size evolution: no evolutionary allometric relationship between male and female stature. J Hum Evol 47(4):253–266.

Haines M. 1998. Health, stature, nutrition, and mortality: Evidence on the 'Antebellum Puzzle' from the Union Army recruits for New York state and the United States. In: Komlos J, Baten J, editors. The biological standard of living in comparative perspective. Berlin: Franz Steiner Verlag. p 155–180.

Haines M, Craig L, Weiss T. 2003. The short and the dead: nutrition, mortality, and the "antebellum puzzle" in the United States. J Econ Hist 63:382–413.

Harbeck, M., Seifert, L., Hänsch, S., Wagner, D. M., Birdsell, D., Parise, K. L., ... & Scholz, H. C. (2013). Yersinia pestis DNA from Skeletal Remains from the 6th Century AD Reveals Insights into Justinianic Plague. *PLoS pathogens*, 9(5), e1003349.

Harris B. 2004. Public health, nutrition, and the decline of mortality: the McKeown thesis revisited. Soc Hist Med 17(3):379–407.

Harris B, Gálvez L, Machado H, editors. 2009. Gender and well-being in Europe: Historical and contemporary perspectives. Burlington: Ashgate Publishing Company.

Henry C, Ulijaszek S. 1996. Long-term consequences of early environment: growth, development and the lifespan developmental perspective. 37th Symposium Volume of the Society for the Study of Human Biology. Cambridge: Cambridge University Press.

Herrmann B, Grupe G, Hummel S, Piepenbrink H, Schutkowski H. 1990. Prähistorische Anthropologie: Leitfaden der Feld- und Lehrmethoden. Berlin: Springer.

Heußner B. 1987. Neue Aussagemöglichkeiten anthropologischer Leichenbranduntersuchungen unter Einbeziehung histomorphometrischer Methoden. Materialhefte zur Ur- und Frühgeschichte Mecklenburgs 2. Schwerin: Mueseum für Ur- und Frühgeschichte.

Hindman H. 2009. The world of child labor: an historical and regional survey. Armonk, N.Y.: M.E. Sharpe.

Huber N. 1967. Anthropologische Untersuchungen an den Skeletten aus dem alamannischen Reihengräberfeld von Weingarten, Kr. Ravensburg. Naturwissenschaftliche Untersuchungen zur Vor- und Frühgeschichte in Württemberg und Hohenzollern. Stuttgart: Müller & Gräff.

IPCC, Solomon S, Qin D, et al. editors. 2007. IPCC fourth assessment report (AR4): climate change, 2007: synthesis report. Contribution of Working Group I to the Fourth Assessment Report of the Intergovernmental Panel on Climate Change. Cambridge: Cambridge University Press. p 467.

Jordan W. 1996. The Great Famine. Northern Europe in the early fourteenth century. Princeton: Princeton University Press.

Kelly J. 2006. The great mortality: an intimate history of the Black Death, the most devastating plague of all time. New York: Harper Perennial.

King A. 1984. Animal bones and the dietary identity of military and civilian groups in Roman Britain, Germany, and Gaul. In: Blagg T, King A, editors. Military and civilian in Roman Britain. Oxford: B.A.R.

King A. 1999a. Animals and the Roman army: the evidence of animal bone. In: Goldsworthy A, Adams C, editors. The Roman army as a community. Vol. S34. Portsmouth: Journal of Roman Archaeology. p 139–150.

King A. 1999b. Meat diet in the Roman world: a regional inter-site comparison of the mammal bones. Journal of Roman Archaeology 12:168–202.

Klasen S. 2002. Warum fehlen 100 Millionen Frauen auf der Welt. Eine ökonomische Analyse. In: Fabel O, Nischik R, editors. Femina oeconomica. Frauen in der Ökonomie. Munich: Rainer Hampp Verlag. p 181–202.

Koepke N. 2008. Regional differences and temporal development of the nutritional status in Europe from 8th century B.C. to 18th century A.D. Ph.D. thesis. Germany: University of Tuebingen.

Koepke N, Baten J. 2005. Climate and its impact on the biological standard of living in north-east, centre-west and south Europe during the last 2000 years. Hist Meteorol 2(1):147–159.

Koepke N, Baten J. 2008. Agricultural specialization and stature in ancient and medieval Europe. Explor Econ Hist 42(2):127–146.

Komlos J. 1989. Nutrition and economic development in the eighteenth century Habsburg monarchy: An anthropometric history. Princeton: Princeton University Press.

Komlos J. 1998. Shrinking in a growing economy? The mystery of physical stature during the industrial revolution. J Econ Hist 58:779–802.

Komlos J, Hau M, Bourguinat N. 2003. The anthropometric history of early-modern France. Eur Rev Econ Hist 7(2):59–190.

Küster HJ. 2006. Zerstörung – Ängste – Gestaltung. Impulse für die Entwicklung von Landschaft durch den Menschen in Mittelalter und Neuzeit. In: Bayer Akademie der Wissenschaften, editor. Natur und Mensch in Mitteleuropa im letzten Jahrtausend. Rundgespräche der Kommission für Ökologie, Band 32, 176S. München: Verlag Dr. Friedrich Pfeil. p S37–S46.

Lao O, Lu TT, Nothnagel M, et al. 2008. Correlation between genetic and geographic structure in Europe. Curr Biol 26(16):1241–1248.

Larsen C. 1997. Bioarchaeology – Interpreting behaviour from the human skeleton. Cambridge studies in biological and evolutionary anthropology 21. Cambridge: Cambridge University Press.

Livi-Bacci M. 1991. Population and nutrition. An essay on European demographic history. Cambridge studies in population, economy and society in past time 14. Cambridge: Cambridge University Press.

Luff, RM. 1993. Animal bones from excavations in Colchester, 1971–85. Colchester archaeological report 12. Colchester: Colchester Archaeological Trust.

Lunn P. 1991. Nutrition, immunity and infection. In: Schofield R, Reher D, Bideau A, editors. The decline of mortality in Europe. Oxford: Clarendon Press. p 131–145.

Mann M, Jones P. 2003. Global surface temperatures over the past two millennia. Geophys Res Lett 30(15):1820.

Manouvrier L. 1893. Détermination de la taille d'après les grands os des membres. Rev mens de l'ec d'Anthrop Paris 2:227–233.

Mascie-Taylor C, Bogin B. 1995. Human variability and plasticity. Cambridge: Cambridge University Press.

May E. 1997. Bemerkungen zur Relevanz von Körpergrößenermittlungen aus kleinen Knochenmaßen. Beitr Archäozoolog u prähist Anthrop 1:134–139.

McEvedy C, Jones R. 1980. Atlas of world population history. London: Penguin.

McKeown T. 1983. Food, infection and population. J Interdiscip Hist 14:227–247.

Menghin W, Planck D, editors. 2005. Menschen, Zeiten, Räume – Archäologie in Deutschland. Stuttgart: K. Theiss Verlag.

Mokyr J, Voth H. 2009. Understanding growth in Europe, 1700–1870: Theory and evidence. In: Broadberry S, O'Rourke K, editors. The Cambridge economic history of modern Europe: Volume 1, 1700–1870. Cambridge: Cambridge University Press.

Moradi A, Baten J. 2005. Inequality in sub-saharan Africa: new data and new insights from anthropometric estimates. World Dev 33(8):1233–1265.

Mummert A, Esche E, Robinson J, Armelagos GJ. 2011. Stature and robusticity during the agricultural transition: evidence from the bioarchaeological record. Econ Hum Biol 9(3):284–301.

Olds K. 2006. Female productivity and mortality in early-20th-century Taiwan. Econ Human Biol 4(2):206–221.

Omran A. 1971. The epidemiologic transition. Milbank Mem Fund Q 49:509–538.

Osmani S, Sen A. 2003. The hidden penalties of gender inequality: foetal origins of ill-health. Econ Hum Biol 1(1):105–121.

Pearson K. 1899. On the reconstruction of stature of prehistoric races. Mathematical contributions to the theory of evolution. Trans R Soc A 192:169–245.

Perkins J, Khan K, Smith GD, Subramanian, S. 2011. Patterns of trends of adult stature in India in 2005–206. Econ Hum Biol 9(1):184–193.

Pfister C. 1988. Klimageschichte der Schweiz 1525 – 1860. Das Klima der Schweiz von 1525–1860 und seine Bedeutung in der Geschichte von Bevölkerung und Landwirtschaft. Bern/Stuttgart: Haupt.

Poskitt E. 1999. Feeding problems. In: Sadler M, Strain J, Caballero B, editors. Encyclopaedia of human nutrition. San Diego: Academic Press. p 156–157.

Quiroga Valle G. 1998. Stature evolution in Spain, 1893–1954: an analysis by regions and professions. In: Komlos J, Baten J, editors. The biological standard of living in comparative perspective: Contributions to the conference held in Munich, January 18–22, 1997, for the XIIth international economic history association. Berlin: Franz Steiner Verlag. p 359–383.

Rey J, Bresson J. 1997. Conséquences à long terme de la nutrition fœtale. Arch Pediatr 4(4):359–366.

Rodríguez L, Cervantes E, Ortiz R. 2011. Malnutrition and gastrointestinal and respiratory infections in children: a public health problem. Int J Environ Res Public Health 8:1174–1205.

Rösing F. 1977. Methoden und Aussagemöglichkeiten der anthropologischen Leichenbrandbearbeitung. Archives u Naturwiss 1:53–80.

Rth HW, editor. 1984. Hessenim Frühmittelalter. Archäologie und Kunst. Ausstellungskat. Frankfurt a.M.: Thorbecke.

Scheidel W. 2010. Approaching the Roman economy. Available at SSRN: http://ssrn.com/abstract=1663560 (accessed November 5, 2013).

Schönwiese CD. 1995. Klimaänderungen. Daten, Analysen, Prognosen. Berlin: Springer.

Scott S, Duncan C. 2002. Demography and nutrition: Evidence from historical and contemporary populations. Oxford: Blackwell Publishing Company.

Seetah, K. 2005. Butchery as a tool for understanding the changing views of animals: cattle in Roman Britain. In: Pluskowski A, editor. Just skin and bones? New perspectives on human-animal relations in the historical past. BAR International Series 1410. Oxford: Archaeopress. p 1–8.

Sen A. 1984. Food entitlements and economic chains. In: Newman L, editor. Hunger in history. Food shortage, poverty, and deprivation. Oxford: Blackwell. p 374–386.

Sherratt A. 1981. Plough and pastoralism. In: Hodder I, Issacs G, Hammond N, editors. Patterns of the past. Cambridge: Cambridge University Press.

Silventoinen K. 2003. Determinants of variation in adult body stature. J Biosoc Sci 35(2):263–285.

Steckel R. 1995. Stature and the standard of living. J Econ Lit 33(4):1903–1940.

Steckel R. 2004. New light on the "Dark Ages": the remarkably tall stature of northern European men during the medieval era. Soc Sci Hist 28(2):211–229.

Steckel R. 2009. Statures and human welfare: recent developments and new directions. Explor Econ Hist 46(1):1–23.

Steckel RH, Prince JM. 2001. Tallest in the world: Native Americans of the Great Plains in the nineteenth century. Am Econ Rev 91(1):287–294.

Steckel R, Rose J. 2002. The backbone of history: Health and nutrition in the western hemisphere. Cambridge: Cambridge University Press.

Stephensen CH. 1999. Burden of infection on growth failure. J Nutr 129(S2):534–538.

Stinson S. 2000. Growth variation: biological and cultural factors. In: Stinson S, Bogin B, Huss-Ashmore R, O'Rourke D, editors. Human biology. An evolutionary and biocultural perspective. New York: Wiley-Liss. p 425–463.

Stinson S, Bogin B, O'Rourke D, editors. 2012. Human biology: an evolutionary and biocultural perspective. New York: Wiley-Liss.

Thüry G. 2001. Müll und Marmorsäulen. Siedlungshygiene in der römischen Antike. Mainz am Rhein: von Zabern.

Thüry G. 2007. Kulinarisches aus dem römischen Alpenvorland. Linzer Archäologische Forschungen XXXIX. Linz: Stadtmuseum/PG Druckerei.

Trotter M, Gleser G. 1952. Estimation of stature from long bones of American Whites and Negroes. Am J Phys Anthropol 10:463–514.

Ulijaszek S. 2006. The international growth standard for children and adolescents project: environmental influences on preadolescent and adolescent growth in weight and stature. Food Nutr Bull 27(S4):279–294.

UNICEF. 1990. Strategy for improved nutrition of children and women in developing countries. A UNICEF policy review. New York: UNICEF.

UNICEF. 1998. The state of the world's children, 1998. New York: Oxford University Press.

van Wieringen J. 1972. Secular changes of growth, 1964–1966. Stature and weight surveys in the Netherlands in historical perspective. Leiden: Netherlands Institute for Preventive Medicine.

Verhoff M, Kreutz K, Ramsthaler F, Schiwy-Bochat KH. 2006. Forensische Anthropologie und Osteologie – Übersicht und Definitionen. Dt Ärzteb 103(12):A-782, B-661, C-641.

Wahl J. 1982. Leichenbranduntersuchungen. Ein Überblick über die Bearbeitungs – und Aussagemöglichkeiten von Brandgräbern. Prähist Zeitschr 57:1–125.

Wahl J, Kokabi M. 1999. Das römische Gräberfeld von Stettfeld I – Osteologische Untersuchungen der Knochenreste aus dem Gräberfeld. Stuttgart: K. Theiss Verlag.

Walker S, Wach, T, Gardner J, et al. 2007. Child development: risk factors for adverse outcomes in developing countries. Lancet 369(9556):145–157.

Ward-Perkins B. 2005. The fall of Rome and the end of civilization. Oxford: Oxford University Press.

Waterlow J. 1972. Classification and definition of protein-energy-malnutrition. Br Me d J 3(5826): 566–568.

Whitehead R. 1977. Protein and energy requirements of young children living in the developing countries to allow for catch-up growth after infections. Am J Clin Nutr 30(9):1545–1547.

WHO. 1995. Physical status: The use and interpretation of anthropometry. Geneva: WHO.

WHO. 2007. Growth reference data for 5–19 years. Available at: http://www.who.int/growthref/en/ (accessed November 5, 2013).

Wiese B, Zils N. 1987. Deutsche Kulturgeographie: Werden, Wandel und Bewahrung deutscher Kulturlandschaften. Herford: Busse Seewald.

WolaDski N. 1953. Graficzna metoda obliczania wzrostu na podstawie ko[ci dBugch. Przeglad Antropologiczny 19:403–404.

Woolgar C. 2006. Meat and dairy products in late medieval England. In: Woolgar C, Serjeantson D, Waldron T, editors. Food in medieval England: Diet and nutrition. Oxford: Oxford University Press. p 88–101.

Wrigley E, Schofield R. 1981. The population history of England 1541–1871: A reconstruction. Cambridge: Harvard University Press.

Chapter 5
The Epidemiological Transition in Practice: Consumption, Phthisis, and TB in the 19th Century

Jeffrey K. Beemer
Department of Sociology, University of Massachusetts Amherst, Amherst, MA

Introduction

The general shift from acute infectious to chronic and degenerative diseases[1] during the late 19th and early 20th centuries marked the beginning of what Omran (1971) described as the *epidemiological transition*. Subsequent declines in mortality over this same period ushered in an era of rising life expectancies that continues to define health trends today. The factors that gave rise to this transitional period are often obscured by the lack of common historical criteria during this period for diagnosing disease and identifying causes of death. Our understanding of the North American epidemiological transition, for example, lacks precision and scope due in large part to the equivocal nature of the available data. With an ever shifting etiological landscape and lack of uniform standards for recording the 19th century causes of death, persistent doubts remain vis-à-vis our ability to provide more comprehensive conclusions about the mortality transition over this period (Woods, 1991; Alter and Carmichael, 1996, 1999; Risse, 1997; Arrizabalaga, 1999; Haines, 2003). McKeown's (1976) nutritional/standard-of-living explanation for declining mortality in the late 18th and 19th centuries provided an earlier generation of historians with a compelling basis for rejecting explanations that posited medical and public health interventions exclusively. McKeown's thesis, nevertheless, has come under considerable doubt due to a number of critical reexaminations, not the least of which were further historical and demographic studies since the mid-1970s that employed more nuanced approaches to existing and new data sources. Szreter (1988), for instance, offers an alternative to McKeown's argument, one that places

---

[1] The distinction between acute infectious and chronic degenerative diseases, while meaningful within the context of Omran's thesis, has given way to more recent etiological understandings that recognize the infectious nature of many diseases traditionally classified as chronic degenerative, for example, cervical and oral cancers and human papillomavirus (HVP), among other inflammatory and bacterial conditions. Given the historical focus of this chapter, I use this distinction being mindful that earlier classifications of chronic degenerative diseases are progressively being identified as infectious within a contemporary etiological framework.

Modern Environments and Human Health: Revisiting the Second Epidemiologic Transition, First Edition. Edited by Molly K. Zuckerman.
© 2014 John Wiley & Sons, Inc. Published 2014 by John Wiley & Sons, Inc.

the role of public health in a more substantial explanatory position and criticizes McKeown's method of parsing out the 19th-century disease categories. What Szreter and others have shown is that the ways in which diseases and causes of death are identified, grouped, disaggregated, and interpreted plays a significant role in isolating the mitigating factors that emerge in telling the story of what is arguably the most important development in the history of human health.

This paper does not attempt to resolve the ongoing disputes surrounding this broader debate, but it does help advance the discussion by focusing more closely on the everyday conditions that reporting personnel faced on the ground. Consequently, we may be able to disentangle some of the broader issues of data reliability and the historical validity of cause-specific mortality trends. I do this by first concentrating on reporting issues at the state level in Massachusetts, analyzing some of the perennial difficulties that state officials faced in implementing the registration system through the first half of this period. I then extend this analysis by providing a localized view of the changing structure of cause-of-death reporting in Northampton and Holyoke, Massachusetts, during the latter half of the 19th century. I examine the impact of changing the 19th-century conceptions of tuberculosis on cause-of-death reporting during the early part of the North American epidemiological transition. Using recorded literal causes of death[2] in Northampton and Holyoke, from 1850 to 1912, I look at the changes in tuberculosis as a reported cause of death throughout this period and examine the terminological shifts leading up and subsequent to the first International Classification of Diseases (ICD) in 1900. Both communities experienced the full effects of the 19th-century industrialization and are significant for their comparative advantage in assessing the impact of various economic, political, legal, and other institutional frameworks that emerged during this period. The stresses of urbanization, rapid population growth, immigration, lagging infrastructures, rising mortality, and so on all contributed to similar sets of circumstances. Although both Northampton and Holyoke shared in these stresses, the degree to which these changes took place was quite different. I examine the historical and demographic similarities and differences between the two towns and make a comparative analysis of changing health trends by focusing on tuberculosis throughout this period.

I use a method of formal decomposition in which recorded literal causes of death are compared and identified through a parsing routine, which resolves the component parts of principal and secondary causes into a classification schema. The results were then cross-validated to the ICD in accordance with the ICD cause-of-death coding rules. The overall aim of this formal decomposition and analysis is to document changes in the leading causes of death over the study period and provide quantifiable data from which to analyze the evolving 19th-century conceptions of disease (Anderton and Leonard, 2004). The scope of the data allows for more comprehensive comparisons of ICD classifications and common cause-of-death terminology over an entire half century of the epidemiological transition.

After providing brief histories of Massachusetts' death registration system and the two towns of Northampton and Holyoke, I compare the 10 leading literal causes of death with the 10 leading ICD-coded causes of death for both towns (note: an ICD-coded cause of death refers to a literal cause of death as coded according to the standards of the 1909 ICD). In doing so, I consider the following questions: Do we find different or similar pictures of tuberculosis given these two distinct ways of organizing causes of death? What do the differences or similarities suggest with regard to Massachusetts' evolving and increasingly complex death registration system? And finally, what effects did changing the 19th-century

---

[2] A literal cause of death is a cause of death as it literally appeared in the death records.

nomenclature have on the reporting of tuberculosis during the early part of the North American epidemiological transition? I argue that Massachusetts' new and evolving system for reporting deaths had an important impact on the changing death narratives through the turn of the century. I begin with the registration history of Massachusetts from the colonial period through the end of the 19th century.

Massachusetts Death Registration

During the early part of this period, from the 1840s through the 1860s, public health officials were just beginning to use cause-of-death statistics to systematically monitor and report on disease trends and epidemic outbreaks. Acquiring a uniform set of aggregate measures for causes of death was not feasible in the early part of the 19th century due to the absence of civil registration systems in much of Europe and the United States. Even after establishing Britain's General Register Office (GRO) in 1837 and Massachusetts' state registration system in 1842 (the first in the USA), maintaining a reliable system of registration remained an ongoing struggle for registration and public health officials throughout much of this period.

Death registration in North America began with the early settlement of Massachusetts in the 17th century. The practice of recording deaths in Europe began in the 16th century, when Thomas Cromwell introduced the Parish Registers in 1538, and the Council of Trent made the registration of births and marriages a part of ecclesiastical law in 1563 (Edge, 1928). It was the early 17th century, however, that marked the first time in either Europe or North America that an official governing body enacted a law requiring town officials to register deaths as a function of secular rather than ecclesiastical authority. On September 9, 1639, the General Court of the Massachusetts Bay Colony ruled that the registration of births, deaths, and marriages was a matter of public record to be administered through local town officials (Gutman, 1959: 61):

> Whereas many judgments have been given in our Courts, whereof no records are kept of the evidence and reasons whereupon the verdict and judgment did pass,…it is therefore by this Court ordered and decreed that hence forward every judgment, with all the evidence, be recorded in a book, to be kept to posterity…that there be records kept of all wills, administrations, and inventories, as also of every marriage, birth, and death of every person within this jurisdiction.

Prior to this statute, the responsibility of recording marriages, births, and deaths was a function of ecclesiastical authority. This new delegation of authority and responsibility was unique in its administrative scope and intent. Secular authority in England and elsewhere did not officially assume this responsibility in any sustained sense until the early 19th century. During the mid- to late 17th century, England implemented a nonecclesiastical system of registration as a substitute for the Parish Registers but returned to the old system by the 18th century (Edge, 1928). This difference between colonial America and England was significant in terms of the public and civil character that the early colonists accorded to vital events. Deaths, births, and marriages were not simply matters of ecclesiastical record as in England and elsewhere but also matters of civil record between individuals and governing authorities. The role that civil registration played became an important first step in establishing the systematic recording of vital events for statistical purposes. Nevertheless, such statistical objectives did not surface for another 200 years. For the colonists, vital registration was a matter of legal record alone, designed primarily for "the just administration of law and the protection of individual rights" (Hetzel, 1997). As Kuczynski (1900) notes, "Massachusetts was the first state in the world which recorded the dates of the actual facts of births, deaths, and marriages rather than the subsequent ecclesiastical ceremonies of baptisms, burials, and weddings; and Massachusetts was the first state in the world which imposed on the citizen the duty of giving notice to the government of all births, deaths [sic], and marriages occurring in his family."

For over 2 centuries, this system of vital registration remained relatively unchanged in terms of its legal function. Those changes that were enacted throughout this period concerned minor provisions to compensate town clerks for their recording duties, penalties for failing to provide such information on the part of informants, and expanding the responsibility for reporting births and deaths to include family members. Massachusetts' early registration system throughout the 17th and 18th centuries functioned as a protocensus. The impulse for gathering vital statistics, however, waned considerably during the 18th century with very little change in the existing laws. Massachusetts' registration laws were intended less with the public's health in mind than for cases of probate (Gutman, 1959). It wasn't until the 1840s, when Lemuel Shattuck took it upon himself to reform the ways in which vital statistics were gathered and applied, that death registration became more than just a legal matter of public record but one that specifically focused on the public's health (Swedlund, 2010).

Massachusetts was unique for being the first state to provide vital registration in 1842 and led the way in public health and sanitation reform in the United States throughout this period (Beemer et al., 2005). Without the use of statistical data, the substantive basis for public health reports would not have been possible. These and other measures provided the necessary information to combat growing epidemics and create public health initiatives in a more systematic way. Here, the role of medical science in conjunction with state authority opened the door for increased regulatory initiatives. This is but one layer in a multilayered social, political, and historical context, which highlights the need for a more complex approach in understanding the changing conceptual landscape of the 19th-century disease. Isolating the complexities that lay beneath these layers requires an account that extends beyond the shifting *advances* of the 19th-century medicine. As Alter and Carmichael (1996) note, "[t]he history of cause of death registration cannot be viewed as simply a part of the development of medicine; rather it reflects a much more complex interaction between the state, the medical community, and the public."

The Registration Act of March 3, 1842, established the requirements and guidelines for implementing a modern system of vital statistics in Massachusetts. Prior to the 1842 Registration Act, the Massachusetts State Legislature in 1835 adopted and confirmed without revision an earlier 1796 registration law, which included the following provisions (Gutman, 1959: 73):

> It shall be the duty of parents to give notice to the clerk of the town or district in which they dwell, of all the births and deaths of their children; and it shall be the duty of every householder to give notice of every birth and death which may happen in his house; and of the eldest person next of kin to give such notice of the death of his kindred; and it shall be the duty of the master or keeper of any almshouse, workhouse, or prison, and of the master or commander of any ship or vessel to give notice of every birth or death which may happen in the house or vessel under his care or charge, to the clerk of the town or district in which such event shall happen.

The responsibility for reporting deaths once again expanded in scope but became more specific in detail. More significant, however, was the added provision that each town rather than parents or kin would be required to pay the registration fee (Gutman, 1959). The shift from a more private to a more public focus began to take shape. Rather than focusing exclusively on issues of probate, vital registration was moving away from the domain of private, legal interests and toward the realm of the public good. Four decades later, the registration

> **1. FORM OF RETURN**
> *Return of the Births, Marriages, and Deaths which have taken place in the town of*
> _____
> **For the year ending May 1, 1842**
> [N.B.—if this blank should not be found sufficient for the purpose, the Town Clerk is requested to lengthen it by the addition of blank paper, ruled to correspond with the columns. The Town Clerk is also requested to sign the Return.]
> _____
> Whole number of Births in the town, for the year ending May 1, 1842, ...
> Whole number of Marriages in the town during the same period, ...
>
> **DEATHS IN THE TOWNS DURING THE ABOVE PERIOD**
>
> | Date of death. | Name of deceased. | Age. | Sex. | Occupation of adult males. | Disease, or cause of death. |
> |---|---|---|---|---|---|
> | | | | | | |

Figure 5.1. Registration blank for Massachusetts, 1842.

system of the Commonwealth of Massachusetts effected a more permanent shift toward the public good with the Registration Act of 1842.

The practical significance of the 1842 Act, according to Gutman (1959), was negligible as its provisions were already addressed in the 1796 legislation. Nevertheless, it established the consolidation of registration requirements into a single centralized system, stating that clerks of "several towns and cities in the Commonwealth" were required to submit annual reports (in May) of births, marriages, and deaths to the Secretary of the Commonwealth, under a penalty of $10.00 for noncompliance. This systematic centralization of information is what Gutman identifies as the beginning of a modern vital statistical regime. The 1842 Act also added the requirement that the Secretary of the Commonwealth "furnish blank forms of return" (blanks), with "suitable instructions and explanations; to receive said returns; to prepare therefrom such tabular results as will render them of practical utility; and make report thereof annually to the Legislature" (Bolles, 1843).

The 1842 Act stipulated that the Secretary of the Commonwealth sends a blank (Figure 5.1) and a letter of instruction to every town and city clerk in Massachusetts. These blanks were changed several times over the years for clarification and in response to pressures from the Secretary himself for a more uniform method of registration. For example, Secretary John A. Bolles (1843), in the first annual report, appealed to the legislature to adopt a more rigorous system of collecting vital statistics similar to that of the "French Code of Registration." Bolles suggested an alternative blank providing more detailed information compared to the official form (note: the surviving death records do not include these blanks for Northampton and Holyoke). These appeals continued throughout the years and were marked with relative success in making changes to the official registration forms. All of the available annual reports also contained nosological tables, which reported aggregate, cause-of-death information by county. Each of these reports noted the relative inconsistency of returns (full reports) of deaths received from some counties. Steps were taken to help increase reporting by increasing the amount paid to clerks and registrars per recorded death, stiffer fines for noncompliance, clearer instructions, and more detailed forms, but the effectiveness of such measures was negligible.

The new and evolving system for reporting deaths in Massachusetts undoubtedly shaped the changing death narratives over the 19th century. The 1842 law was repeatedly amended over the years, leading to the realization of modern death registration in Massachusetts by 1878. The first amendment of this law in 1844 added sextons or other persons having charge of burial grounds or burials to those required to make returns of facts connected with a death to the town clerk. In 1849, the law was revised to include a *proviso* requiring towns with a population of 10,000 or more to appoint a *registrar*, whose exclusive duty was to supervise registrations. The shift in responsibility from town clerks to registrars as towns grew in size was a less radical change than one might have supposed. Several of the legislative revisions over this period were primarily efforts to equalize the compensation and incentives provided to different authorities for recording deaths and to maintain equivalent quality in smaller towns and rural areas. Both types of reporting authorities were required to report the same information to the Commonwealth for review. In fact, differences in the number and quality of undertakers and physicians in different communities were often cited as reasons for differences between submitted reports from clerks and registrars. Gutman (1959) notes that many of the revisions between 1842 and 1855 followed changing practices rather than *vice versa* and were due in large part to pressure from town clerks and registrars.

In 1860, the law was revised again, requiring undertakers to certify all burials and give notice to the city clerk. This amendment also required all physicians to record and provide a cause-of-death certificate but only if a request was made of him within 15 days of the death occurring (Gutman, 1959). Following the Civil War, a fiscal revision in 1866 raised the compensation for clerks in smaller towns, who often had to collect death information personally rather than simply record what was reported by physicians and undertakers. Within a decade of this legislation, however, both Northampton and Holyoke reached a size where a registrar was required. In 1872, town boards of health were empowered to license undertakers and limit private burials. In 1873, a revision again raised fees paid to town clerks for recording of deaths.

It was not until 1878 that death certification moved toward full compliance when legislation provided that no body could be buried until the town clerk issued a burial certificate indicating that the clerk had received all details of the death including a record of the cause of death prepared by a physician. The full language of Public Statute 32, section 3 (Massachusetts State Board of Health, 1889), reads as follows:

> A physician who has attended a person during his last illness shall, when requested, forthwith furnish for registration, a certificate stating, to the best of his knowledge and belief, the name of the deceased, his age, the disease of which he died, the duration of his last sickness, and the date of his decease; and a physician who has attended at a birth of a child dying immediately thereafter, or at the birth of a stillborn child, shall, when requested, forthwith furnish for registration a certificate, stating to the best of his knowledge and belief the fact that such a child died after birth or was born dead. If a physician neglects or refuses to make a certificate as aforesaid, or makes a false statement therein, he shall be punished by a fine not exceeding fifty dollars.

The onus nevertheless remained with town clerks and registrars to ensure that requests were made of physicians to certify all deaths. The new law differed from the 1860 statute in that issuing a permit for legal burial required a physician-certified death certificate stating the cause of death, including the demographic information and length of morbidity, if any. The latter, interestingly enough, was not included on the registration schedules that city and town clerks prepared.

The last significant change in the registration laws took place in 1889 requiring physicians to list a primary and secondary cause of death but only for Civil War veterans. "A physician … shall, in case the deceased was a soldier or a sailor who served in the war of the rebellion, give both the primary and the secondary or immediate cause of death as nearly as he can state the same. If a physician refuses or neglects to make such a certificate he shall forfeit to the treasurer the sum of ten dollars for the use of the town in which he resides" (Massachusetts State Board of Health, 1889). The penalty *proviso* was a regular feature of almost every public statute on death registration since its enactment in 1842. Such penalties applied to almost everyone along the reporting chain—clerks, registrars, superintendents of almshouses, undertakers, and even parents. For example, the 1796 and 1842 registration laws stated that "[p]arents shall give notice to the clerk of their city or town of the births and deaths of their children…Whoever neglects to give such notice for the space of six months after birth or death shall forfeit a sum not exceeding five dollars" (Public Statute 32, section 2.) The full statute, however, only required giving notice to the town clerk with no requirements to provide cause of death.

Reliability and Accuracy

Although Gutman (1959) does not fully address the quality of Massachusetts' cause-of-death reporting, gradual improvement in the general quality of death statistics did take place over this period. The 1860 law not only encouraged registrars and clerks to consult physicians as to the cause of death but empowered them to obtain such reports from physicians if needed. The problem in implementing this provision, however, stemmed largely from its voluntary nature and the reporting practices to which clerks had been accustomed. According to Gutman, clerks and undertakers did not use the powers granted them under the law of 1860, to request physicians to provide a certificate of the cause of death of persons to whom they had attended. Town clerks and registrars had little incentive to verify the accuracy of reported causes of death because verifying reported causes, whatever the source, was not mandated by the law. As they saw it, their primary responsibility in the reporting chain was to record cause-of-death details as they received them. Undertakers were often the first link in the reporting chain and were notoriously unreliable in properly certifying deaths. The Massachusetts State Board of Health polled both physicians and town clerks in 1877 regarding reporting accuracy. Many of the complaints from this poll and throughout this period focused on the unreliability of cause-of-death certifications from undertakers (see Beemer, 2011). For their part, town clerks often consolidated their efforts by gathering much of the vital registration information on a periodic basis, which involved canvassing their communities annually to record births, marriages, and deaths in one fell swoop.

The lack of complete and accurate reporting was regularly cited in both the Massachusetts' annual vital registration reports and the annual State Board of Health reports. Earlier registration reports from 1843 through the 1850s acknowledged the insufficiency of their data to allow for specific conclusions about the health of the state's population but were, nevertheless, improving and exhibited sufficiency with regard to elucidating *general truths* and principles of the *general laws of mortality*. The tone during this period was optimistic. The reports expressed confidence that the information they were receiving on the ground, while lacking in completeness and accuracy, was nevertheless moving toward that goal (Massachusetts State Board of Health, 1877: 233):

> The protracted labor which has necessarily arisen from the heterogeneous intermingling of the whole twenty months, in every conceivable combination, in the returns, has prevented the elucidation of some interesting principles and facts, deducible from the records of registration, inasmuch as it would prolong the undesirable delay already incurred, concerning this Report. It is confidently believed, however, that herein will be found sufficient ground covered, and with sufficient faithfulness, to exhibit very reliable and highly important general truths.

By the 1870s, however, the tone had changed. Concerns surrounding the accuracy of death records were by no means confined to Massachusetts. In 1876, the Secretary of Michigan's State Board of Health, Henry B. Baker, commented on the issue of accuracy as he believed it stood across the country, "No method has yet been found, or at least acted upon, whereby the actual death-rate can be positively ascertained for the United States, or so far as I know, for any single State" (Massachusetts State Board of Health, 1877). In response to Baker's rather dim observation, Charles F. Folsom, Secretary of Massachusetts's State Board of Health, was not eager to offer a vigorous defense of Massachusetts's vital registration system: "It has seemed desirable to ascertain how we stand in reference to so sweeping a criticism, and the result of the inquiry has been that we can only say that we are a little better than some of our neighbors" (Massachusetts State Board of Health, 1877). The inquiry to which Folsom referred was a survey that the Board of Health sent out to "medical correspondents of the Board" (i.e., physicians) and town clerks throughout the state. The following questions were asked of the physicians (I.) and clerks (II.), respectively (Massachusetts State Board of Health, 1877: 233):

> I. Is the registration of deaths and causes of death complete and satisfactory in your town? If not, please suggest any deficiencies of which you are aware, whether *all* deaths are returned to the undertakers, whether the undertakers themselves return them promptly and accurately to the clerks, whether causes of death are reported by the physicians in all cases, etc., etc.?

> II. Will you be so kind enough to inform us whether the registration of deaths and causes of death is complete and satisfactory in your town? If not, please suggest any deficiencies of which you are aware, whether *all* deaths are returned to the undertakers, whether the undertakers themselves return them promptly and accurately to the clerks, in what proportion of cases the *causes of death* are reported by physicians, etc., etc.?

The State Board of Health had regularly included correspondence from physicians and clerks in their annual reports in years past, but such notes from the field were typically unsolicited letters they had received on a wide variety of health-related issues. The difference with the 1877 report was the Board's systematic effort to solicit physicians and clerks from across the state on the specific issue of accuracy and completeness in cause-of-death registration. The responses were overwhelmingly unflattering and demonstrated well the central problem with the current system—the lack of physician input and participation.

Before examining the reported aggregate numbers of those who responded to the survey, it will be useful to look at a few examples of the actual responses that both physicians and clerks returned to the Board. Not all of the returned responses were included in the 1877 report, but a sizable selection was nevertheless published. In selecting the responses, the State Board of Health endeavored to highlight those that articulated the problems associated with the registration system but not to the exclusion of those who thought the system was working well. Here are just a few of the statements from physicians and town clerks in response to the State Board of Health survey (Massachusetts State Board of Health, 1877):

Replies of Medical Correspondents:

> 7. The cause of death is invariably returned on the undertaker's certificate and is given by the friends of the deceased. I am never asked to make out a physician's certificate. (235)

> 39. Many times the physician never sees the return at all. The statement of some member of the family is all the authority. I don't think, during my practice, that fifty per cent [*sic*]. of the deaths were returned in a proper manner. (237)

57. I am not called upon in half my cases to give the certificate. I understand that undertakers get them filled by the family, and I know that many certificates are wholly false as to cause of death. Almost any cough is reported either *consumption* or *lung fever*, and so in other diseases. (239)

Replies of City and Town Clerks:

8. The custom has been to return the deaths at the end of the year, except when the body is carried out of town for burial; then I get a return near the date of death. I seldom get the physician's certificate with the return; the cause of death is usually named, also the name of the physician; but all in the handwriting of the undertaker, and I think it is obtained from some member of the family of the deceased. I think the disease or cause of death is in many cases guessed at, so that my return to the department is not accurate as to the prevailing disease. (250)

31. The law is in no case complied with, either as regards the undertaker, physician, or by the families themselves. At the close of the year, the births and deaths of the year past are collected by going from house to house throughout the town, making the result very unsatisfactory and expensive. (251)

90. I have been clerk of the town for six years in succession, and have employed a man to gather all the information in regard to births and deaths called for in the blanks sent to me for that purpose. I have never received any information from physicians or undertakers. (252)

The aforementioned responses were just a few of the nearly 150 responses included in the 1877 report. While a response rate cannot be given because the report did not indicate the total number of surveys sent out, the responses do provide a good representation of the general sentiment expressed by both physicians and clerks across the state. Overall, only 25% of physicians and 29% of clerks who responded felt that the registration system was adequate. Table 5.1 shows how physicians and town clerks responded to the survey questions. On the question of accuracy in registered causes of death, over 60% of the physicians who responded thought that causes of death were not competently reported. The completeness of registered deaths, that is, without regard to cause, fared only slightly better in their estimation. Of these, 27% reported that the number of deaths recorded was satisfactory compared with 23% who reported that the number was unsatisfactory. The majority of responses, approximately 50%, were deemed indefinite, which meant "… they [did] not answer that particular question, they [had] no suggestions, they [did] not know, or the reply [was] so worded as not to convey a definite statement" (Massachusetts State Board of Health, 1877). On the question of cause-of-death accuracy, the ratio of satisfactory to unsatisfactory among clerks was similar to that reported by the physicians. Roughly, two out of every five physicians and two out of every four clerks thought that causes of death were

Table 5.1. Physician and town clerk responses to 1877 Massachusetts's Board of Health survey.

	Responses	Number of deaths registered	Causes of deaths registered	Row total
Physicians	Satisfactory	54	44	98
	Unsatisfactory	45	118	163
	Indefinite	97	34	131
	Response total	196	196	
Town clerks	Satisfactory	102	50	152
	Unsatisfactory	43	113	156
	Indefinite	117	99	216
	Response total	262	262	

adequately reported. When comparing the two groups on the question of completeness (number of deaths registered), it is not surprising that the clerks were more optimistic in that regard. Accuracy in terms of recorded numbers of deaths would have been more reflective of their responsibilities and capabilities as town clerks than the actual causes themselves.

From the time that the registration system was initiated in 1842, the architects who implemented the system envisioned that physicians' participation was more extensive than it actually was. It soon became clear that such participation would not be forthcoming. Public health officials were not naïve, however, and attempted to hedge their bets. Shattuck (1850) among others wanted physician participation in the new registration system to be mandatory. The legislature, however, resisted any move toward compulsory participation on the part of physicians and did so for over 30 years. The unsuccessful attempt to maintain legislation supporting licensed physicians in the 1830s was still fresh in their minds. The state's interest to not interfere with the free flow of economic opportunity overrode any public health interests.

Beyond the political obstacles against mandatory physician certification of causes of death, there were more mundane obstacles with which to contend. Physicians were not always available to determine a cause of death or given adequate notice that a death had occurred. There were few legal requirements during this period for physicians to document patient morbidity, and patient confidentiality was a professional necessity. Nevertheless, had these obstacles not stood in the way of accurate and reliable cause-of-death reporting, the lack of compensation from the state provided an additional disincentive for physicians to actively participate in a system they felt had no demonstrable benefit to their practice. The broader public health initiative of gathering aggregate statistical data for analyzing statewide health trends was not one that occupied the professional interests of the average physician. As such, most physicians were not inclined to actively provide the necessary information to properly certify causes of death.

19th-Century Northampton and Holyoke

The towns of Northampton and Holyoke exemplified the struggles that many urban, industrial centers experienced during the late 19th century. Over the course of the early mortality transition, both towns changed dramatically with the rise of industrial growth and the influx of immigrant populations from their neighbors to the north in Canada and across the Atlantic in Europe. Both communities experienced the full impact of industrialization but in different degrees.

Northampton was the older of the two communities, having established itself as a permanent settlement nearly 2 centuries before Holyoke. Northampton transitioned from a rural market economy to a robust mix of industry, agriculture, and commerce, quadrupling its population, from 5278 in 1850 to 19,431 in 1910 (Hautaniemi, 2002). Industries began developing in neighborhoods along the Mill River in the Northampton area with associated enclaves of worker housing in areas known as Bay State, Paper Mill Village, and Florence. Northampton's infrastructure was slow to develop, with severe housing shortages and the lack of adequate public works contributing to poor health conditions. The housing shortage quickly became a leading concern for city officials. As late as 1881, the local newspaper reported, "Tenements of all descriptions are difficult to obtain, and rents will be firm at present prices, if not higher, in the spring" (*The Hampshire Gazette and Northampton Courier*, 22 February 1881). Housing shortages were one of several indicators of Northampton's strained capacity to respond to the demands of local industry during this period of economic expansion. There were a small number of industrial mills of moderate size in Northampton, which in 1855 employed approximately 10% of the labor force for the surrounding region

(Hankins, 1954). A number of these firms were destroyed by the 1874 Mill River Flood, which forced many of the larger companies to move their operations to nearby Holyoke (Green, 1939; Benson, 1954; Jacob, 1999). Although Northampton never fully recovered from this to emerge as a major industrial center, it did continue to flourish with a mix of farming and commercial trade as part of a diverse economy (Hautaniemi, 2002). Like other industrializing cities of this period, Northampton experienced the pressures of urban growth, lagging infrastructure, and poor sanitation.

Over this same period, Holyoke residents faced similar conditions but on a much larger scale. Holyoke experienced a far greater industrial expansion and demographic shift than Northampton. By comparison, Holyoke's population grew far more rapidly, from 3249 in 1850 to 59,732 in 1910 (Hautaniemi, 2002). Like Northampton, economic opportunity was the driving force underlying this growth, but unlike Northampton, Holyoke's industrial emergence followed a more deliberate and focused path. Holyoke was one of several planned communities in the region. This was due to the enterprising vision of the Boston Associates. This group of venture capitalists sought to expand the textile industry at various sites along the Merrimack and Connecticut Rivers in Western Massachusetts. Earlier ventures in Lowell and Lawrence, Massachusetts proved lucrative for the Boston Associates, and with the natural resources of the Connecticut River, Holyoke held even greater potential as an industrial mill town. The cotton textile industry became Holyoke's first principal economy as a result, serving as a catalyst for its exploding immigrant population. Much of Holyoke's history emerged out of this expanding textile industry, which exploited the hydraulic power provided by the Connecticut River (Hautaniemi et al., 1999). Similarly, the influx of foreign immigrants, which consisted of Irish, French Canadians, and Eastern Europeans, provided local industry with the requisite supply of cheap labor. Immigration peaked during this period of expansion in the 1870s with the foreign born constituting 52% of Holyoke's residents (Hautaniemi, 2002). Two groups in particular comprised the bulk of Holyoke's labor force, the Irish and French Canadians.

Despite the historical and demographic differences, there were many parallels between these two towns. Both communities experienced intense population pressures. The ensuing strains of urban growth and resulting high-mortality environments were ameliorated only by the later development of public works and a sanitary infrastructure (Hautaniemi et al., 1999; Beemer et al., 2005). Waves of immigration were transnational in scale and affected the composition of the two towns similarly over time. Moreover, the constant influx of poor immigrant labor, fueled by these transnational population movements, sustained remarkable levels of inequality in these developing urban-industrial areas. In such settings of high mortality and dramatic inequality, one would expect to find the full impact of misery and poverty among the working poor.

Not surprisingly, Holyoke's identity as a community in the late 19th century owed much to its immigration history. The first large-scale immigrant population was the Irish, who began arriving in Holyoke in the late 1840s following the Irish Famine of 1846. The Irish were, in large part, responsible for building the infrastructure of Holyoke, of which construction of the dam and canals proved to be the most crucial for Holyoke's emerging industrial economy (Hautaniemi, 2002). Given the circumstances of their immigration, the Irish were more likely to establish permanent roots in the community. Like Northampton, housing was a major issue for newly arriving mill workers as many families were forced to take residence in shanties along the Connecticut River. Holyoke's population density in 1880 was the third highest in the country (Hautaniemi, 2002).

The second largest immigrant population in Holyoke was the French Canadians, who began arriving in large numbers around 1859, shortly after the economic crisis or *panic* of 1857. The influx of French Canadians during this period was an orchestrated endeavor, as was the case with much of Holyoke's economic designs. As Hautaniemi (2002) notes, the French Canadians were directly recruited into Holyoke's work force due to the efforts of Nicholas Prue, an agent for the Lyman Mills who went to Quebec with a specially built wagon resembling a prairie schooner to recruit workers, for which he was paid four dollars a head. Rather than hire skilled men as he was instructed, however, Prue hired mostly girls and young women. The apparent motivation behind this recruiting effort was the assumption that most of the workers would be temporary or seasonal laborers (Hautaniemi, 2002). This was also the intent of many of the newly recruited French Canadians themselves. Again, the recruitment effort factored into the idea of supplemental employment for Canadian families. Green (1939) notes that Prue went from village to village in Quebec and spoke of the prospect of wages that could be sent home or brought back in a few years' time to set up whole families of business in Canada. The women who were hired in this first migration phase via direct recruitment had year-long contracts. However, whether they actually returned to Canada after their contracts ended or simply remitted their earnings is unclear.

The migration patterns of the French Canadians, nevertheless, followed the economic fluctuations that occurred in Holyoke throughout this period. Holyoke underwent a series of economic crises beginning in the 1850s with periodic panics occurring through the end of the 19th century. In 1879, following the panic of 1873, another influx of French Canadians began arriving in Holyoke (Hautaniemi, 2002). This was the largest influx of French Canadian immigrants into Holyoke up to that point. The major difference with this second migration phase was that the immigrants consisted primarily of families. One of the significant contributing factors to this massive influx was the depressed economic conditions in Quebec. As conditions became untenable there, French Canadians emigrated to the United States, and New England in particular, looking for economic opportunity. This shift in immigration most likely represented more of an abandonment of homeland than the earlier recruitment migration in 1859.

The initial rapid population growth of these towns was fueled by younger factory labor immigrants seeking new opportunities in developing sectors. As immigration became a smaller segment of the growing population and population growth stabilized, these urban centers began experiencing an aging population. For some ethnic groups, migration also changed from predominantly younger factory workers to families. This combination of trends accelerated the aging of the population, generated age differences among social groups, and presumably accentuated age–sex-specific changes in mortality over the course of the transition. The shifting composition of the population, particularly in Holyoke, brought with it changing public health concerns and initiatives. Other cities in Massachusetts were undergoing similar changes due to the same dynamics of industrialization, immigration, high population densities, economic opportunity, rising inequality, and poor living conditions, all of which changed the disease ecology in a relatively short period of time. State and local officials in Massachusetts responded with a system of surveillance capable of monitoring epidemic and endemic diseases, without which the charge of public health would not have moved forward.

New England as a whole had higher child mortality levels than other regions of the country and experienced a mortality plateau throughout the latter half of the 19th century. By the time of sustained mortality decline at the turn of the 20th century, Northampton and Holyoke had matured into substantial urban-industrial centers, population growth had slowed, public health infrastructures had been initiated, and the population had aged significantly as mortality and immigration of younger factory laborers declined. The mortality rates for both towns reflected these changes, particularly in Holyoke, which experienced a more dramatic set of demographic changes than did Northampton. Nevertheless, beginning around 1880, both towns experienced a very similar shift in mortality (Figure 5.2). In the

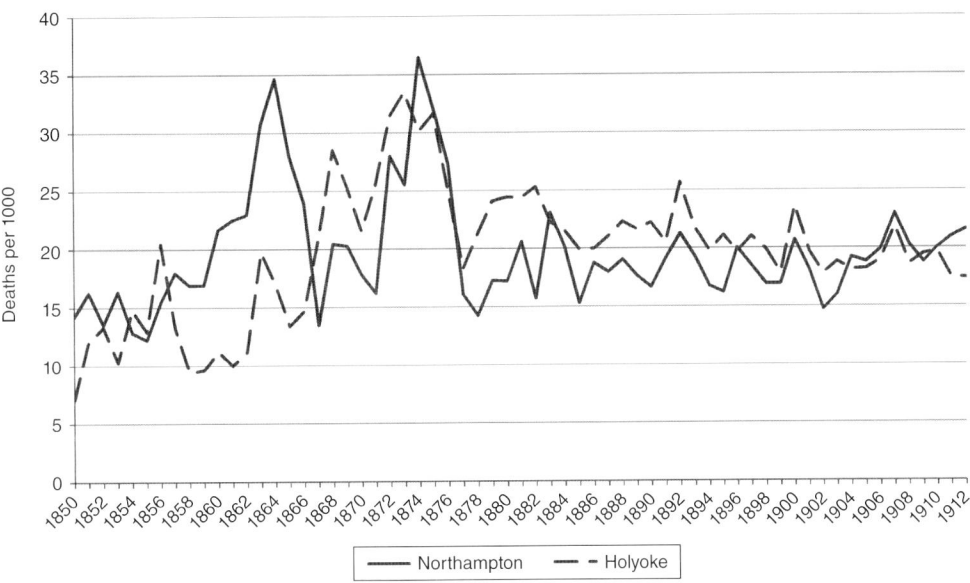

Figure 5.2. Northampton and Holyoke standardized mortality rates, 1850–1912.

earlier decades, Holyoke experienced rising mortality due primarily to compositional shifts in its population. This was less evident for Northampton, but for both towns, compositional shifts coupled with specific disease outbreaks and high child mortality account for a large proportion of rising mortality up through the 1870s. At the end of this decade, however, both towns experienced dramatic declines in their overall mortality. It is at this stage that the mortality plateau becomes evident, and we begin to see an overall stabilization of mortality develop across most disease categories.

Analyzing Grammars of Death

There are several ways of explaining the mortality plateau, none of which are mutually exclusive. Changes in immigration patterns, aging populations, implementation of public health and sanitation infrastructures, and regulatory measures designed to improve accuracy in cause-of-death reporting were all players in this period of mortality stabilization. Given the history of vital registration in Massachusetts, one would expect that changes in the quality of reported causes of death, similar to changes in completeness, would gradually improve. While the changing registration system must always be kept in mind, evolving conceptions of disease and disease diagnoses over the period had a more proximate effect on recorded causes of death than incremental changes in the registration system *per se*. Again, these developments were not mutually exclusive.

To better understand the impact of the epidemiological transition on Northampton and Holyoke during this period, I compare the top 10 leading literal and ICD-coded causes of death and focus on three particular causes—consumption, phthisis, and tuberculosis—noting shifts in nomenclature and frequency. Although it is not surprising that overall mortality trends in Northampton and Holyoke reflect similar trends among certain leading causes, understanding these trends in light of Massachusetts' history of death registration provides some insight into the reasons why these changes took place. Table 5.2 shows the frequency of the 10 leading literal and ICD-coded causes of death by decade from 1850 to 1912.

Table 5.2. Ten leading, parsed literal and ICD-coded causes of death, Holyoke and Northampton, Massachusetts, 1850–1912.

Decade	Frequency	Parsed literal cause	Frequency	ICD 1a-2 code/disease title
1850 $N = 1399$	312	Consumption	315	028 Tuberculosis of the lungs
	78	Dysentery	78	014 Dysentery
	74	Fever	67	007 Scarlet fever
	63	Scarlet fever	66	154 Senility
	53	Old age	53	092 Pneumonia
	42	Lung fever	36	001 Typhoid fever
	36	Typhoid fever	35	064 Cerebral hemorrhage, apoplexy
	35	Dropsy	35	187 Ill-defined organic disease
	31	Fit	34	079 Organic diseases of the heart
	24	Croup	32	104 Diarrhea and enteritis (under 2 years)
1860 $N = 3253$	465	Consumption	570	028 Tuberculosis of the lungs
	138	Dysentery	205	104 Diarrhea and enteritis (under 2 years)
	126	Typhoid fever	164	009 Diphtheria and croup
	120	Cholera infantum	159	092 Pneumonia
	108	Croup	140	014 Dysentery
	108	Lung fever	132	001 Typhoid fever
	101	Old age	113	154 Senility
	88	Scarlet fever	97	007 Scarlet fever
	80	Phthisis	76	079 Organic diseases of the heart
	65	Heart disease	56	006 Measles
1870 $N = 6431$	658	Consumption	879	028 Tuberculosis of the lungs
	515	Cholera infantum	665	104 Diarrhea and enteritis (under 2 years)
	262	Diphtheria	480	009 Diphtheria and croup
	200	Typhoid fever	347	092 Pneumonia
	194	Old age	234	001 Typhoid fever
	191	Smallpox	203	007 Scarlet fever
	189	Croup	194	154 Senility
	163	Pneumonia	192	005 Smallpox
	156	Scarlet fever	180	151 Congenital debility, icterus, and sclerema
	147	Lung fever	153	079 Organic diseases of the heart
1880 $N = 8578$	640	Cholera infantum	1153	028 Tuberculosis of the lungs
	457	Consumption	1001	104 Diarrhea and enteritis (under 2 years)
	436	Phthisis pulmonalis	513	092 Pneumonia
	430	Pneumonia	410	009 Diphtheria and croup
	264	Stillborn	352	001 Typhoid fever
	259	Typhoid fever	249	079 Organic diseases of the heart
	191	Bronchitis	230	061 Simple meningitis
	184	Enteritis	223	151 Congenital debility, icterus, and sclerema

Table 5.2. (Continued)

Decade	Frequency	Parsed literal cause	Frequency	ICD 1a-2 code/disease title
	177	Phthisis	214	154 Senility
	169	Convulsions	210	064 Cerebral hemorrhage, apoplexy
1890 $N = 11{,}330$	769	Cholera infantum	1188	028 Tuberculosis of the lungs
	617	Pneumonia	1183	104 Diarrhea and enteritis (under 2 years)
	490	Stillborn	731	092 Pneumonia
	359	Phthisis pulmonalis	437	151 Congenital debility, icterus, and sclerema
	317	Consumption	434	079 Organic diseases of the heart
	294	Bronchitis	430	061 Simple meningitis
	294	Meningitis	341	064 Cerebral hemorrhage, apoplexy
	278	Premature birth	338	009 Diphtheria and croup
	221	Apoplexy	336	089 Acute bronchitis
	211	Convulsions	300	154 Senility
1900 $N = 13{,}516$	700	Stillborn	1068	104 Diarrhea and enteritis (under 2 years)
	664	Pneumonia	1055	028 Tuberculosis of the lungs
	629	Cholera Infantum	925	079 Organic diseases of the heart
	414	Pulmonary tuberculosis	839	092 Pneumonia
	392	Premature birth	539	151 Congenital debility, icterus, and sclerema
	369	Malnutrition	517	120 Bright's disease
	302	Meningitis	513	064 Cerebral hemorrhage, apoplexy
	275	Senility	471	061 Simple meningitis
	269	Bronchopneumonia	448	154 Senility
	267	Gastroenteritis	307	009 Diphtheria and croup
1910–1912 $N = 4366$	216	Pulmonary tuberculosis	387	104 Diarrhea and enteritis (under 2 years)
	213	Stillborn	322	028 Tuberculosis of the lungs
	210	Gastroenteritis	312	079 Organic diseases of the heart
	201	Pneumonia	306	092 Pneumonia
	164	Bronchopneumonia	216	120 Bright's disease
	150	Premature birth	211	064 Cerebral hemorrhage, apoplexy
	132	Senility	180	151 Congenital debility, icterus, and sclerema
	119	Cerebral hemorrhage	165	091 Bronchopneumonia
	116	Cholera infantum	161	154 Senility
	102	Nephritis	108	061 Simple meningitis

For certain diseases, we see an evolving nomenclature that highlights the consolidation of several causes into a single cause and the partitioning of others into several distinct causes. For example, consumption is the leading literal cause of death, and tuberculosis is the leading ICD-coded cause of death for three consecutive decades, 1850, 1860, and 1870. By the 1880s, however, consumption falls to the second leading literal cause with 457 reported

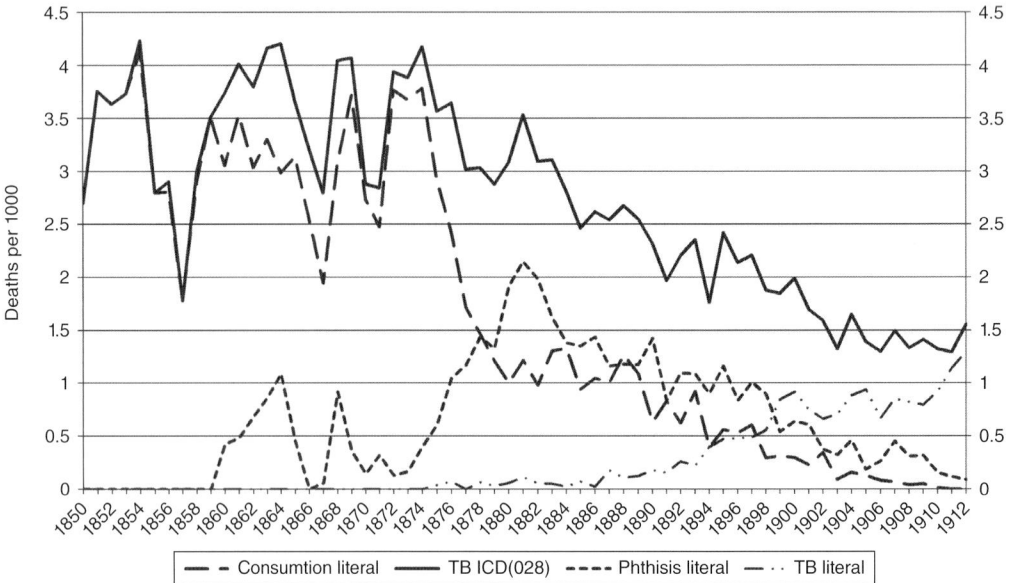

Figure 5.3. Disease-specific mortality rates: all ages by year, Northampton and Holyoke, 1850–1912.

cases and the fifth leading literal cause in the 1890s with 317 reported cases, and by the turn of the century, consumption drops off the list entirely as one of the top 10 causes of death in Holyoke and Northampton. By contrast, tuberculosis is either the first or second leading ICD-coded cause of death throughout the period. As a literal cause of death, tuberculosis does not appear in the death records until 1875 but quickly rises as one of the top 10 causes by the 1900s and becomes the leading literal cause by 1910.

The differences between tuberculosis as a literal cause of death and tuberculosis as an ICD-coded cause of death are due to the way the ICD consolidated causes. The ICD employed a classificatory system that lumped earlier cognates, such as phthisis and consumption, under the contemporary nomenclature *pulmonary tuberculosis*. Phthisis and consumption were terms used to identify tuberculosis in the earlier decades, with consumption as the more common term. As Anderton and Leonard (2004) note, even at this early time, consumption was a lay equivalent to the medically prevalent phthisis, or simply *wasting*. During the last half of the 19th century, such deaths were increasingly identified as phthisis in Western Massachusetts. Figure 5.3 shows the decline of consumption, the rise and fall of phthisis, and the rise of tuberculosis over this period. Consumption began to lose its currency as a cause-of-death term in the mid-1870s, around the same time that tuberculosis began to be recorded as a cause of death. The dramatic decline of consumption, however, cannot be attributed entirely to the increasing use of *tuberculosis*. As Figure 5.3 shows, phthisis begins its ascendency at nearly the same time that consumption begins its descent. Tuberculosis, while the leading cause of death throughout this period, was nevertheless on the decline (see TB ICD(028) in Figure 5.3). If we compare all three of the literal causes of death (consumption, phthisis, and tuberculosis) with the ICD-coded tuberculosis, the picture we get of tuberculosis as the leading 19th-century cause of death in Northampton and Holyoke becomes much more complex and nuanced.

This change in nomenclature was rather dramatic, with reported cases of consumption falling to nearly 75% of its previous levels and phthisis increasing twofold. This shift may reflect a move toward more diagnostic precision, which would closely correspond with the

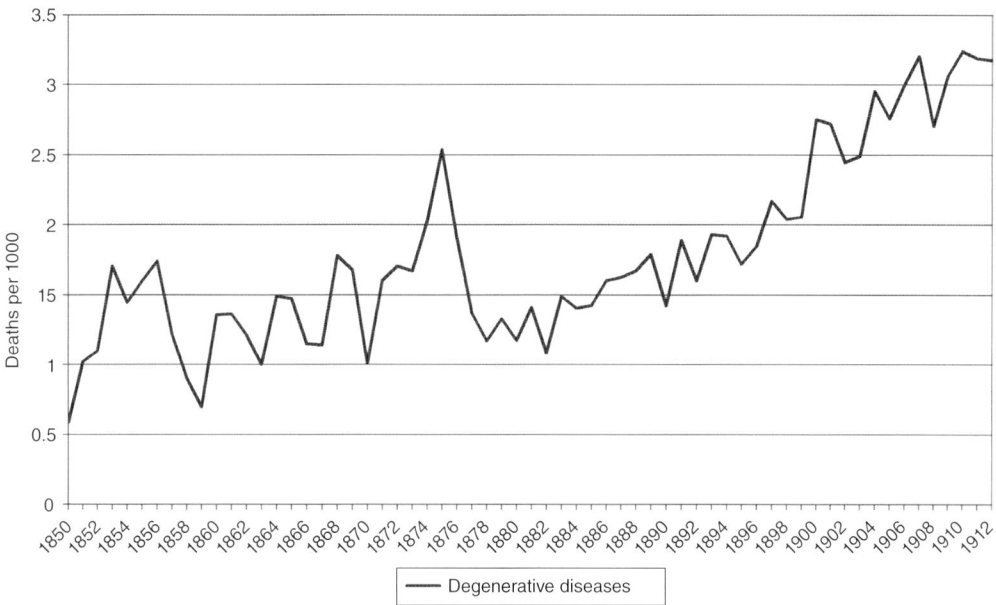

Figure 5.4. Mortality rates for degenerative diseases: all ages by year, Northampton and Holyoke, 1850–1912.

changes taking place in death registration during the mid- to late 1870s, namely, the increased pressure on town clerks and registrars to certify all causes of death from a physician, or it may reflect real declines. As noted earlier, by 1878, town clerks and registrars were required by law to secure a detailed death report from a physician. The 1878 statute required town clerks and registrars to get a physician-certified death report before burial of the deceased could legally take place. Moreover, it is not until the early 1860s that phthisis began to be recorded as a cause of death, which was the same period when the registration law was amended requiring "…all physicians to record and provide a cause of death certificate, but only '*if a request was made of him within fifteen days after the death occurred.*'" It is also important to note that the mortality patterns of both phthisis and consumption began to stabilize at nearly the same time with similar rates in the early 1880s, which corresponded to the overall pattern of stabilization of mortality in Northampton and Holyoke during this same period (see Figure 5.3). When compared to pre-1880 rates, overall and disease-specific mortality rates after 1880 exhibited far less variability and fluctuation. Again, such stabilization is suggestive of a more uniform method of recording and certifying causes of death, but it could also simply be an etiological shift in the overall disease ecology. It is during this period that Northampton and Holyoke are implementing modern public health infrastructures that included water filtration and sewerage systems (Beemer et al., 2005).

This trend was not limited, however, to tuberculosis. In fact, we see it occur in almost all cases and in various disease groupings. While tuberculosis is an endemic disease, the same sorts of conditions responsible for reducing the risk factors for epidemic diseases, like cholera, typhoid, malaria, and smallpox, can be responsible for reducing risk for certain endemic diseases as well. But what about those diseases that were identified as chronic and degenerative? In both Northampton and Holyoke, we begin to see an aging of the population. Epidemic mortality is declining toward the end of this period, while mortality from degenerative diseases is on the rise. Figure 5.4 shows causes of death due to chronic or degenerative diseases

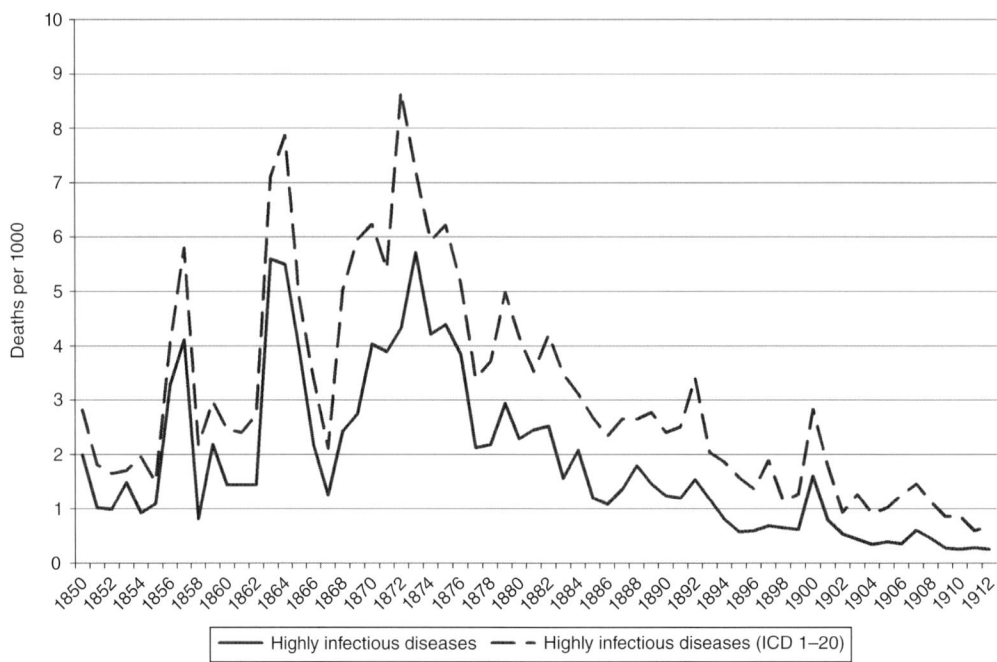

Figure 5.5. Mortality rates: all ages by year, Northampton and Holyoke, 1850–1912 (cholera, diphtheria, malaria, smallpox, and typhoid).

and conditions (note: chronic and degenerative diseases included all cancer-related causes of death, senility, and heart disease). Again, when compared to the pre-1880 rates, we see these cause-specific rates exhibit less variability (i.e., a more stable pattern) after 1880. As Figure 5.5 shows, highly infectious acute diseases, such as cholera, diphtheria, malaria, smallpox, and typhoid, begin to exhibit a more stable pattern after 1880 as well. We would expect to see high variability and fluctuations across time with epidemic mortality but not necessarily with degenerative causes.

The stabilization, while evident, is not as salient among chronic and degenerative causes as it is with acute infectious causes. This may suggest that shifts in the overall disease ecology were more likely etiologically based. In other words, chronic and degenerative diseases should not be subject to the same stabilizing influences that affect acute infectious disease if the stabilization is simply a result of changing etiological conditions. However, if we look at causes of death that were ill defined, namely, those that did not fit with any other category, we see the same sort of pattern emerge here as well (Figure 5.6).

Answering the question of whether we find different or similar pictures of tuberculosis given the distinct ways in which causes of death are organized seems clear. Grouping consumption, phthisis, and their cognates under the ICD classification of tuberculosis (ICD 028) provides us with a picture of steady decline beginning in the mid-1870s. Disaggregating the causes reveals a much more complex development of nomenclature than is present under the ICD classification schema. Of course, the utility of one approach over the other depends on what the research intends to accomplish. Using literal, recorded causes of death for historical analyses of the 19th-century epidemiological trends can be problematic in several respects (see Maudsley and Williams, 1996; Risse, 1997). Some of these shortcomings have already been noted, but one of the limitations not mentioned concerns the qualitative value

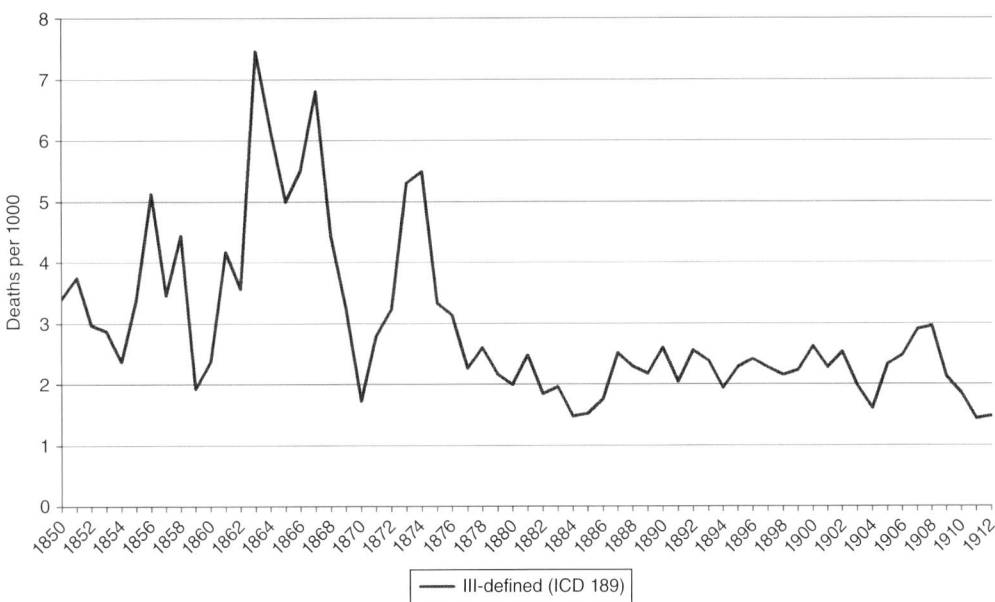

Figure 5.6. Mortality rates: all ages by year, Northampton and Holyoke, 1850–1912 (ill-defined and unknown diseases).

of simple cause-of-death descriptions. Diagnostic accounts throughout this period were often terse, lacking sufficient detail in identifying cases where cause-of-death descriptions (literals) were notoriously vague. The heterogeneous nature of cause-of-death reporting coupled with the minimalist style inherent in recorded causes of death certainly constrains the scope to which the information can be applied. Moreover, detailed death narratives can be less precise and more ambiguous when attempting to identify single or primary causes of death (Anderton and Leonard, 2004). When causes of death are coded in accordance with a particular classification system, like the ICD, loss of information is likely to be extensive but more useful for analyzing a population's health. For instance, using the 1910 ICD standards retrospectively for classifying pulmonary tuberculosis in the decades prior to this classification schema consolidates historically divergent nomenclature (e.g., consumption, phthisis, and tuberculosis) of what was understood to be pulmonary tuberculosis in 1910. Consequently, this attenuates the historical complexity (a loss of information) of the contemporaneous understandings of this disease as it evolved over time. On the other hand, using more contemporary etiological classification standards gives us an understanding of the epidemiological trends within a historical population that better accords with accepted classificatory standards that are meaningful today.

Conclusion

The expanding authority and scope of Massachusetts' emerging death registration system played a subtle but an important role in the history of Northampton and Holyoke's shifting epidemiological environment. We have no way of identifying the reporting agents, physician, or otherwise, because the death records did not contain this information. Nevertheless, we do have evidence of the broader outlines influencing reporting trends throughout this period. The regulatory requirements for reporting causes of death became more specific and restrictive in terms of who and what could be reported toward the last quarter of the 19th century. Public health initiatives were on the rise throughout the state, and the cultural impact of the sanitation movement was felt at the local, national, and international levels. Nevertheless, certain cause-specific trends are difficult to explain on demographic or public health grounds alone. We would expect to see high variability across time with epidemic mortality, smallpox, cholera, etc., but not necessarily with ill-defined or degenerative causes. Holyoke, for example, did experience a smallpox epidemic from 1870 to 1872 that caused higher than normal mortality rates, especially among children. In that decade, smallpox was the 6th leading cause of death in Holyoke and the 24th in Northampton, while tuberculosis and ill-defined causes still remained at the top as leading causes of death.

It is difficult to say with complete certainty what accounts for this contrast between high variability in mortality during the earlier decades followed by a pattern of stabilization after 1880. Generally speaking, changes in adult mortality are seemingly amplified by patterns of development and population growth in emergent urban-industrial centers like Northampton and Holyoke. Whether improvements in recording and certifying causes of death offer a more compelling explanation for Northampton and Holyoke's stabilizing mortality over this later period is at this stage undecided. Changes in Massachusetts's death registration system undoubtedly played a role in how deaths were recorded, thus having an impact on the reliability and validity of recorded causes of death during this period. The evidence also suggests that reported mortality trends were contingent upon sociopolitical and economic contexts in ways that may not directly reflect shifting disease ecologies. Having considered a variety of reasons for Northampton and Holyoke's shifting mortality over this period, the most likely scenario is one that takes into account a combination of the earlier factors working together to bring about these changes.

References

Alter G, Carmichael A. 1996. Studying causes of death in the past: problems and models. Hist Methods 29:44–48.

Alter G, Carmichael A. 1999. Classifying the dead: toward a history of the registration of causes of death. J Hist Med 54:114–132.

Anderton DL, Leonard SI. 2004. Grammars of death: an analysis of nineteenth-century literal causes of death from the age of miasmas to germ theory. Soc Sci Hist 28:111–143.

Arrizabalaga J. 1999. Medical causes of death in preindustrial Europe: some historiographical considerations. J Hist Med 54:241–260.

Beemer JK. 2011. Social meanings of mortality: the language of death and disease in 19th century Massachusetts. Ph.D. dissertation. Amherst: University of Massachusetts-Amherst.

Beemer JK, Anderton DL, Hautaniemi SI. 2005. Sewers in the city: a case study of individual-level mortality and public health initiatives in Northampton, Massachusetts at the turn of the century. J Hist Med Allied Sci 60:42–72.

Benson LW. 1954. Floods and disasters. In: Tercentenary History Committee, Editor. The Northampton book: Chapters from 300 years in the life of a New England town 1654–1954. Northampton: The Tercentenary Committee. p 355–363.

Bolles JA. 1843. Second annual report to the legislature: Under the Act of March, 1842, relating to the registry and returns of births, marriages and deaths in Massachusetts. Boston: Dutton and Wentworth.

Edge PG. 1928. Vital registration in Europe. The development of official statistics and some differences in practice. J R Stat Soc 91:346–393.

Green CM. 1939. Holyoke, Massachusetts: A case history of the industrial revolution in America. New Haven: Yale University Press.

Gutman R. 1959. Birth and death registration in Massachusetts 1639–1900. New York: Milbank Memorial Fund.

Haines MR. 2003. The great modern mortality transition. Soc Sci Hist Assoc Newsl, Winter, 1–22.

Hankins FH. 1954. Economic transition: 1817–1860. In: Tercentenary History Committee, Editor. The Northampton book: Chapters from 300 years in the life of a New England town 1654–1954. Northampton: The Tercentenary Committee. p 77–84.

Hautaniemi SI. 2002. Demography and death in emergent industrial cities of New England. Ph.D. dissertation. Amherst: Department of Anthropology, University of Massachusetts.

Hautaniemi SI, Swedlund AC, Anderton DL. 1999. Mill town mortality: consequences of industrial growth in two nineteenth-century New England towns. Soc Sci Hist 23:1–39.

Hetzel AM. 1997. U.S. vital statistics system major activities and developments, 1950–95. Hyattsville: National Center for Health Statistics. Available at: http://www.cdc.gov/nchs/data/misc/usvss.pdf (accessed October 27, 2013).

Jacob EC. 1999. One morning in May: The Mill River disaster of 1874. Haydenville: Edward C. Jacob.

Kuczynski RR. 1900. The registration laws in the colonies of Massachusetts Bay and New Plymouth. Publ Am Stat Assoc 7 (51):1–9.

Massachusetts State Board of Health. 1877. Eighth annual report of the State Board of Health of Massachusetts. Boston: Albert J. Wright.

Massachusetts State Board of Health. 1889. Massachusetts State Board of Health manuals of statutes relating to the public health. Boston: Wright & Potter.

Maudsley G, Williams EMI. 1996. Inaccuracy in death certification: where are we now? J Public Health Med 18:59–66.

McKeown TH. 1976. The modern rise of population. New York: Academic Press.

Omran AR. 1971. The epidemiological transition: a theory of the epidemiology of population change. Milbank Mem Fund Q 49:509–538.

Risse GB. 1997. Cause of death as a historical problem. Contin Chang 12:175–188.

Shattuck L. 1850. Report of a General Plan for the Promotion of Public and Personal Health. Boston: Dutton & Wentworth.

Swedlund AC. 2010. Shadows in the Valley: A cultural history of illness, death, and loss in New England, 1840–1916. Boston: University of Massachusetts Press.

Szreter S. 1988. The importance of social intervention in Britain's mortality decline, c. 1850–1914: a reinterpretation of the role of public health. Soc Hist Med 1:1–37.

Woods R. 1991. Public health and public hygiene: the urban environment in the late nineteenth and early twentieth centuries. In: Schofield R, Reher D, Bideau A, Editors. The decline of mortality in Europe. Oxford: Clarendon Press. p 233–247.

Part 2
Epidemic Infectious Disease and the Second Epidemiologic Transition

Chapter 6
Agent-Based Modeling and the Second Epidemiologic Transition

Carolyn Orbann, Jessica Dimka, Erin Miller and Lisa Sattenspiel
Department of Anthropology, University of Missouri, Columbia, MO

Introduction

Numerous studies have addressed the characteristics and impact of the second epidemiologic transition (e.g., this volume, Omran, 1971, 1983; Preston, 1976; Higgs, 1979; Preston and Haines, 1991; Schofield and Reher, 1991; Coleman and Salt, 1992; Gage, 1993, 1994, 2005; Barrett et al., 1998). In general, this transition is associated with a shift to a situation wherein the primary causes of death are chronic and degenerative diseases rather than infectious diseases. In his seminal article laying out transition theory, Omran (1971) claimed that there are broad patterns of human morbidity and mortality and that in a wide range of societies, these patterns can be tied to technological innovations and economic shifts. The continued development of Omran's ideas has examined numerous factors, including advances in Western medicine and public health, such as vaccination campaigns, sanitary regimes, and the use of antibiotics to fight infections (e.g., Omran, 1971, 1983; Meeker, 1972; Schofield and Reher, 1991; Troesken, 1999; Beemer et al., 2005; Cutler and Miller, 2005) as well as nutritional changes (McKeown, 1976) and improvements in education, transportation, and animal waste control (Coleman and Salt, 1992).

A variety of methods have been used in these studies of the transition. This chapter focuses on a novel method of studying it, agent-based modeling, that can further add to the body of knowledge about human health over time. We make a case that agent-based computer simulation models can be a powerful tool to use in testing hypotheses about the second transition. These kinds of models are being used more frequently in a variety of disciplines, including epidemiology, demography, ecology, anthropology, and archaeology, and they can be used by researchers with different types of data and a diversity of research questions. We argue that the application of an agent-based model to questions about disease and demographic change in the past provides an additional way to make use of the often incomplete or imperfect data that are commonly encountered by researchers working with historical populations.

The model that we describe in this chapter is designed to address questions about the spread of the 1918 influenza pandemic in a small Newfoundland fishing community. The pandemic was documented throughout the world and resulted in the deaths of at least 50 million people (Johnson and Mueller, 2002). Spread of the pandemic has been linked to the movement of troops at the end of World War I as well as global transportation of goods and people (Crosby, 1989). At the time of the pandemic, most countries in North America and Western Europe were nearing the end of the second transition. For these places, the 1918 pandemic represented one of the last great epidemiologic events until the recent resurgence of infectious diseases (Barrett et al., 1998). For populations that had not yet gone through the second transition, this pandemic was still a major, and probably one of the most significant, single epidemiologic events in history.

The Dominion of Newfoundland and Labrador was hit very hard by the 1918 pandemic. Overall, mortality in the Dominion averaged 7.5 per 1000 population, well above the average of about 5 per 1000 in the United States, Canada, and Western Europe (Mamelund, 2011), but it varied substantially and was as high as 80–90% in some small communities in Labrador (Sattenspiel, 2011). At the time of the pandemic, the Dominion was in the midst of the second transition, especially in the capital city, St. John's, and nearby towns and villages. Archival sources indicate, for example, that the government was implementing policies and programs to increase access to health care; raise awareness of health issues, especially among mothers with young children; and diversify the economy and increase the standard of living (Baker and Pitt, 1984; Overton, 1998; Bishop-Sterling and Webb, 2008). Because of the remoteness of most communities, however, these efforts had only minimal impact on the health of the Dominion's population outside of the capital region.

With an eye toward eventually exploring the entire time period encompassing the second transition as well as the importance of different possible reasons for that transition, the model described here has been designed for a remote community whose status at the time of the pandemic was essentially pretransition. The purpose of this model is to learn about the effects of acute, infectious disease in small, kin-based populations with limited access to medical care. Of particular interest is the way in which gendered behavior patterns, occupational categories, and age-grade behaviors affect epidemic size and spread. Although the model cannot yet directly be used to address questions specific to the transition in Newfoundland or elsewhere, it is being designed with a generalized structure that will allow it to do so in the future. As it is further developed, this model will be used to study the interplay between chronic and acute infectious diseases as well as the effect of medical interventions and other community-level health improvements on the long-term demographic structure of a population. Eventually, results generated by the model will be used to evaluate the relative importance of different factors that are thought to play a role in changing patterns of mortality as a consequence of the second transition. Here, we describe the overall model structure, some of our initial analyses of the model, and the ways in which models such as this one can be used to address questions about the transition.

Background

In general, the second transition followed developments in economic industrialization and public health measures, beginning in the 19th century in the United States, Canada, and Western Europe. Most populations outside of these regions have now also experienced this transition, although not all have passed through it completely. Depending on the time and location of the transition in a population of interest, researchers may face different sets of challenges in data collection that may require different types of analytical methods. For example, because of later adoption of compulsory birth and death registration, data from populations experiencing the second transition more recently may not have the time depth needed to identify significant long-term health trends. In order to compensate for this, researchers may be required to use sophisticated statistical or computer modeling techniques that take full advantage of the existing data while accounting for the limited information from the past.

One of the advantages of studying historical populations is the availability of a wide variety of sources, such as vital statistics, skeletal data, environmental reconstructions, and archaeological data. However, numerous challenges are also associated with these data sources, including damage and decay of sources such as original census or parish documents, photographs, and poorly preserved skeletons; biases in what material is recorded or retained; and changes in language and the understanding of disease processes and diagnoses (Hudson, 1993). Further, it is well known that recording and preservation biases disproportionately affect marginalized and disenfranchised population subgroups, such as women and ethnic or racial minorities (e.g., Hollingsworth, 1968; Macfarlane, 1977; Howell and Prevenier, 2001). Additionally, highly mobile portions of the population may appear and disappear from historic records as they move across a landscape (Kasakoff and Adams, 1995). These concerns should not dissuade researchers from using historical resources, though, as there are ways in which to identify and deal with many potential problems with the resources. One strategy includes building models that can overcome such limitations by incorporating complex historical data to identify important factors affecting the phenomenon of interest. Such models can then be used to test hypotheses about the specific impact of these factors.

The Scope of Agent-Based Modeling

The use of models, in the general sense, is integral to the practice of science. Scientists build models by identifying essential components of the systems under study and examining how alterations in those components may affect system-wide outcomes. Building a model can be as simple as constructing a flowchart of causes and outcomes or as complex as designing a computer simulation with a myriad of processes that can be used to generate quantitative data for statistical analysis. It is up to the researchers to decide what kind of model is most appropriate for the questions that they wish to ask with the data available to them.

Agent-based modeling is an effective tool to use in studying the second transition because of the quantity of supporting historical materials that are often available and that can be used to develop realistic models with which to test hypotheses about the impact of the transition on health and demography. An agent-based model is one that explicitly models a process within a group of distinct "individuals" (which could correspond to individual organisms, households, communities, or other distinct units). Models of this type rely on the model developer understanding the dynamics of the system as a whole. Because of this complexity, which makes them difficult to analyze using mathematical techniques, the vast majority of these models are computer simulations. Agent-based models can incorporate as much data as the programmer desires, including, for example, population movements, land-use data, weather or seasonal data, or marriage patterns. The major constraint is the availability of necessary data within the historic record for the population under study.

A danger with agent-based models is that they can easily become as complex and realistic as one desires, but often, that leads to difficulties in understanding the implications of the results. A savvy researcher thus incorporates essential knowledge about the study population into the model rules but keeps the model simple enough to both address the questions being asked and understand which aspects of the model are most affecting the output. In addition, a good model is designed to be flexible enough to easily incorporate different kinds of rules when research questions change. For example, the model described in this chapter is not set at a particular time of year, but does have the ability to keep calendar time if that becomes an area of interest. Additionally, most agent-based modeling platforms have the ability to render stochasticity (randomness) into a model. The element of randomness is important, especially when modeling human behavior and small populations. Indeed, the most casual observer of human behavior must concede that even under the strictest rules, humans do not always behave in a rigid, rule-driven fashion. Stochasticity in a model is vital for discovering the potential variety of outcomes held within model parameters.

A growing number of agent-based models are being developed to address questions in all subfields in anthropology and in other social sciences. Cultural anthropologists have used them to study the transmission of religious beliefs (Tomlinson, 2009), interpersonal violence (Younger, 2011), and marriage practices (Billari et al., 2007). Archaeologists have used agent-based models to test hypotheses about ancient land-use patterns and site-formation processes (Kohler and Gumerman, 2000). One of the most common uses of simulation in biological anthropology is to evaluate genetic change over time by modeling mutations or changes in gene frequencies in human or primate populations (e.g., Barbujani et al., 1995). Computer simulations are also commonly used by anthropologists interested in health and epidemic disease to study how culturally mediated behaviors affect disease transmission and other aspects of health in populations of anthropological interest (e.g., Herring and Sattenspiel, 2007; Carpenter and Sattenspiel, 2009; O'Neil and Sattenspiel, 2010).

Newfoundland at the Time of the Second Epidemiologic Transition

Before describing the structure and analysis of our model, we set the stage with an overview of the nature of life in Newfoundland during the early 20th century. A wealth of historic information is available to explore the impact of epidemics and health initiatives in Newfoundland, including detailed mortality records, census information, parish records, hospital records, and a variety of historical accounts of epidemics and other health-related issues. To ensure that the model adequately represents the population being studied, ethnographic and demographic information is used to design the structure of a model fishing village, or an outport, that is based on a particular community on the island—St. Anthony, a small town in the Northern Peninsula (Figure 6.1).

The island of Newfoundland, together with Labrador, became the easternmost province of Canada in 1949 (Rodgers and Witney, 1981). At the time of the 1918 pandemic, Newfoundland and Labrador formed an independent dominion within the British Commonwealth. The total population of the Dominion was about 255,000, but only about 4000 of those people lived in Labrador. The remainder lived on the island of Newfoundland itself, with 40% residing in or near the capital city, St. John's, on the northeastern edge of the Avalon Peninsula (Sattenspiel, 2011). Almost all of the remaining island residents lived in a few towns situated slightly inland or in outports scattered in harbors and inlets along the coast. The small size of these communities and their relative geographic isolation, both from each other and from the rest of the world, make the study of infectious disease spread within Newfoundland communities ideally suited to computer simulation approaches, which are better able to deal with the randomness inherent in epidemic processes operating in small populations.

Settlement of the island of Newfoundland occurred in stages. Indigenous peoples colonized the island around 5000 years ago, followed by several waves of different groups (Rankin, 2008). European contact began with the Vikings as early as the 11th century but was sporadic until the 16th century, when travel to the region began to increase as a

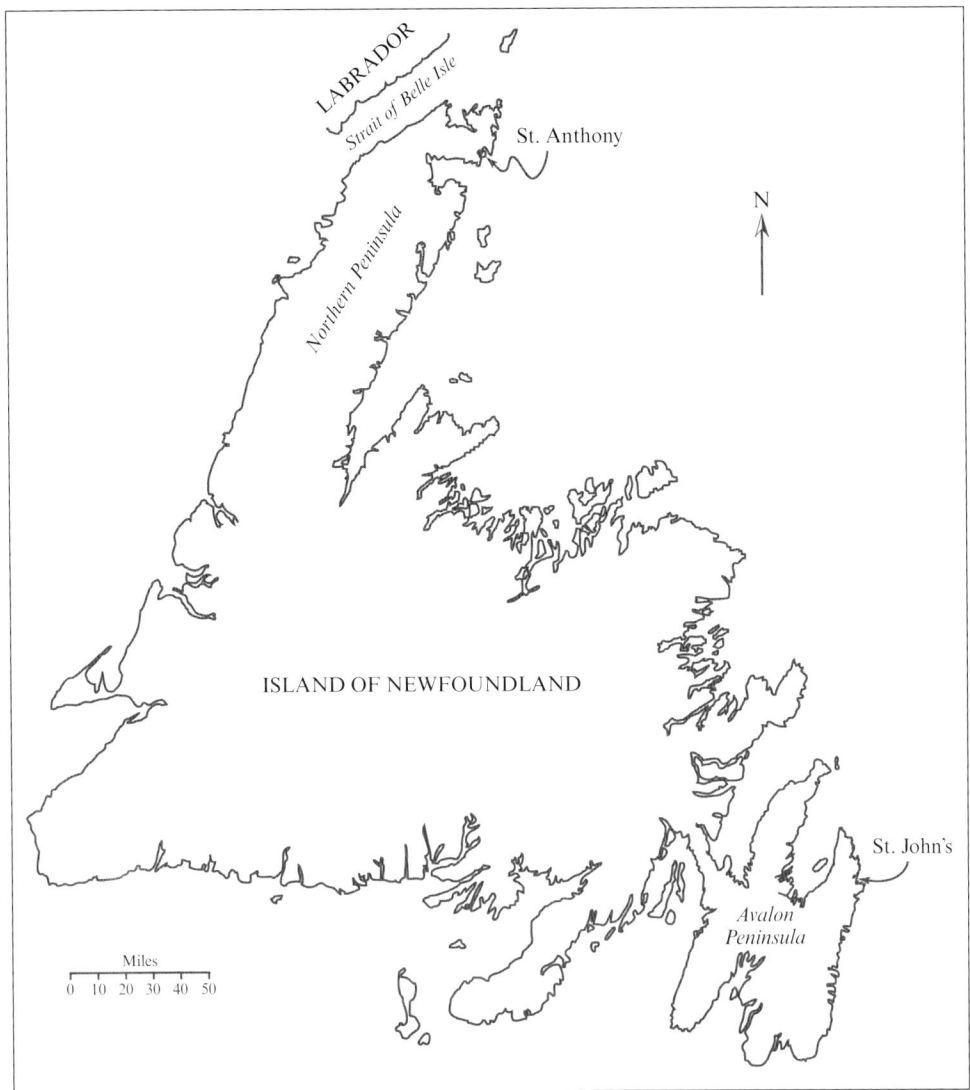

Figure 6.1. Map of Newfoundland showing the location of St. Anthony, the Northern Peninsula, St. John's, and the Avalon Peninsula.

consequence of seasonal French, English, and Portuguese fishing expeditions. Although English policies forbade permanent settlement to facilitate better control of resources, many fishermen chose to settle anyway, frequently in outports (Harris and Warkentin, 1974; Prentice et al., 1988). By the mid-18th century, as an increasing number of women and families contributed to the population, the number of permanent residents became larger than that of temporary residents (Handcock, 1977; Thornton, 1985). Most of the settlements remained on the coast, although some families moved a short distance inland for greater protection from the elements during the winter. Outports typically consisted of a small number of families, although larger populations could be supported in some harbors (Firestone, 1967; Harris and Warkentin, 1974; Smith, 1987).

Ethnographers have noted consistent and enduring patterns to the social structure and daily life found in Newfoundland outports, at least prior to Confederation with Canada in 1949 (Queen and Habenstein, 1974). These patterns were built around the primary subsistence and economic activity of fishing. Small crews of related males (typically fathers, sons, and brothers) fished together, establishing the patrilineal extended family as the basic organizational unit of the community (Firestone, 1967; Nemec, 1972; Queen and Habenstein, 1974). Upon marriage, couples would reside with the husband's parents before eventually moving to independent households nearby; the youngest son and his family would often remain with the parents and inherit the house (Firestone, 1967). Fishing gear, berths, and boats would be inherited jointly by male siblings, who would continue to fish together until their sons were old enough to become part of new father–son crews (Firestone, 1967; Nemec, 1972; Davis, 1983). Seasonal fishing activities were supplemented by income from lumbering, sealing, farming, mining, and other employments (Newfoundland Colonial Secretary's Department, 1923; Murray, 1979).

Although there was a sexual division of labor, Porter (1985) notes that the boundary for this division was at the shoreline. In addition to tasks such as housework; preparing meals; caring for animals, children, elderly, and the sick; gardening; making butter and picking berries, which were sold for cash or credit with merchants; and making clothing, many women frequently participated in the fishery activities (Murray, 1979; Porter, 1985; Prentice et al., 1988). Shore crews cleaned and processed fish, and women could perform virtually any of the related tasks. In some cases, wives also had authority over hiring and supervision of labor (Queen and Habenstein, 1974; Murray, 1979; Porter, 1985). Overall, women's work in the fishery and other areas is estimated to have contributed at least half of the family income (Murray, 1979). Other forms of employment, particularly for young and unmarried women, included nursing, teaching, and domestic service (Kivimaki, 1937; Murray, 1979; Porter, 1985; Beaton and Walsh, 2004).

At the community level, outports were characterized by egalitarianism, most likely because individuals generally maintained similar levels of income, occupation, and education (Davis, 1983; Porter, 1985). The few status distinctions recognized by residents included clergy, merchants, and outsiders, and these differences could be seen, for example, during visits when these individuals would be entertained in the relatively finer front rooms of houses (Pocius, 1979). Generally speaking, all other townspeople could enter and exit each other's homes without invitation, gathering in kitchens, which were the centers of activity of houses and, by extension, the community (Porter, 1985). Women also engaged in social and voluntary associations, including church or exercise groups and service organizations (Murray, 1979; Davis, 1983).

The early 20th century in Newfoundland was a time of relative prosperity compared to the hardscrabble past on the island. Successive governments worked hard at diversifying the economy by promoting expansion of the pulp and paper industry, for example. Attempts were also made to improve the overall standard of living by funding government organizations such as the Child Welfare Agency and building small health clinics and hospitals in regions outside of St. John's (Bishop-Sterling and Webb, 2008). Although these efforts resulted in some improvements in the Dominion's well-being, its economic base remained unstable. Ultimately, this instability played a major role in Newfoundland and Labrador's shift from a self-governing Dominion within the British Empire to its 1949 incorporation as a province of Canada.

Historically, like most parts of the world in the early 20th century, Newfoundland experienced many health problems. These issues included chronic conditions such as tuberculosis and malnutrition and repeated epidemics of several infectious diseases. The largest of these were the 1918 influenza epidemic, with over 2000 deaths, and the 1916 measles epidemic, with over 500 deaths; these two epidemics alone were responsible for the deaths of nearly 1% of the population. Data from the Newfoundland Annual Reports of the Registrar General of Births, Marriages, and Deaths (Newfoundland House of Assembly, 1910–1942) indicate that influenza epidemics occurred again in 1935 and 1941, and unusually large mortality levels from measles reoccurred in 1924, 1930, and 1937. In addition to these two diseases, epidemics of pertussis occurred in 1921, 1925, 1931, 1935, and 1940, and there was an epidemic of scarlet fever in 1924. Health conditions were exacerbated by poverty and poor sanitation and hygiene. Furthermore, the relative isolation of communities and poor transportation systems hindered attempts to improve health in the communities by the government and organizations such as the Grenfell Mission, a privately run hospital and medical mission that provided health care to the residents of the Northern Peninsula of Newfoundland and the central and southern parts of Labrador.

The Government of Newfoundland and Labrador made several significant attempts at improving health and well-being during the first half of the 20th century (Baker and Pitt, 1984; Overton, 1998). A permanent medical health officer with responsibilities to oversee all health issues in the Dominion was appointed in 1905, and a public health laboratory for the Medical Health Officer opened in 1906. Initiatives put into place by the governor, Sir William MacGregor (1904–1909), included the formation of a volunteer antituberculosis association, construction of a new tuberculosis sanatorium and a modern laboratory, creation of a Department of Health and Education, implementation of educational programs focused on teaching mothers about caring for children, and the improvement of home sanitation, sewers, and water supply systems. Many of these initiatives were not put in place until after the completion of MacGregor's governorship, but his efforts guaranteed continuing attempts at improving the health situation of the Dominion. Unfortunately, Newfoundland's population was scattered across a wide geographic area, with essentially no roads and limited rail service, making most communities outside St. John's and surrounding areas very difficult to reach. This geographic distribution of settlements ensured that the island's health-care infrastructure, especially in the outports outside St. John's and the H-shaped Avalon Peninsula on the eastern side of the island, would remain inadequate for decades to come.

A 1916 measles epidemic, the 1918 influenza pandemic, and other problems associated with activities during World War I led to a number of additional health improvements (Baker and Pitt, 1984). A large number of veterans returned with active cases of tuberculosis, prompting the government to open a tuberculosis hospital in St. John's to serve the military population. The 1916 epidemic, which was a serious problem among the servicemen, stimulated the government to open a special facility, also in St. John's, to deal with infectious diseases. This hospital moved to larger facilities later that year and was used throughout the 1918 influenza epidemic, finally closing in May 1920. Effort and funds were also spent on improving child welfare, primarily through the initiation of a community nursing service, the establishment of milk stations, and the distribution of clothing to families in need. Although these improvements were helpful, most efforts were centered in St. John's and the more densely populated areas surrounding it. Health care throughout most of the rest of the Dominion remained woefully inadequate.

This history of health reforms and improvements in Newfoundland provides support for the view that the island was in the midst of the second transition, but it also suggests that any changes are likely to be observed most strongly in the St. John's region where they were implemented earliest. Data on island-wide mortality from the most common causes also suggest that Newfoundland was experiencing the transition at this time. Deaths from infectious causes other than tuberculosis declined somewhat, deaths from tuberculosis dropped precipitously, deaths from more chronic conditions (e.g., cancer, paralysis/apoplexy, and old age) increased slightly, and infant mortality declined substantially. Comparable data for the city of St. John's are available but require more detailed analyses before it is possible to assess the relative timing of the transition in the city versus the Dominion as a whole. The attempts made to institute many of the health reforms island-wide had varying success and timing of implementation, however; thus, the island presents an ideal location in which to study health changes as the second transition evolved over time and space.

The Model

Details about life in Newfoundland described in the previous section are used to design an ideal community within which influenza spreads. In general, agent-based models consist of a population of agents, the environment or space in which they interact, and a set of rules governing the nature and frequency of interactions (Epstein and Axtell, 1996). In common with most agent-based and microsimulation models in epidemiology, the model we describe here also incorporates an epidemic process, the SEIR process, that takes place as a consequence of agent interactions on the model space (Figure 6.2). The SEIR process includes four distinct disease states—susceptible (S), exposed (or latent) (E), infectious (I), and recovered (R)—and has been used extensively and with considerable success in both mathematical modeling and computer simulations of directly transmitted infectious diseases, such as influenza and measles (see Anderson and May, 1992; Hethcote, 2000; Keeling and Rohani, 2007; Sattenspiel, 2009).

Influenza is spread as a consequence of respiratory contact between susceptible and infectious individuals. Transmission occurs fairly readily, but is not certain; thus, a key parameter of simulation models is the probability that transmission occurs following contact. In influenza, once an individual is infected, he or she enters a latent phase that lasts from a few hours to up to 3 days. After this phase, the individual becomes infectious and capable of transmitting the virus to others. This phase can begin before a person is aware he or she is sick and lasts for 3–5 days in adults and up to 7 days in children (Heymann, 2004). Recovery confers permanent immunity to the specific strain of the virus; however, the influenza virus changes structure often, and so immunity during subsequent outbreaks of

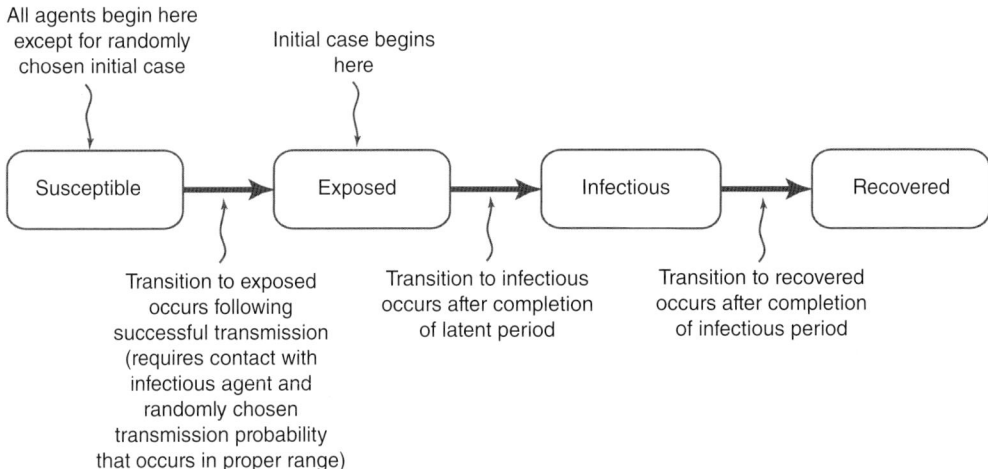

Figure 6.2. The SEIR framework. Disease states are represented by boxes; dark arrows indicate transitions between disease states.

the disease is often imperfect or absent (depending on the degree of similarity between the particular circulating viruses).

Our influenza simulation model was developed using the programming language Java and the Repast Simphony simulation libraries and packages,[1] which are designed for social science modeling. Repast software facilitates model creation using Java and provides built-in options for generating visual data output. The model described here was written using the direct coding option in Java. The fundamental structures of the model are a space and a group of agents that represent individual people. The agents move on the space according to predetermined rules, but not in a predetermined fashion. That is, there are guidelines based on sex, age, and occupation that influence where an agent can move during a particular time, but in almost all cases, there is a chance that those guidelines will not be followed. Disease transmission takes place throughout the entire model timeline, as described in the following.

Basic Model Structure

The model space is a 100 × 100 bounded grid with clusters of cells that represent "buildings" scattered over the space. Agents are only allowed to occupy a cell that is part of a building, and agent movements are constrained so that only one agent can occupy a particular cell at a time. The town of St. Anthony is being used to develop the model space, but it is important to note that the model focuses on the general issue of epidemic spread within a generic, small, kin-based fishing community rather than the specific issue of epidemic spread in St. Anthony itself. The geography of the model does not directly reflect the geography of the town, in that the layout of structures and distances between them do not match. However, in order to ensure that the model is a realistic representation of a small Newfoundland fishing community, the number and types of many structures known to have existed within St. Anthony were reproduced in the model structure. Because of the way that agents move on the space (described later), the distances between the structures are irrelevant, and so accurate georeferencing is not necessary. The virtual town was given 84 houses, corresponding to the number found in the 1921 census of St. Anthony. It is known from archival sources that a hospital and orphanage were located in the town; a school and two churches were also programmed into the simulation. These structures are based on information about St. Anthony derived from newspaper articles, parish registration information, memoirs, and other archival sources.

Agents in the model are idealized, but, again, in order to ensure that the model is suitably realistic, their characteristics were chosen on the basis of information in the 1921 census of St. Anthony, which lists all residents of the community, their age, occupation, residence, and family membership. Thus, the model population of 503 residents reflects the historical population of the community in terms of age and sex structure, family distributions, and occupations. While the most common occupation among the men of St. Anthony was fishing, the 1921 census also contains references to pastors, doctors, nurses, and servants. These occupations were distributed through the model population to reflect their distribution within the real village. Currently, the time length of the simulation runs is only long enough for a single flu epidemic to run its course (100 days); thus, fertility and background mortality have not been included.

Time is split into 4-h increments in this model. The model initializes on a Monday morning at 6 a.m. A typical model run is 600 ticks (100 days) using current parameter values. Agent behaviors are linked closely to day of the week and time of day; however, the model is not set in a particular month of the year. Future iterations of this model may take this into account in order to test hypotheses about seasonal employment changes or other seasonal questions.

[1] http://repast.sourceforge.net

Behaviors and Movement

While the model population is small (503 agents) for a model of this type, agent behaviors are complex and modeled on historical and ethnographic evidence from St. Anthony and similar fishing villages along the north coast of Newfoundland. Most male agents over the age of 15 work on fishing boats from Monday to Saturday. In the simulation, agents work with relatives (as assigned by grouping individuals whenever possible by surname listed in the 1921 census) with between 4 and 8 males per boat. A few male agents were assigned to be pastors/schoolteachers (pastors were assumed to also serve as schoolteachers during the week), servants, and doctors rather than fishermen. These agents all move to the appropriate building (school, house, hospital) each weekday. Doctors and servants also do their usual activities on Saturdays; pastors do whatever their wives do on Saturdays.

The behavior of adult female agents is somewhat dependent on the presence and age of any children in the family. If preschool-aged children (<5 years) are present, at least one female agent over the age of 15 (usually the stay-at-home mother) stays with them. The preschool children always accompany their caretaker on any movement during the week. Most of the time, these caretakers stay in their residences, but they do have a chance to visit another family between the hours of 10 a.m. and 2 p.m., Monday through Saturday. Female agents with older or grown children and unmarried female agents over the age of 15 are either fisherwomen or work as nurses, teachers, or servants. Fisherwomen, who assist in processing the catch near the shore, either travel to their family boat every day (Monday–Saturday) (though they go to the boat later in the day than the males) or have a small chance of visiting another family. Nonfisher female agents move to the building associated with their profession on weekdays. Nurses go to the hospital for work on Saturdays; other nonfisher female agents do whatever the stay-at-home mother does on Saturdays.

All agents between the ages of 5 and 15 go to school 5 days a week. Orphans move to school from the orphanage during schooltime. On Saturdays, children 10 years and older have a chance of going to the boats to help in fishing activities. Otherwise, they and all the younger schoolchildren have a chance to go to the school building to "play" with other children. School-aged children remaining at home on Saturday mornings may go visiting with their mother.

On Sundays, all families attempt to go to one of two church services, moving together as a group. If they do not attend church, they may visit another family. If they do neither, they stay at home, where they may receive visitors. Pastors always go to church on Sundays, but no other agent is guaranteed a spot in either service. The church capacity is such that it is impossible for all agents to go to church every Sunday.

Due to the small size of the village and the short travel times, agents in this model essentially "teleport"—pick a new location, disappear from the current one, and reappear in the new location—from one place to another rather than moving along pathways or in vehicles. This type of movement has been assumed because the space is a social space rather than a geographic space. The order in which agents move is random and changes for each time tick. All agents' locations are tracked through a series of lists so that the location of any agent at any point during a simulation run is known.

The Disease Process

At model initialization, all agents are susceptible to the disease being modeled. Before the first tick (Monday 6 a.m.), an agent is chosen at random from all the agents in the simulation and given the disease. This agent and any others who later become infected spend 1 day (6 ticks) exposed before the disease status changes to infectious. Once an agent is infectious, it has a chance to transmit the disease to susceptible neighbors it encounters following its movement during a time step. These neighbors must be located within the same building and in adjacent cells to the north, south, east, or west (termed "von Neumann neighbors"). The chance of infection (transmission probability) is set by the user in a parameter file and can be changed easily to fit the specific research questions. In our simulations, agents remain infectious for 5 days (30 time ticks of the simulation), after which they move into the recovered category. Recovered agents engage in their normal scheduled behaviors throughout the rest of the simulation. Susceptible agents may also become infected if they move to a new location that is adjacent to an infectious agent. As in the process following movement of infectious agents, the probability of infection is determined by the transmission parameter set by the user. If transmission occurs, the newly infected agent moves through the disease states described earlier.

Currently, an infectious agent's movement pattern does not change throughout the disease process, and all agents recover from the disease, but future iterations of the model will include altered behaviors when an agent is infectious, as well as disease-related mortality. The number of possible transmissions from a single infectious agent is regulated only by the schedule of movement and contact with other agents.

Analysis of the Model

Although the ultimate reason to develop an agent-based model is to address specific research questions, before this can be done it is necessary to fully test the model at various stages during its development to see how it works and to make sure that it generates reasonable results given the assumptions it is built upon. These early tests also provide baseline information with which to evaluate changes occurring in later stages of the model. To illustrate the testing process, we compare results from the analysis of the first two stages of model development. The two stages we consider are (i) a random movement model, wherein agents move randomly on the landscape, and (ii) a directed movement model, wherein the movement of agents occurs to and from specific buildings in accordance with the demographic characteristics of particular agents. The model is still being developed, and several significant features of the 1918 influenza epidemic in Newfoundland, including disease-related mortality, changes in mobility during an epidemic, and use of the hospital, have not yet been incorporated. Nonetheless, the comparison of results from the first two stages has led to the identification of important trends in the patterns of epidemics generated within the model population, some of which have implications for understanding the potential effect infectious disease epidemics may have had on populations undergoing the second transition.

General Model Behavior

Simulations of the model have been run using a variety of parameter sets. In general, for both the random and directed movement models, these simulations generate typical epidemic curves that build slowly, come to one or a few well-defined peaks, and then die out (Figure 6.3). Because of the stochastic nature of the models, the epidemic curves vary from one simulation to another, even if all parameters are held constant. In addition, it is always possible for the disease to die out quickly without spreading substantially. In order to distinguish these situations of rapid epidemic extinction from other small epidemics, we have defined an epidemic to be a situation wherein >5% of the agents become infected. Examination of the simulation results shows that epidemic extinction can occur even when the parameter values are set to maximize disease spread, although the frequency of extinction depends on the specific parameter values chosen for the model.

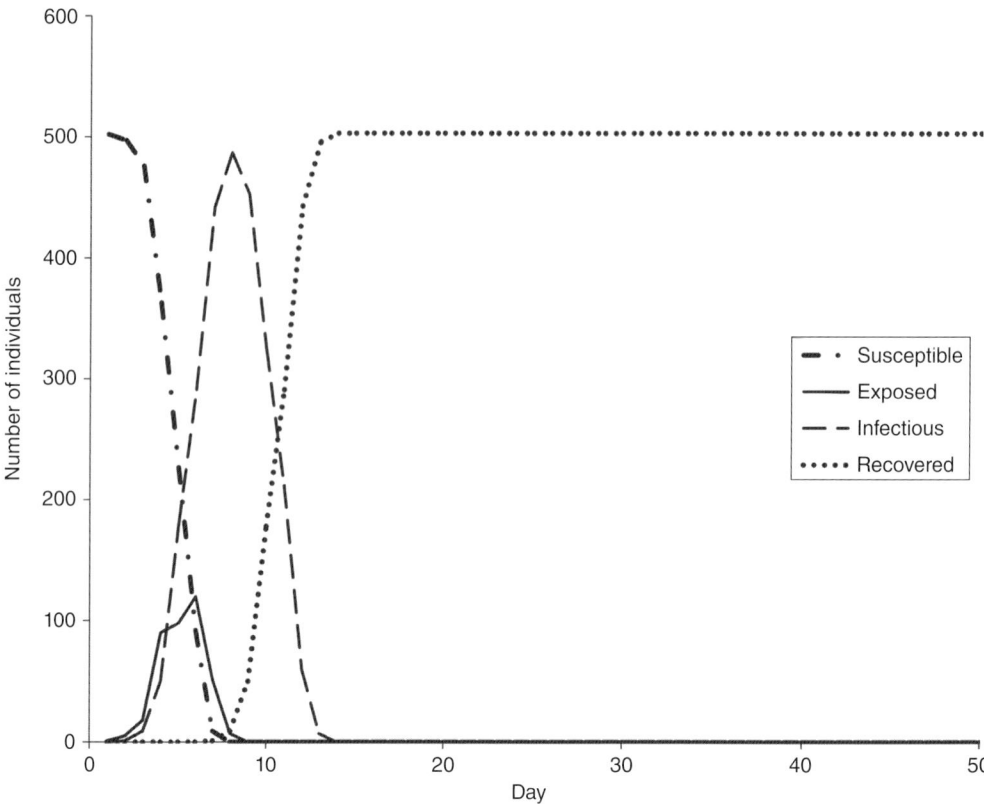

Figure 6.3. A typical simulation run resulting in an epidemic (using the directed move model). Population size = 503, transmission probability = 1.0, latent period = 1 day, and infectious period = 5 days.

In summary, the model generates simulations that reflect what is known about real-world epidemics, both in terms of their general shape and in their potential for disease extinction due to stochastic factors. These results are not totally realistic, however. The current iteration of the model produces epidemics in which almost all of the population is infected, a circumstance that is not often observed in actual infectious disease outbreaks. This model behavior is a consequence of more frequent transmission per contact than occurs in the real world as well as a lack of mortality. Both of these features will be adjusted in subsequent versions of the model.

Sensitivity Analyses

Sensitivity analysis is a strategy used to assess the impact of variation in essential model parameters; for this model, these parameters include transmission probability, length of the latent and infectious periods, and population size. The value of this analysis is that it identifies which of the parameters are the strongest determinants of overall epidemic outcomes. The sensitivity analyses for the model described here initially varied the values of each of the four parameters mentioned earlier, varying one parameter while holding the rest constant (Table 6.1).

Each single-parameter analysis consisted of 100 runs of the simulation for each value of the parameter being varied. These analyses were completed with models using both random

Table 6.1. Parameter values used in the sensitivity analyses.

Parameter	Value when held constant	Range when varied singly	Range when varied jointly
Latent period	1 day	1–7 days, increment 1 day	N/A
Population size	503	Target of 50–500, increment 50	Target of 50–500, increment 50
Transmission probability	1	0.01, 0.05, 0.075, 0.10, 0.13, 0.17, and 0.2–1.0; increment 0.1	0.1–1.0, increment 0.1
Infectious period	5 days	1–12 days, increment 1 day	N/A

and directed movement. Data on peak time of an epidemic, peak number and/or proportion of infectious agents, final epidemic size, and time of the last infectious agent were averaged for two data sets: (i) across all 100 runs and (ii) across all runs for which an epidemic occurred. These data were then compared for all values of the relevant parameter. Because we are primarily interested in general impacts on epidemic patterns, we limit our discussion of these results to epidemic runs only.

In general, the sensitivity analysis of the directed movement model revealed that variation in the latent and infectious periods and population size had negligible effects on the probability that an epidemic occurs and on the proportion ultimately infected. Latent period and infectious period did affect epidemic timing and number infectious at the peak of an epidemic, however. For example, with longer latent periods, both the peak and the end of an epidemic were delayed. As the infectious period increased from 1 to 6 days, the peak number of infectious agents also increased until essentially all agents were infectious at the peak, a value that remained stable for longer infectious periods. The peak and end times of an epidemic increased linearly through the entire range of variation of the infectious period length. It is important to note, however, that even at values thought to reflect reality, the chance of an epidemic occurring was near certainty and the total proportion of agents infected was over 90% for most runs. The chance of experiencing an epidemic was lowest when population size was smallest, but because population density was held constant and family groupings were preserved in populations of all sizes, the overall impact of changes in population size was not significant.

Unlike the other parameters, variation in transmission probability did have measurable effects on the probability of an epidemic and the proportion ultimately infected, as well as on the timing of an epidemic, although the effects were only moderate in size. Initially, transmission probability was varied between 0.1 and 1.0. The resulting epidemics reached over 90% of the population in most runs. To further test the sensitivity of the model to the transmission probability, simulations were also run with the transmission probability set to 0.01, 0.03, 0.05, and 0.075. For all but the lowest of these probabilities, the majority of runs produced an epidemic, and most epidemics still reached over 80% of the population.

One of the most interesting outcomes of the sensitivity analysis is that the results using the directed movement model differ substantially from the random movement model for all parameters tested. These differences apparently relate to the clustering of agents at social nodes (e.g., schools, boats, or houses) that occurs as a consequence of the directed movement behaviors. This clustering facilitates contact and therefore disease transmission, which results in significantly larger and faster epidemics than in models where agents engage only in random movement and contact. Thus, even though the directed move model leads to unrealistically high rates of infection within populations, the comparison of results from both models, which differ only in the nature of movement (and the resulting contact patterns), highlights the importance of population clustering at social nodes in the transmission of infectious diseases.

Human populations in the process of industrialization or urbanization experienced increased clustering that was often the result of changes in occupation (e.g., rural agriculture to factory work), increased population density, or even changes in social roles (e.g., when public schooling for children was implemented or as women began to work outside the home). These changes in behavior had consequences, including increases in diseases such as influenza or measles (the so-called crowd diseases) that are associated with people living in higher population densities, and increases in food- and waterborne diseases as the result of contamination and decreasing diversity of resources. Eventually, in many places, public health measures were passed in order to provide for workplace safety and clean food and water. It has been argued (e.g., McKeown, 1976; Colgrove, 2002) that these measures are in some way responsible for the emergence of the second transition in industrializing societies. Indeed, our model demonstrates the increased risk of disease that individuals face in highly clustered populations.

Our model also demonstrates that changes in human behavior can alter the paths through which epidemic disease can move into or out of subgroups of human populations. For example, the introduction of compulsory school attendance resulted in schools becoming a more important site at which children interacted. Schools facilitated not only more frequent contact, but also longer periods of contact. Thus, they became more important social nodes than they had been and took on a more central role within the community. Model results showed that simulated epidemics beginning with schoolchildren peaked faster with a higher number of infectious agents than simulated epidemics beginning with other types of agents. This is likely due to the fact that the close contact within schools increased the chances that children became infected once the epidemic entered the school. This model behavior suggests that the introduction of compulsory schooling may have played a significant role in the process by which epidemics spread horizontally within age groups and into new families.

Advantages of Agent-Based Modeling in Studies of the Second Epidemiologic Transition

While the results presented here are preliminary, there are important lessons to be learned from the use of an agent-based model to look at disease in a population experiencing the second transition. In the following discussion, we outline the specific ways in which the choice of an agent-based model to address our research questions proved to be a strength in terms of making use of available historical information, in dealing with the problems of a small population, and in comparison to other modeling approaches.

Historic and Ethnographic Background

This project relied on multiple types of primary and secondary documentation. A traditional epidemiologic study might have focused solely on those data that directly recorded morbidity or mortality information. While it is likely that any researcher interested in the health of this historic population would have been attracted by the first-person accounts, photos, newsletters, and other ephemera, it is difficult to weave that information into a traditional demographic or epidemiologic study. Inferences derived from those kinds of data, however, can be built into a model and can play a strong role in defining population dynamics. For example, ethnographic data indicated that extended family groups owned boats, so in the model, the boats were organized according to the similarity of surnames. Historic recollections contained information about the importance of visiting among women, so the likelihood of daily visiting too was adjusted to reflect that as a priority. Memoirs and government documents informed decisions about school attendance. Historic and contemporary photos of the community were used to get a general feel about distances, building sizes, and travel times, which led to the decision not to incorporate travel into the model. All of these types of data build a rich context for the model that adds authenticity to the patterns of disease transmission during simulation runs.

Another advantage of using historic data is its incorporation into the evaluative process. As described earlier, sensitivity testing is the standard means for testing a model's strength and reliability. If a researcher has access to disease records from the time period being modeled, it may be possible to further validate the model through a comparison with those data. However, those data should typically not be used to build the model (depending on the research questions), as they increase the risk of circular logic within the model.

Modeling a Historic Population

While having diffuse and varied types of historic data on which to build a model is a clear strength, as mentioned previously, there is further strength in choosing a real-world historic population to work with (vs. a generic one). In many places, there are census records, parish records, or other individualized records that can be used to reconstruct the population under study. These make it possible to not only know the age and sex structure of the population, but in some cases to know the family structures, diversity of occupations, and rate of immigration and emigration. With these data available, the model can reflect historical reality to a much greater degree. The incorporation of cultural values, activity patterns, or other information from the background research can lead to a closer concordance between the model and the population under study. The importance here is not that the model population closely approximates the real one, but that these kinds of information can be used within the model to understand *which* kinds of patterns are most important to disease transmission in that population. Is it something having to do with the age and sex structure of the population? Is it an issue of seasonality or epidemic timing? Or, perhaps, do age- and sex-structured behaviors play a large role in determining the outcome of the epidemic? As discussed, we have seen the importance of structured activities in spreading disease both horizontally across peer groups and then vertically into family groups. It is these activities (e.g., school, church) that provide the areas of most likely transmission. Basing model activity patterns on hypothetical and idealized patterns rather than knowledge of the historical activity patterns may have resulted in imperfect and possibly incorrect assessments of the role these social institutions played in the spread of disease in this small community.

The Use of an Agent-Based Model

The final point of discussion is the choice to use an agent-based model to study the second transition rather than an alternative type of model. Many modeling strategies can make use of historic contextual data, as we have done in the construction of this model. The agent-based approach has certain strengths that make it well suited for this specific topic. The population of interest for this study is small, kin-based, and geographically isolated. While this may not be the case for all human populations that experience the second transition, it certainly is not an uncommon phenomenon. Even today, many American small towns or rural hamlets could still be described as geographically isolated and primarily composed of kin groups. Worldwide, there are innumerable small communities with limited population movement that are organized around kin networks, especially within developing countries. An agent-based model is ideal for these communities as this type of model can handle the randomness associated with the behavior of individuals and can easily incorporate essential kin-based behaviors. The results generated by an agent-based model may show a wide range of potential outcomes, primarily because of the effects of small, random iterations of agent actions. Models that operate on the population level may lose the level of detail that provides the full range of both common and rare outcomes.

Conclusion

This paper has sought to make a case for the use of models, specifically agent-based computer simulation models, for the study of the second transition. We posit that these techniques make fullest use of available historical and contextual data to generate quantitative disease data that the historical record may not contain. Modeling allows for experimentation on simulated human populations that would be unethical and infeasible in a laboratory setting. It allows for evaluation of interventions, testing of hypotheses about routes of transmission and the importance of social structures, and discovery of the ways in which culture, history, and biology may have combined in order to produce patterns of health that we see in the past. Increases in computing power and the availability of modeling software packages have made this type of research more accessible than ever before, and we propose further use of these methods in the study of the second epidemiologic transition.

References

Anderson RM, May RM. 1992. Infectious diseases of humans: Dynamics and control. Oxford: Oxford University Press.
Baker M, Pitt JM. 1984. A history of health services in Newfoundland and Labrador to 1982. In: Smallwood JR, Editor. Encyclopedia of Newfoundland and Labrador. Vol. 2. St. John's: Newfoundland Book Publishers. p 864–875.
Barbujani G, Sokal RR, Oden NL. 1995. Indo-European origins: a computer-simulation test of five hypotheses. Am J Phys Anthropol 96(2):109–132.
Barrett R, Kuzawa CW, McDade T, Armelagos GJ. 1998. Emerging and re-emerging infectious diseases: the third epidemiologic transition. Annu Rev Anthropol 27(1):247–271.
Beaton M, Walsh J. 2004. From the voices of nurses: An oral history of Newfoundland nurses who graduated prior to 1950. St. Johns: Jesperson Publishing.
Beemer JK, Anderton DL, Leonard SH. 2005. Sewers in the city: a case study of individual-level mortality and public health initiatives in Northampton, Massachusetts, at the turn of the century. J Hist Med Allied Sci 60:42–72.
Billari FC, Prskawetz A, Aparicio Diaz B, Fent T. 2007. The "Wedding Ring": an agent-based marriage model based on social interaction. Demogr Res 17:59–82.
Bishop-Sterling T, Webb JA. 2008. The twentieth century. In: Newfoundland Historical Society, Editor. A short history of Newfoundland and Labrador. Portugal Cove-St. Philip's: Boulder Publications. p 103–140.
Carpenter C, Sattenspiel L. 2009. The design and use of an agent-based model to simulate the 1918 influenza epidemic at Norway House, Manitoba. Am J Hum Biol 21(3):290–300.
Coleman D, Salt J. 1992. The British population: Patterns, trends, and processes. Oxford: Oxford University Press.
Colgrove J. 2002. The McKeown thesis: a historical controversy and its enduring influence. Am J Public Health 92(5):725–729.

Crosby AW. 1989. America's forgotten pandemic: The influenza of 1918. Cambridge: Cambridge University Press.
Cutler DM, Miller G. 2005. The role of public health improvements in health advances: the twentieth-century United States. Demography 42:1–22.
Davis D. 1983. The family and social change in the Newfoundland outport. Culture 3(1):19–32.
Epstein JM, Axtell R. 1996. Growing artificial societies: Social science from the bottom up. Washington: Brookings Institution Press.
Firestone M. 1967. Brothers and rivals: Patrilocality in Savage Cove. Newfoundland. Social and Economic Studies, ARDA Project No. 1016(5). St. John's: Memorial University of Newfoundland.
Gage TB. 1993. The decline of mortality in England and Wales 1861 to 1964: decomposition by cause of death and component of mortality. Popul Stud 47:47–66.
Gage TB. 1994. Population variation in cause of death: level, gender, and period effects. Demography 31:271–296.
Gage TB. 2005. Are modern environments really bad for us? Revisiting the demographic and epidemiologic transitions. Yearb Phys Anthropol 48:96–117.
Handcock WG. 1977. English migration to Newfoundland. In: Mannion JH, Editor. The peopling of Newfoundland: Essays in historical geography. Social and Economic Papers, No. 8. St. John's: Institute of Social and Economic Research, Memorial University of Newfoundland. p 15–48.
Harris RC, Warkentin J. 1974. Canada before confederation: A study in historical geography. New York: Oxford University Press.
Herring DA, Sattenspiel L. 2007. Social contexts, syndemics, and infectious disease in northern Aboriginal populations. Am J Hum Biol 19(2):190–202.
Hethcote HW. 2000. The mathematics of infectious diseases. SIAM Review 42:599–653.
Heymann DL, Editor. 2004. Control of communicable diseases manual. 18th ed. Washington: American Public Health Association.
Higgs R. 1979. Cycles and trends of mortality in 18 large American cities, 1871–1900. Explor Econ Hist 16:381–408.
Hollingsworth TH. 1968. The importance of the quality of the data in historical demography. Daedalus 97(2):415–432.
Howell M, Prevenier W. 2001. From reliable sources: An introduction to historical methods. Ithaca: Cornell University Press.
Hudson RP. 1993. Concepts of disease in the West. In: Kiple KF, Editor. The Cambridge world history of human disease. Cambridge: Cambridge University Press. p 45–52.
Johnson NPAS, Mueller J. 2002. Updating the accounts: global mortality of the 1918–1920 "Spanish" influenza pandemic. Bull Hist Med 76:105–115.
Kasakoff AB, Adams JW. 1995. The effect of migration on ages at vital events: a critique of family reconstitution in historical demography. Eur J Popul 11:199–242.
Keeling MJ, Rohani P. 2007. Modeling infectious diseases in humans and animals. Princeton: Princeton University Press.
Kivimaki A. 1937. Nursing with the Grenfell Mission. Am J Nurs 37(6):593–598.
Kohler TA, Gumerman GG, Editors. 2000. Dynamics in human and primate societies: Agent-based modeling of social and spatial processes. Oxford: Oxford University Press.
Macfarlane A. 1977. Reconstructing historical communities. Cambridge: Cambridge University Press.
Mamelund S.-E. 2011. Geography may explain adult mortality from the 1918–20 influenza pandemic. Epidemics 3:46–60.
McKeown T. 1976. The modern rise of population. New York: Academic Press.
Meeker E. 1972. The improving health of the United States, 1850–1915. Explor Econ Hist 9:353–373.
Murray HC. 1979. More than fifty percent: Woman's life in a Newfoundland outport 1900–1950. Canada's Atlantic folklore and folklife series, No. 3. St. John's: Breakwater Books.
Nemec TF. 1972. I fish with my brother: The structure and behaviour of agnatic-based fishing crews in a Newfoundland Irish outport. In: Andersen R, Wadel C, Editors. North Atlantic fishermen: Anthropological essays on modern fishing. Newfoundland Social and Economic Papers, No. 5. St. John's: Institute of Social and Economic Research, Memorial University of Newfoundland. p 9–34
Newfoundland. Colonial Secretary's Department. 1923. Census of Newfoundland and Labrador, 1921. St. John's: Colonial Secretary's Department.

Newfoundland. House of Assembly. 1910–1942. Annual Report of the Registrar General of Births Marriages and Deaths. St. John's: House Assembly Newfoundland.

Omran AR. 1971. The epidemiologic transition. Milbank Mem Fund Q 49:509–538.

Omran AR. 1983. The epidemiologic transition theory: a preliminary update. J Trop Pediatr 29:305–316.

O'Neil C, Sattenspiel L. 2010. Agent-based modeling of the spread of the 1918–1919 flu in three Canadian fur trading communities. Am J Hum Biol 22(6):757–767.

Overton J. 1998. Brown flour and beriberi: the politics of dietary and health reform in Newfoundland in the first half of the twentieth century. Newfoundland and Labrador Studies 14:1–27.

Pocius GL. 1979. Hooked rugs in Newfoundland: the representation of social structure in design. J Am Folk 92(365):273–284.

Porter M. 1985. "She was skipper of the shore-crew:" notes on the history of the sexual division of labour in Newfoundland. Labour/LeTravail 15:105–123.

Prentice A, Bourne P, Brandt GC, Light B, Mitchinson W, Black N. 1988. Canadian women: A history. Toronto: Harcourt Brace Jovanovich.

Preston SH. 1976. Mortality patterns in national populations. New York: Academic Press.

Preston SH, Haines MR. 1991. Fatal years: Child mortality in late nineteenth-century America. Princeton: Princeton University Press.

Queen SA, Habenstein RW. 1974. The family in various cultures. 4th ed. Philadelphia: JB Lippincott.

Rankin L. 2008. Native peoples from the Ice Age to the extinction of the Beothuk (c. 9,000 years ago to AD 1829). In: Newfoundland Historical Society, Editor. A short history of Newfoundland and Labrador. Portugal Cove-St. Philip's: Boulder Publications. p 1–22.

Rodgers RH, Witney G. 1981. The family cycle in twentieth century Canada. J Marriage Fam 43(3):727–740.

Sattenspiel L. 2009. The geographic spread of infectious diseases: Models and applications. Princeton: Princeton University Press.

Sattenspiel L. 2011. Regional patterns of mortality during the 1918 influenza pandemic in Newfoundland. Vaccine 29S:B33–B37.

Schofield R, Reher D. 1991. The decline of mortality in Europe. In: Schofield R, Reher D, Bideau D, Editors. The decline of mortality in Europe. Oxford: Oxford University Press. p 1–17.

Smith PEL. 1987. Transhumant Europeans overseas: the Newfoundland case. Curr Anthropol 28(2):241–250.

Thornton PA. 1985. Newfoundland's frontier demographic experience: the world we have not lost. Newfoundland and Labrador Studies 1(2):141–162.

Tomlinson B. 2009. A proximate mechanism for communities of agents to commemorate long dead ancestors. JASSS 12(1):7.

Troesken W. 1999. Typhoid rates and the public acquisition of private waterworks, 1880–1920. J Econ Hist 59:927–948.

Younger S. 2011. Leadership, violence, and warfare in small societies. JASSS 14(3):8.

Chapter 7
Does Exposure to Influenza Very Early in Life Affect Mortality Risk during a Subsequent Outbreak? The 1890 and 1918 Pandemics in Canada

Stacey Hallman[1] and Alain Gagnon[2]
[1] *Department of Sociology, Western University, London, Ontario, Canada*
[2] *Département de Démographie, Université de Montréal, Montreal, Quebec, Canada*

Introduction

The second epidemiologic transition in North America involved a gradual shift from epidemic infectious diseases to those of a chronic and degenerative during the 19th and early 20th centuries (Omran, 1971; Barrett et al., 1998). While Omran identified this long and progressive transition as "the Age of Receding Pandemics," it was briefly interrupted by two major pandemics of infectious disease, the 1890 Russian influenza pandemic and the 1918 Spanish pandemic. The 1918 pandemic was particularly lethal; it is estimated to have killed between 40 and 100 million people worldwide (Johnson, 2003), far more than the 9 million soldiers killed during the First World War (Patterson, 1986; Quinn, 2008; Storey, 2009). As explained by Crosby (1989: 311), "nothing else—no infection, no war, no famine—has ever killed so many in so short a period." Despite the vast number of casualties, the Spanish pandemic rapidly left the public consciousness and often receives only a cursory mention in histories of the First World War (Crosby, 1989). This is even more applicable to the 1890 pandemic, which is all but forgotten in the collective memory and is seldom the topic of scholarly interest (Herring and Carraher, 2011a). The lack of attention given to both pandemics may have influenced the theorizing about the stages and progression of the epidemiologic transition. As Barrett and colleagues (1998: 262) suggest, "had the historical precedents of influenza been given closer consideration, previous projections for the continued decline in infectious diseases might not have been so optimistic."

The Spanish flu pandemic is also notable for its unusual pattern of death: it killed unprecedented numbers of young adults between the ages of 20 and 40. This pattern has been well noted in the literature (Crosby, 1989; Luk et al., 2001) and confirmed recently in India, a country with a less developed statistical system (Hill, 2011). However, debate remains as to the exact reason for this unusual occurrence. Several scenarios have been proposed. First, it has been suggested that the excess deaths among young adults could be due to a negative interaction between tuberculosis and influenza. This is because young adult males were at greatest risk of tuberculosis, and those individuals with tuberculosis infection and lung damage from the disease were at greater risk of dying from influenza (Noymer and Garenne, 2000; Noymer, 2009). Contrastingly, the idea has been advanced that deaths among young adults resulted from an overactive immune response that caused the lungs to fill with fluid, known as a cytokine storm, causing death from drowning. As young adults are at the height of immunocompetency, this would have placed them at highest risk from the pandemic (Loo and Gale, 2007). Third, acquired immunity has also been hypothesized as a means by which older adults could have been protected through previous infection from a strain similar to the 1918 flu that would have been in circulation prior to 1890 (Ahmed et al., 2007). All of these explanations have been based on the general finding that young adults between the ages of 20 and 40 were at greatest risk, often with the highest unexpected mortality being found among those aged 25–29. However, it is generally agreed that key elements of the modern environment, including the rapid movement of people, the gathering of previously more isolated and thus more susceptible individuals, and the unsanitary conditions of the war, were the elements that contributed to the astounding spread and impact of this disease.

This paper presents the results of preliminary research into the age distribution of deaths among young adults during the 1918 pandemic. We show that among various cities and regions in Canada, it was not merely an entire age range that was at greatest risk, but that mortality centered on the specific age of 28. To develop this, we use the Developmental Origins of Health and Disease Hypothesis (DOHaD) (e.g., fetal origins or Barker hypothesis) (Barker, 2006) and concepts of scarring during critical periods of development (Preston et al., 1998) and original antigenic sin (Francis, 1953; Kim et al., 2009) to explain why individuals at this particular age were at heightened risk. Is it possible that mortality risk in 1918 was conditioned by exposure to influenza 28 years earlier, during the other forgotten influenza pandemic in 1890? Was it not only the modern environment in 1918, in the midst of the second transition, that allowed the spread of the influenza, but were certain people at greater risk because of the environmental conditions at the time of their birth?

Background

Theoretical Framework

Understanding the complex interrelationship between early life conditions and adult mortality has become an area of concern in the development of strategies for the protection of population health. Following the work by Barker and others on the DOHaD hypothesis, it is now established that inadequate nutrition during growth *in utero* results in compensatory physiological restrictions that increase the risk of metabolic and cardiovascular disease in later life (Barker, 2006). In the critical period model, which is a component of the DOHaD hypothesis, exposures acting during specific windows of time have irreversible effects on health. There is also evidence that maternal exposure to a virulent epidemic disease can divert resources to the maternal immune response at the detriment of fetal maturation, leading to higher cardiovascular disease prevalence and mortality in the offspring at older ages (Almond and Mazumder, 2005; Myrskylä et al., 2010). More directly, the concept of scarring during critical periods of development posits that infection with airborne infectious diseases in critical periods of development can harm the lung tissue, causing scarring and leading to greater susceptibility to (and mortality from) later airborne infectious diseases (Preston et al., 1998; Bengtsson and Lindstrom, 2003). Critical period models also connect prenatal undernutrition to infectious death in adolescence. What has yet to be addressed is the possibility that exposure to a virulent agent such as influenza early in life may lead to increased susceptibility from later outbreaks of the same disease. It is unknown, for instance, whether sequential pandemics of different strains of the same disease are related, whether risk of death during a pandemic relates not only to health and immune status but also to exposure during the last major pandemic.

We first hypothesize that exposure to influenza during early development results in physiological impairments that increase risk of death later in life from airborne infectious diseases. According to the DOHaD hypothesis, we should expect to find a mortality peak at the age of 28 years during the pandemic of 1918 since 28 years separate this pandemic from the previous one in 1890. As this hypothesis posits detriments to long-term health based on the gestational environment, those individuals who were exposed to a severe manifestation of the Russian influenza strain *in utero* in January 1890 would have experienced developmental impairments that would have increased their mortality in 1918, at the age of 28.

It is also possible that an immune response to the 1890 flu was the cause of higher mortality in 1918, rather than physiological sequelae from maternal infection or exposure early in life. According to the "original antigenic sin" model (Francis, 1953; Kim et al., 2009; Ma et al., 2011), the first influenza virus strain encountered during childhood conditions the immune response to that specific variant. Having built antibodies to this initial strain, the immune system will respond inadequately to a highly virulent and antigenically novel one. We posit as a second hypothesis that *commitment* to a specific strain depends on both virulence and on the age at which it is first encountered. Since maternally derived antibodies provide protection against influenza in the first 6 months after birth (Beaudry et al., 1995; Munoz, 2003), infants do not have the capacity to mount an immune defense prior to that age. If the antigenic sin hypothesis holds true, then people born a year or two before 1890 were at a higher risk of death during the 1918 pandemic because they first encountered (and developed an immune response to) the 1890 strain at the youngest possible age and at the time of its maximum virulence. Important to this hypothesis, the pandemic strain in 1890 (thought to be either a subtype of influenza A/H2 or H3N8) had a quite dissimilar surface antigen than the strain in 1918 (influenza A/H1N1) (Dowdle, 1999; Taubenberger and Morens, 2006; Kolte et al., 2008; Valleron et al., 2010). Thus, this second overreaching hypothesis implies higher death tolls at ages slightly older than 28. Scarring from infection in the first years of life could also cause deaths at age 28 as well as at slightly older ages. Using the information available to us, it would not be possible to separate the effects of scarring from original antigenic sin or gestational impairments, but scarring would not be expected to result in such a clear peak in mortality at age 28 (due to the higher risk of mortality at all ages following the first infection).

The 1890 and 1918 Pandemics in Canada

The age of exposure to the 1890 pandemic strain can be known with high precision because of the short duration of this pandemic, known as the *Russian flu*. Through newspaper accounts, we have determined that its transmission in Canada was as swift as elsewhere (Valleron et al., 2010; Le Goff, 2011), leaving an impression of simultaneity for the pandemic around the globe. The first wave of the pandemic began in 1889, on the eve of the New Year, December 31, 1889, when a few cases were reported in North America in the Toronto Globe newspaper. No cases were reported for Canada at this time (but see Patterson, 1986, who reports cases in December in Montreal). Then, on January 8, 1890, the disease began to rage in most of Eastern and Central Canada. However, by January 15, few new cases in Canada were reported in the Globe and practically none in the following months. Maris (2011) reports that the pandemic in Canada had peaked by the first week in January, although it continued along the Eastern Coast of the United States until the end of the month. In Toronto, the highest numbers of deaths occurred between the end of January and mid-February, consistent with Patterson's assertion that influenza deaths are normally first noticed to be elevated approximately 4 weeks after the start of an epidemic (Patterson, 1986; Ancestry.com, 2010). The disease is believed to have entered Canada from the port of Halifax, Nova Scotia, and to have spread westward to the rest of the country along the transportation networks, but it may also have entered the country through the railways linking New York and Chicago with Ontario (Thompson, 2011).

The second wave of the pandemic emerged in the spring of the following year, 1891, when new cases were reported in New York City. However, no mention of the affliction is found for Canada in the Toronto Globe in 1891, although the disease could have very well been present in a milder form. Indeed, Patterson (1986) does report cases in Canada in both the second wave (January to June 1891) and in the third wave (September 1891 to February 1892). It appears from studies conducted in England and Wales (Smith, 1995; Langford, 2002) that this pandemic strain was endemic for many years after 1890. More detailed research establishing the exact mortality curves from death records in Ontario from the 1890 flu will help to establish the true extent of this pandemic in Canada. However, all indications at present, including the case study of the Russian pandemic in Hamilton by Herring and Carraher (2011b), suggest that the spread of the flu in Canada was similar to other parts of the world. The Russian influenza was followed by a much milder pandemic in 1899–1900, reported to have been present in Ontario, that has been theorized as varying little from the previous strain, only on the H antigen, and was not nearly as virulent as either the 1890 or 1918 pandemics (Patterson, 1986; Dowdle, 1999). As is typical of any influenza variant, virulence most likely decreased over time from its pandemic peak until it was fully replaced by a novel strain in 1918.

Similarly, the 1918 Spanish flu pandemic was also of short duration. While the exact geographic origins of the pandemic have not been established (theories suggest either the battlefields of France or an American training base in Kansas; see Oxford et al., 2002; Barry, 2004; Humphries, 2005) in Canada, the disease is thought to have entered the country through a military training base in southern Ontario and from there to have spread westward with the Siberian Expeditionary Forces (Humphries, 2005). It appears to have first struck Canada during the second global wave in the fall of 1918. Deaths were most numerous in October, although the epidemic in Canada continued throughout November and December, followed by a less severe resurgence during the third wave in the spring of 1919.

The Russian and Spanish influenza pandemics are similar in that they are both described as completely global and reliant on rapid trade and transit for their almost instantaneous spread (Patterson, 1986; Crosby, 1989; Le Goff, 2011). Yet, there are two major differences between the 1890 and 1918 pandemics: the case-fatality rate and the age structure of mortality. Valleron and colleagues (2010: 8778) report that the Russian influenza had a clinical attack rate of 60% (interquartile range of 45–70%), while the case-fatality rate ranged from 0.1% to 0.28%. This suggests that approximately 60% of any given population would have contracted the flu during the pandemic, but of those who became ill, only 0.1% would die from the disease, resulting in relatively few deaths being attributed to the pandemic. This is significant for this study since high morbidity and low mortality in 1890 means that many individuals would have been exposed early in life and would have survived to meet the pandemic in 1918, thus giving us a sample of susceptible individuals. It is typical of influenza epidemics and pandemics to infect far more people than those who die (Patterson, 1986; Glezen and Couch, 1997), but those who usually die are those most at risk: infants, the elderly, and the immunocompromised. This results in the typical *U*-shaped mortality curve, where the two highest rates of death are among the very young and the very old (Crosby, 1989; Valtat et al., 2011). While the 1890 pandemic followed the classical influenza pattern (Morens and Fauci, 2007; Valtat et al., 2011), the 1918 pandemic is known for the *W*-shaped age distribution of deaths, wherein young adults were at an unexpectedly high risk of death from the disease. In contrast to the case-fatality rate of 0.1–0.28% in 1890, the 1918 pandemic had a case-fatality rate greater than 2.5% (Taubenberger and Morens, 2006), while the clinical attack rate ranged from 20% to 60% (Morens and Fauci, 2007).

Materials and Methods

For our analysis, we utilized the registered death records from September to December 1918 for the Ontario cities of Toronto, Ottawa, London, and Hamilton and the counties of Welland and Lincoln in southern Ontario. We focus on Ontario since the microfilmed death records for the pandemic period have been digitized and are available online.[1] However, we also utilized the indexes of the registered death records for Winnipeg, Manitoba, and Vancouver, British Columbia, in order to compare experiences in major Canadian cities covering a large geographic distance. From the data available in the complete death records for Ontario, information on sex, age at death, date of death, and cause of death were recorded. The death records for Manitoba and British Columbia are not available online, but the indexes containing sex, date of death, and age at death are available through the Vital Statistics Agency of Manitoba[2] and the British Columbia Archives.[3] Therefore, for the cities and counties in Ontario, we have information on both total deaths for the 4-month period and deaths specifically from pandemic-related causes (e.g., influenza, pneumonia, and bronchitis). These three causes of death have been grouped together as *influenza related* since death during the pandemic was generally caused by secondary bacterial infections of the lungs (Harder, 1918; Crosby, 1989). Selecting individuals who only had a cause of death of *influenza* would miss a substantial proportion of pandemic-related deaths. For Winnipeg and Vancouver, however, we are limited to an analysis of deaths from all causes. As population totals are as of yet unavailable for these cities in 1918, we are unable to calculate rates of death. However, as a comparison, the total numbers of deaths were recorded for each city for September 1918, before the epidemic had entered Canada (see Table 7.1 for the populations and densities for each city from the 1921 census). As can be seen in Figure 7.1, the mortality pattern in Toronto was drastically different during the epidemic than for the preceding, nonepidemic month.

Death registration is subject to different forms of error for various reasons. Although death records for Ontario are deemed to have become complete by around 1920 (Emery, 1993) due to careful legislation requiring a death certificate before a grave can be dug, it is possible that the deaths of some people were not recorded. Those individuals not present in the records are probably a select group (e.g., immigrants or the impoverished), but as this is

Table 7.1. 1921 population size and population density (Canada, 1924).

	Toronto	Ottawa	Vancouver	Winnipeg	Hamilton	London	Welland and Lincoln
1921 Population (census)	521,893	107,843	117,217	179,087	114,151	60,959	115,293
1921 Population density (persons per km^2)	5766.77	979.50	1454.72	2815.83	925.80	357.11	57.22

[1] www.ancestry.ca
[2] http://vitalstats.gov.mb.ca/index.html
[3] http://www.bcarchives.gov.bc.ca/index.htm

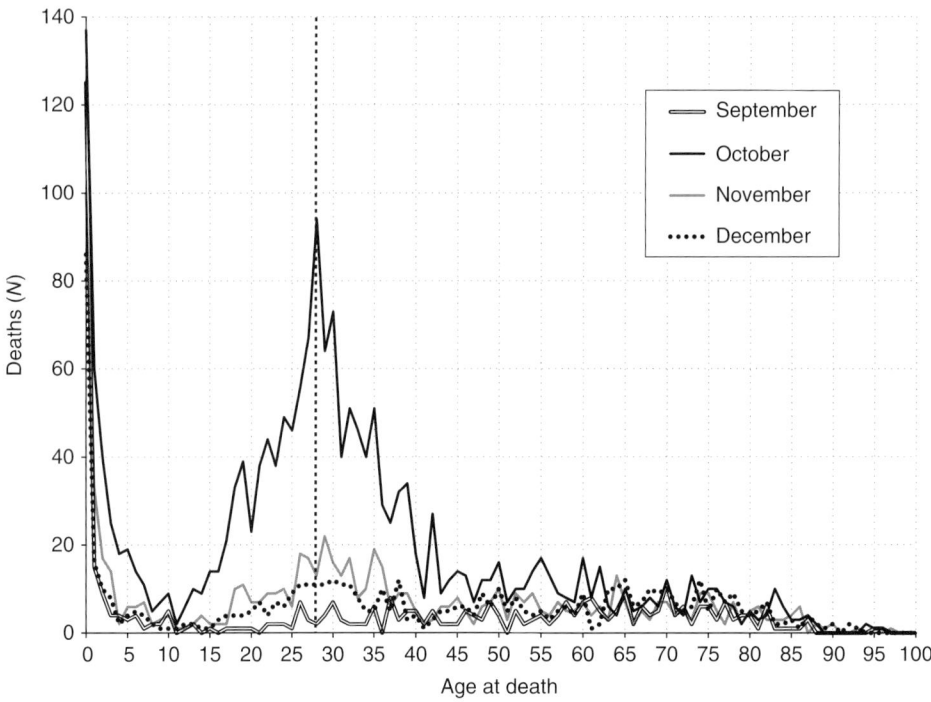

Figure 7.1. Number of deaths by age from all causes in the City of Toronto, September to December 1918 (September, $n = 441$; October, $n = 1885$; November, $n = 731$; December, $n = 618$. Total, $n = 3675$). The vertical line indicates age 28.

not a nominal analysis, individuals not missing at random are not of a large concern as it is unlikely that they would overrepresent any specific age.

The death records themselves have been transcribed multiple times, beginning with the initial individual who filled out the record. From there, the records were sent to the Registrar General for Ontario where they were transcribed into a comprehensive register (Emery, 1993). Much later, the records were transcribed onto microfilm and then digitized and made available on the Internet. At any stage, inaccuracies in transcription may have occurred. The extent to which this may have influenced the results will be assessed in a future records linkage project meaning to confirm exact date of birth of the individuals who died during the pandemic in Ontario.

Results

The results for the total number of deaths by age recorded during the deadliest wave of the Spanish flu in Toronto in October 1918 and in the surrounding months are depicted in Figure 7.1. Strikingly, mortality peaked at age 28, precisely for the generation born at the time of the peak of the previous influenza pandemic, in January 1890.

Also apparent in Figure 7.1 is the elevated number of deaths among young adults aged 20–40, which was noted in contemporary accounts and which has been reinforced by more modern analyses (Harder, 1918; Oertel, 1919; Crosby, 1989; Luk and Gross, 2001; Phillips and Killingray, 2003; Taubenberger and Morens, 2006). Due to available population data, however, these previous investigations have collapsed yearly ages at death into age-groups (i.e., 20–24, 25–29, 30–34) and, as a result, were not able to reveal if there was a noteworthy peak at any specific age. Yet, in Toronto, death totals at age 28 were higher than in any single age between the ages of 20 and 25 or between 31 and 35, which were abnormally high as well. Figure 7.1 also shows a secondary peak in deaths at the age of 30, which would represent those individuals who met the 1890 pandemic at the earliest ages to mount an immune response. This can also be seen for deaths from pandemic-related causes in other cities in Ontario.

It is possible that some of our results are influenced by age heaping at ages 28 and 30, with correspondingly less deaths attributed to age 29 and to age 31 (see Figure 7.1 for heaping evident in older ages). Age heaping may occur in historical data because age declaration often tends to be rounded up or down to the nearest number that ends in 0, 2, 5, or 8, especially among the elderly or the uneducated. The level of such bias can be estimated with Myers's summary index (Myers, 1940), which measures the amount of preference for a specific terminal digit while accounting for the effect of mortality, such that there would be expected to have more individuals in a specific decade with a terminal digit of 0, decreasing from 1 to 9 (e.g., there are more people aged 80 than aged 89, simply due to the increased mortality rate with age). Using this method, a summary index with a value close to 0 represents no heaping and 90 would indicate all deaths being reported at the same terminal digit (Hobbs, 2004). In our data, the summary index for all cities and regions pooled together for the 4-month period is 4.21, a relatively low figure. Redistribution of the deaths between the ages of 25 and 34 would not be of such a magnitude as to reduce the pattern evident. Yet, it is possible that deaths at age 29 were slightly underreported, causing deaths at the ages of 28 and 30 to be increased.

To show that this pattern is due to pandemic-related causes of death, Figure 7.2 shows deaths from influenza, pneumonia, and bronchitis for the cities of Toronto, Ottawa, Hamilton, and London and the counties of Welland and Lincoln, Ontario. Again, the deaths are elevated for those aged approximately 25–34 with the peak coming at the age of 28. The same pattern of ages at death can also be seen when looking at two major Canadian cities outside of Ontario (according to the 1921 census, the populations of the four major cities we analyzed were as follows: Toronto = 521,893; Ottawa = 107,843; Vancouver = 117,217; and Winnipeg = 179,087) (Canada, Dominion Bureau of Statistics, 1921). Figure 7.3 shows the age distribution of deaths from all causes from September to December 1918. Among young adults, the most deaths occurred at age 28 in Winnipeg, and there are peaks at both age 28 and 30 for Vancouver, with age 30 having the higher number of deaths.

To highlight the experience of young adults, Figure 7.4 shows the deaths at each age from 15 to 45 as the percentage of the total mortality. It displays mortality from all causes from every city combined as well as the pooled flu fatalities from all cities and regions in Ontario, the only province from which we had cause-of-death data readily available, for the 4-month period of September to December 1918. Clearly, the age of 28 accounted for the highest percentage of total young adult mortality during this period, with a secondary peak at age 30. This is more pronounced for those individuals whose listed cause of death was pandemic related, revealing that this phenomenon was specifically the result of the 1918 influenza epidemic.

Discussion

Our research into selected Canadian cities confirms the finding that young adults, specifically those between the ages of 25 and 35, were at greatest risk of death from the 1918 Spanish influenza pandemic (Crosby, 1989; Luk and Gross, 2001; Loo and Gale, 2007). Previous investigators have proposed mechanisms that would explain the higher death tolls

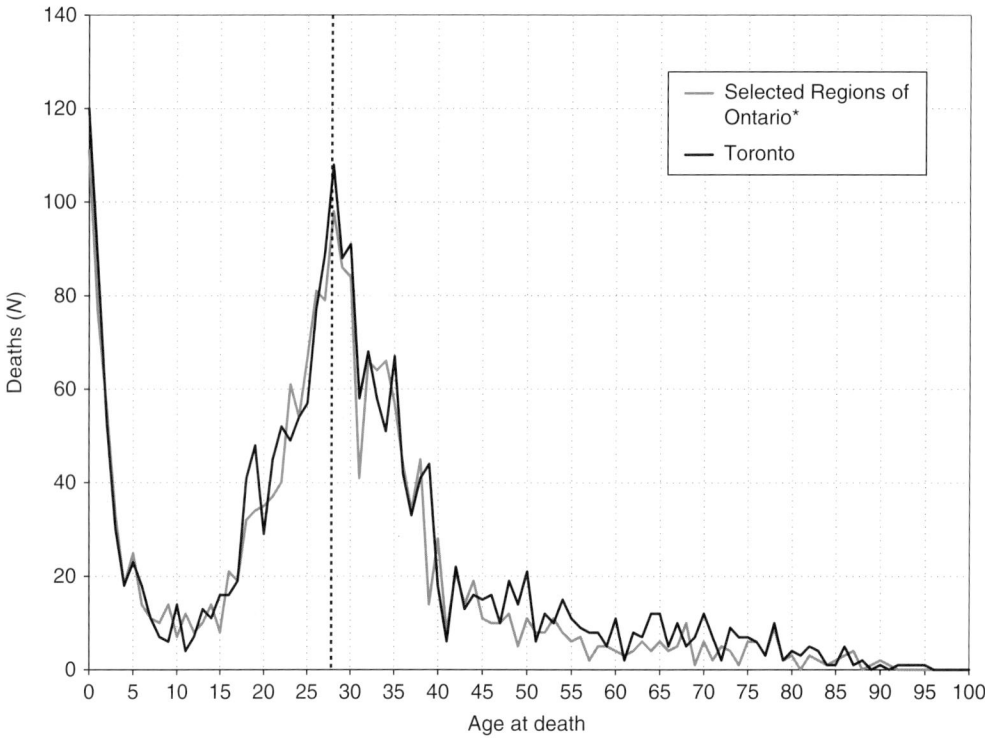

Figure 7.2. Deaths from influenza, pneumonia, and bronchitis, September to December 1918 in selected Ontario cities. Toronto $n = 2195$. The vertical line indicates age 28.
*Includes London ($n = 290$), Hamilton ($n = 536$), Ottawa ($n = 640$), and Lincoln and Welland Counties ($n = 550$).

among young adults, such as tuberculosis or an overactive immune response (i.e., cytokine storm). But, none of these mechanisms can account for the concentration of deaths at a specific age. Significantly, we found that those at the ages of 28 and 30 were most at risk. Future research will need to use dates of birth (through linking birth and death records) to determine exactly the timing of potential exposure during *in utero* development and infancy, to the month or even to the day. This will help to distinguish whether the observed patterns are the product of scarring during critical periods of development of very young individuals (i.e., deaths at age 28) or the result of early exposure to an antigenically different strain of influenza, our main working hypotheses (i.e., deaths at ages slightly older than 28).

For the time being, we note that even though our study does not allow us to distinguish between the two scenarios for the increased death toll at age 30, the situation for age 28 is much clearer. As these individuals would have been *in utero* or the majority still exclusively breastfeeding, and thereby protected by the mother's antibodies and not producing their own, at the time of the 1890 pandemic (Beaudry et al., 1995; Munoz, 2003), their response to the 1918 flu would have resulted from an insult during a critical period of development and not primarily from an antigenic sin.

As seen in Figure 7.1, mortality was increasing from around the age of 10 to the peak at 28. According to our reading of the theory of antigenic sin, this could be the result of exposure to less and less virulent forms of the 1890 flu that continued to circulate in the population in the years after 1890. Likewise, mortality was declining after the age of 28,

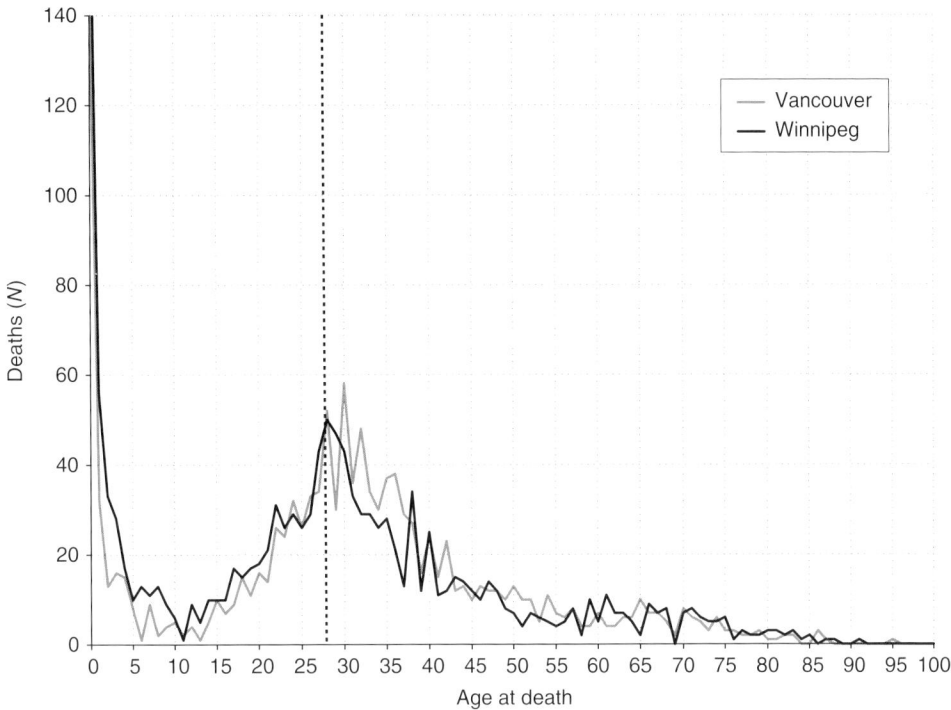

Figure 7.3. Deaths from all causes, Winnipeg ($n = 1193$) and Vancouver ($n = 1141$), September to December 1918 (excluding infants). The vertical line indicates age 28.

which may have been the result of children who had encountered the 1890 pandemic at slightly older ages and who had presumably already developed antibodies to either another influenza strain or to a less virulent form of the 1890 virus. Their response would then have been less specific to the 1890 pandemic variant, allowing for better protection from the 1918 strain because they were less compromised. Patterson (1986) reports that the last flu epidemic to hit North America occurred in 1873–1874. The individuals born during this period would have been 16–17 in 1890 and 44–45 in 1918, while anyone born between 1874 and 1889 would probably have been exposed to this flu variant early in life, through the circulation of seasonal influenza strains. Alternatively, if we suppose that an H1N1-like virus was circulating and drifting during the years leading up to the 1890 antigenic shift (Palese, 2004), then it is possible that those born around perhaps 1878 were exposed early in life to a strain that was closer antigenically to the 1918 strain than those perhaps born in 1888. If so, the susceptibility and mortality of those aged 30 would have been higher in the 1918 pandemic than for those aged 40. At the other end of the spectrum, some elderly people (60–65+) may have been spared in 1918 by earlier exposure to an influenza virus similar to the 1918 strain that circulated prior to 1890 (Luk and Gross, 2001), although mortality at older ages was also high during this pandemic (Crosby, 1989).

The second transition is associated with industrialization and urbanization. However, these were also exactly the conditions that aid in the spread of airborne infectious diseases. Influenza spreads quite easily in the winter, since people cluster in heated indoor environments and hot, dry air weakens the mucosal surfaces of the nasal passages, making it easier for the virus to enter the lungs (Kaslow and Evans, 1997). Through rapid migration, major Canadian cities, such as Toronto, were overcrowded with cramped living conditions,

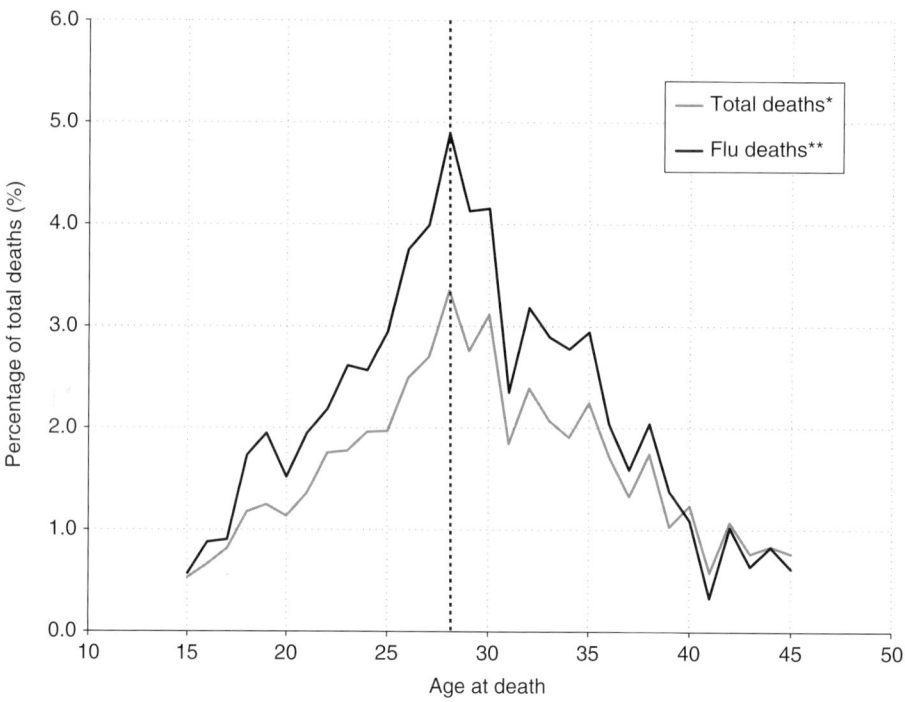

Figure 7.4. Percentage of deaths from all causes and from influenza from ages 15 to 45, September to December 1918. The vertical line indicates age 28. *Deaths from all causes, from Toronto, Ottawa, Hamilton, London, Welland and Lincoln Counties, Winnipeg, and Vancouver. **Deaths from influenza, bronchitis, and pneumonia from Toronto, Ottawa, Hamilton, London, and Welland and Lincoln Counties.

especially among immigrants and the most impoverished members of society (Mercier, 2006). Not only were these conditions conducive to the spread of influenza, they also helped to propagate tuberculosis infections (Sherman, 2006). And, as hypothesized by Noymer and Garenne (2000) and Noymer (2009), tuberculosis infection may have intensified deaths among young adults from the flu in 1918, since both diseases targeted these populations and previous lung damage from tuberculosis allowed for earlier and more intensive infections with influenza and the ensuing secondary bacterial infections of the lungs. However, the early life exposure to influenza is the preferable hypothesis, as it is not clear as to how prior infection with tuberculosis could lead to higher death tolls at specific ages.

The very modern and novel Great War further brought about the conditions of spread for this pandemic. Technological advances and an increasingly interconnected world enabled the 4-year-long stagnant war and the horrific conditions in the trenches throughout northern Europe that characterized it. Many soldiers came from rural areas, such that they would have had limited previous exposure to crowd diseases, therefore simulating the effects of *virgin soil* epidemics, and their immune systems would have been greatly strained when hit by the highly virulent influenza virus in 1918. In this way, the modern development of trench warfare not only allowed the spread of the flu among the troops but was also directly responsible for its rapid, global spread (Crosby, 1989). Beyond the cold, wet, and disease-ridden conditions faced by the soldiers, the war led directly to food and energy rationing and stressed civilian populations, complete with the concomitant immune-compromising effects of deprivation and psychosocial stress (Livi-Bacci, 1991). Thus, it was modern conditions that allowed for both decreases in infant and maternal mortality and the resulting extensions of the average life expectancy that are characteristic of the second epidemiologic transition but also facilitated the widespread devastation of the 1918 influenza pandemic. In particular, the reduction in exposure to epidemic diseases early in life starting at the end of the 19th century, coupled with the further reductions in mortality in the first quarter of the 20th century that occurred with the transition, could have made the early life signal of the 1890 pandemic *stronger* and more decipherable in 1918. In previous historical periods, early life effects were likely blurred by overall higher levels of infant and adult mortality (Gagnon and Mazan, 2009). When overall mortality decreases, it becomes possible to detect more subtle changes in typical mortality patterns. In this way, perhaps the epidemiologic transition was the necessary historical prerequisite for a clear manifestation (and thus understanding) of early life exposure to influenza on later life mortality from the same disease.

Conclusion

Previous research has been valuable in highlighting the alarming experience of young adults during the 1918 flu. Those who are generally the healthiest and most productive in society, who had just survived 4 years of global war, were dying in unprecedented numbers. This could not be adequately explained in 1918, and a universally approved hypothesis has yet to be proposed. Consistent with current literature, we propose that individuals who died in 1918 cannot be analyzed in isolation: their deaths were a product of their experiences in the world around them, extending from conception to the last moments of their lives. The exact age at which they met the previous pandemic strain of influenza in 1890 could have been the deciding factor as to whether they survived the pandemic of 1918.

Even though there have been vast improvements in health and life expectancy in certain parts of the world in the 20th century, it cannot be universally asserted that the modern world is invariably good for human health. Instead, modern environments created the conditions that allowed for the deaths of up to 100 million individuals in the 1918 Spanish influenza pandemic. There is little reason to believe that this could not happen again; it could be hypothesized that in conditions where there is overcrowding, poor nutrition, poor sanitation, and high levels of endemic infectious diseases, those epidemics and potential pandemics will not be far behind. Our analysis demonstrates that previous experience early in life with the various forms of the influenza virus may affect the age distribution of susceptible individuals in future pandemics and should be addressed in order to guide our responses to outbreaks and to inform our preventative policies.

References

Ahmed R, Oldstone MBA, Palese P. 2007. Protective immunity and susceptibility to infectious diseases: lessons from the 1918 influenza pandemic. Nat Immunol 8:1188–1193.
Almond D, Mazumder B. 2005. The 1918 influenza pandemic and subsequent health outcomes: an analysis of SIPP data. Am Econo Rev 95:258–262.
Ancestry.com. 2010. Ontario, Canada, Deaths, 1869–1938 and Deaths Overseas, 1939–1947 [database on-line]. Provo: Ancestry.com Operations Inc.
Barker D. 2006. Adult consequences of fetal growth restriction. Clin Obstet Gynecol 49:270–283.
Barrett R, Kuzawa CW, McDade T, Armelagos GJ. 1998. Emerging and re-emerging infectious diseases: the third epidemiologic transition. Annu Rev Anthropol 27:247–71.
Barry JM. 2004. The site of origin of the 1918 influenza pandemic and its public health implications. J Transl Med 2:3.

Beaudry M, Dufour R, Marcoux S. 1995. Relation between infant feeding and infections during the first six months of life. J Pediatr 126:191–197.

Bengtsson T, Lindstrom M. 2003. Airborne infectious diseases during infancy and mortality in later life in southern Sweden, 1766–1894. Int J Epidemiol 32:286–294.

Canada, Dominion Bureau of Statistics. 1924. Sixth census of Canada, 1921 [Sixieme recensement du Canada, 1921]. Ottawa: F.A. Acland.

Crosby A. 1989. America's forgotten pandemic: The influenza of 1918. New York: Cambridge University Press.

Dowdle WR. 1999. Influenza A virus recycling revisited. Bull World Health Organ 77(10):820–828.

Emery G. 1993. Facts of life: The social construction of vital statistics, Ontario 1869–1952. Montreal/Kingston: McGill-Queen's University Press.

Francis T. Jr. 1953. Influenza: the new acquaintance. Ann Intern Med 39:203–211.

Gagnon A, Mazan R. 2009. Does exposure to infectious diseases in infancy affect old-age mortality? Evidence from a pre-industrial population. Soc Sci Med 68(9):1609–1616.

Glezen WP, Couch RB. 1997. Influenza viruses. In: Evans AS, Kaslow RA, editors. Viral infections of humans: Epidemiology and control, 4th ed. New York: Plenum Medical Book Company. p 473–505.

Harder T. 1918. Some observations on the more severe cases of influenza occurring during the present epidemic. Lancet 192:871–873.

Herring DA, Carraher S. 2011a. The Russian influenza pandemic (1889–90) and the margins of memory. Paper presented at the 39th Annual Meeting of the Canadian Association for Physical Anthropology, Montreal, Quebec, October 29, 2011.

Herring DA, Carraher S, editors. 2011b. Miasma to microscopes: The Russian influenza in Hamilton. Hamilton: McMaster University, Faculty of Social Sciences.

Hill K. 2011. Influenza in India 1918: excess mortality reassessed. Genus 67:9–29.

Hobbs F. 2004. Age and sex composition. In: Siegel J, Swanson D, editors. The methods and materials of demography. New York: Elsevier Academic Press. p 125–173.

Humphries MO. 2005. The horror at home: the Canadian military and the great influenza pandemic of 1918. J Can Hist Assoc 16:235–260.

Johnson NPAS. 2003. The overshadowed killer: influenza in Britain in 1918–1919. In: Phillips H, Killingray D, editors. The Spanish influenza pandemic of 1918–19: New perspectives. New York: Routledge. p 132–155.

Kaslow RA, Evans AS. 1997. Epidemiologic concepts and methods. In: Evans AS, Kaslow RA, editors. Viral infections of humans: Epidemiology and control, 4th ed. New York: Plenum Medical Book Company. p 3–58.

Kim JH, Skountzou I, Compans R, Jacob J. 2009. Original antigenic sin responses to influenza viruses. J Immunol 183:3294–3301.

Kolte IV, Skinhøj P, Keiding N, Lynge E. 2008. The Spanish flu in Denmark. Scand J Infect Dis 40(6–7):538–546.

Langford C. 2002. The age pattern of mortality in the 1918–19 influenza pandemic: an attempted explanation based on data for England and Wales. Med Hist 46:1–20.

Le Goff JM. 2011. Diffusion of influenza during the winter of 1889–90 in Switzerland. Genus 67:77–99.

Livi-Bacci M. 1991. Population and nutrition: An essay on European demographic history. New York: Cambridge University Press.

Loo Y, Gale M. 2007. Influenza: fatal immunity and the 1918 virus. Nature 445:267–268.

Luk J, Gross P, Thompson WW. 2001. Observations on mortality during the 1918 influenza pandemic. Clin Infect Dis 33:1375–1378.

Ma J, Dushoff J, Earn DJD. 2011. Age-specific mortality risk from pandemic influenza. J Theor Biol 288:29–34.

Maris NK. 2011. The impact of influenza: A global perspective. In: Herring DA, Carraher S, editors. Miasma to microscopes: The Russian influenza in Hamilton. Hamilton: McMaster University, Faculty of Social Sciences. p 30–39.

Mercier ME. 2006. The social geography of childhood mortality, Toronto, 1901. Urban Geogr 27(2):126–151.

Morens DM, Fauci AS. 2007. The 1918 influenza pandemic: insights for the 21st century. J Infect Dis 195:1018–1028.

Munoz FM. 2003. Influenza virus infection in infancy and early childhood. Paediatr Respir Rev 4:99–104.

Myers RJ. 1940. Errors and bias in the reporting of ages in census data. Transactions of the Actuarial Society of America 41(104, Part II):395–415.

Myrskylä M, Mehta N, Chang V. 2010. Exposure to the 1918 pandemic and later mortality by cause: Evidence from the NHIS data. In: Program Summary of the Population Association of America 2010 Annual Meeting (Dallas), Session 49: The demographic impact of pandemics. Available at: http://paa2010.princeton.edu/ (accessed October 30, 2013).

Noymer A. 2009. Testing the influenza–tuberculosis selective mortality hypothesis with union army data. Soc Sci Med 68:1599–1608.

Noymer A, Garenne M. 2000. The 1918 influenza epidemic's effects on sex differentials in mortality in the United States. Popul Dev Rev 26:565–581.

Oertel H. 1919. Anatomical and bacteriological findings in the recent epidemic pneumonia. Can Med Assoc J 9:339–344.

Omran AR. 1971. The epidemiologic transition: a theory of the epidemiology of population change. Milbank Mem Fund Q 49(4):509–537.

Oxford JS, Sefton A, Jackson R, Innes W, Daniels RS, Johnson NPAS. 2002. World War I may have allowed the emergence of "Spanish" influenza. Lancet Infect Dis 2:111–114.

Palese P. 2004. Influenza: old and new threats. Nat Med 10:S82–S87.

Patterson KD. 1986. Pandemic influenza 1700–1900: A study in historical epidemiology. Totowa: Rowman & Littlefield.

Phillips H, Killingray D, editors. 2003. The Spanish influenza pandemic of 1918–19: New perspectives. New York: Routledge.

Preston SH, Hill ME, Drevenstedt GL. 1998. Childhood conditions that predict survival to advanced ages among African-Americans. Soc Sci Med 47:1231–1246.

Quinn T. 2008. Flu: A social history of influenza. London: New Holland Publishers.

Sherman IW. 2006. The power of plagues. Washington, DC: ASM Press.

Smith FB. 1995. The Russian influenza in the United Kingdom, 1889–1894. Soc Hist Med 8:55–73.

Storey WK. 2009. The First World War: A concise global history. Toronto: Rowman & Littlefield Publishers, Inc.

Taubenberger J, Morens DM. 2006. 1918 Influenza: the mother of all pandemics. Emerg Infect Dis 12:15–22.

Thompson S. 2011. The Russian flu rushes to Hamilton. In: Herring DA, Carraher S, editors. Miasma to microscopes: The Russian influenza in Hamilton. Hamilton: McMaster University, Faculty of Social Sciences. p 40–49.

Valleron AJ, Cori A, Valtat S, Meurisse S, Carrat F, Boelle P. 2010. Transmissibility and geographic spread of the 1889 influenza pandemic. Proc Natl Acad Sc U S A 107:8778–8781.

Valtat S, Cori A, Carrat F, Valleron AJ. 2011. Age distribution of cases and deaths during the 1889 influenza pandemic. Vaccine 29S:B1–B6.

Part 3
Regional and Temporal Variation in the Second Epidemiologic Transition

Chapter 8
The Second Epidemiologic Transition in Western Poland

Alicja Budnik
Instytut Antropologii, Uniwersytet Im. Adama Mickiewicza, Poznań, Poland

Introduction

Diseases, be they endemic or epidemic, have accompanied humanity throughout our entire history. Importantly, they are a critical component of human evolution because they influence demographic and genetic changes in human populations (Mielke and Fix, 2007). The type and frequency of diseases present in human populations have changed along with shifts in the cultural, economic, and demographic characteristics of these groups. Large-scale, population-wide changes in the patterns of health and disease, characteristic for various periods of human history, were identified as *epidemiologic transitions* by Omran (1971, 1977, 1983). Epidemiologic transition theory focuses on the complex changes in patterns of health and disease that occur over time and the interactions between these patterns and their demographic, economic, and sociologic determinants and consequences (Omran, 1971). These issues were initially raised by McKeown and colleagues (McKeown and Brown, 1955; McKeown and Record, 1962), among others, in their work describing changes in patterns of health and demographic dynamics of historical populations, as well as their interrelationships and causes.

Epidemiologic transition theory, whether in its classic formulation (Omran 1971, 1977, 1983) or in modified versions which consider changing patterns of health and disease in a broader evolutionary context (e.g., Armelagos et al., 1996, 2005; Barrett et al., 1998; Armelagos and Barnes, 1999; Armelagos, 2009), elicits considerable debate (e.g., Mackenbach, 1994; Wolleswinkel-van den Bosch et al., 1997; Gage, 2005; McKeown, 2009). It seems, however, that it provides a very useful framework for interpreting the many health problems which characterize economically developed countries.

Given the larger evolutionary framework suggested by Barrett, Armelagos, and colleagues, Omran's original transition is now often termed the *second* epidemiologic transition. It is characterized by a gradual decline in mortality caused by acute, epidemic infectious diseases and an increase in deaths resulting from degenerative conditions. It occurred the earliest in Western Europe, where it began around the mid-19th century (Omran, 1971;

Modern Environments and Human Health: Revisiting the Second Epidemiologic Transition,
First Edition. Edited by Molly K. Zuckerman.
© 2014 John Wiley & Sons, Inc. Published 2014 by John Wiley & Sons, Inc.

Barrett et al., 1998), but was delayed elsewhere in Europe, such as in Holland, where it started in the 1870s (Wolleswinkel-van den Bosch et al., 1997). These large decreases in mortality in high-income developed countries were associated with the development of advances in sanitation, improved hygiene, and medical advances, among other factors (Stephan and Henneberg, 2001; Saniotis and Henneberg, 2011). Control of infectious disease drastically reduced infant mortality and more moderately reduced adult mortality, but this increase in life expectancy (i.e., the demographic transition) allowed degenerative diseases, such as stroke, cancer, and cardiovascular disease, to become more common as causes of death. This was further exacerbated by the air and water pollution associated with industrialization and by increases in sedentism.

In Poland, virtually no research has been conducted on epidemiologic transitions (Budnik, 2008), and the second transition is especially poorly documented (but see Liczbińska and Budnik, 2007). This may be partially due to the fact that until 1918 Poland was not a sovereign political unit with unified statistical documentation. Instead, Central and Eastern Poland were under Russian rule, Southern Poland was part of the Austro-Hungarian Empire, and Western Poland was ruled by Prussia. Only Western Poland produced reliable health and demographic statistics. Here, I attempt to demonstrate that the changes in mortality levels and in causes of deaths that occurred in the second half of the 19th century and at the beginning of the 20th century in Western Poland, under Prussian rule, fit into the classic pattern of the Western European epidemiologic transition as defined by Omran (1971). Several expectations can be generated. Given the well-documented linkage between living conditions and epidemiologic characteristics found, it is expected that the effects of the second transition would be more pronounced in industrializing cities of the region and occur earlier than in rural areas. Additionally, in rural areas, peasant villages based on agriculture are predicted to have better biodemographic indicators, including life expectancy, infant mortality, and values of the Biological State Index (Henneberg and Piontek, 1975; Henneberg, 1976), and a lower prevalence of degenerative diseases than villages in heavily industrialized regions. In particular, it is expected that the effects of improved sanitation and modernization should be reflected in decreasing normal mortality (see Gage, this volume) and infant mortality specifically, with a reduction in deaths from acute, epidemic infectious disease. Budnik and Liczbińska (2006) have already documented urban–rural differences in biodemographic indices and causes of deaths for the historical Poznań Province, Great Poland, or Wielkopolska, which constituted a large part of the Prussian Polish territory. Therefore, the aim of this work is to describe changes in patterns of health and disease occurring during the 19th and the early 20th centuries in Western Poland. Detailed aims include estimation of a number of biodemographic indicators related to the level, structure, and causes of mortality and their relationships to such cultural variables as economic strategies, levels of urbanization and industrialization, and improvements in sanitary and hygienic conditions.

Background

Between 1772 and 1795, the territory of Poland was occupied by Prussia, Austria, and Russia and divided among them. The occupation was due to a weakening of the Polish state by invasions from Russia, Prussia, and Austria and to complex political machinations of European powers at the time. This situation lasted for 120 years, during which time the western part of Poland, including present-day Eastern Pommerania and the city of Gdańsk, Great Poland (Wielkopolska), and Silesia, a large region covering part of present-day Western Poland, was under Prussian rule (Davies, 1981; Gierowski, 1984). The present study includes urban and rural populations from all three of these occupied regions, each of which has a distinctive history and character.

Eastern Pommerania

Eastern Pommerania, also known as Gdańsk Pommerania, has been a part of Poland since the reign of Mieszko the First, a prince of the Piast dynasty and the founder of the Polish statehood (10th century). After a stormy history, during which Polish and German interests in that territory were intertwined in numerous ways, the region became incorporated into the Kingdom of Prussia at the end of the 18th century and remained so until 1919 when the Treaty of Versailles returned its major part to Poland.

During the 19th century, most of the region belonged to the Prussian Gdańsk Regency (Regierungsbezirk Danzig) (Słownik Historii Polski, 1973; Davies, 1982; Labuda, 1993). At the beginning of the century, the surface area of the Gdańsk Regency was close to 8000 km^2, with a population of 709,312 individuals. This equates to a population density of 89 persons per km^2 (Statistisches Jahrbuch, 1911). Rural areas of the Regency were farmed despite rather poor soil quality, varying value of the land, and a sometimes inadequate intensity of agricultural production (Szumowski and Wajda, 1957; Wajda, 1964; Szukalski, 1988). These shortcomings often resulted in food scarcity and frequent winter and late spring famines in some areas (Budnik, 2005). Within Eastern Pommerania, Gdańsk, a medieval city which was historically strongly linked with Poland, remained for a long time under German and later specifically Prussian influence. In 1920, after World War I, it became an artificial political entity, the *Free City of Gdańsk*, which, though to a certain extent integrated with Poland, was under control of the League of Nations. The city was unequivocally returned to Poland after World War II in 1945. It flourished from the end of the 15th century to the 17th century as a dynamic commercial center and a seaport. In the second half of the 19th century and at the beginning of the 20th century, Gdańsk was a large city with extensive harbor and port facilities and a shipbuilding industry. Shipyards employed several thousand workers producing not only ships, but also locomotives and industrial machinery (Słownik Historii Polski, 1973; Chwalba, 2000).

Great Poland

Great Poland (Wielkopolska) is a historical center of the Polish state. In the 10th century, Mieszko the First established a significant center of government in this region, which impelled economic development there for the next millennium (Budnik et al., 2004). At the end of the 18th century, Great Poland fell under Prussian rule and remained under it until the end of World War I, with a short interlude between 1807 and 1815 when it belonged to the Duchy of Warsaw. During the 19th century, the major part of the region was known as the Poznań Regency (Regierungsbezirk Posen) in the Kingdom of Prussia. The population density of this Regency in 1905 was 72 persons per km^2 with 1,260,000 total inhabitants residing on over 17,500 km^2 of the surface area (Statistisches Jahrbuch, 1911). Poznań (Posen)—a sizeable city since the times of Mieszko the First—was its capital (Jakóbczyk, 1973; Słownik Historii Polski, 1973; Davies, 1981; Chwalba, 2000). Great Poland has been known since the 19th century as a farming region characterized by high agricultural productivity, essentially the *breadbasket* of Poland. This is not so much the result of its natural conditions, because the soils are of average quality for the country (Gorzelak, 1989), but because of its sociocultural history. Polish peasants and landowners very decisively opposed discrimination by the Prussian government during the 19th century by strenuously improving and modernizing their agricultural practices and technologies, such as by introducing modern machinery and extensive chemical fertilizing.

By the mid-19th century, agriculture in Great Poland possessed a solid organizational structure with a modern framework of land ownership by individual farmers, a highly productive labor force, and high levels of practical knowledge. Intensive use of machinery and fertilizers produced harvests double those found in other regions of Poland (Borowski, 1962; Buszko, 1984). Poznań also emerged as a strong commercial and political center in the 19th century, with dynamically developing heavy industries, chemical industries, and agrofood processing (Łuczak, 1965; Topolski and Trzeciakowski, 1994).

Silesia

Silesia was the most industrialized among the three regions. This is especially true of its Eastern part, Upper Silesia. Silesia became a part of Poland in the early middle ages, at the end of the Mieszko the First's reign. Since the beginning of the 14th century, however, its rulers changed repeatedly, from Czechs to Austrians to Prussians. Despite this, the region continued to have large numbers of Polish inhabitants and maintained links with Poland. Around the mid-18th century, Silesia became a part of Prussia, and Upper Silesia became the Opole Regency (Regierungsbezirk Oppeln). After World War I, as a result of Polish uprisings, a part of Upper Silesia was returned to Poland in 1922. The rest, together with other Silesian areas, was returned to Poland only in 1945 (Słownik Geograficzny, 1880–1914; Słownik Historii Polski, 1973; Buszko, 1984; Gierowski, 1984; Chwalba, 2000). Silesia is rich in natural resources, especially large seams of coal. For this reason, since the end of the 18th century, and with particular intensity since the mid-19th century, Upper Silesia became heavily industrialized and developed mining and metallurgy industries. Many deep mines and foundries were built, transforming rural villages into industrial urban centers. Environmental degradation from mining and metal works in many areas, as well as new economic opportunities presented by industry, caused many changes in the lifestyles of rural communities, as many inhabitants abandoned agriculture in favor of industrial labor (Słownik Geograficzny, 1880–1914; Słownik Historii Polski, 1973; Puch, 1993; Chwalba, 2000; Budnik and Liczbińska, in press). This also accelerated urbanization, although unevenly across the territory. For instance, in the Bytom District, the most densely populated area of Upper Silesia, there were 900 inhabitants per km^2, while in other areas of Silesia, the population density was nine times lower (Słownik Geograficzny, 1880–1914). At the beginning of the 19th century, population density there was 154 persons per km^2 for the entire Opole Regency, of which Upper Silesia was a part; the surface area of this Regency was 13,200 km^2 and was populated by 2 million inhabitants (Statistisches Jahrbuch, 1911). This was much higher density than in the other two Regencies. Bytom (Beuthen), a typical mining town in Upper Silesia, is among the oldest cities in the territory. It was founded in 1254, but its rapid growth dates to the mid-19th century when mining developed and large-scale heavy industry became established in the city (Słownik Geograficzny, 1880–1914; Słownik Historii Polski, 1973; Drabina, 2000).

Materials and Methods

In this study, original Prussian demographic and medical statistics on the numbers of deceased by age, causes of death, living population size, and numbers of live births were obtained for the years 1875–1913 from the Preussische Statistik (1876, 1878–1908, 1912, 1913). The Preussische Statistik is the official publication of the Royal Statistical Office in

Berlin (Königlich Preussischen Statistischen Bureau) and has been produced since 1861 in separate specialist issues containing demographic, medical, economic, and climate data. Several researchers have confirmed that these data are highly reliable (Budnik, 2005; Budnik and Liczbinska, 2006). The parameters of the included data and the quality of said data expanded after 1875 when uniform state civil registries were established in Prussian provinces, including the Polish territories. From 1875 onwards, the state administration determined precisely how individual demographic and medical data had to be registered, as well as taking responsibility for systematically collecting and reporting the data. For instance, medical practitioners became obligated to report standardized causes of death to the state offices. The offices then categorized these data according to state instructions and reported them as summary statistics (Preussische Statistik, 1877, 1880; Klotzke, 1980; Budnik, 2005).

The data used in this study include the following: Pommeranian rural agricultural villages and the port city of Gdańsk in the Gdańsk Regency ($n = 357,933$ deceased; $n = 228,955$ live births); typical agricultural villages and the city of Poznań in Wielkopolska in the Poznań Regency ($n = 684,967$ deceased; $n = 411,517$ live births); and the Opole Regency, including the mining region of Upper Silesia and Bytom (total $n = 845,827$ deceased). Altogether, data on 1,888,727 deceased and 640,472 live births were collected and included in the analysis.

Analysis

Death Rates

Using these data, several parameters were calculated. These include crude death rates and infant death rates (Holzer, 1980). Crude death rates were calculated as the number of deaths in a given period divided by the total population size for this period. Infant death rates, those of children less than 1 year of age, were estimated as a ratio of the number of deaths before the end of the first year of life to the number of births in a given period. Second, biometric functions of life tables were estimated for both stationary and stable population models (Pressat, 1966; Acsádi and Nemeskéri, 1970). Stationary population models assume zero natural increase and stable death and birth rates, stable models do not require that these rates are equal, thus allowing for a nonzero natural increase. Because of the stable rates in both models, the age and sex structure of the population also remains stable. The biometric functions obtained for the stable population models were corrected for nonzero growth rates, which were calculated as a difference between numbers of births and deaths divided by the population size in a given period. The most synthetic biometric function of a life table is the life expectancy, e_x. This standard function indicates the number of years in the future to be lived by an individual who reached age x. In other words, it provides the number of years remaining to be lived by an average person reaching the age x in a given population.

Indices of the Opportunity for Natural Selection

Indices of the opportunity for natural selection were also calculated. The data employed in this study enabled detection of the operation of natural selection via one of its components, differential mortality. Three different indices were used to accomplish this, as each provides somewhat different information. The oldest of them is the Index of the Opportunity for Natural Selection through Differential Mortality, or Crow's I_m index (Crow, 1958; Spuhler, 1976). It considers only a fragment of the entire information on mortality—subadult deaths—but is widely used. The index is calculated as a ratio of the proportion of children who did not survive to the age of reproductive maturity (P_d) to those who survived to that age (P_s):

$$I_m = \frac{P_d}{P_s} \qquad (8.1)$$

As can be seen, this index completely ignores adult mortality, even if they died early in their reproductive life span before producing many children.

The second is the Biological State Index (I_{bs}). This index is less commonly used but is more precise and provides more complete information about the influence of mortality on processes of natural selection (Henneberg and Piontek, 1975; Henneberg, 1976; Stephan and Henneberg, 2001; Budnik et al., 2004; Budnik and Liczbińska, 2006; Saniotis and Henneberg, 2011). The I_{bs} index expresses the complete opportunity for natural selection, estimated through the mortality of both subadults and those adults who die before the end of their natural reproductive life span. The opportunity for natural selection is calculated as the magnitude of actual mortality moderated by the reproductive value of individuals, which changes with age (e.g., the loss of a dying child is a complete reproductive loss; the death of a female in her 30s is approximately a 50% reproductive loss; the death of an elderly person is not a reproductive loss, genetically).

$$I_{bs} = 1 - \sum_{x=0}^{\omega} d_x s_x \qquad (8.2)$$

Here, d_x denotes the fraction of persons dying at age x, ω the age of the oldest individual in the population, and s_x is a measure of the reproductive loss resulting from the death of an individual at age x. The variable s_x expresses a probability that a person at age x has not produced their total possible number of offspring (i.e., the total fertility rate) as predicted by fertility models in non-Malthusian populations. For instance, a subadult (0–14 years) will have $s_x = 1$, a postmenopausal woman (>45 years) will have $s_x = 0$, while an individual aged 30 will have $s_x = 0.5$. In this study, values of s_x are estimated for an *archetype of fertility*, based on fertility in populations prior to the advent of effective birth control (Henneberg, 1975, 1976; Stephan and Henneberg, 2001).

The third is the Potential Gross Reproductive Rate (R_{pot}), which is roughly analogous to the Biological State Index but considers only the differential mortality of reproductive-age adults. The R_{pot} can be considered a component of the Biological State Index and is especially useful in situations where child mortality cannot be reliably estimated. The R_{pot} is calculated by Henneberg (1975, 1976) as

$$R_{pot} = 1 - \sum_{x=15}^{\omega} d_x s_x \qquad (8.3)$$

The variables are the same as those in the Biological State Index, but fractions of deceased are calculated taking into account only adult deaths.

By definition, values of both the I_{bs} and R_{pot} indices vary between 0 and 1, making interpretation simple. The value of I_{bs} for a given population expresses the average probability of persons born into this population being able to pass their genetic material on to the next generation. The value of R_{pot} provides information about the fraction of the reproductive span completed by an average adult before death in a given population.

In cases where data required for the specific calculations for these indices were incomplete, estimates from the literature were used (see Borowski, 1968, Liévin, 1880; Puch, 1989; Budnik and Liczbińska, 2006).

Cause of Death

Analyses of causes of death are fraught with difficulties (see McKeown and Record, 1962; Gage, 2005; Beemer, this volume), and the data employed in this study are not without their limitations. For instance, Prussian statistical yearbooks used the medical knowledge existing at the time they were compiled, meaning that the classification of causes of deaths varies between periods. Where possible, causes of deaths were grouped so as to approximate present-day classification in the ICD system. As discussed by Wolleswinkel-van den Bosch et al. (1996) and Beemer (this volume), such procedures can produce acceptable results.

Percentage frequencies of particular causes of death within the total number of deaths were calculated for particular locations and periods. In addition, where possible, death rates from specific causes per 1000 living population were estimated. The so-called childhood diseases such as smallpox, scarlet fever, measles, rubella, diphtheria, croup, and whooping cough together with typhoid fever and diarrhea were categorized here as acute, epidemic infectious diseases. Tuberculosis was considered separately. Degenerative diseases such as cancers, cardiovascular diseases, stroke, and other diseases of the brain were also included.

Results and Discussion

Total Mortality and Infant Mortality

Crude death rates calculated for the studied populations are shown in Table 8.1. Crude death rates were lower in rural areas than in the cities, but in both cases, they decreased over time, especially at the beginning of the 20th century. This change is likely linked to living conditions, which were poor throughout Polish cites in the 19th century and at the beginning of the 20th century but improved afterwards. Multiple Polish cities show this trend; for instance, data from the Preussische Statistik (1877, 1908) reveal that Poznań and Gdańsk had a crude death rate of over 31 per 1000 population (/1000) between 1873 and 1877, declining to approximately 23 between 1903 and 1908. In Wrocław (Breslau), the crude death rate declined from 32.1 in 1875 to 21.5 in 1906, while data from the Statistisches Jahrbuch (1911) reveal that the crude death rate in Szczecin (Stettin) reached a low of 18.4 only in 1909.

Significant differences in crude rates between rural and urban settlements were also found in the parts of Poland ruled by Russia. For instance, in the second half of the 19th century, crude death rates in the city of Warsaw reached 3.7%, while in rural areas of the surrounding

Table 8.1. Crude death rates (per 1000 population) for studied areas of Western Poland in the 19th and at the beginning of the 20th century.

Population	1873–1877[a]	1878–1882[b]	1883–1887[c]	1888–1892[d]	1903–1908[e]
Gdańsk Regency					
Gdańsk city	31.1	31.0	29.1	27.3	22.3
Rural areas	26.9	30.9	27.4	24.5	23.2
Poznań Regency[f]					
Poznań city	31.5	29.2	28.8	26.9	24.5
Rural areas	22.3	23.6	26.4	22.9	17.8
Silesia[g]	31.1	30.1	30.7	29.3	25.4

[a–e]For Silesia, respectively: 1871–1875, 1876–1880, 1881–1885, 1886–1890, 1901–1905.
[f]From Budnik and Liczbińska (2006).
[g]From Borowski (1968).

district, the Warsaw Governorship, they were only 2.6% (Janczak, 1994; Budnik and Liczbińska, 2006). Leon (2008) and others have also observed differences in crude death rates related to the size of urban settlements for other parts of Western Europe.

The observed high urban mortality rates had multiple causes. Many of the changes associated with the beginning stages of industrialization and urbanization were initially detrimental for large parts of the population. Destruction of the natural environment, the lack of modern sanitation and of clean water supplies, poor hygiene and dangerous working conditions, poverty, inadequate diet, overcrowding, inadequate health services, and the stresses of daily existence produced extremely difficult living conditions, especially for the urban working class. Instances of widespread epidemics of infectious diseases and physical exhaustion were common. These dynamics have been described for other early industrializing nations, such as the United Kingdom and the United States (Condran et al., 1984; Barrett et al., 1998; Armelagos et al., 2005; Leon, 2008; Armelagos, 2009), and the situation was similar in Poland.

Overcrowding, lack of adequate housing, and high housing prices also plagued all of the Polish urban populations included in this study. Many urban inhabitants lived in basements and in attics, which were often overcrowded, damp, and poorly ventilated or heated. In historical Poznań, for example, often as many as 12 people lived in one room. No wonder, because the lowest rent for a modest apartment amounted to a fortnight's wages for a common laborer. Housing problems were substantial and persistent in many Polish cities—for instance, even as late as in 1900, nearly 45% of all inhabitants of Poznań lived in overcrowded accommodations (Łuczak, 1965; Jakóbczyk, 1973; Budnik and Liczbińska, 2006). In Gdańsk during the second half of the 19th century, population density exceeded 6000 per km^2, while in surrounding rural areas, it was just a few scores of persons per km^2 (Friedrich, 1895; Budnik and Liczbińska, 1997). The situation was also similar in Upper Silesia (Zieliński, 1984).

Changes of crude death rates in Gdańsk over an extended time period are shown in Figure 8.1. Until 1871, death rates were very high, exceeding as much as 58 deaths per 1000 persons in some years, such as in 1848. This was partially due to frequent cholera epidemics (Liévin, 1880), which spread easily in the large, densely populated city, especially as it lacked efficient sewer systems or clean water supplies. Importantly however, the poor quality of regional roads and transportation systems (Brandsfäter, 1879) prevented the epidemics from spreading into surrounding rural areas. Cholera epidemics stopped in the late 19th century due to the introduction of modern sewage and water supply systems in 1871; the last epidemic, which occurred in 1866, killed as many as 1450 people, raising the crude death rate to 5.0%. However, it spread into the surrounding rural areas so weakly that at the distance of 30 km from the city, only 37 people died, 25 times less than in the city (Topographisch-Statistische Handbuch, 1869; Budnik and Liczbińska, 1997). The improved sanitary systems have also been linked to a subsequent reduction in the spread and mortality of other acute epidemic infectious diseases in the late 19th century, specifically waterborne conditions, resulting in reduction of the average crude death rate by over 9% (Figure 8.1). The systems installed in Gdańsk were among the most advanced in Europe at the time and became an example followed by such cities as Frankfurt on the Mein, Basel, and Trieste (Masłowski, 2011).

Analysis of the mortality of infants revealed that this statistic is an even more sensitive index of improved hygiene and sanitation in Poland than are crude death rates. Infant death rates for Gdańsk and Poznań are shown in Table 8.2 and Figure 8.2. The decrease in infant death rates was in all instances very marked but occurred somewhat more slowly than in some other European populations. For instance, in the Netherlands, infant mortality declined in urban areas from an average of 218 per 1000 population between 1875 and 1879 to 172 from 1895 to 1899, while in rural areas it declined from 178 to 144, respectively

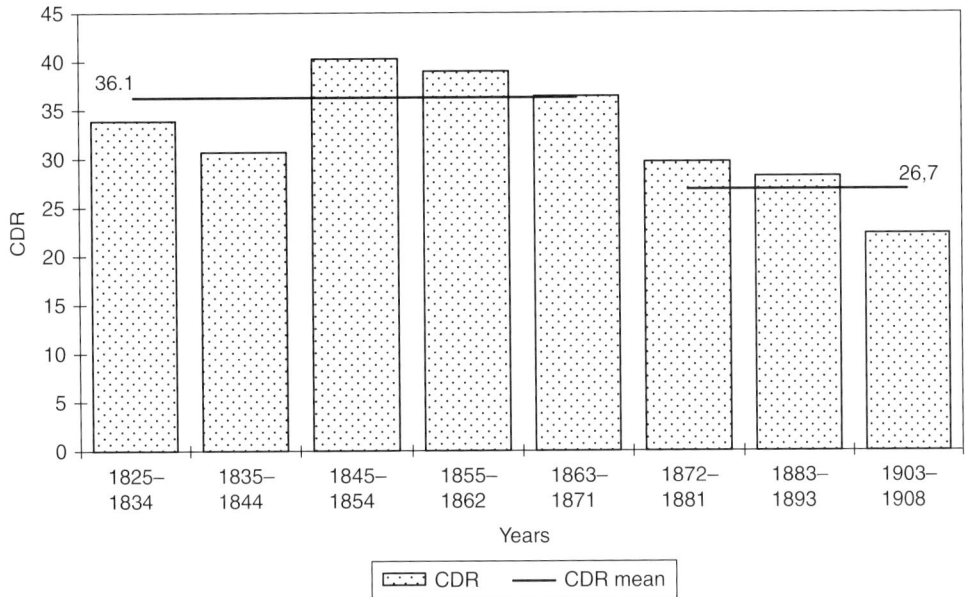

Figure 8.1. Crude death rates (per 1000 population) in Gdańsk in the 19th century and beginning of the 20th century (CDR for years 1825–1871) (Liévin, 1880).

Table 8.2. Infant mortality rates (per 1000 population) in studied areas of Western Poland in the 19th and at the beginning of the 20th century.

Population/ years	1875–1877	1878–1882	1883–1887	1888–1892	1893–1901	1902–1906	1907–1913
Gdańsk Regency							
Gdańsk city	196.8	204.9	191.8	197.8	211.3	223.5	185.9
Rural areas	158.5	163.6	152.8	161.7	171.5	221.4	207.0
Poznań Province[a]							
Poznań city	277.0	280.5	262.2	258.4	263.1	219.1	190.6
Rural areas	208.4	209.4	219.4	201.9	209.1	188.5	181.8
Rural Silesia[b]	1870–1879	1880–1889			1890–1899		
	257.7	206.5			220.3		

[a]From Budnik and Liczbińska (2006).
[b]Calculations based on Puch (1989).

(Wolleswinkel-van den Bosch et al., 2000). In England and Wales, it ranged from 170 to 140/1000 between 1885 and 1900, while in Sweden, it dropped progressively during the 19th century, declining from 216.1/1000 from 1761 to 1770 to 198.7 between 1801 and 1810, 183.4 between 1811 and 1820, 146.0 between 1851 and 1860, and 100/1000 between 1896 and 1900 (McKeown and Brown, 1955; Lynch and Greenhouse, 1994). Major differences in urban–rural mortality rates persisted, however. In the second half of the 19th century in Swedish cities, the rate was 193/1000, while in rural areas it was 119 (Preston et al., 1981; Bogin, 1988). As in Sweden and the United Kingdom, gastrointestinal infections were the major cause of infant deaths in the Polish material analyzed here. This was especially clear in

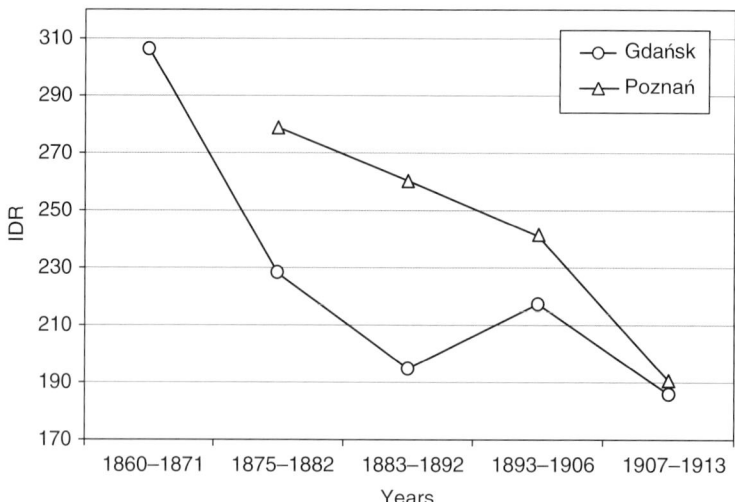

Figure 8.2. Infant mortality rates (infant deaths per 1000 live births) in historical Gdańsk and Poznań cities.

the cities where, before the construction of sewers and water supplies piped to individual buildings, infant mortality rates were especially high. It is striking that in Poznań during industrialization, infant mortality was much higher than in Gdańsk, a port city. In Poznań, however, the first sewers were put into use only in 1896 (Topolski and Trzeciakowski, 1994; Kaniecki, 2004; Budnik and Liczbińska, 2006). After that date, infant mortality started to decline but remained higher in the city than in rural villages. Persistently higher levels of infant mortality in the city, especially during summer months, were largely related to poor hygiene, specifically what has been labelled the *urban-sanitary diarrheal effect* wherein poor sanitary conditions, combined with warm temperatures, greatly exacerbated the spread of diarrheal diseases (Wolleswinkel-van den Bosch et al., 2000).

Life Tables

Here, the values of only one biometric function of the life table are presented, but it is the function of greatest informative value—life expectancy (e_x). Figure 8.3, Figure 8.4, Figure 8.5, Figure 8.6, and Figure 8.7 show differences in values of life expectancy in relation to the region, the level of urbanization, and the ecological situation of each population in the middle of the last decade of the 19th century. In each Regency, significant urban–rural differences are seen, with cities being most disadvantaged (see Figure 8.3, Figure 8.4, and Figure 8.5).

Among rural populations, the highest life expectancies are seen in agricultural Wielkopolska (Poznań Regency) where the local environment remained relatively unpolluted despite economic modernization. Newborn life expectancy there was just over 46 years, which is 8 years higher than in rural Pommerania (Gdańsk Regency) and 9.5 years more than in rural Silesia (Figure 8.6).

Overall, all cities considered here had lower life expectancies than did rural villages (Figure 8.7).

The lowest life expectancies were found in the mining city of Bytom (Silesia). A newborn there could expect to live only 28.3 years, while a 20-year-old adult could expect to live on for less than 32 years. The situation was not much better in the progressively industrializing city of Poznań, while the city of Gdańsk had the best life expectancies: 30.1 for a newborn and 35.7 years for a 20-year-old. However, newborn life expectancies in large cities of

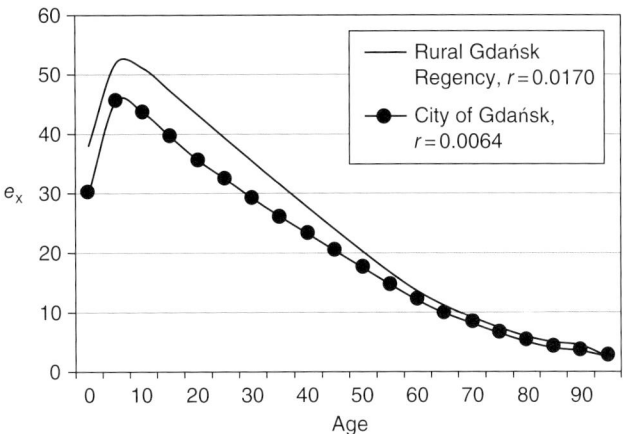

Figure 8.3. Life expectancies by age and by area of residence in Gdańsk Regency in the second half of the 19th century (r—rate of natural increase for the stable population model used to calculate life expectancies).

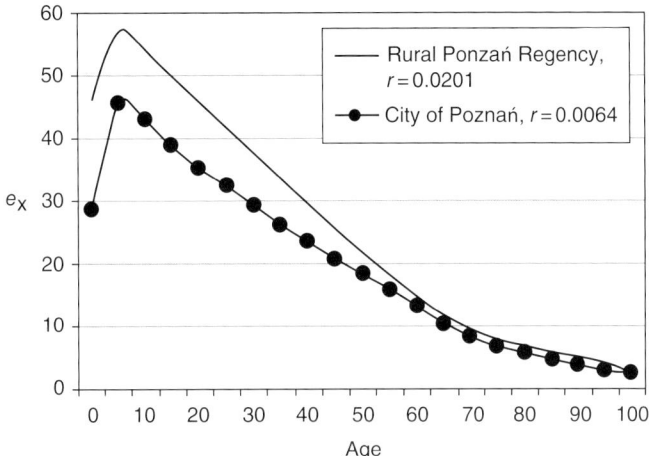

Figure 8.4. Life expectancies by age and by area of residence in Poznań Regency in the second half of the 19th century (r—rate of natural increase).

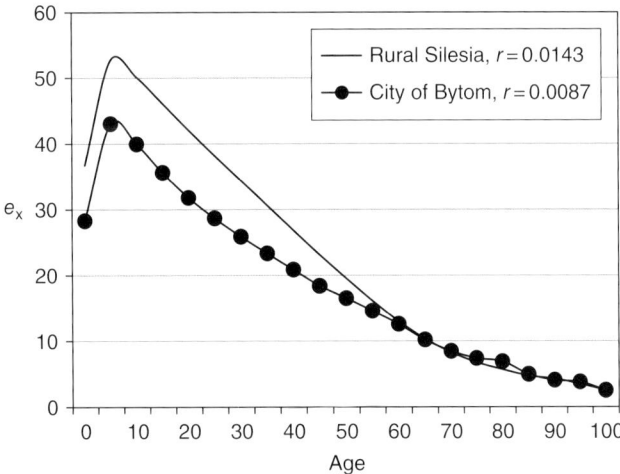

Figure 8.5. Life expectancies by age and by area of residence in Opole Regency (Silesia) in the second half of the 19th century (r—rate of natural increase).

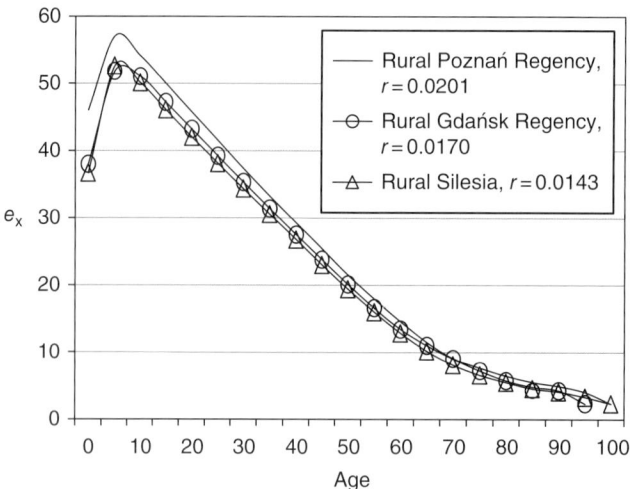

Figure 8.6. Life expectancies by degree of industrialization of the region in the second half of the 19th century. Rural populations (r—rate of natural increase).

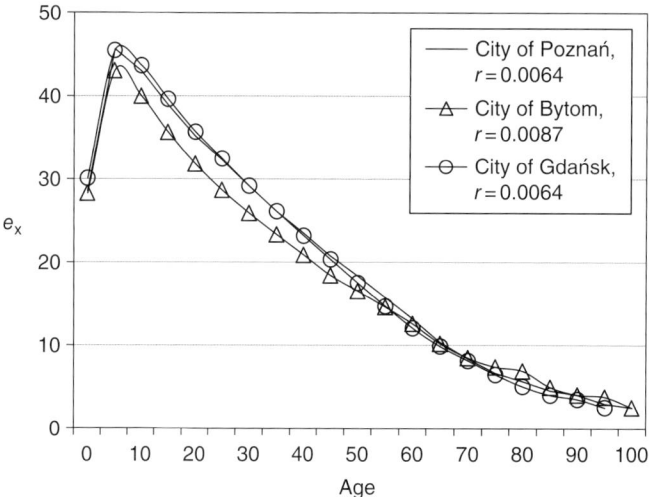

Figure 8.7. Life expectancies by degree of industrialization of the region in the second half of the 19th century. Urban populations (r—rate of natural increase).

England and Wales in the last three decades of the 19th century exceeded those in Gdańsk by 4–17 years. For instance, in Liverpool, newborn life expectancy was approximately 36 years and 46 years in Bristol (Szreter and Mooney, 1998); towards the end of the first half of the 19th century, it was 40.2 years for the entirety of England and Wales (Preston et al., 1981; Bogin, 1988), but it rose to 47.6 years between 1870 and 1910 (Vielrose, 1961). Elsewhere in Europe during approximately the same period, newborn life expectancy was on average 37.8 years in Austria, in Holland nearly 46 years, in Sweden 47.5 years, and more than 48 years in Denmark (Vielrose, 1961; Preston et al., 1981; Bogin, 1988). In France at the turn of the 19th century, newborn life expectancy reached 47 years (Pressat, 1966).

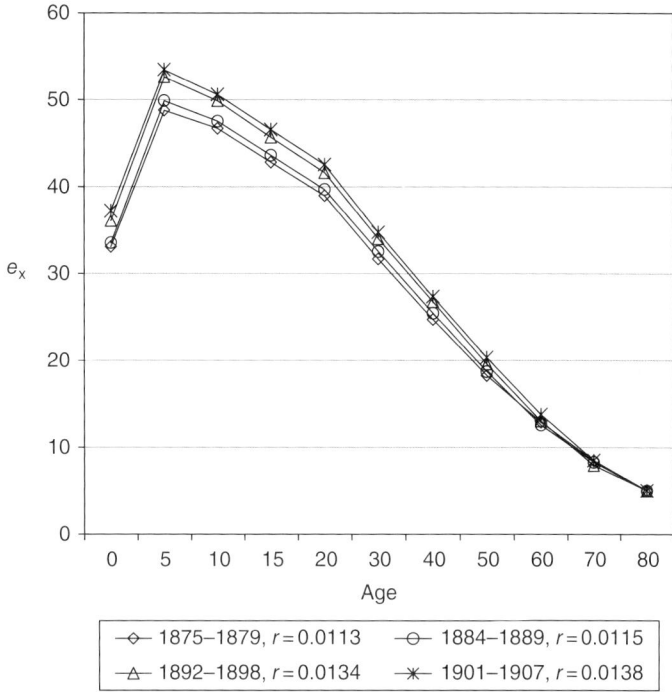

Figure 8.8. Changes in values of life expectancies throughout time in Opole Regency (Silesia, r—rate of natural increase).

In the areas studied in Poland, life expectancy increased substantially during the 19th century. Figure 8.8, Figure 8.9, and Figure 8.10 depict these changes. Figures from the Gdańsk Regency illustrate this, as well as the persistence of urban–rural differences (Figure 8.9 and Figure 8.10) (see also Budnik and Liczbińska, 2006 for a discussion of the same in the Poznań Regency). Throughout the last quarter of the 19th century, newborn life expectancy increased in Gdańsk by 11.5 years—from 27.8 years in 1875 to 1879 to 39.3 years between 1900 and 1908—while life expectancy at age 20 increased by 6.6 years—from 33.8 to 40.4 years. Respective values for rural regions of the Gdańsk Regency were 10.7 years (from 34.3 to 45.0 years) and 5.6 years (from 41.2 to 46.8 years). This means that at the beginning of the 20th century, a person reaching age 20 in the city of Gdańsk could expect to live, on average, to the age 60.4 years but in the surrounding rural areas, they could expect to live to nearly 67 years.

Changes in Mortality Caused by Acute, Epidemic Infectious Diseases and Degenerative Diseases

As aptly noted by McKeown (2009), traditional distinctions between infectious and noninfectious, degenerative disease—Omran's *man-made* diseases of civilization—may be difficult to maintain at present. This is not only because some infectious diseases may have a chronic character but also because the etiology of some degenerative diseases traditionally recognized as being noninfectious, such as stomach ulcers, cervical cancer, and a number of inflammatory disorders, is increasingly recognized as being infectious. It must be stressed that in the present study, due to the limitations of historical sources, infectious diseases were considered to be mainly childhood diseases, such as typhoid fever, smallpox, and diarrhea

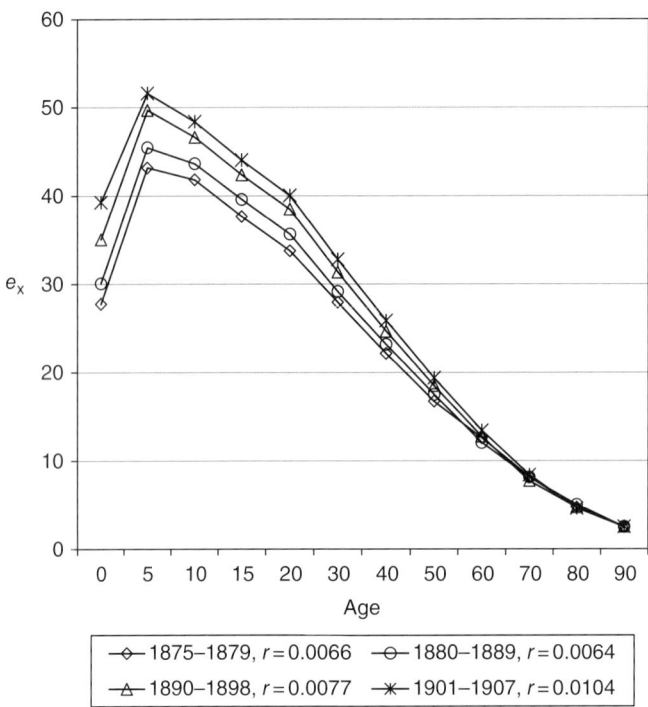

Figure 8.9. Changes in values of life expectancies throughout time in historic city of Gdańsk (r—rate of natural increase).

(in the vast majority of affected children), that is, diseases of an acute, epidemic nature. Tuberculosis, which is clearly an infectious disease, but also of a chronic and wasting nature, was considered separately. Degenerative diseases considered here include conditions like cancer as well as those that could result in sudden death, like heart attacks and strokes.

Figure 8.11, Figure 8.12, and Figure 8.13 illustrate changes through time in the proportion of deaths caused by acute infectious diseases and degenerative diseases. Tuberculosis and diarrhea are not included in these categories. It can be seen that in all Regencies, deaths caused by acute infectious diseases declined while the proportion of deaths from other causes increased. However, the rate of change varied. Acute infectious diseases were the major cause of death in Poland for a very long time. Their significant decline occurred only at the end of the 19th century and beginning of the 20th century. Most notably, smallpox mortality declined dramatically, and deaths attributed to typhoid fever diminished exponentially. Declining smallpox mortality coincided with the introduction of vaccination, while that of typhoid fever was due to improvements in hygiene and sanitation related to the construction of sewer systems and access to clean water, as mentioned earlier. In comparison, substantial declines in deaths caused by acute infectious diseases such as typhoid fever, scarlet fever, measles, and smallpox after 1880 in association with improved sanitation have also been noted for the Netherlands (Wolleswinkel-van den Bosch et al., 1997, 1998). In the USA, in Philadelphia, mortality caused by typhoid fever declined only after 1907 when the city started to distribute filtered water to its inhabitants (Condran et al., 1984).

Mortality from smallpox in Poland clearly demonstrates this overall trend. Artificial inoculation with smallpox by introduction (variolization) of the pus from pox pustules of a patient into the body of a healthy person originated in India approximately 2000 years BCE and was practiced for several centuries with variable success (Stocki, 1964a, 1964b). In

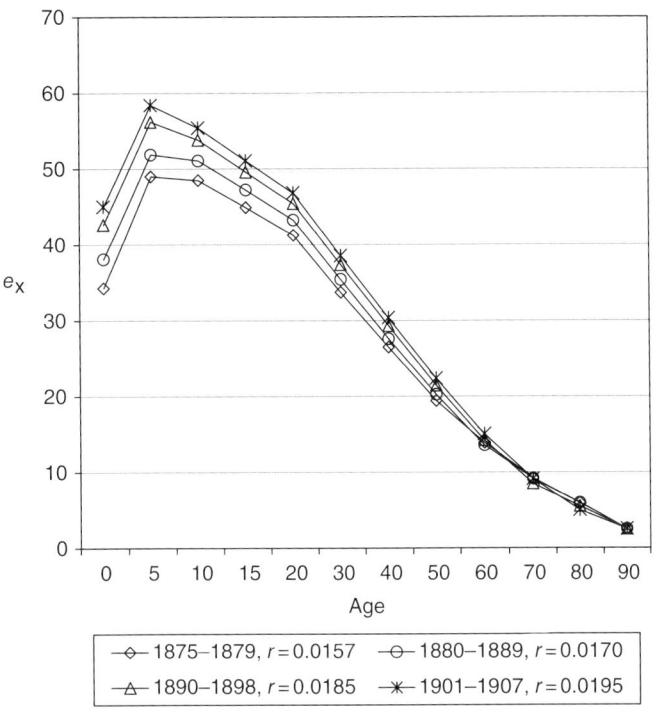

Figure 8.10. Changes in values of life expectancies throughout time in Gdańsk Regency (r—rate of natural increase).

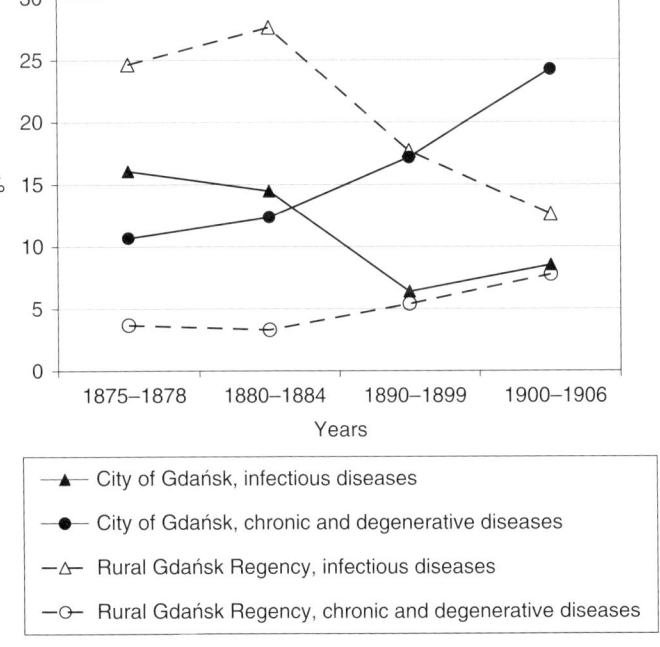

Figure 8.11. Changes through time in the frequencies of cause of death from both infectious and degenerative conditions in the historical Gdańsk Regency.

154 *Modern Environments and Human Health*

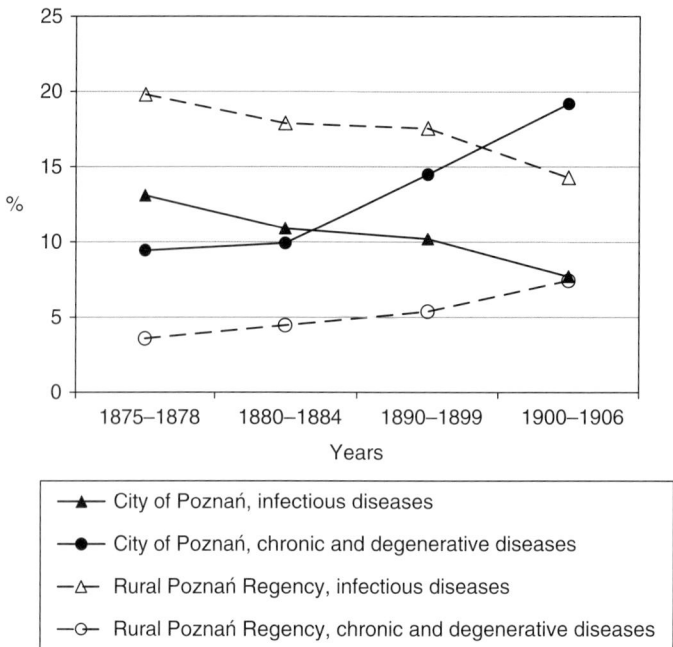

Figure 8.12. Changes through time in the frequencies of cause of death from both infectious and degenerative conditions in the historical Poznań Regency.

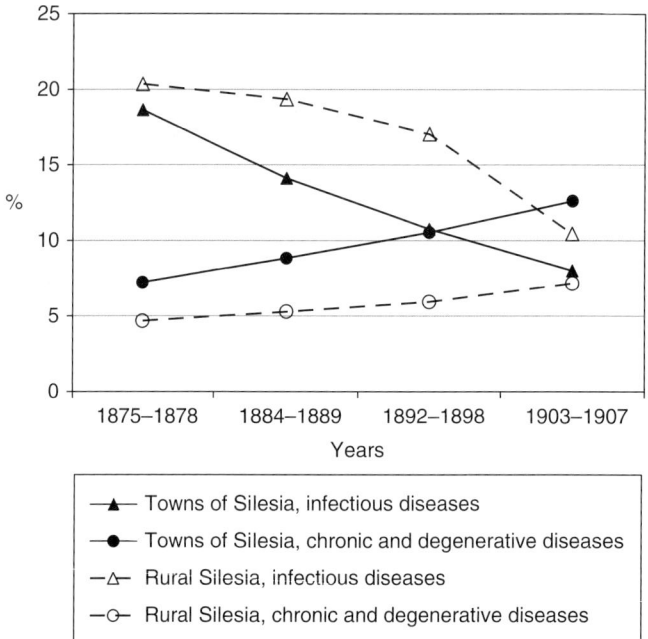

Figure 8.13. Changes through time in the frequencies of cause of death from both infectious and degenerative conditions in the historical Opole Regency (Silesia).

Poland, with the consent of the King Stanislaus Augustus, inoculations were first carried out in Warsaw in 1769; vaccination, developed by Edward Jenner in 1796, did not become popular in Poland for several decades. In Prussia, including the parts of Poland under Prussian rule, compulsory inoculation began in 1874 (Bogdanowicz, 1964). Results suggest that these practices had little initial effect, but produced a substantial reduction in mortality over time. For instance, between 1875 and 1878, in the villages of the Poznań Regency, 190 persons died of smallpox, despite inoculation, meaning that smallpox was responsible for 0.31% of all annual deaths there. In the same time period, in the city of Poznań, only five people died of smallpox, amounting to 0.09% of all annual deaths, or a mortality rate of 0.03 per 1000 persons. In the Gdańsk Regency, smallpox accounted for 0.08% of all deaths in rural areas, or 0.01 per 1000 population annually; in the city of Gdańsk, it amounted to 0.05% of all deaths and 0.02 per 1000 population. After 1878, there were many years without a single smallpox death, particularly in rural areas. For instance, in rural areas of the Poznań Regency, only 0.005% of the annual number of deaths was attributed to smallpox—one death per one million inhabitants. The situation was similar in rural areas of the Gdańsk Regency, where smallpox produced 0.003% of all deaths, or 0.8 deaths per 1 million population. Urban areas marked higher mortality but still at reduced levels; in the city of Gdańsk, for instance, smallpox represented 0.07% of all deaths, or 0.02 deaths per 1000 population (Preussische Statistik 1877–1880, 1894, 1895, 1897, 1900). After 1903, Preussische Statistik abandoned reporting of smallpox deaths because the disease had been eliminated from the territory of Prussia, including its Polish provinces. It seems therefore that inoculations produced very significant effects, though their full impact took some time. Despite a number of past controversies surrounding the effectiveness of smallpox inoculations, ultimately, other authors have concluded the same—that the procedure had the desired effect, only delayed (McKeown and Record, 1962).

The prevalence of tuberculosis varied in the material studied here. Overall, however, there were more deaths attributed to tuberculosis in cities than in rural villages (Figure 8.14).

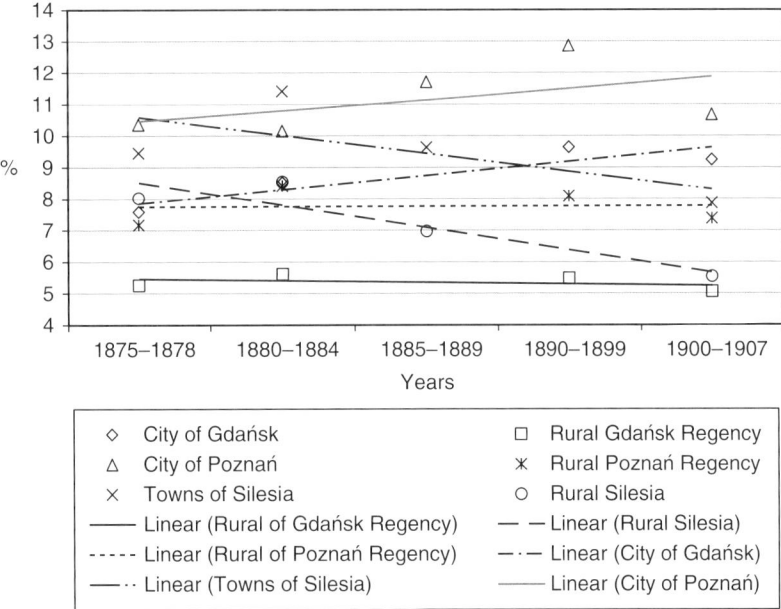

Figure 8.14. The share of deaths caused by tuberculosis in the total number of deaths in the Regencies studied.

Changes noted throughout time were minimal. This agrees with earlier observations for the Netherlands (Wolleswinkel-van den Bosch et al., 1997).

Throughout high-income, developed nations, improvements in sanitation, in hygiene, and to a certain extent in living conditions, as well as in medical treatment according to some authors (e.g., Condran et al., 1984; Armelagos, 2009), contributed to the decline in mortality from acute, epidemic infectious diseases and in infant mortality during the second epidemiologic transition. In turn, this resulted in the increase in life expectancy found in the transition. As noted by Armelagos (2009), life expectancy is related to the increase in the frequency of degenerative diseases. Here, analysis of data from Poland during the second transition demonstrates a progressive increase in the number of deaths caused by these conditions in both rural and urban populations. This increase is especially clear in the cities where, it seems, it was a result of changes associated with industrialization that were detrimental to health. This increase was much less pronounced and slower in rural areas (Figure 8.11, Figure 8.12, and Figure 8.13). In Polish cities, the change in patterns of mortality resulting from the gradual replacement of acute, epidemic infectious diseases by degenerative diseases as the major cause of deaths started after 1884. In Bytom (Silesia), which was much smaller than Gdańsk and Poznań and which intensely developed economically only at the very end of the 19th century, the change started later, around 1892–1899. It must be added that correcting for life expectancy, using measures of newborn life expectancy, did not alter the observed changes in patterns of disease in any of the urban populations studied here. This confirms that the occurrence of the second epidemiologic transition was related to industrialization and associated processes of urbanization in historical Polish territories rather than other processes.

All considerations regarding causes of deaths and their changes through time were hitherto based on the relative share of particular diseases in total mortality. McKeown (2009) has warned that, paradoxically, such relative frequencies of specific causes within the total mortality may not correctly reflect changes in the actual risk of death. It is possible that an increase in the share of specific causes of death in total mortality can occur even when the risk of death from specific causes declines, given that the risk of death from other causes may decrease faster (McKeown, 2009). In Table 8.3, rates of deaths resulting from specific causes are shown for the city of Gdańsk and the rural areas of Gdańsk Regency for those years for which complete information was available. As can be seen, the decline of deaths caused by acute, epidemic infectious diseases and the increase—though smaller—in deaths caused by degenerative diseases remains the same. The relationship between urban–rural patterns also remains the same.

Opportunities for Natural Selection

The opportunity for natural selection has been defined by Crow (1958; see also Spuhler, 1976) as the total premature mortality, specifically the deaths of subadults and to a certain extent young adults, for a given population. Values of the indices for the opportunity for natural selection through differential mortality in the areas of Poland under study are presented in Table 8.4 and Table 8.5. Several items deserve mention. Selection acts especially strongly through eliminating subadults from a population, as evidenced by high values for the Crow's index and very low values for the Biological State Index. Theoretically, values for the Biological State Index can range from 0 to 1. For instance, the value of 0.424 for Gdańsk from 1875 to 1879 reflects so poor an adaptation of this population to environmental conditions that less than half of newborns would have had a chance to achieve adulthood and reproduce. Nearly 60% of newborns would have no chance to produce offspring.

Table 8.3. Deaths by cause (per 1000 population) in the historic Gdańsk Regency.

	City of Gdańsk			Rural areas		
Causes of death/year	1880	1885	1906	1880	1885	1906
Infectious diseases	6.28	3.40	1.80	8.74	5.89	2.56
Chronic and degenerative diseases	3.35	3.77	5.12	0.86	0.93	1.58
Tuberculosis	2.10	2.21	1.95	1.62	1.46	1.03

Table 8.4. Changes in the opportunity for natural selection through differential mortality through time in Gdańsk Regency.

	City of Gdańsk			Rural Gdańsk Regency		
Years	R_{pot}	I_{bs}	Crow's index, I_m	R_{pot}	I_{bs}	Crow's index, I_m
1875–1879	0.854	0.424	1.012	0.908	0.496	0.831
1880–1889	0.874	0.457	0.912	0.927	0.546	0.697
1890–1899	0.896	0.528	0.697	0.939	0.603	0.556
1900–1908	0.906	0.586	0.546	0.946	0.630	0.503

Table 8.5. Changes in the opportunity for natural selection through differential mortality through time in Opole Regency (Silesia).

Years	R_{pot}	I_{bs}	Crow's index, I_m
1875–1879	0.893	0.489	0.827
1884–1889	0.896	0.491	0.825
1892–1898	0.912	0.523	0.744
1901–1907	0.916	0.535	0.713

The Potential Gross Reproductive Rate is also a part of the Biological State Index. As discussed later, it measures the opportunity for selection through the premature death of reproductive-age adults. Its values also range from 0 to 1, and these can be interpreted in a way similar to those of the Biological State Index. For instance, a value for the Potential Gross Reproductive Rate of 0.854 means that about 85% of adults could achieve full reproductive success, while some 15% was deprived of this success by premature mortality.

Values of all three indices show that mortality in urban areas provided greater opportunities for natural selection to act than did mortality in rural areas (Table 8.4). Overall, this suggests that urban populations were less adapted to their environments than were rural populations and that this was largely the result of the unprecedented changes in population density, levels of hygiene and public health, and overall living conditions that accompanied industrialization and urbanization in 19th-century Poland. However, all of these values indicate increasing levels of adaptation through time; the Biological State Index and the Potential Gross Reproductive Rate values increase while the Crow's index values decrease through time. This is similar to other situations found in industrializing Europe; for instance, Korpelainen (2003) also found decreasing values for the Crow's index for industrializing and urbanizing Finland.

Conclusion

All parameters of mortality studied here improve their values through time. Historical records suggest that these changes are directly associated with and caused by improved sanitation, hygiene, and control of infectious diseases. This is augmented by results demonstrating significant urban–rural differences in mortality, indicating that rural populations enjoyed a substantially better ecological situation in relation to determinants of disease. However, rural areas were not homogenous in their ecological situations and epidemiologic indicators; populations relying on agriculture had lower mortality than did populations relying on mining and heavy industry, as seen in Silesia. Indices of the opportunity for selection through differential mortality confirm these differences and demonstrate their impacts upon the adaptive success of these populations throughout time in the face of changing environments.

With the decrease in deaths caused by acute, epidemic infectious diseases in Poland, there was also an increase in mortality caused by degenerative diseases. This changeover of mortality patterns was especially marked in the cities, where it occurred earlier than in rural areas. This documents that Poland underwent a *classic* version of the second epidemiologic transition, initiating in the 1880s in Gdańsk and Poznań and in the 1890s in Silesia.

References

Acsádi GY, Nemeskéri J. 1970. History of human life span and mortality. Budapest: Akadémiai Kiadó.
Armelagos GJ. 2009. The paleolithic disease-scape, the hygiene hypothesis, and the second epidemiological transition. In: Rook GAW, editor. The hygiene hypothesis and Darwinian medicine. Basel: Birkhäuser Verlag. p 29–43.
Armelagos GJ, Barnes K. 1999. The evolution of human disease and the rise of allergy: epidemiological transitions. Med Anthropol 18:187–213.
Armelagos GJ, Barnes K, Lin J. 1996. Disease in human evolution: the re-emergence of infectious disease in the third epidemiological transition. AnthroNotes 18:1–7.
Armelagos GJ, Brown PJ, Turner B. 2005. Evolutionary, historical and political economic perspectives on health and disease. Soc Sci Med 61:755–765.
Barrett R, Kuzawa C.W, McDade T, Armelagos GJ. 1998. Emerging and re-emerging infectious diseases: the third epidemiologic transition. Annu Rev Anthropol 27:247–271.
Bogdanowicz J. 1964. Kilka danych dotyczących historii szczepień ochronnych przeciw ospie prawdziwej. Pediatr Pol 39:609–610.
Bogin B. 1988. Rural-to-urban migration. In: Mascie-Taylor CGN, Lasker GW, editors. Biological aspects of human migration. Cambridge: Cambridge University Press. p 90–129.
Borowski S. 1962. Rozwarstwienie wsi wielkopolskiej w latach 1807–1914. Poznań: Rada Naukowo-Ekonomiczna przy Prezydium WRN w Poznaniu.
Borowski S. 1968. Emigracja z ziem polskich pod panowaniem niemieckim w latach 1815–1914. Przeszl Demogr Pol 2:139–167.
Brandsfäter FA. 1879. Land und Leute des Landkreises Danzig: eine topographisch-historisch-statistische Schilderung. Danzig: Verlag von Theodor Bertling.
Budnik A. 2005. Uwarunkowania stanu i dynamiki biologicznej populacji kaszubskich w Polsce. Studium antropologiczne. Poznań: Wydawnictwo Naukowe UAM.
Budnik A. 2008. Przejścia epidemiczne na ziemiach polskich. In: Dzieduszycki W, Wrzesiński J, editors. Epidemie, klęski, wojny. Poznań: SNAP. p 53–66.
Budnik A, Liczbińska G. 1997. Mortality in the populations of Danzig (Regierungsbezirk Danzig) in the second half of the nineteenth century. Anthropol Rev 60: 13–24.

Budnik A, Liczbińska G. 2006. Urban and rural differences in mortality and causes of death in historical Poland. Am J Phys Anthropol 129:294–304.
Budnik A, Liczbińska G, Gumna I. 2004. Demographic trends and biological status of historic populations from Central Poland: the Ostrów Lednicki microregion. Am J Phys Anthropol 125:369–381.
Buszko J. 1984. Historia Polski 1864–1948. Warszawa: Państwowe Wydawnictwo Naukowe.
Chwalba A. 2000. Historia Polski 1795–1918. Kraków: Wydawnictwo Literackie.
Condran P, Williams H, Cheney RA. 1984. The decline in mortality in Philadelphia from 1870 to 1930: the role of municipal services. Pa Mag Hist Biogr 108:153–177.
Crow JF. 1958. Some possibilities for measuring selection intensities in man. Hum Biol 30:763–775.
Davies N. 1981. God's playground: A history of Poland. Volume II: 1795 to the present. Oxford: Oxford University Press.
Davies N. 1982. God's playground. A history of Poland. Volume I: The origins to 1795. New York: Columbia University Press.
Drabina J. 2000. Historia Bytomia 1254–2000. Bytom: Towarzystwo Miłośników Bytomia.
Friedrich E. 1895. Die Dichte der Bevölkerung im Regierungbezirk Danzig. Danzig: AW Kafemann.
Gage T. 2005. Are modern environments really bad for us? Revisiting the demographic and epidemiologic transitions. Yearb Phys Anthropol 48: 96–117.
Gierowski JA. 1984. Historia Polski 1764–1864. Warszawa: Państwowe Wydawnictwo Naukowe.
Gorzelak E. 1989. Przesłanki regionalizacji polityki rolnej. Wieś Współczesna 4–6: 60–73.
Henneberg M. 1975. Notes on the reproduction possibilities of human prehistorical populations. Przegląd Antropologiczny 41: 75–89.
Henneberg M. 1976. Reproductive possibilities and estimations of the biological dynamics of earlier human populations. J Hum Evol 5:41–48.
Henneberg M, Piontek J. 1975. Biological state index of human groups. Przegląd Antropologiczny 41:191–201.
Holzer JZ. 1980. Demografia. Warszawa: Państwowe Wydawnictwo Ekonomiczne.
Jakóbczyk W. 1973. Dzieje Wielkopolski. Tom II. Lata 1793–1918. Poznań: Wydawnictwo Poznańskie.
Janczak JK. 1994. Statystyka ludności Królestwa Polskiego w drugiej połowie XIX wieku. Przeszłość Demograficzna Polski 19:47–116.
Kaniecki A. 2004. Poznań. Dzieje miasta wodą pisane. Poznań: Wydawnictwo Poznańskiego Towarzystwa Nauk.
Klotzke Z. 1980. Ludność obwodu Urzędu Stanu Cywilnego Luzino w latach 1874–1918. Przeszłość Demogra Pol 12:65–104.
Korpelainen H. 2003. Human life histories and the demographic transition: a case study from Finland, 1870–1949. Am J Phys Anthropol 120(4):384–390.
Labuda G, editor. 1993. Historia Pomorza. Poznań: Poznańskie Towarzystwo Przyjaciół Nauk.
Leon DA. 2008. Cities, urbanization and health. Int J Epidemiol 37:4–8.
Łepkowski T, editor. 1973. Słownik Historii Polski. Warszawa: Wiedza Powszechna.
Liczbińska G, Budnik A. 2007. Zjawisko przejścia epidemiologicznego w wybranych metropoliach polskich i niemieckich z przełomu XIX i XX wieku [abstract]. Przegląd Antropologiczny – Anthropological Review Suppl 5:91.
Liévin AKL. 1880. Ueber die Sterblichkeit in Danzig in den Jahren 1863 bis 1879. Danzig in naturwissenschaftlicher und medizinischer Beziehung. Danzig.
Łuczak C. 1965. Życie gospodarczo-społeczne w Poznaniu 1815–1918. Poznań: Wydawnictwo Poznańskie.
Lynch KA, Greenhouse JB. 1994. Risk factors for infant mortality in nineteenth-century Sweden. Popul Stud 48(1):117–133.
Mackenbach JP. 1994. The epidemiologic transition theory. J Epidemiol Community Health 48:329–332.
Masłowski A. 2011. Eduard Wiebe – honorowy obywatel Gdańska. Available at: http://ibedeker.pl (accessed November 5, 2013).
McKeown RE. 2009. The epidemiologic transition: changing patterns of mortality and population dynamics. Am J Lifestyle Med 3:19–26.

McKeown T, Brown RG. 1955. Medical evidence related to English population changes in the eighteenth century. Popul Stud 9:119–141.
McKeown T, Record RG. 1962. Reasons for the decline in mortality in England and Wales during the nineteenth century. Popul Stud 16:94–122.
Mielke JH, Fix AG. 2007. The confluence of anthropological genetics and anthropological demography. In: Crawford MH, editor. Anthropological genetics: Theory, methods and applications. Cambridge: Cambridge University Press. p 112–140.
Omran A.R. 1971. The epidemiologic transition. A theory of the epidemiology of population change. Milbank Mem Fund Q 49:509–538.
Omran AR. 1977. A century of epidemiologic transition in the United States. Prev Med 6:30–51.
Omran AR. 1983. The epidemiologic transition theory. A preliminary update. J Trop Pediatr 29:305–316.
Pressat R. 1966. L'analyse démogrphique, méthodes, résults, applications. Paris: Press Universitaires de France.
Preston SH, Haines MR, Panuk E. 1981. Effects of industrialization and urbanization on mortality in developed countries. International Population Conference, Manila, 1981. Proceedings 2:233–253.
Preussische Statistik. Die Bewegung der Bevölkerung. Die Geburten, Eheschliessungen und Sterbfälle. Herausgegeben in zwanglossen Heften vom Königlichen Statistischen Bureau in Berlin. 1876, 42; 1878, 45; 1879–1908, 48–220; 1912, 238; 1913, 245. Berlin: Königlichen Statistischen Bureau.
Preussische Statistik. Medicinalstatistik. Beiträge zur Medicinalstatistik des preussischen Staates und zur Mortalitätsstatistik der Bewohner desselben. Herausgegeben in zwanglosen Heften vom Königlichen Statistischen Bureau in Berlin. 1877, 43; 1878, 46. Berlin: Königlichen Statistischen Bureau.
Puch EA. 1989. Dynamika biologiczna populacji wiejskich różnych regionów Polski (II połowa XVIII i XIX wiek). Ph.D. dissertation. Poznań: UAM.
Puch EA. 1993. Dynamika biologiczna polskich społeczności wiejskich z różnych systemów ekologiczno-kulturowych w XVIII i XIX wieku. Przegląd Antropologiczny 56:5–36.
Saniotis A, Henneberg M. 2011. Medicine may be constructing human bodies in the future. Med Hypotheses 77:560–564.
Słownik geograficzny Królestwa Polskiego i innych krajów słowiańskich. 1880–1914. Warszawa: nakładem Filipa Sulimierskiego i Władysława Walewskiego.
Spuhler JN. 1976. The maximum opportunity for natural selection in some human populations. In: Zubrow EBW, editor. Demographic anthropology: Quantitative approaches. Albuquerque: University of New Mexico Press. p 185–226.
Statistisches Jahrbuch für den Preussischen Staat 1910. 1911. Berlin: Verlag des Königlichen Statistischen Landesamts.
Stephan CN, Henneberg M. 2001. Medicine may be reducing the human capacity to survive. Med Hypotheses 57:633–637.
Stocki E. 1964a. Ospa na przestrzeni wieków. Wiad Lek 17(6):517–524.
Stocki E. 1964b. Ospa prawdziwa i sposoby zapobiegania jej w zarysie historycznym. Wiad Lek 17(7):605–611.
Szreter S, Mooney G. 1998. Urbanisation, mortality and the standard of living debate: new estimates of the expectation of life at birth in 19th-century British cities. Econ Hist Rev 51:84–112.
Szukalski J. 1988. Cechy charakterystyczne środowiska przyrodniczego Kaszub, jego walory i ochrona. Kaszuby. Pomorze Gdańskie 18: 53–85.
Szumowski Z, Wajda K. 1957. Pomorze Gdańskie i Wielkopolska w początkach okresu imperializmu. Studia i Materiały do Dziejów Wielkopolski i Pomorza 3: 9–31.
Topographisch-Statistische Handbuch fur Regierungbezirk Danzig bei der Königlichen Regierung. 1869. Danzig.
Topolski J, Trzeciakowski L. 1994. Dzieje Poznania 1793–1918. Warszawa/Poznań: Wydawnictwo Naukowe PWN.
Vielrose E. 1961. Elementy ruchu naturalnego ludności. Warszawa: Państwowe Wydawnictwo Ekonomiczne.
Wajda K. 1964. Wieś pomorska na przełomie XIX i XX wieku. Poznań: Wydawnictwo Poznańskie.
Wolleswinkel-van den Bosch JH, Poppel FW, Mackenbach JP. 1996. Reclassifying causes of death to study the epidemiologic transition in The Netherlands, 1875–1992. Eur J Popul 12:327–361.

Wolleswinkel-van den Bosch JH, Looman CWN, Poppel FWA, Mackenbach JP. 1997. Cause-specific mortality trends in The Netherlands, 1875–1992: a formal analysis of the epidemiologic transition. Int J Epidemiol 26:772–781.

Wolleswinkel-van den Bosch JH, Poppel FWA, Tabeau E, Mackenbach JP. 1998. Mortality decline in The Netherlands in the period 1850–1992: a turning point analysis. Soc Sci Med 47(4):429–443.

Wolleswinkel-van den Bosch JH, Poppel FWA, Loomana CWN, Mackenbach JP. 2000. Determinants of infant and early childhood mortality levels and their decline in The Netherlands in the late nineteenth century. Int J Epidemiol 29(6):1031–1040.

Zieliński A. 1984. Górny Śląsk i Zagłębie w dawnych opisach. Wiek XIX. Katowice: Wydawnictwo Śląsk.

Chapter 9
The Timing of the Second Epidemiologic Transition in Small US Towns and Cities: Evidence from Local Cemeteries

Lisa Sattenspiel and Rebecca S. Lander
Department of Anthropology, University of Missouri, Columbia, MO

Introduction

Populations through Western Europe and the United States underwent a demographic transition, marked by a shift from a situation of high fertility and high mortality prior to the Industrial Revolution to one of low fertility and mortality throughout the early 19th and 20th centuries. Omran (1971, 1983) has suggested that most of the decline in mortality is due to reduced deaths from epidemic infectious diseases, leading to the idea that an epidemiologic transition occurred as well. More recent conceptualizations of this shift have led to the recognition that there have been other major, global shifts in patterns of human health and disease throughout the history, leading to a relabeling of Omran's transition as the *second* epidemiologic transition.

Although the declines began early in the 19th century or before in some parts of the world (Schofield and Reher, 1991), the rates of decline escalated in the late 19th century following widespread acceptance of the germ theory of disease. The recognition of *germs* as a cause of many diseases stimulated the development of sanitation and sewer systems and other public health measures in most large cities of Western Europe and the USA, and these social changes strongly correlate with the increased rapidity of declines in mortality (Ellms, 1928; Meeker, 1972; Higgs, 1979; Okun, 1996; Troesken, 1999; Cutler and Miller, 2005). In a quantitative assessment of the impact of sanitary facilities, Cutler and Miller (2005) estimated that in major American cities between 1900 and 1936, nearly half of the reduction in overall mortality and three-quarters of the reduction in infant mortality could be traced to the construction of city sewer systems and more readily available clean water. Although other social changes have been proposed as playing a role in the late 19th- and early 20th-century mortality declines, it is clear that sanitary improvements were among the most important advances.

The improvements in sanitary facilities occurred rapidly throughout the large cities of the USA between the end of the 19th century and the beginning of the 1920s (Melosi, 2000). Melosi noted, however, that although sewer systems were built quickly, treatment facilities to improve water quality did not always accompany sewer systems immediately. Nevertheless,

---

Modern Environments and Human Health: Revisiting the Second Epidemiologic Transition,
First Edition. Edited by Molly K. Zuckerman.
© 2014 John Wiley & Sons, Inc. Published 2014 by John Wiley & Sons, Inc.

even in cities with sanitary facilities but no major emphasis on the quality of drinking water, between 1900 and 1913 the typhoid death rate dropped by more than 55%, since the large population in these cities made typhoid such a substantial problem that inadequate treatment could still significantly improve the situation (Ellms, 1928; Okun, 1996). Due to water treatment and other public health measures, mortality rates in the large cities were reduced to near modern levels by the early 1920s, long before the post-WWII advent of widespread antibiotics and other modern health practices.

The situation outside of the large cities during the early 20th century is much less well understood, however. Though the second transition occurred throughout high-income, developed countries in the 19th and 20th centuries and continues in low-income, developing ones today, the analysis of the decline in mortality rates associated with it has been clearly biased towards the study of large cities and of national-level data sources (e.g., Preston, 1976; Higgs, 1979; Preston and Haines, 1991; Coleman and Salt, 1992; Gage, 1993, 1994, 2005). Problematically, major public health initiatives were not implemented uniformly across the country, and the timing of community improvements in smaller towns and cities was not likely to be the same as in larger cities, meaning that metropolitan- and national-level data cannot be clearly extrapolated to the epidemiologic situation in these smaller areas. For example, most rural areas of the USA did not implement milk pasteurization, water purification practices, sanitary sewage disposal, and other measures until the 1930s or later (Higgs, 1973). Higgs also noted that before the 1920s fewer than 50 counties in the entire USA had established full-time county health units, and according to the US Census Bureau (1990), over half of residents in the USA at the turn of the 20th century lived in communities with fewer than 10,000 people. Therefore, increased knowledge of the demographic history of residents of rural areas and small cities and towns will fill out the picture of when and how the second transition came to pass throughout the USA. Additionally, it may help to assess more effectively the relative importance of different factors influencing mortality changes at this time and may aid in resolving some of the disagreements about which factors had the most significant influence on the second transition, both overall and in particular times and places.

Studying local-level changes in demography is not without difficulties, though. One of the major problems is that relevant demographic data may be very hard to find. As was the case with the implementation of public health initiatives, official policies requiring the registration of deaths were implemented at various times across and within the USA. In urban areas, nearly all deaths were registered on a regular basis by the turn of the 20th century, but death registration was not customary in many small communities until after 1910 (Higgs, 1973). Furthermore, even if death registration data existed for a region of interest, data for small cities and towns are often not reported in aggregate census and vital statistics reports, and locating the original data sources can be a difficult undertaking.

A direct result of this lack of data is that direct estimates of changes in mortality during the second transition and the impact of community improvements on infectious disease mortality are often difficult or impossible to generate. As Sattenspiel and Stoops (2010) demonstrated, it may be possible, however, to observe the signal of these changes in mortality by analyzing a data source—local cemetery headstones—that is easily accessible in even the smallest communities. This kind of analysis can help to extend our understanding of the local tapestry of changes in demography during the recent past so that we can begin to build a more balanced picture of events throughout Western Europe and North America.

In their study, Sattenspiel and Stoops (2010) analyzed data from nearly 6000 gravestones of individuals who died and were buried between 1900 and 1990 in the Columbia Cemetery, located in the City of Columbia, Missouri. Examination of the time series of burials by age indicated a marked decline in burials of individuals under age 45 beginning in the early to mid-1920s, while burials of older individuals declined somewhat later. This drop in mortality of younger individuals was shown to coincide with a number of public health and sanitation improvements, including the implementation of TB testing of cows and the construction of city sewage facilities. Furthermore, analyses showed that, at least at the state level, declines in water- and milk-borne diseases such as infantile diarrhea, typhoid, diphtheria, and tuberculosis also occurred at this time. Because of this, Sattenspiel and Stoops suggested that cemetery records provided a signal of the timing of changes associated with the second transition in Columbia and that this timing was somewhat later than that observed in both eastern seaboard cities and the large Missouri cities of St. Louis and Kansas City.

One of the limitations of Sattenspiel and Stoops' study is that there are two large cemeteries in Columbia. The Columbia Cemetery was established in 1820; the Memorial Park Cemetery was established in 1928, nearly a century later. At the time of their study, sampling of headstones at the Memorial Park Cemetery was not complete enough to include in a systematic study. Thus, Sattenspiel and Stoops based their conclusions only on data from the Columbia Cemetery, under the assumption that the late establishment of the Memorial Park Cemetery would ensure that the picture derived from the Columbia Cemetery adequately reflected community-wide patterns during the first few decades of the 20th century, when local infrastructure and policy changes generally associated with the second transition were taking place. The number of different sections of the cemetery that have been sampled and the overall sample size of data collected from the Memorial Park Cemetery are now sufficient to allow the assumption about the representativeness of the Columbia Cemetery to be evaluated and to see if the signal of the second transition observed in the Columbia Cemetery data continues to be apparent when data from both cemeteries are included. Hence, in this paper, we discuss both the patterns observed in the newer cemetery and those observed when both cemeteries are combined. We discuss these results and their implications in light of the prior results from the Columbia Cemetery as well as in regards to larger issues of whether and how local cemeteries can contribute to knowledge of the timing and extent of changes in mortality associated with advances in sanitation, health care, and other factors that correlate with the second transition elsewhere.

Materials and Methods

The present population of Columbia, Missouri, is about 110,000 people, but during the first half of the 20th century, when many of the advances associated with the second transition occurred, it ranged from about 5500 to just over 30,000 people. The Columbia Cemetery, located near the center of modern-day Columbia, is $0.14\,km^2$ in size and is composed of a number of historically distinct units, including, for example, a historic cemetery containing the remains of many of the original inhabitants of the city, a Jewish cemetery, and a section serving residents who were economically disadvantaged. None of these units are physically separated from the rest, and most burials are marked by aboveground gravestones. The density of gravestones is relatively high throughout; exceptions include the northern end, which has largely been reserved for new burials, and the southern end, the resting place of the disadvantaged residents. The latter section putatively has numerous burials, but there are few gravestones, most of which are of poor quality and highly eroded. This cemetery was and is open to all residents of Columbia, and for all intents and purposes, it was the only cemetery in town until the establishment of the Memorial Park Cemetery at the end of the 1920s, although a few small church and family cemeteries can be found in the surrounding region (almost all of which are outside the limits of even present-day Columbia).

The Memorial Park Cemetery is about $0.18\,km^2$ in size, but about 25% of the area remains undeveloped. It contains approximately 32,000 burials overall, 95% of which

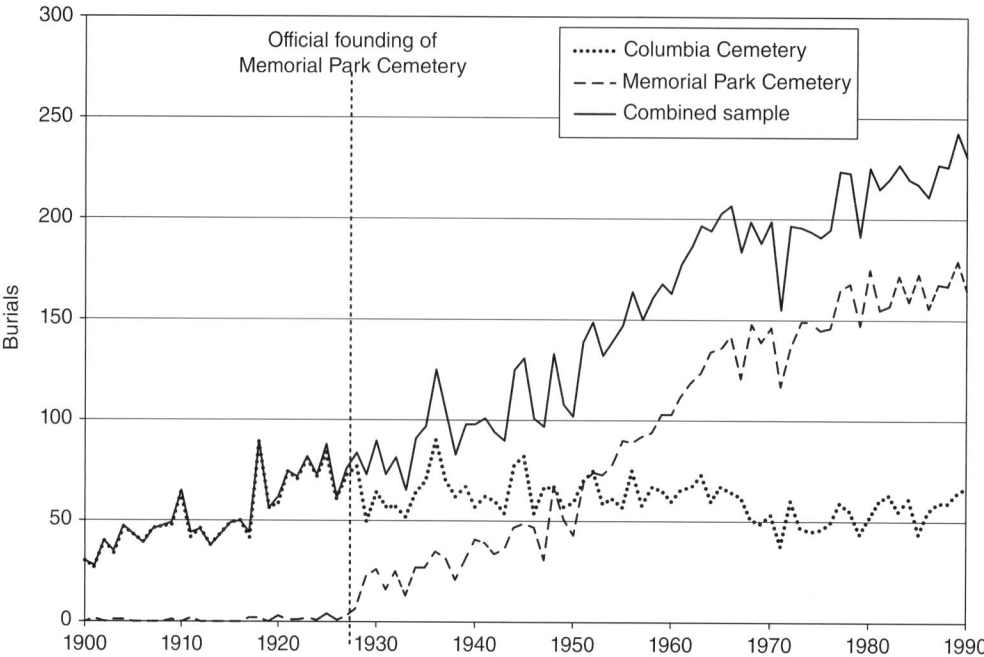

Figure 9.1. Raw number of burials in the Columbia Cemetery, Memorial Park Cemetery, and combined cemetery samples.

include legible markers. Within two decades of its establishment, it became the most common place of burial for Columbia's residents (Figure 9.1). It, too, was and is open to all residents of Columbia. At the time it was founded, it was outside of the city limits, but it is now well within the boundaries on the west side of the city. It also consists of several distinct units, including four sections with a preponderance of infant burials, a veterans' area, and a section that is a traditional burial place for many of Columbia's African American residents, but all sections are open to anyone desiring space within them. The types of burials at the cemetery are also more variable than at the older Columbia Cemetery and include a small number of aboveground tombs and mausoleums, three large columbaria holding the remains of cremated individuals, wall vaults within the grounds of the cemetery, and numerous small gardens with monuments and individualized gravesites or memorials in addition to traditional graves or crypts for ashes. Aside from wall vaults and memorials, aboveground markers are rare at the cemetery.

Data used in this study include information from about 7000 headstones and markers in the Columbia Cemetery and over 10,000 markers in the Memorial Park Cemetery. The existing data represent about 95% of the visible burials in the Columbia Cemetery and around 35% of the burials in the Memorial Park Cemetery. Birth date, death date, name, age (if given), and other information such as relationship to other individuals or military service were collected from each of the headstones. When a specific burial was recorded more than once, which is a common situation at the Columbia Cemetery, but rare at the Memorial Park Cemetery, the data were crosschecked for accuracy, and when any discrepancies occurred, these were resolved through on-site checks of the actual stones. In both samples, a substantial number of burials did not possess sufficient information for the analyses performed; because of this, final sample sizes are under 6000 for the Columbia Cemetery and slightly over 7500 for the Memorial Park Cemetery. The combined sample used for most analyses consists of

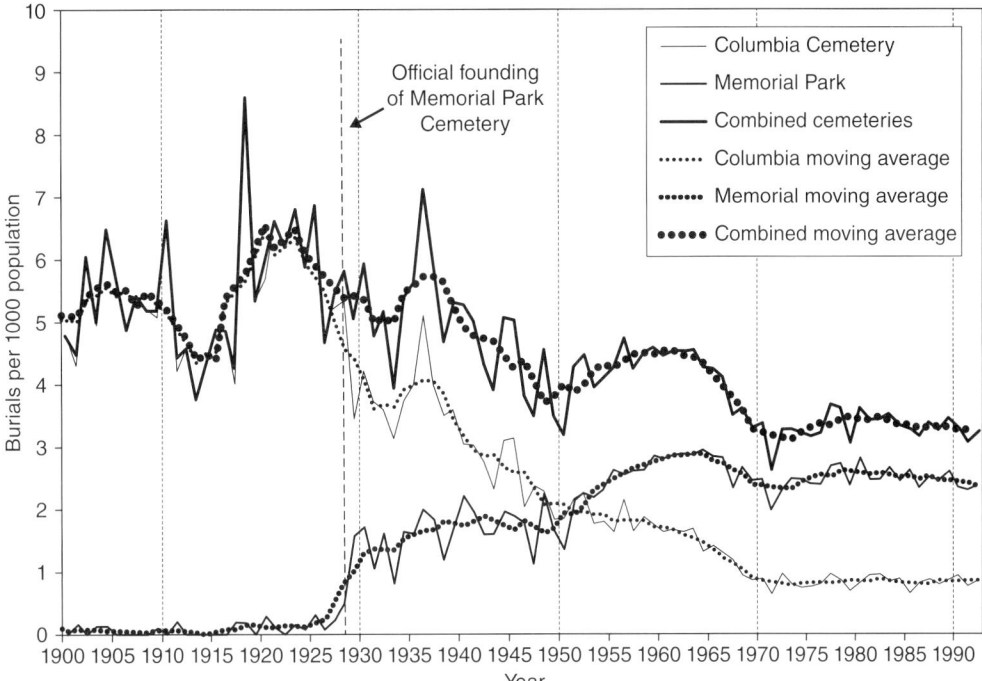

Figure 9.2. Number of burials in the Columbia Cemetery, Memorial Park Cemetery, and combined cemetery samples standardized by population size. *Solid lines* show burials per year; *patterned lines* show 5-year moving averages of the burials.

data from 13,447 burials, with the earliest recorded death occurring in 1806 for the Columbia Cemetery and in 1870 for the Memorial Park Cemetery. It is likely that the earliest burials in both cemeteries represent reburials from elsewhere, since they predate the establishment of the cemeteries, but extant records are not available to verify this.

A small percentage of headstones recorded age at death, but in most cases, it was necessary to calculate this using birth and death dates. An algorithm that corrected for potential age biases due to lack of complete information on birth/death days and months was used to determine the age at death for all individuals with at least year of birth and year of death information (see Sattenspiel and Stoops, 2010). Analyses were conducted using the resulting age distributions from both cemeteries individually and from the combined sample.

General methods of analysis were similar to those used in Sattenspiel and Stoops (2010). Because the City of Columbia experienced substantial growth during the time period under study, burial counts were standardized by age and population size using age distributions and total counts published in US Census reports whenever possible. Census data for the City of Columbia were not published prior to 1930, and the 1930 data used 10-year age-classes for adults rather than the 5-year classes that became the standard in later censuses. Thus, standardization in the early years of the 20th century involved some estimation using a combination of data from Columbia, Boone County, and the State of Missouri. Both the Columbia Cemetery and Memorial Park Cemetery served the population of Columbia predominantly; thus, the same standardization procedure and population figures were used to derive standardized burial numbers and rates for both cemeteries individually and for the combined cemetery. These data and a 5-year running average of the burial numbers are shown in Figure 9.2.

Following the methods used in Sattenspiel and Stoops (2010), ANOVAs were run to study the impact of changes within age-groups across time periods and changes within time periods across age-groups. The analyses directed at changes within age-groups across time periods divided burials into two sets of four 20-year time periods: (i) 1910–1929, 1930–1949, 1950–1969, and 1970–1989, and (ii) 1905–1924, 1925–1944, 1945–1964, and 1965–1984. The analyses directed at changes within time periods across age-groups used 10-year time periods between 1910 and 1989 or 1905 and 1984. The use of offset time periods allowed for finer resolution of the specific times when changes were most likely to have occurred. For all statistical analyses, data were aggregated into five age-groups: infants (0–1), children (1–14), young adults (15–44), older adults (45–64), and elderly (65–74). In order to avoid the use of open age intervals, burials of the oldest individuals (75+) were not included in the statistical analyses. Further statistical analyses directed at determining when and in which age-groups changes occurred were provided through *post hoc* analyses using Tukey's test (Hinkle et al., 1994).

Other sources of data are available to help gain more knowledge of the lives (and deaths) of Columbia's residents during the early 20th century. In compliance with a 2004 law, death certificates of individuals who died more than 50 years ago are publically available online from the Missouri State Archives.[1] Although Missouri did not require the filing of death certificates until 1911, some public death records of select counties from before this time are also available at this site. Unfortunately, pre-1911 records for Boone County (including the City of Columbia) are not available, but a small number of individuals buried at the Columbia Cemetery or Memorial Park Cemetery died in neighboring counties with available records and may be able to be located in this database. However, to date, a search has been completed only for death certificates dating between 1925 and 1935.

The Missouri death certificates usually include the following information: full name of decedent and spouse, cause of death, place and date of death, place and date of burial, and residing address. Thus, it is possible to use these certificates to verify essential demographic data in cases where recording errors at the cemeteries or during database entry may have been made. Death certificates can also be used to collect information on cause of death. However, it is rare to find death certificates for infants and young children, especially in the pre-WWII years, and the Missouri death records database does not include anyone dying outside the State of Missouri who may have been buried in the cemeteries. This, combined with the short time span of records that have been searched at present limits the utility of cause of death information available from the death certificates.

Data on the prevalence of different infectious diseases are available at the state level in the Annual Reports of the Missouri State Board of Health (1911–1942), but data at the county or city levels have not been located. The state level data are discussed briefly in the introduction to this chapter and in more detail in Sattenspiel and Stoops (2010). Local newspapers can be used to identify whether and when serious epidemics occurred, but these events are not systematically reported. It is also possible to search for obituaries of individuals recorded in the database, but the size of the cemetery samples and the fact that obituaries are available for only a proportion of deaths limits the utility of this data source as well. Because of their limitations, these two data sources are not included in the study at the present time. Local newspapers are a good source of information on local improvements in infrastructure and health initiatives, however, and several reports of these have been uncovered. In addition, two general histories of the City of Columbia, Havig (1984) and Crighton (1987), can be used to provide additional information on important events in Columbia's past.

---

[1] http://www.sos.mo.gov/archives/resources/deathcertificates/

Results

Observation of the time series of burials presented in Figures 9.1 and 9.2 shows that changes in the burial patterns throughout the city prior to the 1940s mirror changes in burials in the Columbia Cemetery, indicating that the impact of the Memorial Park Cemetery at this time was minimal and supporting Sattenspiel and Stoops' (2010) assumption that data from the Columbia Cemetery prior to the 1940s are likely to be representative of the city as a whole. Standardized burial rates at the Columbia Cemetery decline precipitously between 1930 and 1950 and then decline more gradually until 1970, when they stabilize at a relatively low rate. Standardized burial rates at the Memorial Park Cemetery increase moderately from its establishment in 1928 until 1950; they then increase more rapidly over the next two decades before stabilizing in the mid- to late 1960s. Rates of burial at the Columbia Cemetery drop below those of the Memorial Park Cemetery in the early 1950s, and as a result of the continued decline in burial rates at the Columbia Cemetery and the marked increase in burial rates at the newer cemetery, the overall pattern of burials in the combined cemeteries after 1950 mirrors the pattern observed in Memorial Park alone.

As was the case in the Columbia Cemetery data alone, the four time periods—1910–1929, 1930–1959, 1950–1969, and 1970–1989—capture much of the interesting variation observed in the time series of standardized burial rates, but the addition of the Memorial Park sample to the data set changes the pattern, especially in the second half of the 1900s. As Figure 9.2 shows, burial rates in Columbia fluctuated around a relatively constant value between 1900 and 1930 and declined markedly between 1930 and 1950, a pattern that is largely reflected in the Columbia Cemetery data and was noted by Sattenspiel and Stoops (2010). However, the contribution of the Memorial Park Cemetery indicates that in the city as a whole, rather than declining steadily as was observed in the data from the Columbia Cemetery alone, burial rates increase slightly between 1950 and 1960 and then decline sharply between 1960 and 1970 before stabilizing for the final time period.

While recognizing that the establishment of the Memorial Park Cemetery likely influenced the patterns they observed in the Columbia Cemetery data, Sattenspiel and Stoops (2010) presented arguments designed to explain long-term declines in the burial rates in Columbia extending throughout the 20th century. Data from both cemeteries combined clearly indicate that such long-term declines were an artifact of the increasingly unrepresentative nature of the burials from the Columbia Cemetery. This points to the significance of underlying assumptions about the representativeness of burial samples, both skeletal and cemetery, for understanding the history of human populations within a region across time. We return to this important point in the following discussion section. The focus of Sattenspiel and Stoops was on explaining the mortality patterns in the early decades of the century, however, and the present data indicate that their assumption that the Columbia Cemetery data reflects the population of Columbia at that time is reasonable.

As was apparent in the earlier study, the mean rates of burial in the combined sample vary markedly within all age-groups and at all times (Figures 9.3 and 9.4). A comparison of Figures 9.3 and 9.4 indicates that the rapid post-1920 declines in mortality of the two oldest age-classes observed by Sattenspiel and Stoops using the Columbia Cemetery burials only were not representative of what was happening in Columbia as a whole; rather, declines are gradual and apparent primarily in the oldest age-class. Infant burials show slightly lower rates of decline in the combined sample, but a major peak in infant deaths during the 1920s shows up in both samples, largely because the burials are found primarily in the Columbia Cemetery, and there is an additional peak in the 1940s that was not especially marked in the Columbia Cemetery sample alone. It has been suggested that these peaks might reflect

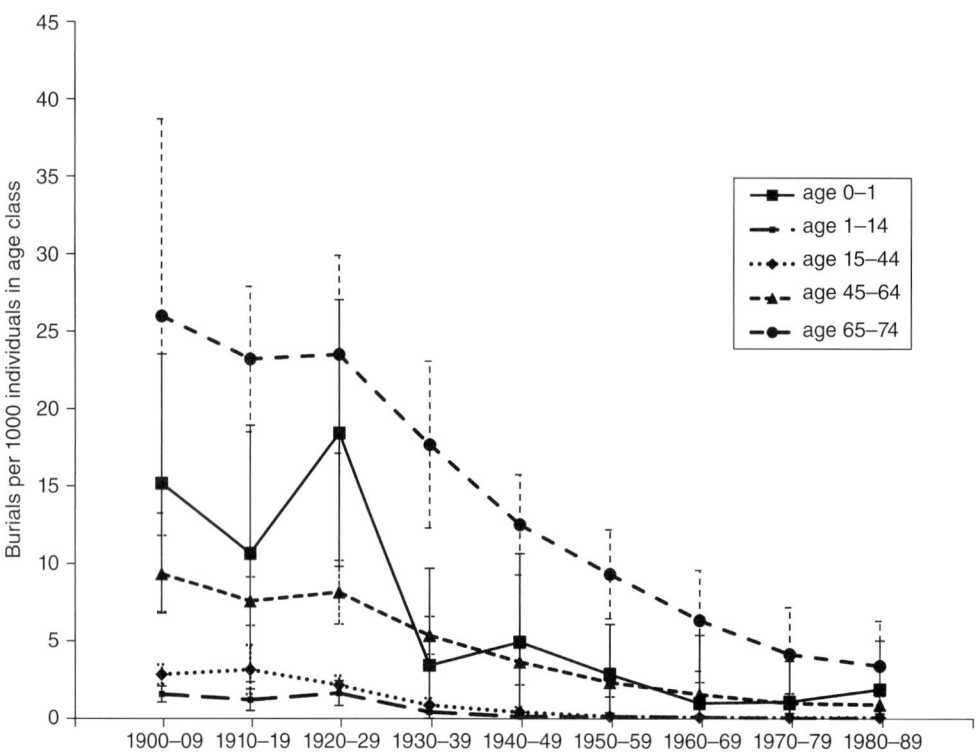

Figure 9.3. Means and 95% confidence intervals for burial numbers standardized by population for each age-group in the Columbia Cemetery sample only. (From Sattenspiel and Stoops, 2010, Figure 4b.)

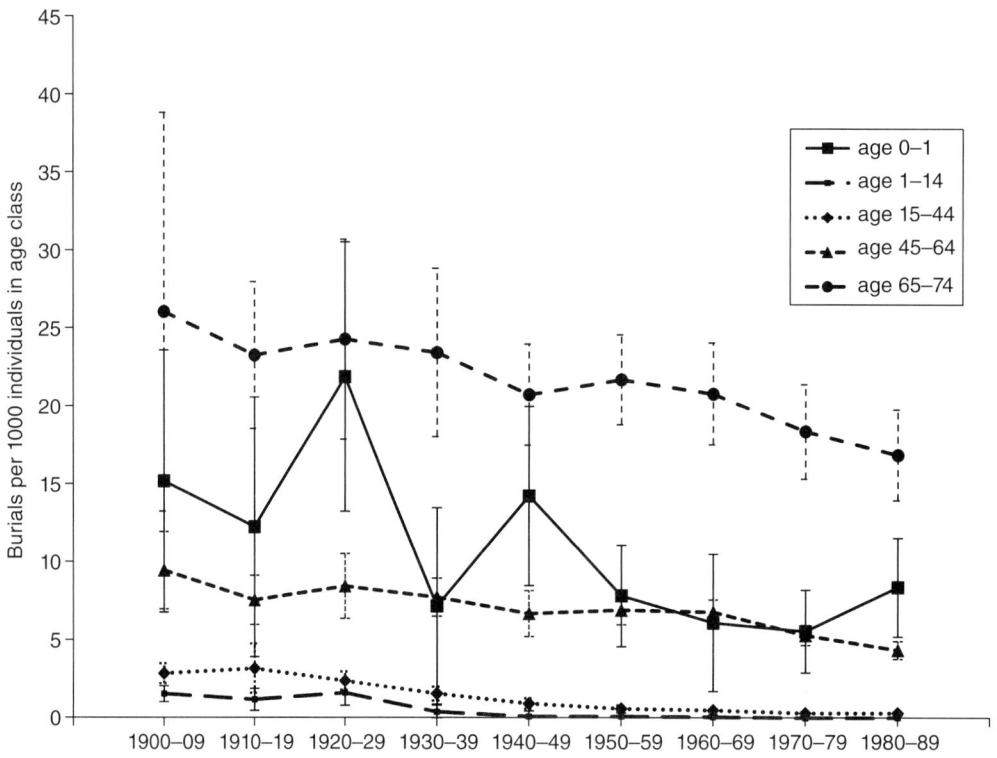

Figure 9.4. Means and 95% confidence intervals for burial numbers standardized by population for each age-group in the combined cemetery sample.

the *Great Migrations* of African Americans from the South to the northern Midwest and Western States during the first half of the 20th century (de la Cova, personal communication), but no evidence has been found to suggest an influx in the population of Columbia due to these events. Determining whether these infant deaths occurred disproportionately among African Americans is also difficult, since only a very small proportion of the death certificates of infants under age 1, no matter their ancestry, are available in the State of Missouri online database.

Because the impact of the Memorial Park Cemetery was minimal until the middle of the 20th century, the marked difference in mortality patterns between those dying below and above age 45 is still apparent in the combined sample (Figure 9.5a and b). Thus, the cornerstone of this study—that data from local cemetery headstones may provide an indication of the timing of changes in mortality that accompanied the second transition—remains intact. In particular, *beginning* in the 1920s, there was a marked decline in mortality of individuals under age 45 that continued until stabilizing around 1970. As discussed earlier, this contrasts with the experiences of large cities in the United States, where declines in mortality began much earlier. One major difference between the pattern observed in the combined sample and that of the Columbia Cemetery alone (Sattenspiel and Stoops 2010; Figure 5) is that the decline in burial of individuals older than 45 largely disappears once the burials from the Memorial Park Cemetery are added to the sample, suggesting even more strongly that events in the first few decades of the 20th century differentially affected the survivorship of the young.

Statistical analyses of the combined data set confirm this conclusion. The analyses of differences within age-groups across time suggest that the major changes in the mortality of individuals aged 1–44 were occurring primarily in the late 1920s and early 1930s and were completed by 1935, although the exact timing of the transition within this time period was not as well demarcated for children aged 1–14 as it was for young adults aged 15–44. As Figure 9.4 suggests, mortality rates in individuals aged 45–64 showed slight decreases in the early years of the century, but changes were fairly gradual and occurred most rapidly after 1975. Although Figure 9.4 indicates gradually declining mortality for the oldest individuals throughout the century, none of these differences were statistically significant from one decade to the next. Finally, because of the significant and unusual fluctuations in the mortality of infants, it was not possible to pinpoint whether and when their patterns of mortality changed most dramatically; statistical results merely indicated either that the 1920s were different (using traditional decades) or that there were no discernible trends (using 10-year periods beginning in years ending in five). It is not yet clear why the 1920s stand out for infants, but it seems to be an unusual and a negative event (in terms of mortality) rather than something systematically associated with community-level changes.

The analyses comparing different age-groups within time periods indicate that ages 1 to 44 have similar rates and patterns of burial throughout the time period under study. In general, all other age-groups follow distinct patterns, although, as is also indicated by Figure 9.4, in time periods other than when the infant peaks occurred, infants and individuals aged 45–64 had similar rates of burial; however, it is likely that the reasons for these similar rates are different.

Using data from the Columbia Cemetery alone, Sattenspiel and Stoops (2010:16) concluded that "major changes in the mortality experience of the people of Columbia began in the early to mid-1920s and affected infants first, followed closely by changes in the mortality of children aged 1–14 and young adults aged 15–44." The additional infant burials present in the combined sample cloud the picture for this age-class, but in general, new results suggest that improvements began somewhat later than originally posited—in the mid- to late 1920s, rather than in the early 1920s. The analyses confirmed the earlier result

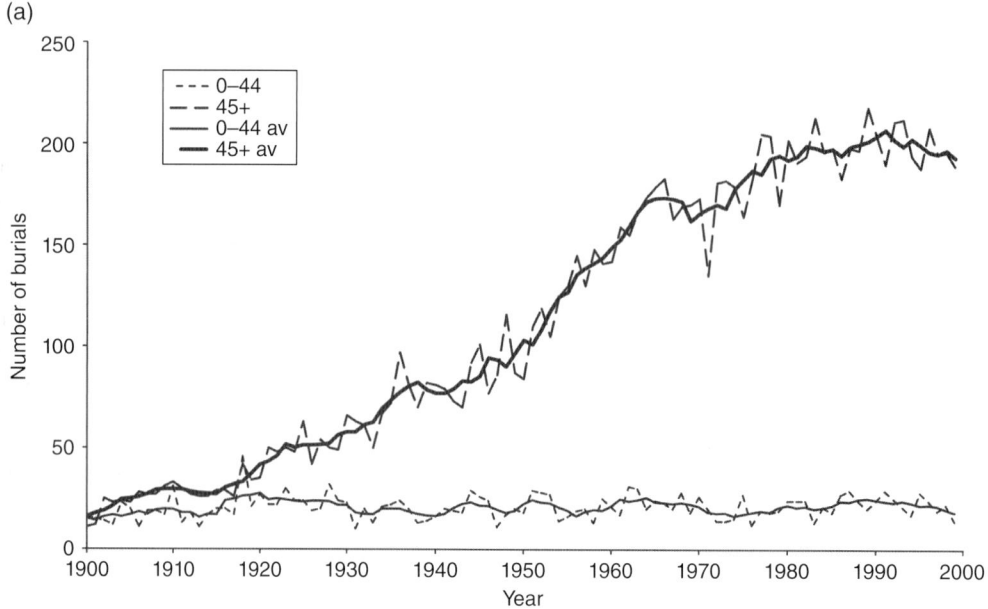

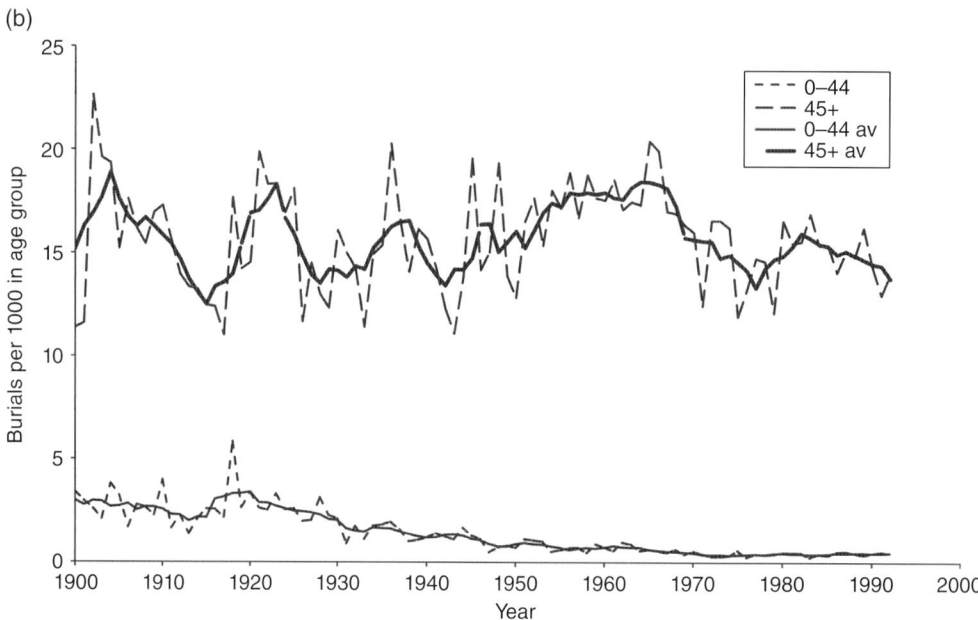

Figure 9.5. Burials across time comparing individuals under age 45 at time of death with those over age 45. (a) Raw burial numbers, (b) burial numbers standardized by population and age. *Solid lines* show 5-year moving averages.

that children aged 1–14 and young adults aged 15–44 formed a single uniform group that could not easily be distinguished and that the mortality of all individuals under age 45 deviated markedly from that of individuals over age 45 beginning in the 1920s. The mortality of adults over age 45 experienced a slight decline throughout the 20th century, but the trend was slow and individual time periods could not, in general, be distinguished from one another.

These analyses reinforce the primary conclusion of Sattenspiel and Stoops (2010) that data from Columbia's cemeteries provide a strong signal of events that affected the mortality of children and young adults, but not older adults, during the first few decades of the 20th century. Given the wealth of material detailing the impact of changes associated with the second transition on the mortality of people in other parts of the United States and the nearness in time of the changes in Columbia to the timing of the second transition elsewhere, a first hypothesis for these changes in Columbia is that they are evidence of Columbia's second transition and that they are caused by factors similar to those observed elsewhere. A full discussion of historical data pertinent to this question is provided in Sattenspiel and Stoops. Using newspaper reports and Havig's (1984) and Crighton's (1987) histories of the city, Sattenspiel and Stoops showed that there was a major push in Columbia in the early to mid-1920s to improve water sanitation and cleanliness of the milk supply, including the construction of a sewage plant that opened in 1923 and multiple cow testing ordinances. Thus, major health initiatives that have been linked to the second transition elsewhere can also be tied to Columbia, and, consistent with the evidence from analysis of the cemetery burials, these changes were implemented slightly later in Columbia than in the large cities.

Correlating such changes with the timing of mortality declines does not prove that the cleaner water and milk made available after these changes caused the mortality declines, however. One way to show the existence of a causal connection is to look at the specific causes of death before, during, and after the time period of interest. Remembering that the mortality declines occurred primarily in children and young adults at this time, analyses directed at their causes of death could provide more evidence for the impact of these community improvements. The Missouri State Archives online database was searched to find the death certificates for all individuals buried in the two Columbia cemeteries during the time of the apparent second transition. Unfortunately, time constraints limited this activity to the time period from 1925 to 1935. A total of 709 individuals were included from the Columbia Cemetery, of which 542 had accessible death certificates. There were 174 burials in the Memorial Park Cemetery, with death certificates identified for 142 of these. If community infrastructure and public health improvements were an important factor leading to reductions in the mortality of individuals under age 45, then it would be expected that deaths from diseases related to contaminated water and milk (e.g., dysentery, typhoid, cholera, tuberculosis, diphtheria) would show evidence of a decline as the improvements were put in place.

Infectious diseases accounted for the deaths of 169 of the individuals with identifiable death certificates, but only six of these deaths were due to water- or milk-borne diseases. The majority of deaths from infectious disease were a consequence of influenza, pneumonia, and/or tuberculosis. Thus, the data are insufficient to directly test the impact of the community improvements, although, as discussed in Sattenspiel and Stoops (2010), statewide data do show a clear decline in mortality from these diseases (Missouri State Board of Health, 1911–1942).

Discussion

As the analyses here and in other studies have shown, determining when, how, and why populations made their way through the second transition is a daunting task. For large cities and larger population aggregations, such as counties or states, the quantity of age-specific data on mortality is usually sufficient to identify whether and when declines in mortality may have occurred during the late 19th or early 20th century. For smaller communities such as Columbia, for which mortality rates and population figures may not be available at the resolution needed (e.g., by age and sex), even the task of pinpointing the existence and timing of these mortality changes may be difficult or impossible. As this study has shown, it may be possible to observe the changes in data derived from cemetery markers, which at least provide age- and sex-specific numbers of deaths across time; however, as discussed further in the following paragraphs, there are several constraints on the selection of data that must be kept in mind.

Even when the existence and timing of mortality declines can be shown in a data set, identifying the factors that led to those declines is a complex problem. Many factors have been proposed, including not only community-level improvements in sanitation, water treatment, and safety of the milk supply (e.g., Ellms, 1928; Meeker, 1972; Higgs, 1979; Okun, 1996; Troesken, 1999; Cutler and Miller, 2005) but also a transition to motorized transport and consequent reduction in animal waste in public places (Coleman and Salt, 1992), a reduction in insect-borne diseases as a result of vector control methods (Riley, 1986; Coleman and Salt, 1992), and improvements in the educational levels and the health and well-being of women (Coleman and Salt, 1992) as well as the availability of domestic child care (Santow, 1999). All of these factors have been correlated with mortality declines in a variety of locations, but correlation does not mean they play a causal role. What is needed to prove cause is information on exactly which causes of death were responsible for the observed declines as well as a clear functional relationship between the changing causes of death and proposed factor(s). Such a relationship has been shown for at least some large cities in the United States. For example, as mentioned earlier, using typhoid fever as a marker for waterborne disease, Cutler and Miller (2005) traced large reductions in overall and infant mortality in 10 US "death registration states" and several large cities to the construction of sewer systems and provision of cleaner drinking water. Similarly, Hauteniemi et al. (1999) and Beemer et al. (2005) linked the late 19th- and early 20th-century declines in mortality from gastrointestinal diseases to improving water supplies in several Massachusetts towns.

Proving a causal relationship between hypothesized causes and mortality declines is not easy, however, especially for small towns and cities like Columbia. Even in the large populations studied by Cutler and Miller (2005), mortality rates from typhoid fever at the turn of the 20th century were less than 50 per 100,000 people, and by 1920, they declined to under 5 per 100,000 people. In a city the size of Columbia at the turn of the century ($N = 5651$), at the largest of these rates, only about three deaths from typhoid fever per year would be expected, making it much harder to detect the presence of significant declines over time. In addition, challenges associated with verifying cause of death for all individuals because of incompleteness in the death certificate database makes the task of showing a connection between particular causes and observed trends even more difficult.

One solution to this is to proceed as Cutler and Miller (2005) did—by basing the analyses on a large enough sample size to facilitate identification of a causal link. However, if the goal is to assess the role of community improvements in *small* towns and cities, this can only be achieved through multiple studies of the type presented here. Although it may not be possible to link community improvements in Columbia, Missouri directly to changes in specific causes of death that would be tied to those improvements, if similar correlations between mortality changes and infrastructure improvements can be observed for other small communities outside of the large cities, the basis for assuming a causal link becomes stronger. Without such a set of "repetitions" of the experiment, the results from this study should only be viewed as correlative and not construed as evidence for a causal connection, however logical and attractive that may seem.

One of the major issues that must be dealt with in conducting studies such as the one described here is the degree to which the sample being analyzed is representative of the population from which it was derived. The large size of both cemeteries included in this study, especially relative to the underlying population size of Columbia, ensured that their burials represented a cross section of Columbia's population, and the sample sizes derived from the cemeteries also guaranteed that a cross section was included in the data analyzed. Nonetheless, even when coverage is relatively complete, cultural practices and temporal changes can lead to unanticipated biases in a sample. For example, when the Memorial Park sample was added to the database, it became clear that conclusions from the previous study using only the Columbia Cemetery data did not accurately reflect the real patterns of mortality in the oldest individuals, at least after the 1940s, when burials in the Memorial Cemetery surpassed those in the Columbia Cemetery. Because the annual number of deaths of younger individuals was small by this time, the bias is not observed in their burials.

It is also important to remember that even though a community may have one or two major cemeteries, in many parts of the country there are numerous small private and church-based cemeteries in which people may be buried. This is definitely the case for Columbia, and it is possible that, for this reason, some sectors of the community are not well represented in the existing database. However, most of these outlying cemeteries are likely to draw their inhabitants from locations near the cemeteries and thus outside the city limits. They are also likely to be relatively small and to contain a similar distribution of burials by age and sex as in Columbia's cemeteries. Although it is possible that some biases result from omitting these cemeteries, the size of the database used in this study and the likelihood that systematic sex and age biases are absent from the outlying cemetery populations suggest that the existing database is sufficient to generalize about Columbia's population. It is important to remember these cemeteries, however, and if the questions to be addressed deal with differences among families, ethnic groups, or other social characteristics, it is essential to make sure that the database being used includes a representative sample of the groups of interest. Careful research on the patterns of use of the different cemeteries in a region can identify potential biases and allow one to account for them when drawing conclusions.

When dealing with demographic analyses of skeletal populations, these issues are potentially more serious. Results from the present study show clearly that the Columbia Cemetery sample was not representative of the underlying population during the second half of the 20th century, but it required the sampling of both cemeteries to illustrate how significantly the biased sample affected the conclusions about the underlying patterns of mortality. Skeletal biologists do not usually have the time and resources to sample multiple cemeteries to the degree that is possible when recording information off of cemetery markers, and constraints on exhumation and analysis of human remains further complicate the situation. Results from the present study suggest research on the demographic history of skeletal populations should proceed with even more caution than is observed in current practice, because conclusions derived from skeletal biologists' limited samples may differ significantly from actual mortality patterns in unanticipated ways.

The availability of supporting information also makes studying the impact of changes associated with the second transition in small cities and towns challenging. Despite repeated efforts, adequate census data for the City of Columbia and even Boone County have not been located for the first four decades of the 20th century. For most decades, data were available for some age-groupings, but the groupings were not consistent from census to census. Thus, a number of estimations had to be made, and these estimations usually had to be based to a significant degree on statewide age distributions, which may have differed from those present in Columbia. Although the limited data on Columbia's or Boone

County's age distributions were used in the estimations whenever possible, it is likely that these procedures resulted in estimated age distributions that were not entirely representative of the actual age distributions.

It has also not been possible to locate local-level cause of death data other than for the subsample of individuals for whom death certificates can be located. Many of the death certificates for infants and young children are not available from the early decades of the 20th century, because, although Missouri state law required that deaths be registered beginning in 1911, the implementation of this law varied from place to place. In addition, registration of births was not required in the state until 1927 (Linder and Grove, 1947). This makes it difficult to assess whether there were changes in cause of death for the youngest as a consequence of the second transition. As described earlier, death certificates are available for about three-quarters of the individuals buried in the two Columbia cemeteries, but the small size of Columbia's population means that the total number of deaths among individuals below the age of 45 was small, making it difficult to assess whether rates declined significantly. In addition, the analysis of the death records may have missed detecting a potential decline because data collection efforts did not extend far enough before the proposed declines began, and so a good baseline with which to estimate pretransition death rates for Columbia is not yet available.

Conclusions

In spite of the data issues and caveats associated with this study, the analyses presented here help to fill in the overall picture of how and when the second epidemiologic transition occurred across communities within the USA. Results suggest that these events were not uniformly experienced at the same time and in the same fashion in all parts of the USA, although it appears that, as in the large cities, sanitary improvements and other health-related initiatives are likely to have been significant influences on changing mortality patterns in Columbia and other small towns and cities. The research that has been done with large data sets from major USA cities sheds significant light on the impact of the second transition in the USA, but without including a sample of smaller communities such as Columbia in the body of research on this important event, the resulting picture is incomplete. The addition of smaller communities like Columbia allows a more nuanced and complex understanding of exactly how and when changes in mortality related to the second transition occurred throughout the USA.

References

Beemer JK, Anderton DL, Leonard SH. 2005. Sewers in the city: a case study of individual-level mortality and public health initiatives in Northampton, Massachusetts, at the turn of the century. J Hist Med Allied Sci 60:42–72.

Coleman D, Salt J. 1992. The British population: Patterns, trends, and processes. Oxford: Oxford University Press.

Crighton JC. 1987. A history of Columbia and Boone County. Columbia: Computer Color-Graphics.

Cutler DM, Miller G. 2005. The role of public health improvements in health advances: the twentieth-century United States. Demography 42:1–22.

Ellms JW. 1928. Water purification. 2nd ed. New York: McGraw-Hill.

Gage TB. 1993. The decline of mortality in England and Wales 1861 to 1964: decomposition by cause of death and component of mortality. Popul Stud 47:47–66.

Gage TB. 1994. Population variation in cause of death: level, gender, and period effects. Demography 31:271–296.
Gage TB. 2005. Are modern environments really bad for us? Revisiting the demographic and epidemiologic transitions. Yearb Phys Anthropol 48:96–117.
Hauteniemi SI, Swedlund AC, Anderton DL. 1999. Mill town mortality. Soc Sci Hist 23:1–39.
Havig AR. 1984. From Southern village to Midwestern city: Columbia, an illustrated history. Woodland Hills: Windsor Publications.
Higgs R. 1973. Mortality in rural America, 1870–1920: estimates and conjectures. Explor Econ Hist 10:177–195.
Higgs R. 1979. Cycles and trends of mortality in 18 large American cities, 1871–1900. Explor Econ Hist 16:381–408.
Hinkle DE, Wiersma W, Jurs SG. 1994. Applied statistics for the behavioral sciences, 3rd ed. Boston: Houghton Mifflin.
Linder FE, Grove RD. 1947. Vital statistics rates in the United States 1900–1940. Washington, DC: US Government Printing Office.
Meeker E. 1972. The improving health of the United States, 1850–1915. Explor Econ Hist 9:353–373.
Melosi MV. 2000. The sanitary city: Urban infrastructure in America from colonial times to the present. Baltimore: The Johns Hopkins University Press.
Missouri State Board of Health. 1911–1933, 1935–1942. Annual Report. Jefferson City: MO State Board of Health.
Okun DA. 1996. From cholera to cancer to cryptosporidiosis. J Environ Eng 6:453–458.
Omran AR. 1971. The epidemiologic transition. Milbank Mem Fund Q 49:509–538.
Omran AR. 1983. The epidemiologic transition theory: a preliminary update. J Trop Pediatr 29:305–316.
Preston SH. 1976. Mortality patterns in national populations. New York: Academic Press.
Preston SH, Haines MR. 1991. Fatal years: Child mortality in late nineteenth-century America. Princeton: Princeton University Press.
Riley JC. 1986. Insects and the European mortality decline. Am Hist Rev 91:833–858.
Santow G. 1999. The mortality, epidemiologic and health transitions: their relevance for the study of health and mortality. In: Health and mortality issues of global concerns: Proceedings of the symposium on health and mortality, Brussels, 19–22 November 1997. New York: United Nations. p 39–53.
Sattenspiel L, Stoops M. 2010. Gleaning signals about the past from cemetery data. Am J Phys Anthropol 42:7–21.
Schofield R, Reher D. 1991. The decline of mortality in Europe. In: Schofield R, Reher D, Bideau D, editors. The decline of mortality in Europe. Oxford: Oxford University Press. p 1–17.
Troesken W. 1999. Typhoid rates and the public acquisition of private waterworks, 1880–1920. J Econ Hist 59:927–948.
U.S. Census Bureau. 1990. 1990 Population and housing unit counts: United States. Washington, DC: GPO.

Chapter 10

Industrialization and the Changing Mortality Environment in an English Community during the Industrial Revolution

Peter M. Kitson
*Cambridge Group for the History of Population and Social Structure,
University of Cambridge, Cambridge, UK*

Introduction

The notion of a *second epidemiologic transition*, taking place at the same time as the Industrial Revolution, has been widely used as a means of conceptualizing the changes in mortality and morbidity that took place in Britain and in other high-income developed nations during the late 18th and 19th centuries. However, the methodologies that have been used to study changing patterns of mortality in preindustrial societies have only rarely been applied to communities that were at the vanguard of industrialization. Here, infant and child mortality rates derived from a family reconstitution study for an industrializing community in the West Midlands of England are discussed within the context of an epidemiologic transition. The results presented here suggest that the concept of a straightforward transition from one epidemiologic regime to another is difficult to sustain, as infant and childhood mortality patterns here parted from national trends and worsened during the early 1800s.

Background

The Second Epidemiologic Transition in an English Context

Scholars have often located transitions between different phases in the economic and social development of England to between 1700 and 1850. At its core lies the characterization of English society at the start of this period as one where most of the population resided in settlements of less than 5000 inhabitants and derived their living either directly from agriculture or indirectly through the manufacture and retail of goods produced from raw materials derived from plant and animal husbandry (Wrigley, 1988, 2010). However, this period witnessed the emergence of societies based around other organizational paradigms. Rapid population growth combined with increasing urbanization and the growth of employment in large-scale capital- and energy-intensive forms of production led to the emergence of industrial society. Britain is generally seen as having been at the forefront of these changes, which were geographically concentrated in particular regional economies within Britain. *Modernization* is often used to characterize this transition, the Industrial Revolution, and this concept remains entrenched in scholarly approaches to understanding economic, social, cultural, and demographic change during this period. The notion of escape from Malthusian restraints to growth is one example; the idea that the period witnessed an epidemiologic transition is another.

Omran's (1971) so-called *classic* formulation of an epidemiologic transition identified multiple phases. The first is the *Age of Pestilence and Famine*, characterized by high and fluctuating mortality rates and a low average life span. The second, the *Age of Receding Pandemics*, features declining mortality due to progressively less frequent epidemics and increased life expectancy. In the third, the *Age of Degenerative and Man-Made Diseases*, anthropogenic, degenerative diseases replace acute, epidemic infectious conditions as the major source of mortality and morbidity. Omran suggested that the reduction in crisis mortality was especially beneficial to children and reproductive-age women. The probability of children surviving childhood increased as a consequence of improvements in living standards, nutrition, and the development of improved sanitation; the greatest improvement in survival was for those aged between 1 and 4. Meanwhile, females experienced lower levels of mortality compared to men for the whole duration of their lives, rather than it being restricted to the years over the age of 50, as before. Declining mortality further led to rapid population growth during this period. While some aspects of Omran's scheme have been challenged, the decline in infectious disease coincident with the Industrial Revolution still holds currency. As two additional epidemiologic transitions have now been identified (Barrett et al., 1998), Omran's classic transition is now known as the second epidemiologic transition.

Three causal factors underpinned Omran's second phase. The first, termed *ecobiological*, referred to the complex relationship between disease vectors, environmental factors, and host resistance. He noted that the changing virulence of certain types of disease might not be related to medical advances and that other aspects of vector–host interactions could explain increasing survival rates and declining morbidity. He did, however, note that these relationships were highly complex and still poorly understood. The implication is that the changing epidemiologic environment could be driven by factors independent of human agency, be they social, economic, or technological. Omran's second set of factors was labeled *socioeconomic* and included a set of explanations related to economic and social change. These explanations are oriented towards medical and biological factors but biased towards those that could not be attributed to medical care and advances. The benefits of economic development express as an increased standard of living and, more specifically, improved nutrition and hygiene. His third group of factors is related to a series of specifically medical and public health developments, such as improvements in sanitation technology, vaccination, and the development of germ theory. He suggested that this third set was introduced at a relatively late stage—typically the late 19th century—within the Western societies that first underwent this transition. The overall tenor of Omran's theory emphasizes factors such as the public health measures, improvements in hygiene and nutrition, and exogenous change in the virulence of disease as the key drivers of the improving mortality environment in 18th- and 19th-century Western societies.

Omran's essay has had a considerable impact, comparable to that of Malthus's *Essay on the principle of population*. While it had focused attention upon the mortality aspect of the demographic transition, some have suggested that it was insufficiently epidemiologic in that it emphasized changing causes of death rather than evolving patterns of disease and infection. It has also been criticized for downplaying the role of developments in medical science in improving mortality (Caldwell, 2001). Notwithstanding these conceptual criticisms, Omran's theory faces other problems in its application to the English epidemiologic regime, since it only has had limited success in providing a framework for understanding the falling volatility in death rates in England between the 16th and 19th centuries. Three specific areas of the theory are problematic when applied to England during this period. These are: the attribution of rapid population growth to mortality change; the role of public health and improvements in nutritional status in improving mortality; and the extent to which mortality change can be treated as an exogenous variable, working independently from changes endogenous to human society.

The pioneering work undertaken at the Cambridge Group for the History of Population and Social Structure from the mid-1960s onwards resulted in the reconstruction of the population history of England from the 16th through to the 19th centuries to a degree that had not hitherto been achieved for any other European country. These efforts culminated in the publication of two volumes that stressed the role of fertility as the *prime mover* of the English population from the 16th to the mid-19th centuries. In particular, changing patterns and levels of nuptiality—the propensity to marry—played an important role in determining population dynamics (Wrigley and Schofield, 1981; Wrigley et al., 1997). This was particularly the case for the century or so after 1730, when population growth was at its most rapid; during this time, falling ages at marriage—especially for females—resulted in higher aggregate fertility and more rapid population growth. Subsequent analysis indicated that marital fertility also increased, though it has been suggested that this is due to a decline in the stillbirth rate. Wrigley suggested that the decline in the endogenous component of the infant mortality rate, as well as declining maternal mortality, indicates an overall improvement in maternal health, which may have contributed to the declining stillbirth rate (Wrigley, 1998). Maternal health may have been ameliorated through improvements in either nutritional status or the epidemiologic environment, though Woods has argued that the magnitude of the change in marital fertility is too great to be accounted for by the stillbirth rate alone (Woods, 2005). Though some scholars continue to argue that improvements in mortality drove change during the 18th century (Razzell, 1994; Hatcher, 2003), the balance of scholarly opinion favors the role of fertility as the key driver of population change. However, not all would subscribe to the relationship between real wages and entry into marriage that, according to Wrigley and Schofield, drove the level of fertility within the population. Other factors, such as the weakening of *traditional* modes of behavior during industrialization as a consequence of the growth of that proportion of the population that was dependent upon waged labor, are often cited as alternative explanations for rapid population growth during the 18th century.

The second problem with Omran's theory lies in the timing of mortality change. Overall, improvements in life expectancy at birth in England were comparatively limited between the 16th century and 1870; indeed, it did not exceed 45 until 1870. However, this bald statement obscures some important changes, which are in turn the consequence of relying upon data from national units and not paying attention to the shifting pattern of mortality risk over the life course. For instance, England appears to have become free from the operation of the classic Malthusian positive check (caused by unsupportable demands placed upon agricultural output by rapid population growth) by the mid-17th century (Appleby, 1978; Galloway, 1998; Kelly and Ó Grada, 2009). In addition, crisis mortality caused by disease became a less prominent feature of the epidemiologic regime with the disappearance of plague from London by the final quarter of the 17th century, for example. However, this did not translate into significant improvements in overall mortality; if anything, life expectancy at birth fell over the course of the 17th century, falling from 42.7 to 35.49 between 1581 and 1721 (Wrigley et al., 1997: 614).

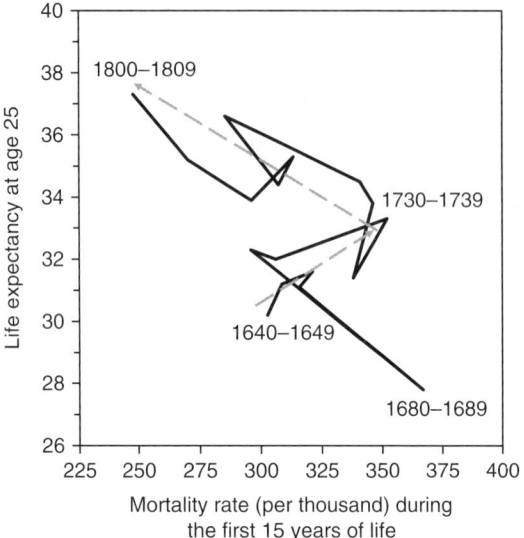

Figure 10.1. Comparing trends in infant and childhood mortality and life expectancy at age 25 in England, 1640–1809. Adapted from Wrigley et al., 1997, Table 6.10, p. 250–251, and Table 6.19, p. 290.

Figure 10.1 further highlights another peculiarity of mortality patterns during this period. This graph traces how estimates of national rates of total infant and childhood mortality ($_{15}q_0$) and life expectancy at age 25 (e_{25}) derived from the family reconstitution studies of 26 English communities changed between 1640 and 1837. On one hand, the probability of surviving to age 15 declined during the 17th and early 18th centuries. However, there was also a steady overall improvement for life expectancy at age 25 throughout the period, though neither of these processes can be described as uniform. Even this summary measure of infant and child mortality obscures important changes. Conventionally, infant mortality is decomposed into two elements. The endogenous component includes the first month of life and is the consequence of deaths caused by prematurity, the trauma associated with birth, or fatal congenital defects. Exogenous infant mortality, which occurs during the remaining first year of life, includes deaths caused by diseases contracted after birth. This distinction is arbitrary, but has proved to be an extremely useful tool in exploring the subtle changes that took place to infant survival chances in past societies (see Perry, this volume). In particular, it allows for some comparison of the severity of the disease environment faced by infants through the exogenous, as well as being a very general indicator of the health status of the mother during pregnancy through the endogenous component. Endogenous infant mortality remained stable during the 17th century but then declined significantly over the 18th century. The exogenous component, however, worsened during the 17th century but then remained at a similar rate throughout the 18th and early 19th centuries. The most significant worsening of mortality within this part of the life span took place between the ages of 1 and 5, again suggesting that the epidemiologic environment worsened for children during the 17th and 18th centuries.

Reliance upon national aggregates also obscures important regional divergences from the national trend. London's mortality regime experienced some dramatic shifts over the 17th and 18th centuries. Evidence from both the bills of mortality and family reconstitution studies of the London Quakers suggests that infant mortality peaked in the second quarter of the 18th century, with rates sometimes exceeding 350 per 1000 live births (Landers,

1993a), in contrast to communities within rural England, where rates could fall below 100 per 1000 (Wrigley et al., 1997). Throughout the early modern period, London acted as a classic *urban sink* or graveyard, with a regular and significant excess of deaths over births, manifested by both higher mortality rates and the endemicization of certain types of disease, such as smallpox. The city's growth could only be sustained by significant levels of in-migration. However, this peculiar epidemiologic regime began to moderate itself to the extent that the metropolitan area was manifesting a surplus of births over deaths by the early 19th century, well before the maturation of the public health movement.

The situation in London by the early 19th century stands in marked contrast to that of the rest of England, where the process of change is very poorly understood. Dobson (1988, 1997) has highlighted areas in southeastern England, where the combination of physical geography—low-lying, poorly drained estuarine and fenland environments—and warm summers could have led to periodic outbreaks of insect-borne infectious disease, especially malaria. Shifts in the distribution of population to less salubrious environments in response to new economic opportunities could also have led to increased mortality and the wider distribution of pathogens through commercial and transportation networks. This is especially so for industrializing communities. The environmental problems caused by overcrowding and unsanitary conditions associated with rapid population growth and urbanization do seem to have produced a very poor epidemiologic environment during the late 18th and early 19th centuries. Overall life expectancies in Manchester, Liverpool, and Glasgow were very low in comparison with the rest of Britain by the mid-19th century (Szreter and Mooney, 1998). The failure to provide important public goods such as sanitation and decent housing conditions created the preconditions for comparatively high levels of infant and childhood mortality, compounded by the regular occurrence of infectious diseases such as cholera. Treating Britain or England as a homogenous entity overlooks the divergent patterns and shifts in mortality patterns, and hence the epidemiologic regime, while the examination of mortality trends for different age groups demonstrates the manner in which different age groups of the population experienced different shifts in their survival chances during the 18th and 19th centuries.

The third critical response to Omran's theory arises from his tendency to treat mortality as an exogenous factor. The fact that mortality patterns varied over space and time in a context characterized by considerable economic and social change leads to the suggestion that shifts in the virulence of disease might also be related to these changes. This is especially the case where the growth of the urban segment of the population was being driven by the rapid expansion of a single metropolitan center—in this case London—which in turn was associated with its increasing integration with both national and international trading networks and significant migrant flows into the capital. The idea that understanding the changing epidemiologic regime may best be framed within an approach that places the spatial structuring of society at its heart has been most rigorously explored by Landers (1993a). London's increasingly prominent role in global trade, its reliance upon internal trade to supply its economy with food and fuel, and sustained population growth fueled by immigration resulted in the creation of high exposure potential to infection and the development of endemic disease pools, perhaps most notably with smallpox. The product was an initial worsening of mortality, but in the longer term, a rising share of the population acquired immunity through prior exposure. Through this mechanism, the virulence of these diseases fell, while short-term instability in urban death rates fell in comparison to those within the countryside, as these diseases became endemic rather than prone to periodic episodes of higher mortality caused by epidemics.

Another dimension to this third critical response has emerged from other scholars who have suggested that certain institutions may have played an important role in the moderation of the epidemiologic regime. The English system of poor relief, typified by payment of relief for the poor in rural as well as urban areas, ensured that high food prices were not accompanied by short-term migration to local towns. A comparison to France, where high food prices were usually accompanied by rural to urban migration and more pronounced mortality within urban centers, is instructive in this regard. The readiness to relieve the poor in their own homes, rather than insisting upon institutional forms of relief such as hospitals and workhouses, also had the potential to render beneficial consequences for mortality and the disease environment (Smith, 2011). However, the role of other political institutions may have also played a significant role in worsening the mortality regime. In particular, the failure by local and national government to act in ways to address the problems caused by urbanization and rapid economic development has been attributed to the narrowness of the electoral franchise. Before the advent of universal suffrage, the wealthier elements of society who possessed the right to vote were reluctant to accept higher taxation to pay for improvements in public health that would disproportionately benefit other segments of society (Landers, 1993b; Szreter, 1988).

It is therefore apparent that the changing epidemiologic environment in 18th- and early 19th-century England is best approached through understanding its evolving spatial structuring, most notably through the flows of goods and people within the economy and society. There are, however, some very significant gaps in our knowledge. We have a good understanding of the mortality regime for London and the rural communities and market towns of England. The onset of civil registration from 1837 onwards ensures that we have a fairly secure empirical basis for understanding changes to the epidemiologic environment during the second half of the 18th century (Woods, 2000). However, we know little about how patterns of mortality shifted in those parts of the country that experienced the most rapid population growth and socioeconomic change before 1837. For instance, the evolution of mortality patterns within the cotton and woollen textile-producing towns of Lancashire and Yorkshire, the mining regions of northern England and South Wales, and the iron-producing areas of the West Midlands between 1700 and the mid-19th century is largely unknown; they had clearly worsened significantly by the early 19th century, but little is known of the pathways by which this took place. It is also now believed that many of these locations had already experienced significant levels of economic development prior to 1750 (Shaw-Taylor and Wrigley, 2008).

The research presented here attempts to go some way towards addressing this imbalance. This chapter will explore how the mortality regime changed in a sizable industrial community in the English West Midlands between the 16th and early 19th centuries. This in turn forms part of a wider project, the purpose of which is to explore the impact upon demographic behavior of a whole bundle of social and economic processes that can be linked under the convenient shorthand of *industrialization* or *modernization*. The key question here is to explore both the nature and timing of changes in mortality patterns in relation to rapid population growth and the changing socioeconomic structures within a particular community, as well as its evolving spatial relationships with other population centers, and thus to infer changes in the epidemiologic regime. The principal reason that it has proved so difficult to study change in such locations relates to both the methodologies available for studying demographic behavior in premodern societies and the peculiar issues with the source material for studying this type of community. The next section introduces the location selected for this case study approach and describes the sources and methods selected for its study.

Materials and Methods

The study of the epidemiologic regime within preindustrial European societies can be approached in many ways, with a variety of different methodologies. The approach adopted here is to infer epidemiologic change by studying changing patterns of mortality, in terms of the estimation of age-specific mortality rates and the study of the volatility of burial registration. This is due to the poor availability of cause-of-death data before the 19th century, especially before the advent of reliable nosological classification (see Beemer, this volume). It should be noted that morbidity data for historical populations is almost entirely absent. The study of mortality rates depends upon the exploitation of registers of the ecclesiastical rituals of baptism, marriage, and burial that were maintained by the legally established state Church from the 16th century onwards. Since WWII, these registers have been widely used for studying the demographic regime in Europe before the advent of compulsory civil registration of births marriages and deaths and have contributed greatly to our understanding of the contours of population change before 1850.

The data recorded by these registers have been used in two distinctive ways. Aggregative analysis relies upon using frequency counts of events for each month and year to explore long-run trends in the numbers of baptisms and burials and hence general population trends. It has also been used to explore the relationships between short-run fluctuations in both burials and prices. The development of inverse projection techniques from the 1970s onwards have also allowed these data to be used as inputs into a system of demographic accounts and thus to generate population-level estimates of fertility and mortality (Lee, 1974; Oeppen, 1993). While the collection of frequency counts needed to perform aggregative analyses is comparatively rapid, it is not usually possible to calculate vital rates, since the total population *at risk* cannot be specified. In addition, inverse projection techniques tend to be applied to larger populations since many of its underlying assumptions work best when large numbers are used.

A second approach involves using ecclesiastical registers to reconstruct family histories using nominative record linkage techniques. The most common framework used in these cases is family reconstitution, a method that has been utilized for over 50 years for the study of demographic behavior in preindustrial societies (Wrigley et al., 1997). The demographer Louis Henry devised the technique during the 1940s and 1950s as a means of studying *natural fertility*, which refers to the levels of fertility found in populations that did not use artificial means of birth control within marriage. While it is fair to say that this technique is primarily oriented towards the study of fertility within marriage, it has been expanded and developed to calculate a whole range of fertility, nuptiality, and mortality measures and statistics.

The technique functions by reconstructing family life histories through linking a particular marriage to the baptism of children resulting from that union, and to the subsequent burials of the parents and children. This is done using the names of the individuals recorded in the ecclesiastical register. Parents can also be linked to their own baptism, allowing their age at marriage and at burial to be calculated by using the date of baptism as a proxy for date of birth. A series of logical rules are then applied to ensure that reconstituted families are plausible. For instance, baptism records cannot be made to women aged over 50, due to the biological near impossibility of this. Once the data has been *cleaned* in this way, the product is a sample of the local population that can be subjected to a series of demographic calculations. The adoption of observation rules to determine which families are *at risk* of bearing children or dying allows the calculation of demographic measures such as age-specific marital fertility rates or infant and child mortality rates. After some modification to deal with the vagaries of English parish registers, it has been used extensively to reconstruct demographic behavior in pre- and early industrial England.

Table 10.1 provides an example of how the process of family reconstitution functions by looking at the family of Henry and Jane Instone, who were married in October 1819.

Table 10.1. An example of a reconstructed family history produced by family reconstitution.

Date	Event type	Event description
October 10, 1819	Marriage	Henry INSTON and Jane MARSH were married
March 05, 1820	Baptism	Henry, son of Henry INSTANT, laborer of Gornal Wood, and Jane his wife, was baptized
December 09, 1821	Baptism	Ann, daughter of Henry INSTANCE, laborer of Gornal Wood, and Jane, his wife, was baptized
March 07, 1824	Baptism	Joyce, daughter of Henry INSTONE, laborer of Graveyard, and Jane, his wife, was baptized
January 23, 1825	Baptism	Jane, daughter of Henry INSTANT, laborer of Graveyard, and Jane, his wife, was baptized
January 23, 1825	Burial	Jane, daughter of Henry INSTANT, laborer of Graveyard, and Jane, his wife, was buried
January 01, 1826	Baptism	Samuel, son of Henry INSTANT, laborer of Graveyard, and Jane, his wife, was baptized
August 10, 1828	Baptism	John, son of Henry INSTAN, laborer of Pensnett, and Jane his wife, was baptized

Source: Sedgley Family Reconstitution Database.

This union resulted in six children, baptized between March 1820 and August 1828. From this, we can infer that the bride was probably pregnant at marriage, though the interval of six months does not preclude the possibility of a premature birth. One of these children—Jane—was buried the same day that she was baptized, on January 23, 1825. Since the family have children born both before and after this event, it is fair to assume that they remain *in observation* for the entire time and thus can be added to both the numerator and denominator for the calculation of infant mortality statistics from the date of the first birth up to and inclusive of the date of the last birth in the reconstructed family history. If the parents can be linked to their own baptisms, then both their age at marriage and subsequent age at death can be calculated, and they too can contribute to the calculation of adult mortality rates and age-specific marital fertility rates.

However, the methodology of family reconstitution has been critiqued. Many of these critiques revolve around the extent to which the *reconstitutable minority*—the subset of the local population for whom demographic statistics may be calculated—are representative of the whole population. For instance, families with a lower propensity to migrate will figure more prominently within the reconstitutable minority. If migration or persistence within a community is related to socioeconomic status, such as landholding, then the *stayers* may not be representative of the whole community. Since English village communities were entwined within strong but highly localized migration fields, the degree of population turnover was quite considerable, ensuring that the size of the reconstitutable minority is often small. There is also the possibility that migration censoring can influence the estimation of mortality and age at marriage. Simply put, if migration is a competing risk to marriage or death, those that marry later or live longer have a cumulatively greater risk of migration by virtue of their longevity, but will be absent from the reconstitutable minority, artificially reducing life expectancy and age at marriage (Ruggles, 1992). In practice, this does not seem to have been a major problem, at least with age-at-marriage estimates, since migration and marriage were closely associated events within the life cycle and not competing risks (Wrigley, 1994).

However, perhaps the strongest criticism of family reconstitution emerges from the practical constraints that have prevented its application to urban and industrializing communities. Very few reconstitution studies have been performed using data from highly urbanized or rapidly growing communities, due to their size as well as the large migration rates often found within these contexts. These issues were particularly pressing when reconstitutions were performed manually, with pen and paper. However, even with modern computing power, the problems of identifying links with large numbers of records and the frustratingly inconsistent spelling of surnames found in premodern registers renders family reconstitution a process that is both labor and time intensive. Communities with populations between 500 and 5000 are usually the subjects of reconstitution studies; any smaller, and the volume of reliable demographic data that can be reconstructed tends to be too small to sustain rigorous analysis; any larger, and the process becomes very time-consuming. This has led to the exclusion of large towns and other urban settings, though some recent studies have applied family reconstitution to the very large London parishes of St James Clerkenwell and St Botolph Aldgate (Newton, 2011a, 2011b).

An additional restraint is that family reconstitution relies heavily upon the quality of the raw material. The quality of the burial registration is the main practical limitation upon whether a community can be made the subject of a family reconstitution study. If a burial entry routinely records only the name of the deceased, there is insufficient information to discriminate between the different individuals to whom that record might refer to. However, if the names of the relatives of the deceased are recorded, the possible set of individuals to whom the burial can be attributed shrinks. Additionally, periods of underregistration caused by clerical negligence can prevent the applicability of reconstitution.

One final point is that religious nonconformity can also have a serious impact upon the quality of parochial registration. This problem is mitigated somewhat if individuals permanently abscond from the ritual observances of the state church, since they cannot then form any part of the *reconstitutable minority*. However, the movement of individuals between different denominations can become a source of error in the calculation of vital rates. This is because the observation rules function on the basis of events recorded within the pages of the parish register; the recording of individuals elsewhere would censor some of these events without altering the observation rules, leading to the underestimation of infant and child mortality rates, for example. Nonconformity with the Church of England was relatively common from the late 17th century onwards; however, by the early 19th century, it was increasingly associated with urbanized and industrializing communities as the administrative geography of the established church failed to keep up with the shifting distribution of population in the English landscape.

Therefore, the practical restraints of size, quality of registration, and extent of nonconformity have limited the application of family reconstitution to communities that are relatively small, rural, and largely agricultural. These locations are hardly at the forefront of the modernization paradigm. It also greatly explains why our knowledge of demographic behavior in industrializing regions of England between 1780 and 1837 is relatively patchy. The research presented here will seek to redress this balance by exploring changing patterns of mortality in one industrializing community that experienced very rapid population growth during the 18th century.

This study focuses on the large industrializing parish of Sedgley, in what would later become known as the *Black Country* of the West Midlands due to the continuous pall of smoke emitted from the region's countless kilns and furnaces. Located 13 miles northwest of Birmingham, it was sandwiched between Dudley and Wolverhampton, two important centers of industrial development in their own right. As with many parishes in the West Midlands, Sedgley was an amalgam of several hamlets and villages, of which there were nine principal settlements: Sedgley, Cotwall End, Gospel End, Upper Gornal, Lower Gornal, Coseley, Ettingshall, Woodsetton, and Brierley. There was considerable variation between these different villages; Sedgley, Cotwall End, Gospel End, and Woodsetton were relatively small communities that exhibited a much greater focus upon agriculture, though in truth farming was pursued throughout the parish. These villages were situated within the parts of the parish that lay directly over the coal measures found within this part of England, and Coseley was situated upon the outcrop of what was called *the ten-yard seam*, a reference to its width. The ready availability of coal ensured an extensive supply of cheap energy, and the region possesses an extensive history of industrial activity. It was well known for the manufacture of iron implements and hardware and Sedgley in particular for the production of nails and screws. By the early 19th century, several ironworks and large coal-mining concerns had also established themselves within the parish, especially within the townships of Coseley, Ettingshall, Upper Gornal, and Lower Gornal.

Sedgley is endowed with a particularly good parish register, with excellent burial registration and few apparent periods of underregistration. Moreover, the compilers of the register routinely recorded the occupation of the men included within the register, either when they were buried or when a member of their family was being interred. These numbers are very heavily weighted towards the end of the period, rather than the start; over 25% of all baptisms in the database take place between 1821 and 1837. In response to such rapid population growth, a series of daughter chapels to the mother church of Sedgley All Saints were founded, at Lower Gornal (1823), Coseley (1830), and Ettingshall (1835), all of which kept their own registers. The parish of Sedgley was also home to many different denominations of nonconformists; registers survive for several groups of Protestant dissenters from the late 18th century onwards. All of the events from these registers have been transcribed and added to the computer database, which contains a very substantial number of events: 38,202 baptisms, 9331 marriages, and 26,652 burials between November 1559 and December 1837.

It is difficult to track the course of population growth for individual English communities before the advent of the first national census in 1801, at which point the total population of the parish was just under 10,000 people. However, some earlier estimates of population drawn up by the ecclesiastical hierarchy can give some sense of its size in the late 17th and mid-16th centuries. In 1563, 126 households were recorded for the parish, equating to a total population of around 600 (Dyer and Palliser, 2005). This had roughly trebled by 1676, when another ecclesiastical census recorded 1200 people aged over the age of 15, giving an estimated population of 1750 (Whiteman, 1986). However, it is clear that at some point during the 18th century, a very rapid period of growth began, as the population almost quintupled between 1676 and 1801, with 9874 recorded in the first census of 1801; this then trebled to the population of 29,447 enumerated by the 1851 census (Wrigley, 2011).

The reasons for this extended period of rapid population growth can be found within the changing local economy. The parish registers record occupations for the fathers of children baptized in the parish for many extended periods before 1813, when the registration of such details became compulsory. These data can be used to generate estimates of the proportions of men employed in each sector of economic activity. The key trends were an early increase in the proportion of the population employed in manufacturing nails; 37% of men belonged to this sector between 1588 and 1595, rising to 53% between 1695 and 1702. However, this was followed by a fall to a mere 20% of men by 1813 to 1820. In its stead, the coal-mining sector grew steadily, employing around 4% of men at the end of the 16th century to 31% between 1813 and 1820. Additionally, other parts of the manufacturing sector, such as iron making, grew considerably, from 15% at the end of the 16th century to 43% by the second decade of the 19th century.

It is therefore possible to discern two distinctive phases to Sedgley's economic development. The first, running from the mid-16th century through to the end of the third quarter of the 18th century, was one of steady population growth based around increased demand for the manufactured goods produced in small-scale workshops and forges, as well as the growing demand for the raw materials of coal and iron to fuel these enterprises. The pattern of industrial development was also distinctive, since it is misleading to suppose that men described in the 16th century as *nailers* were engaged in the manufacture of nails to the exclusion of all other forms of economic activity. Evidence from inventories of the possessions of individuals drawn up at their death during the 17th century demonstrates that nailers were often extensively engaged in animal and crop husbandry, while those given agricultural occupational titles such as *yeoman* or *husbandman* often had significant evidence of involvement in industrial activities such as nail making or agricultural implement manufacture (Roper, 1960; Rowlands, 1975). This was part of a broader pattern of by-employment to be found throughout the Black Country, as individuals took advantage of slack periods of the agricultural calendar to engage in manufacturing while retaining some security through retaining access to the means of food production, for consumption or for sale (Rowlands, 1975).

The second phase of economic development, by way of contrast, was a sudden and dramatic break from this long-established pattern. The development of new technologies permitting the production of iron using coal through the coking process resulted in a massive expansion of the numbers of men employed in iron making and to a lesser extent in coal mining. This also led to the development of large industrial concerns in both these sectors. For instance, a rating of all the property in the parish for the purposes of local taxation compiled in 1826 listed "Park Fields Furnaces, Casting Houses, Building Offices &c. &c. &c." to be rateable at an annual value of £205, whereas a typical dwelling was worth only around £1 per annum (Dudley Archives and Local History Service PR/SED/II/3/1, p. 133). These new capital-intensive forms of production, such as coal mining and iron manufacture, resulted in an increased reliance upon wage labor. The older systems of workshop production using family labor continued into the 19th century and even beyond (Bodey, 2008), but the beginnings of mechanization in nail manufacture in the early 18th century, combined with the dynamic new opportunities in the rapidly growing sectors, resulted in a concerted process of proletarianization and a move away from more *traditional* small-scale modes of production.

This second phase of growth was characterized by rapid population growth, the relative decline of the long-established industrial sector, and the emergence of industrial sectors based around a new technological paradigm. It was associated with urbanization, and the manifold associated social and economic problems. A quote from one of the reporters for the second report of the parliamentary Children's Employment Commission in 1843 conveys a sense of these disamenities:

> Lower Gornal is approached by a lane…the same steep lane descends winding, to the extent of nearly another mile. The invariable sludge occasioned by the rains and water from several springs above, is rendered bestial by the casting forth, both from doors and windows, of everything which would in ordinary cases be deposited on a dung-heap or dust-hole, or carried away by drainage. Low hovels, hutches, and workshops, resembling little black dens, thickly line the lane or main-way, to the extent of perhaps three-quarters of a mile, and in some places are so crowded as to have two or three houses packed close together, with scarcely room to pass between, and sometimes rendering it difficult to open a door…Compared with this, some of the worst streets of Wolverhampton would really appear civilized, if not respectable.

The parish of Sedgley underwent a series of profound and convulsive changes during the late 18th and early 19th centuries. Rapid population growth, radical technological change, and significant growth in religious nonconformity make this community a microcosm of changes to the demographic and epidemiologic environment in England during the 18th and 19th centuries. It is this type of rapidly changing community that is least understood, precisely because the underlying changes challenge the conventional constraints under which historical demography operates. The absence of vital statistics collected by the state requires reliance upon the exploitation of ecclesiastical registration to explore changing mortality patterns. However, rapid population growth challenges the quality and completeness of ecclesiastical registration, due to the strain placed upon existing parochial networks as well as the growth of religious nonconformity. As the following discussion shows, an exploration of mortality change within this community can enrich our understanding of the changing epidemiologic environment and can also shed light upon the notion that a transition took place within this context during this time frame.

Results and Discussion

Changing Patterns of Mortality in Sedgley, c.1580–1837

The first port of call when analyzing the changing mortality patterns in a community is to explore broad trends in the annual frequencies of burials. A simple frequency count reveals that by 1600, there were around 20 burials a year recorded in the parish register of Sedgley. This increased slowly over the course of the 17th century to reach an average of 75 per year by 1700. There are a couple of mortality *spikes*—years with a higher-than-average number of burials—but on the whole, this community seems to have been free from severe instances of crisis mortality before the 18th century.

This slow growth in the annual numbers of burials continues until the 1720s, when around 80 burials per year were registered. However, there is a prolonged period of 5 years where the average number of burials reaches 140 per annum; 1729, with 274 events, represents the worst year in this series. Annual frequencies of burials resume their steady annual growth trend, reaching 100 events per year by the 1750s, despite a peak in 1741. By the first decade of the 19th century, 200 events per year were recorded, reaching 400 by the 1830s. One final peak in the number of burials remained: in 1832, 614 burials were recorded in the parish due to an outbreak of cholera during August and September.

This summary demonstrates the chronology of population growth within this community, as well as how the population remained prone to periods of heightened mortality into the 1830s. However, it is also possible to explore trends in the volatility of burial registration after making certain assumptions. Between the late 16th century and 1824, the burial registers for this community record the parents of buried children with a great degree of consistency, allowing for infant and child burials to be distinguished from that of adults. Most burial registers in England cease to record this information after the introduction of *pro forma* registers by the 1812 Parochial Registers Act; however, the Sedgley registers continue to record age at burial and the parents of the deceased. This can be used to identify the age range during which the parents of children are identified in the burial register. A very clear cutoff point emerges for the age of 20; only 1 of 12 buried individuals recorded as being aged 19 does not record the parent, while 14 out of 30 burials do not record the parent when the deceased is aged 20. There is a clear transition up to age 23–25, after which recording of parents becomes very rare. The clear suggestion is that only those under the age of 20 have the identity of their parents recorded at their burial, and so the absence of parental information can be used as an indirect indication of the age of the deceased.

This insight can be used to explore the volatility in annual burial series. Figure 10.2 shows how standard deviations in the annual numbers of burials change over time. The annual

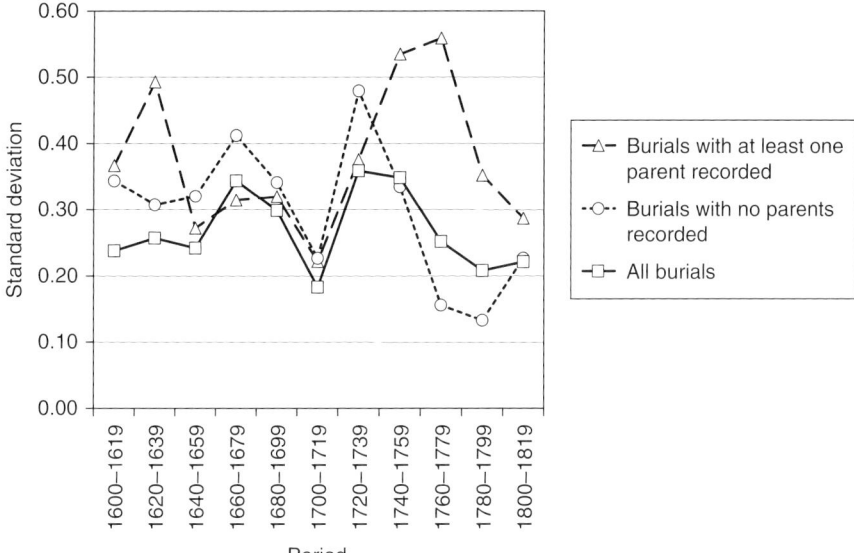

Figure 10.2. Comparing the annual volatility for burials with at least one parent recorded with burials where no parents were recorded in Sedgley, 1600–1819. Source: Sedgley family reconstitution database. Notes: Standard deviations of the detrended fluctuations in burials are reported for each 20-year period. The annual numbers of burials were themselves detrended by dividing the number of burials for each year by a 9-year mean of annual burials around that point and then taking the logarithm of the resulting value.

frequencies have been detrended by taking the logarithm of the annual number of burials divided by a 25-year moving average centered upon that year, and standard deviations of the resulting values have been calculated for each 20-year group from 1600 to 1819. The data itself has been split into all burials, burials with no parents recorded, and burials with parents recorded. This gives a rough indication of how the annual volatility in infant and child burials varied in comparison with adult burials over this period.

After some noticeable peaks during the 17th century, the figure shows that the first 20 years of the 18th century are noticeable for the comparative lack of volatility and the degree of convergence between the different series. However, the years between 1720 and 1739 witnessed a significant increase in volatility for infant and child burials, as well as adult burials. The increase for the latter is short-lived, however, as burial volatility for infants and children remains high for most of the remainder of the 18th century, before starting to fall after 1780. The standard deviation for adult burials falls to comparatively low levels at the same time that it is high for infants and children. This divergence suggests that there was some form of transition in the epidemiologic environment during the early 18th century, leading to increased burial volatility for all age groups, but that this was then increasingly confined to infant and child mortality by the second half of the 18th century. As these changes precede the rapid population growth and economic change experienced by Sedgley after 1780, the suggestion must be that its mortality patterns changed as a result of other changes to the economy and society of England.

It should be noted that the extended period of high mortality manifested by Sedgley during the late 1720s was common to many parts of midland and northern England (Gooder, 1972). In contrast, comparatively few communities in southern England appear to have experienced localized crisis mortality during these years. The variability in the onset of periods of crisis mortality during this time is marked, with some commencing in August and September 1727 and others experiencing the upward surge in mortality in 1728 and 1729 and even into the opening months of 1730. This prolonged period of high mortality also included a deficient harvest year in 1728. All of this suggests that there was not one single cause of high mortality during this period; rather, a series of waves of infectious disease, the precise identification of which is problematic in the absence of reliable cause-of-death data, is probably the underlying cause. It is possible that susceptibility to infectious disease may have been exacerbated by high food prices and a worsening of the nutritional status of those most exposed to changes in the cost of living (Wrigley and Schofield, 1981).

Analysis of burial volatility in this way, while useful, can only indicate change over time. The calculation of mortality rates provides a much more secure basis for the study of change over time, and this is where family reconstitution comes into its own. It is a particularly powerful tool for the study of infant and child mortality, since the observation rules ensure that only children born to families for whom it can be shown that they remain in residence in the parish contribute to the calculation of these rates. They allow the distribution of mortality by age to be traced over time and space, since they allow different communities to be compared. The results for Sedgley that follow are provisional, in that the family reconstitution study upon which they are based has not yet been completed. In particular, it is not yet possible to report adult mortality statistics, or maternal mortality, as not all of the linkage operations have been completed. However, sufficient reconstruction of family life histories has now progressed for infant and child mortality rates to be calculated from the mid-17th century to the early 19th century, and these findings are reported in the following table.

Table 10.2 reports overall infant and childhood mortality rates up to the age of 15 ($_{15}q_0$), as well as a series of other estimates for other time divisions during this part of the life span. The infant mortality rate—measuring mortality in the first year of life—is represented by $_1q_0$, while $_4q_1$ is the mortality rate between the ages of 1 and 4. Two other measures, $_5q_5$ and $_5q_{10}$, measure the mortality in the 5 years of life between 5 and 10 years, respectively. These rates are reported as the number of deaths per 1000 head of population. Two other measures are reported in this table: the endogenous and exogenous infant mortality rates. These two rates may be calculated by plotting the cumulative total of deaths over the first year of life on a graph where the horizontal axis has been subjected to the logarithmic transform $\log^3(d+1)$, where d is the number of days elapsed since birth. The points representing the cumulative number of deaths usually lie close to a straight line, which can then be interpolated backwards to the vertical axis. The point of interception is taken to represent the endogenous mortality rate. The exogenous rate is simply the value gained by subtracting the endogenous mortality rate from the total infant mortality rate (Bourgeois-Pichat, 1951; Wrigley et al., 1997). For the purposes of comparison, the infant mortality statistics for Dudley registration district between the years 1839 and 1845 inclusive are included at the bottom of the table; this was the registration district into which Sedgley was incorporated after 1837.

This table casts considerable light upon the changing climate of infant and childhood mortality from 1650 to approximately 1850. Two very striking trends in infant mortality are noticeable. First, there is a significant reduction in the endogenous mortality rate from the late 17th century to the early 19th century. The rate halves, which is broadly in line with the estimates of national trends in endogenous infant mortality calculated from 26 family reconstitutions. Here, the rate fell from 78 per 1000 between 1580 and 1599 to 33 per 1000 between 1825 and 1837 (Wrigley et al., 1997: 236). Conversely, however, exogenous infant mortality notably worsened. The rate for the period from 1650 to 1699 is comparatively low

Table 10.2. Infant and child mortality rates for the parish of Sedgley, 1650–1837.

Period	Infant and child mortality						
	Endogenous infant mortality (1000d_x)	Exogenous infant mortality (1000d_x)	$_1q_0$	$_4q_1$	$_5q_5$	$_5q_{10}$	$_{15}q_0$
1650–1699	81	59	141	77	31	16	244
1700–1749	78	70	148	111	40	22	289
1750–1799	58	54	112	109	36	27	258
1800–1837	34	131	165	124	34	21	309
1839–1845[a]	42	123	165	—	—	—	—

Source: Sedgley Family Reconstitution Database; Annual reports of the Registrar-General.
[a] Dudley registration district.

in comparison to other English communities, but not implausibly so; Sedgley's location would have spared it some of the environmental problems associated with towns and villages situated in low-lying areas in close proximity to waterways, while its comparatively low population density would have ensured the absence of higher rates of infant mortality associated with urbanization. However, by the 19th century, the exogenous infant mortality rate had more than doubled, to 131 per 1000. These figures are also broadly comparable to those from 1839 to 1845 for the whole of the registration district in which Sedgley was located, suggesting that the results are broadly reliable.

Huck (1994, 1997) has also calculated infant mortality rates for the parish of Sedgley, using the ages from the burial register and a correction for the underregistration of infant burials. His estimates are broadly in line with the ones presented here, even if his are slightly higher. It should be noted however that his figures do not seem to include the baptism or burial data from the nonconformist congregations that are included within this database. Additionally, baptism and burial registrations from the Anglican *daughter* chapelries of the *mother* church of Sedgley—Coseley, Lower Gornal, and Ettingshall—do not seem to have been used for his estimates. These omissions go a long way in accounting for the rates reported by Huck and those presented here.

The figures for childhood mortality are also instructive and indicate a worsening over the period. Mortality between the ages of 1 and 5 worsened from 77 per 1000 to 124 per 1000 between the second half of the 17th century and the first third of the 19th century. It is harder to discern trends for the age groups between 5 and 10 or 10 and 15; mortality in these groups appears to remain relatively constant over time. Accordingly, early child mortality appears to increase between 1650 and 1699 and from 1700 to 1749, without an accompanying increase in infant mortality. However, the trend of worsening early childhood mortality continued into the 19th century, especially during the early years of the century.

Figure 10.3 confirms the general sense of worsening mortality during the early 19th century. It shows annualized infant mortality rates for the parish of Sedgley between 1813 and 1837 and the rates calculated from the figures returned by the *Annual Reports of the Registrar-General* between 1839 and 1845. At the start of this period, the infant mortality rate was typically around 115–125 per 1000, with the occasional year of exceptionally poor early life chances, such as 1815 when the rate surged to 201 per 1000. Notwithstanding these fluctuations, there is a continual tendency for mortality to increase through the second and third decades of the 19th century, until settling down to between 160 and 200 per 1000. There is perhaps a slight decline through to the start of civil registration in 1837, but there is a very strong sense in which the figures from the reconstitution map onto the infant mortality rates

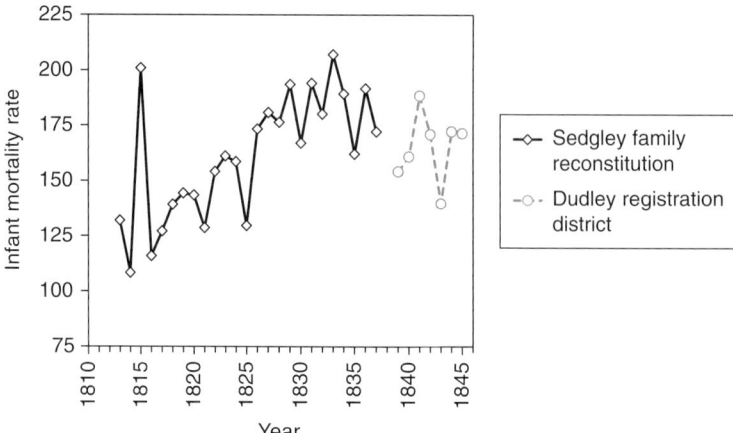

Figure 10.3. Annual variations in the infant mortality rate for the parish of Sedgley (1813–1836) and Dudley registration district (1838–1844). Source: Sedgley family reconstitution database; Annual reports of the Registrar-General.

calculated from the early years of civil registration. Again, the evidence suggests that infant mortality increased during the early years of the 19th century. This was of course a time of rapid population growth, but it is interesting that it was not in lock step with worsening infant mortality. It is, for example, noteworthy that population growth seems to have been most rapid in Sedgley from the final quarter of the 18th century and especially so from 1785 onwards. There was a lag of around 25 years before rapid population growth began to have a serious impact upon infant life chances. This raises the interesting possibility that population density had to pass some critical threshold for the disease environment to worsen.

It is also useful to compare Sedgley's infant mortality rates with that of other communities with a similar trajectory of rapid population growth and urbanization during the late 18th and early 19th centuries. Comparisons can be drawn with the parishes of Birstall, Gedling, and Methley between the late 17th and early 19th centuries and the data for the registration districts in which they were allocated after 1837. The Yorkshire textile-producing parish of Birstall possessed an infant mortality rate of 128 for the years 1675 to 1749. However, registration districts that contained Birstall possessed a combined rate of 170 per 1000 between 1838 and 1844. Methley, a parish in the same county possessing a sizable mining sector, had an infant mortality rate of 134 between 1675 and 1749; the registration district containing this parish had a rate of 164 per 1000 by 1838 to 1844. Finally, Gedling in Nottinghamshire, a stronghold of the framework knitting industry, exhibited a rate of 112 in 1675 to 1649; its registration district had an infant mortality rate of 159 per 1000 by 1838 to 1844. The values for $15q0$ in all three parishes for 1675 to 1649 and their subsequent registration districts in 1838 to 1844 were as follows: Birstall, 256 and 333; Methley, 258 and 309; and Gedling, 225 and 302 (Wrigley et al., 1997). These rates are very similar to the 309 per 1000 reported for Sedgley in Table 10.2 for the years of 1800 to 1837. This confirms that Sedgley's worsening in infant mortality had analogues elsewhere in contemporary England. Clearly, the processes of industrialization produced negative externalities in the epidemiologic environment that were keenly felt by involved communities.

The evidence presented here suggests that mortality change in Sedgley during the 2 centuries or so before 1837 was not a simple, linear process. Broader national trends, such as the decline in endogenous infant mortality and the worsening of exogenous infant mortality and early childhood mortality, occurred in Sedgley. Additionally, the added instability in the mortality environment during the 18th century suggests that the changing force of mortality is not necessarily related to environmental conditions within the community of study, but might also be linked to broader changes in English society. Further research is needed to explore the simultaneity of changes in different industrial communities and assess how they compared with nonindustrial parishes. However, it is clear that rapid growth and industrialization was not favorable to the life chances of infants and children, though the timing of these changes suggests that there was not a straightforward relationship between population growth and worsening infant mortality.

Conclusion

The evidence presented here makes the notion of a straightforward *transition* in the epidemiologic environment during the 18th and 19th centuries problematic. Infants and children constituted a vulnerable group that took the brunt of worsening, not improving, disease and mortality conditions, and the fluctuations that this group experienced over time are worthy of comment. Mortality worsened between the late 17th and the early 18th centuries, before experiencing a further worsening during the early 19th century. In this respect, Sedgley seems to have been following national trends to one point and diverging from them during the early 19th century. The story that unfolds is one of complex shifts in the force and patterns of mortality during this period, rather than a simple linear shift between one regime and another. The role of socioeconomic change, in both leading to the greater spatial incorporation of communities and the externalities resulting from rapid industrialization and population growth, is clearly important but whose impact could vary according to circumstance.

Another important feature of the results presented here is the manner in which the worsening of infant mortality in Sedgley did not march in lockstep with the overall trends in population growth in the parish. Evidence from baptism registrations in the parish register suggests that rapid population growth began in the 1780s, yet on the evidence of Figure 10.2, it is only after 1815 that infant mortality seems to markedly worsen. The broad similarity of worsening mortality in infancy and childhood in other English parishes such as Gedling, Birstall, and Methley suggests that the path that Sedgley followed was by no means unique. It may well be that the spatial integration of the economy around the rapidly growing industrial sectors and the rapid expansion of population in the new industrial towns (fueled in part by migration) combined with poorly articulated local government structures created a cluster of circumstances that created the preconditions for a worsening epidemiologic environment in several parts of England.

It would therefore appear that, in the same way that the application of Omran's model of the epidemiologic transition to England is problematic, its application to developments within this particular community during industrialization is difficult to square with empirical evidence. Changes in the mortality patterns suggest that the epidemiologic environment could vary considerably through time and space. They also suggest that communities at the forefront of the modernization paradigm were most prone to a worsening of overall conditions, rather than benefiting from the higher wages and more regular employment often found in these locations. These economic opportunities encourage the significant migration flows that fed into population growth. Moreover, the lopsided nature of population growth during the 18th and 19th centuries ensured that more and more of the population lived in insalubrious environments. In this sense, economic development does not seem to have been particularly beneficial to the epidemiologic environment from the point of view of infants and children.

References

Appleby AB. 1978. Famine in Tudor and Stuart England. Liverpool: Liverpool University Press.
Barrett R, Kuzawa CW, McDade T, Armelagos GJ. 1998. Emerging and re-emerging infectious diseases: the third epidemiologic transition. Annu Rev Anthropol 27:247–271.
Bodey H. 2008. Nailmaking. Oxford: Shire Publications.
Bourgeois-Pichat J. 1951. La mesure de la mortalité infantile. Population 6:233–248, 459–80.
Caldwell JC. 2001. Population health in transition. Bull World Health Organ 79:159–160.
Dobson MJ. 1988. The last hiccup of the old demographic regime: population stagnation and decline in late seventeenth and early eighteenth-century south-east England. Contin Chang 4:395–428.
Dobson MJ. 1997. Contours of death and disease in early modern England. Cambridge: Cambridge University Press.
Dyer AD, Palliser DM, editors. 2005. The diocesan population returns for 1563 and 1603. Oxford: Oxford University Press.
Galloway PR. 1998. Basic patterns in annual variations in fertility, nuptiality, mortality, and prices in pre-industrial Europe. Popul Stud 4:275–303.
Gooder A. 1972. The population crisis of 1727–1730 in Warwickshire. Midl Hist 1:1–22.
Hatcher J. 2003. Understanding the population history of England 1450–1750. Past Present 180:83–130.
Huck P. 1994. Infant mortality in nine industrial parishes in northern England, 1813–1836. Popul Stud 48:513–526.
Huck P. 1997. Shifts in the seasonality of infant deaths in nine English towns during the 19th century: a case for reduced breastfeeding? Explor Econ Hist 34:368–386.
Kelly M, Ó Gráda C. 2009. The old poor law: resource constraints and demographic regimes. University College Dublin Centre for Economic Research Working Paper Series. Available at: http://ssrn.com/abstract=1667843 (accessed November 5, 2013).
Landers J. 1993a. Death and the metropolis: Studies in the demographic history of London. Cambridge: Cambridge University Press.
Landers J. 1993b. From Colyton to Waterloo: Morality, politics and economics in historical demography. In Wilson A, editor. Rethinking social history: English society, 1570–1920 and its interpretation. Manchester: Manchester University Press. p 97–127.
Lee WR. 1974. Estimating series of vital rates and age structures from baptisms and burials: a new technique, with application to preindustrial England. Popul Stud 28:495–512.
Newton G. 2011a. Infant mortality variations, feeding practices and social status in London between 1550 and 1750. Soc Hist Med 24:244–259.
Newton G. 2011b. Family reconstitution in an urban context: some observations and methods, Online working paper. Available at: http://www.geog.cam.ac.uk/people/newton/UrbanFamilyReconstitution.pdf (accessed November 5, 2013).
Oeppen JE. 1993. Back projection and inverse projection: members of a wider class of constrained projection models. Popul Stud 47:245–267.
Omran AR. 1971. The epidemiologic transition: a theory of the epidemiology of population change. Milbank Mem Fund Q 49:501–538.
Razzell PE. 1994. Essays in English population history. London: Caliban.
Roper JS. 1960. Sedgley probate inventories 1614–1787. Woodsetton: Privately Printed.
Rowlands MB. 1975. Masters and men in the west midland metalware trades before the industrial revolution. Manchester: Manchester University Press.
Ruggles S. 1992. Marriage, migration and mortality: correcting sources of bias in English family reconstitutions. Popul Stud 46:507–522.
Shaw-Taylor L, Wrigley EA. 2008. The occupational structure of England c.1750–1871: A preliminary report, Online working paper. Available at: http://www.hpss.geog.cam.ac.uk/research/projects/occupations/introduction/summary.pdf (accessed November 5, 2013).
Smith RM. 2011. Social security as a developmental institution? The relative efficacy of poor relief provisions under the English Old Poor Law. In Bayly CA, Rao V, Szreter SRS, and Woolcock M, editors. History, historians and development policy. Manchester: Manchester University Press. p 75–102.

Szreter SRS. 1988. The importance of social intervention in Britain's mortality decline c.1850–1914: a re-interpretation of the role of Public Health. Soc Hist Med 1:1–38.

Szreter SRS, Mooney G. 1998. Urbanization, mortality, and the standard of living debate: new estimates of the expectation of life at birth in nineteenth-century British cities. Econ Hist Rev 51:84–112.

Whiteman A. 1986. The Compton census of 1676: A critical edition. London: Oxford University Press.

Woods R. 2000. The demography of Victorian England and Wales. Cambridge: Cambridge University Press.

Woods R. 2005. The measurement of historical trends in fetal mortality in England and Wales. Popul Stud 59:147–162.

Wrigley EA. 1988. Continuity, chance and change: The character of the industrial revolution in England. Cambridge: Cambridge University Press.

Wrigley EA. 1994. The effect of migration on the estimation of marriage age in family reconstruction studies. Popul Stud 48:81–97.

Wrigley EA. 1998. Explaining the rise in marital fertility in England in the 'long' eighteenth century. Econ Hist Rev 51:435–464.

Wrigley EA. 2010. Energy and the English industrial revolution. Cambridge: Cambridge University Press.

Wrigley EA. 2011. The early English censuses. Oxford: Oxford University Press.

Wrigley EA and Schofield RS. 1981. The population history of England 1541–1871: A reconstruction. Cambridge: Cambridge University Press.

Wrigley EA, Davies, RS, Oeppen JE, Schofield RS. 1997. English population history from family reconstitution 1580–1837. Cambridge: Cambridge University Press.

Part 4
Marginalized and Underrepresented Communities in the Second Epidemiologic Transition

Chapter 11
Short Women and Their Stagnating Growth: A Study of Biological Welfare and Inequality of Women in Postcolonial India

Aravinda Meera Guntupalli
Centre for Research on Ageing, University of Southampton, Southampton, UK

Introduction

Omran (1971, 1982) formulated the theory of epidemiologic transitions to describe changing patterns of health and disease preceding and following the industrial revolution in high-income developed western nations. His *classic* model is characterized by a decline in mortality, an increase in life expectancy, and a major shift in the causes of death from infectious to degenerative and noncommunicable diseases (NCDs). At different stages and at different intensities, factors such as improvements in nutrition, advances in medical technology, and developments in sanitation and hygiene contributed to these trends (Omran, 1971; Olshansky and Ault, 1986). When Omran framed the epidemiologic transition, he postulated three stages: the "age of pestilence and famine," the "age of receding pandemics," and the "age of degenerative and manmade diseases" [sic]. Moreover, he identified and discussed the interaction between changing patterns of health and disease with demographic, biological, economic, and sociological factors.

Omran (1971, 1982) also put forward a few points on the heterogeneity of the model. To illustrate the nonhomogeneous nature of the epidemiologic transition, he presented three distinct models to distinguish the period, time, and mode of transition: the *western* model followed by England, Wales, and Sweden; the *accelerated* model followed by Japan; and the *delayed* model followed by Chile and Ceylon, among other nations. Even though some degree of heterogeneity within the transition was addressed by classifying the transition into these models, they are restricted to a unilateral macroperspective. Unfortunately, such macro- and national-level perspectives ignore inequalities within countries and between different groups (e.g., social groups).

Several scholars have critiqued this standardized theory of epidemiologic transitions. For example, McKeown (2009: 19) stated that the theory as proposed by Omran (1971, 1982) "has been useful in providing an overarching perspective on changing demographic patterns" but that it oversimplifies patterns of mortality and morbidity due to different conditions. Moreover, Armelagos et al. (2005) present a variety of evidence to show that the scope and impacts—not just the timing and pace—of the epidemiologic transition has varied greatly within and between countries. This variation, they observe, reflects the influences of economic factors on the disease process. However, they raise the important point that epidemiologic transition theory is not a unilinear evolutionary model and state that the variation that some scholars see as problematic they view as an object of further study.

Global health patterns during the 19th to 21st centuries reflect more than a simple transition from an epidemiologic regime of greater mortality and morbidity from infectious diseases to one of degenerative and NCDs. The risk of having an infectious disease among vulnerable people, such as undernourished groups or those of low socioeconomic position (SEP), in low- and middle-income countries (LMICs) has not improved significantly during this time. Moreover, people born in resource-poor settings are exposed to childhood undernutrition, a lack of vaccination and child health care facilities, and fewer years of education.

Children are particularly affected by a persistently heavy burden of infectious disease and environmental hazards in LMICs. For example, in India between 2001 and 2003, 1.5 million children (0–4 years of age) died, constituting approximately 19% of total deaths in the country. Endogenous mortality from perinatal conditions caused 33% of deaths, followed by respiratory infections (22%); diarrheal disease (14%); and other infectious and parasitic diseases (10.5%); together, these account for nearly 80% of child mortality. These deaths do not represent children from the growing Indian middle class or the upper class but instead mostly represent children of low SEP in rural areas or urban slums (Office of the Registrar General, 2009). Furthermore, statistics reflect that of the children who survive, many experience morbidity from conditions such as anemia, as well as stunting. For instance, in 1998 and 1999, approximately 50% of Indian children under age 3 had experienced stunting. The situation has not improved with time either; nearly 45% of same-age children experienced stunting in 2005 and 2006 (International Institute for Population Sciences, 1995, 2000, 2007).

Cause of death data on adults complicates this picture. India is characterized by high middle-age (25–69 years) mortality. Most deaths arise from cardiovascular diseases (25%); chronic obstructive pulmonary disease (COPD), asthma, other respiratory diseases (10%); tuberculosis (10%); and malignant and other neoplasms (9%). Among males and females, cardiovascular disease is the leading cause of death, accounting for one in every four deaths and one in every five deaths, respectively (Office of the Registrar General, 2009). Among those 70 and older, the picture is similar: cardiovascular disease accounts for 26% of deaths, followed by senility (16%); COPD, asthma, other respiratory diseases (14%), and diarrheal diseases (9%). These figures suggest a population experiencing a regime of higher mortality from degenerative and NCDs. Together, they indicate that India is in the unfortunate position of having experienced few of the health benefits of economic growth and instead the full range of health costs that it and industrialization and urbanization can bring.

Reddy (1989) argued that India reached the second stage of the epidemiologic transition around 1950. Based on current mortality from cardiovascular disease and cancer, one can argue that some population segments in India are experiencing the third stage of the transition. However, people of low SEP are still struggling to meet the economic and dietary requirements, and mortality data suggests that many are still experiencing the second stage of the transition. Moreover, during their life, an individual could experience infectious disease and undernutrition in the early part of the life course and degenerative disease in the later part of the life course if he or she experiences the nutritional transition after reaching adolescence. The model of the nutrition transition (Popkin, 2002) maps on to the stages of the epidemiologic transition. It entails a first stage characterized by a high proportion of undernutrition, which is concurrent with the first stage of the epidemiologic transition; a stage of receding famines, concurrent with stage two; a phase focusing on famine alleviation and prevention, coinciding with a focus on the control of infectious disease; and a final stage

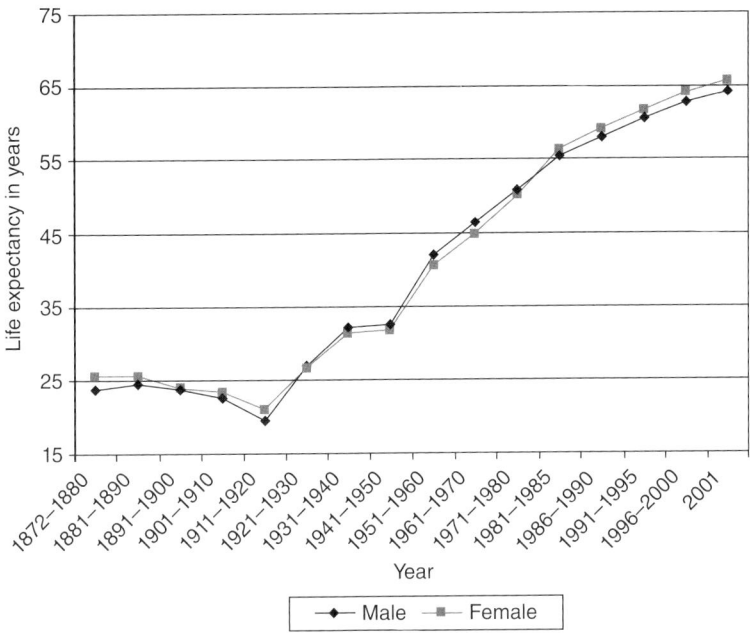

Figure 11.1. Male and female life expectancy in India between 1872 and 2001 (Data from Mayer (1999)).

of a predominance of nutrition-related NCDs, which is concurrent with stage three of the epidemiologic transition. This last stage is associated with what seems to be a convergence on the Western diet, which is high in saturated fat, sugar, and refined foods and low in fiber, among modern societies, including India. Many see this dietary pattern as associated with high levels of chronic and degenerative diseases and reduced disability-free time (Popkin, 2002).

Interesting patterns of longevity are also evident in India. Life expectancy increased steadily from 1950 to 2001 in India for both males and females (Figure 11.1.). However, contrary to the female advantage in life expectancy reported in the Western countries in the 20th century, male life expectancy in India was higher than female life expectancy in some periods. This reflects both discrimination against females and the existence of high (though declining) maternal mortality. Between 1993 and 1997, 480 women died during pregnancy and childbirth for every 100,000 live births; between 1998 and 2002, 390 women died, declining to 200 per 100,000 between 2008 and 2012. For comparison, 12 women died from these causes in the United Kingdom between 2008 and 2012 (World Bank, 2012). Women's mortality and longevity are key components of epidemiologic transition theory. Omran's (1971) original formulation of the theory of epidemiologic transitions consisted of five propositions about patterns of mortality and life expectancy. The third is on the mortality patterns of women and children. This states that "during the epidemiologic transition the most profound changes in health and disease patterns obtain among children and young women." This phenomenon is largely a result of declining infant and maternal mortality and a consequent drop in fertility. Intriguingly though, while maternal mortality is declining in India, the country marks yet another departure from this model: women experienced lower life expectancies than do men between 1940s and 1980s, and both women and children still suffer high mortality. In part to explore this dynamic, this study focuses on the health of females in India during the 20th century, specifically between 1950 and 1975, using anthropometric data, specifically adult stature. The study focuses on female health for three reasons. First, there is limited information on male stature after 1950 as national surveys, such as the National Family Health Surveys, focused exclusively on maternal and child health. Second, by focusing on female health, health inequality in vulnerable groups, such as women and girls, can be examined. Third, after identifying the factors that have influenced the welfare and life expectancy of Indian females, factors that have contributed to these health inequalities in the context of epidemiologic transitions can be discussed.

Other chapters in this volume primarily focus on the 18th to 20th centuries as developed countries experienced the epidemiologic transition concurrent with the industrial revolution. LMICs, however, experienced the second and third stages of the transition in the 20th and 21st centuries. Additionally, it is difficult to obtain historical datasets to study the first phase or the early second phase of the transition in LMICs; vital statistics data is nonexistent (see Gage, 2005). This study is not free of these issues. For instance, the study period was narrowed to post-1950 for three reasons. First, data on indicators such as GDP and poverty does not exist previous to 1950. Second, India experienced strong improvements in life expectancy after 1950 and entered the second phase of the transition in the early 1950s (Reddy, 1989). Finally, India was declared independent in 1947 and became a republic in 1950. Hence, the study period represents a phase of major changes in political conditions, economic growth, and public health. To circumvent issues with data availability and address issues of health and inequality, this study follows Komlos and Baten (1998) and supplements conventional indicators of welfare, such as GDP per capita, with biological welfare measures, namely individual-level adult female stature.

Aims of the Study

The overarching aim of the study is to investigate the impact of economic factors on stature, specifically assessing whether improvements in the economy reached different sections of the Indian population, and do so using female stature. Important questions hardly addressed by previous research—due to the lack of available data—are investigated here. How did biological welfare and inequality develop in India immediately after independence? Did population growth hinder biological welfare, particularly in the postcolonial period? Did the stature of Hindus, Muslims, Christians, Sikhs, and Jains develop differently? Did the increase in GDP in the nonagricultural sector improve the welfare of Indians?

Background

Political and Economic Change in the Postcolonial Period

How did economic indicators change in India during between 1950 and 1975? The Indian economy transformed from a nearly stagnant phase before 1950 to a slow, Hindu, phase of growth after independence. Progress, however, was interrupted by unfavorable environmental and political conditions, namely a border conflict with China in 1962, the Indo-Pakistan conflicts of 1965 and 1972, and severe droughts in 1965, 1966, 1972, and 1974 that produced a declining GDP in those years. Moreover, the oil price shock of 1973 had a drastic impact on economic development.

The extent of economic change during this period, and how it may have affected health, has generated considerable scholarly interest (Virmani, 2006). In part, this is because there are multiple possible outcomes. Health may have improved, as predicted by epidemiologic transition theory. However, it may not have. For instance, according to the Kuznets hypothesis, which holds that as a country develops, there is a natural cycle of economic inequality driven by market forces, which at first increases inequality and then decreases it after a certain average income is attained, an economic expansion could have triggered higher inequality in India and not necessarily have improved welfare (Guntupalli and Baten, 2006). Additionally, population growth might also have outweighed the impact of increasing GDP on welfare of Indians. Virmani (2006) has examined postcolonial economic performance in India, with a focus on fluctuations and breakpoints. Specifically, he examined whether breakpoints resulted from the drought-induced fall in GDP that occurred in 1965 and 1966, which was compounded by drought-related stagnant agricultural growth. He concluded that no structural break occurred in GDP growth from agriculture, once rainfall variations were accounted for. However, no studies have addressed the impacts of these events on biological welfare.

Economic Growth, Population Growth, and Welfare in Postcolonial India

India has long been faced with insalubrious conditions, tied in part to population size. In 1968, for example, India had 14% of the world's population on 2.4% of the world's total land. At the same time, however, longevity measures demonstrate that health conditions were improving, as life expectancy increased from 32 years in 1950 to 51 years in 1968. Meanwhile, although the availability of goods and services increased, per capita consumption did not increase due to contemporary rapid population growth (Chandrasekhar, 1968; Sivasubramonian, 2004). Sivasubramonian (2004) has shown that aggregated real GDP doubled within two decades starting from 1950, but based on GDP per capita estimates, it took nearly three decades for per capita GDP to double. This has been linked to high population growth and suggests that population growth slowed improvements in individual welfare. Nevertheless, GDP only reflects an aggregate picture. The aim here is to distinguish between growth in the agricultural and nonagricultural sectors to see which sector contributed most to an improvement in biological welfare. Ravallion and Datt (1995) used household surveys for the years 1958 to 1990 to measure the effect of agricultural growth on rural poverty and on the rural labor market. Even though the sample size was small—20 surveys—these showed that agricultural growth had no impact on the share of consumption of the poor. Eswaran and Kotwal (1994) have also argued that India's urban-focused industrialization process provided few benefits to the rural poor. This chapter explores the groups that benefited from improvement in the nonagricultural sector.

To incorporate the simultaneous economic expansion driven by growth in the nonagricultural sector and population growth, this study uses the net domestic product per capita (NDP) in the nonagricultural sector. Hence, the population explosion arguments of Chandrasekhar (1968) and Sivasubramonian (2004) can be evaluated.

It is also expected that Indian population growth could have a direct, negative impact on stature, as it has been argued that population growth can hamper GDP growth and decrease nutritional status (Chandrasekhar, 1968; Sivasubramonian, 2004). However, Rodgers (1989) argued that there is little empirical evidence to support the proposition that population growth is a major constraint on poverty alleviation. He further argued that population growth is both a cause and a consequence of poverty and that unraveling the pattern of causation may be an almost hopeless task. Furthermore, Ahlburg (1996) found no association between population growth and changes in poverty. Using stature as a proxy for welfare, for the first time, the impact of population growth rates on biological welfare in postcolonial India can be tested.

Political, Religious, and Ethnic Divisions in India

The political situation in India between the 1950s and 1970s was fairly stable. Wilkinson (2000) defined India in the 1950s and 1960s as a consociational state, a state that has major internal divisions along ethnic, religious, or linguistic lines, yet nonetheless manages to remain stable due to consultation among the elites of its major social groups. Except for the partition years of 1946–1947, India did not experience religious violence[1] during the democratic regime (1950s–1960s) of the first Prime Minister, Jawaharlal Nehru, and was generally stable in terms of ethnic and religious conflict.

The caste system

The Hindu caste system has been in use for several centuries and still plays an important role in contemporary Indian welfare, inequality, and the political system. It is an endogamous system wherein caste is fixed at birth and marriage occurs within a caste. There are four different major levels: upper and dominant castes, backward castes, scheduled castes, and scheduled tribes. Within each of these categories are the actual castes within which people are born and marry; membership is unchanging and lifelong. The system is hierarchical, with the upper and dominant castes generally having a higher SEP and more political power compared to lower castes especially, during the study period.

The majority of the Hindu population belongs to the scheduled castes, tribes, and other backward castes. Less than 20% of the population is upper caste (Varshney, 2000). Especially prior to independence, caste played a substantial role in deciding occupation, income, access to resources, and SEP. Following independence, some states tried to reduce disparities and conflicts between the castes through policy changes, particularly those targeting the welfare of lower-caste women and children. For instance, Kerala initiated major policies of land redistribution and compulsory education, especially after the communist party's first democratic victory in 1957.

The Hindu caste system and its impact on biological welfare and stature of Indians were discussed extensively by Guntupalli and Baten (2006). However, the present study further explores caste and religious inequality in postcolonial India in relation to health and welfare.

Methods and Materials

Though the study of welfare and inequality in India has attracted much attention during the last decades, numerous are limitations posed by the availability of datasets. Most studies of the postcolonial period focused on poverty and inequality have used macrolevel (state-level) aggregated income-based measures (Özler et al., 1996; Datt and Ravallion, 1998; Besley and Burgess, 2000;Ravallion, and Datt, 2002). However, these do not allow assessment of welfare in relation to region, caste, and religious group. Instead, here, stature data allows analysis of individual-level welfare and, through this, regional and group-level variation in welfare.

Research Design

This study revolves around four main hypotheses:

Hypothesis 1: Improvement in the nonagricultural state net domestic product increased stature.
Hypothesis 2: The Indian caste system created particularly high inequality, which, compounded by religious inequality, resulted in differential stature between caste and religious groups.

---

[1]Polity IV data did not show any change in democracy and autocracy in this period. The Polity index value was 9 in 1950–1975. Hence, the available political data could not be used for comparison purposes.

Hypothesis 3: The period of 1964–1966, which featured a declining GDP and agricultural shortfalls, also featured declining stature.

Hypothesis 4: Increasing population pressure and land inequality affected biological welfare in a negative way.

Anthropometric Data on Welfare and Inequality

Stature is a sensitive measure of environmental quality and reflects net nutrition during infancy, childhood, and adolescence. Adult stature is associated with the quality of food, disease burden and exposure to infections, and access to resources during growth and development (Komlos, 1996; Baten, 2000; see Koepke, this volume).

Two major sources were used here to study welfare and inequality. Anthropometric data were taken from the National Family Health Survey-2 (NFHS-2) (International Institute for Population Sciences, 2000). Macroeconomic data were derived from Besley and Burgess (2000). Both these datasets were merged to study the impact of macrolevel poverty, development expenditure, and population growth on individual-level biological welfare and inequality, using stature as a proxy for biological health and welfare.

Anthropometric Data

The NFHS-2 was conducted in 1998 and 1999 in all 20 states of India. It was a household survey with an overall target sample size of approximately 90,000 women, all of whom were married, divorced, or widowed at the time, and between 15 and 49 years of age. The NFHS-2 is a demographic and health survey conducted as part of the Demographic and Health Survey (DHS) program, which is funded by the United States Agency for International Development (USAID) and the United Nations Children's Fund (UNICEF). Data collectors in this study were given extensive, specialized training in collecting anthropometric data.

The reliability, representativeness, and robustness of the data have to be considered. The target sample was set considering the size of the state (Table 11.1). The distribution of age in NFHS-2 is relatively close to the age structure of the population (Table 11.2). However, the most important question is the impact of survivor bias on the sample. It is possible that poor and undernourished girls experienced higher mortality in early life compared to well-nourished girls that are taller. This bias could be reflected in disparities in stature between SEP or religious groups. Plotting the caste or religion-specific samples by birth cohort can be helpful to address this question. If mortality were selectively higher among certain caste and religious groups, a slight underrepresentation of those groups in some years of the study period would be expected. For this purpose, the share of a caste and religious group by birth cohort was plotted. This shows that the share of all castes and religious groups stayed more or less constant over time (Figures 11.2 and 11.3). If mortality had caused selectivity, we would have observed a lower share of scheduled castes and tribes with a lower SEP during the study period. Moreover, the percentage of the sample representing scheduled tribes and castes was close to that demonstrated by census data, though the percentage of scheduled castes is slightly higher (Varshney, 2000).

The sample used here comprises data on the stature of 60,562 women aged 24–49 belonging to all of the states in India (refer to Table 11.1 and 11.2). Women younger than 24 years were excluded, given the potential for continued growth. Due to insufficient observations (<30 observations per year), the birth-years 1948 and 1975 were excluded from the sample. Only 51,678 observations were used for the regression analysis after removing the smaller states such as Sikkim, Meghalaya, for which macroeconomic data are not available (refer to Table 11.1).

Table 11.1. Composition of the NFHS-2 sample by major states of India.

States	Frequency	Percent
Andhra Pradesh	5721	9.45
Assam	1500	2.48
Bihar	5808	9.59
Goa	91	0.15
Gujarat	3194	5.27
Haryana	1259	2.08
Himachal Pradesh	420	0.69
Jammu	550	0.91
Karnataka	3400	5.61
Kerala	2526	4.17
Madhya Pradesh	4647	7.67
Maharashtra	6133	10.13
Manipur	128	0.21
Meghalaya	93	0.15
Mizoram	47	0.08
Nagaland	84	0.14
Orissa	2327	3.84
Punjab	1524	2.52
Rajasthan	3060	5.05
Sikkim	27	0.05
Tamil Nadu	4682	7.73
West Bengal	5452	9.00
Uttar Pradesh	6801	11.23
New Delhi	823	1.36
Arunachal Pradesh	57	0.09
Tripura	207	0.34
Total	**60,562**	**100.00**

Table 11.2. Composition of the NFHS-2 sample by age.

Age	Frequency	Percent
23–24	5307	8.3
25–29	15,321	24.1
30–34	13,623	21.4
35–39	12,018	18.9
40–44	9811	15.4
45–49	7608	11.9
Total	**63,688**	**100.0**

The NFHS-2 also provides information about the place of childhood residence, which enables investigation of rural–urban differences in biological welfare. Other important indicators taken from this database included region, caste, and religious affiliation. Extreme statures (below 100 and above 220 cm) were removed, as these are likely due to either data

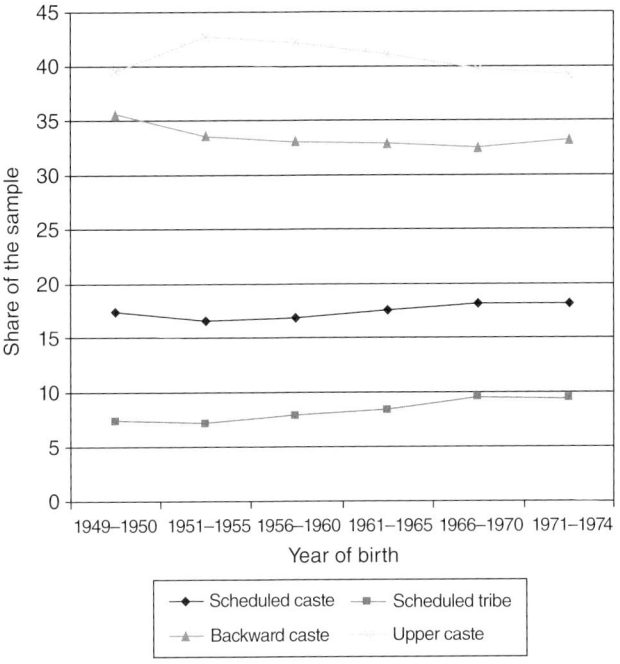

Figure 11.2. Composition by major castes over time.

Figure 11.3. Composition by major religions over time.

entry errors or genetic factors contributing to dwarfism. Overall, the sample used here comprises the stature of women representative of the entire Indian population.

In the sample, women's stature improved steadily from the 1st percentile (138.2 cm) to the 99th percentile (165.6 cm). Moreover, kurtosis statistics show positive kurtosis

Table 11.3. Summary statistics showing the distribution of stature in the sample.

Percentiles	Height
1	1382
5	1423
10	1443
25	1478
50	1515
75	1554
90	1589
95	1612
99	1656
N	51,678
Skewness	0.22
Kurtosis	4.68

(Table 11.3). The stature distribution turns out to have heavy tails and a range from 114 to 199 cm. The major Indian states did not follow a similar distribution pattern in stature. Some states like Madhya Pradesh, Rajasthan, and Gujarat have heavier tails on both sides.

Macroeconomic Data

The macrodata for explanatory variables were taken from Besley and Burgess (2000). Poverty and growth data were initially compiled by Ozler et al. (1996). The explanatory variables used for this study are nonagricultural GDP, development expenditures, population, and poverty.

The database contains detailed statistics on a wide range of poverty and welfare indicators for India. The data are presented separately on the state level from 1950 to 1974. Various sources like the Census of India, the National Sample Survey, and statistical abstracts were compiled to create the annual state-level indicators of inequality, consumption, and welfare used here. In this study, headcount index was used to analyze the relation between poverty and biological welfare. Moreover, the percentage of people owning land was used from this database to study the impact of land inequality on welfare.

The data for poverty, NDP (both for the agricultural and nonagricultural sector), and population were available only for 15 prime states, although stature data was available for all of the Indian states. Therefore, all of the empirical models used here focus on the following 15 prime states of India by combining individual stature data with the state-level socioeconomic data: Andhra Pradesh, Assam, Bihar, Haryana, Karnataka, Kerala, Gujarat, Madhya Pradesh, Maharashtra, Rajasthan, Orissa, Punjab, Tamil Nadu, Uttar Pradesh, and West Bengal.

To make the regional comparisons simple, all 24 states (Figure 11.4) in the sample were classified into the following six regions based on their geographical and cultural clustering (Table 11.4).

Both datasets are combined to study the impact of land inequality, development expenditure, growth in the nonagricultural sector, and population growth on the stature of individuals. The impact of the 1964–1966 drought years was tested using control variables

11 Short Women and Their Stagnating Growth

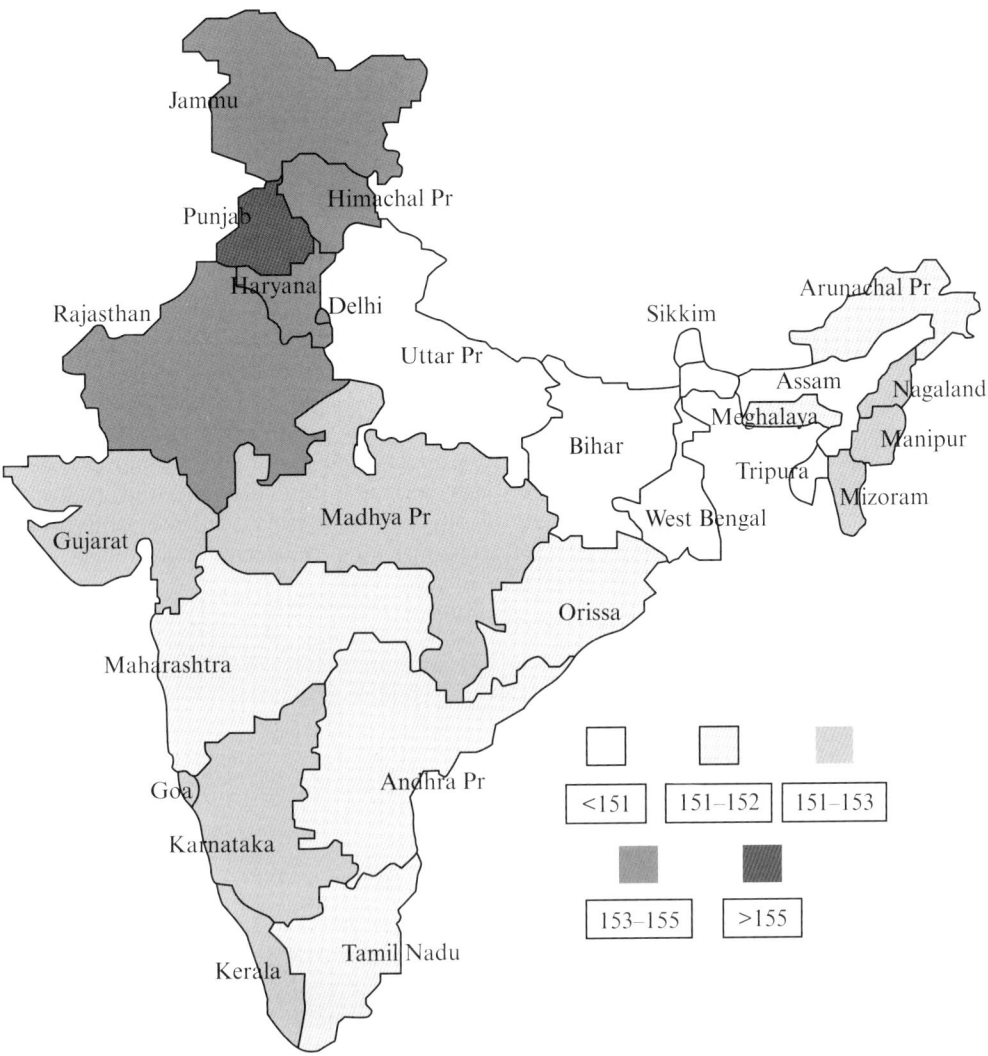

Figure 11.4. States in India.

Table 11.4. Regions of India and the states included within them.

Region	State
Northern India	Delhi, Haryana, Himachal Pradesh, Jammu and Kashmir, Punjab, Rajasthan.
Central India	Madhya Pradesh
East India	Bihar, Orissa, West Bengal
South India	Andhra Pradesh, Tamil Nadu, Kerala, Karnataka
West India	Goa, Gujarat, Maharashtra
South India	Andhra Pradesh, Tamil Nadu, Kerala, Karnataka
Northeast India	Arunachal Pradesh, Assam, Manipur, Moghalaya, Mizoram, Nagaland, Sikkim

for the years 1964, 1965, and 1966. In the regression model, multicollinearity was tested to check the association between variables such as caste and education. The covariates that were included in the final model did not suffer from multicollinearity bias.

Caveats of the Data

Mortality Biases in Cohort Studies

There are a number of possible mortality biases in the sample. Overall, shorter people have a higher risk of dying at a younger age (Waaler, 1984). Therefore, there is the possibility that surviving individuals—those present in the data set used here, for instance—might have been from a slightly taller segment of the population. As the whole analysis is based on stature data for women aged 24–45, maternal mortality and gender discrimination could also bias the sample. For instance, regional differences in gender discrimination might create interstate stature differences, meaning that states in which women were commonly discriminated against were likely to have shorter women (Guntupalli and Moradi, 2008). By using state categories, this can be controlled for to a certain extent. On the whole, the depiction of survivors by caste and religion does not support the presence of selective mortality (Figures 11.2 and 11.3), indicating that the stature data used in the sample are reliable.

Place of Childhood Residence

It was ultimately not possible to extract exact information on how long respondents lived in their childhood place of residence. Nevertheless, childhood place of residence provides better information about geographic origins in comparison to the current place of residence because current residence can be affected by both interstate and rural–urban migration. Moreover, there is uncertainty about the reliability of this measure. Therefore, it was only used for descriptive statistics.

Reduced Stature with Age

Reduced stature in older individuals can also introduce biases in stature series. Limiting the sample to individuals under the age of 45 largely precludes this source of bias. Moreover, analysis of the data set reveals that women aged 45–49 did not exhibit a significantly lower stature compared to younger women. As there was no minimum height requirement to participate in the survey, no bias was introduced to the data for understanding secular trends in stature.

Adolescent Growth

Continued growth can also introduce biases into stature series. Limiting the sample to those under the age of 24 helps to limit this bias. This is important as analysis of the potential for late adolescent growth beyond age 20 in the larger, original dataset revealed that women aged 21–23 were significantly shorter compared to the rest of the sample and were evidently still growing.

Analytical Methods

This study uses quantile regression (QR) instead of a standard Ordinary Least Square (hereafter OLS) method. This is because OLS focuses on the mean and does not provide any information on other parts of the distribution (Buchinsky, 1998; Koenker and Hallock, 2001). Moreover, OLS is strongly influenced by outliers that deviate positively or negatively from the average as the residuals are squared. In QR, different weights are assigned depending on where an observation lies based on the percentile estimated. For quantiles above the median, higher weight is placed on the residuals above the quantile than on the residuals below it. This pushes the minimization above the median. For quantiles below the median, less weight is placed on residuals above the quantile. This means that rather than restricting analysis to a particular quantile of the sample, all observations are used. Consequently, QR can provide powerful insights into the impact of the explanatory variables on different parts of the distribution of the dependent variable.

The quantile regression function is written as

$$Q_\theta\left[\mathbf{y}|\mathbf{X}\right] = \mathbf{X}\boldsymbol{\beta}_\theta \qquad (11.1)$$

The data matrix is $\mathbf{X}$, the dimension $n \times k$, the parameter vector $\boldsymbol{\beta}_\theta$ ($k \times 1$), and the dependent variable $\mathbf{y}$ ($n \times 1$). N represents the number of observations. The parameter θ identifies the quantile we are interested in.

Additional information is obtained by assuming that development expenditure affects all quantile groups in a different way. This means that it is not assumed that improved GDP in the nonagricultural sector improves the living standards of women belonging to lower parts of the stature distribution in a similar way to those in the upper parts. It is very likely that very poor women do not benefit from development in the nonagricultural sector (e.g., lack of skills and mobility). Instead, if development provided additional employment opportunities, it is likely that people with semiskilled status that belong to the middle part of the stature distribution benefit. Hence, impact of various macroeconomic indicators on various parts of the distribution is assessed to get a wider picture of various quintiles.

For this reason, we estimate several quantile regressions:

$$Q_{10}\left[\mathbf{y}|\mathbf{X}\right] = \mathbf{X}\mathbf{b}_{10} \qquad (11.2)$$

$$Q_{25}\left[\mathbf{y}|\mathbf{X}\right] = \mathbf{X}\boldsymbol{\beta}_{25} \qquad (11.3)$$

$$Q_{50}\left[\mathbf{y}|\mathbf{X}\right] = \mathbf{X}\boldsymbol{\beta}_{50} \qquad (11.4)$$

$$Q_{75}\left[\mathbf{y}|\mathbf{X}\right] = \mathbf{X}\boldsymbol{\beta}_{75} \qquad (11.5)$$

$$Q_{90}\left[\mathbf{y}|\mathbf{X}\right] = \mathbf{X}\boldsymbol{\beta}_{90} \qquad (11.6)$$

Results and Discussion

Levels and Trends of Welfare

Female stature in India increased slightly from 1949 to 1974 (Figure 11.5). We see a slight optimistic trend of 0.8 cm (negligible), and interestingly, that stature declined during economic stagnation in the 1960s. We can conclude that Indian stature was faintly greater compared to the pre-1950 period when stature increased by 0.7 cm. Also, we see that from the mid-1960s, the stature series suggests that the biological welfare of Indian women was more or less stagnant. In sum, we can say that stature and therefore biological welfare did not improve significantly in the two and half decades after independence, although we see a slight positive growth until the 1960s. But we need to see whether all

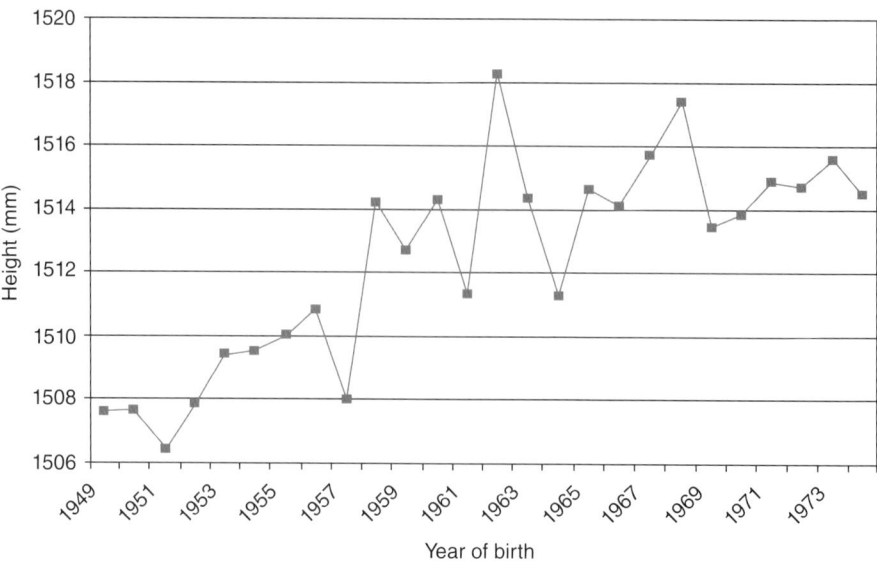

Figure 11.5. Stature in India, 1949–1974.

regions developed in a similar way to get a clear picture of stature and therefore welfare during this period.

Stature and Biological Welfare by Regions

Most investigations of regional variation in biological welfare have been limited to the late 20th century or to the colonial period. For instance, Brennan et al. (1995) found significant interstate differences in stature in India, though mostly for the 19th and early 20th centuries. Datt and Ravallion (1998) and Sahn (2003) studies of interstate differences show that economic development and investment in infrastructure may play a key role in interstate differences. Focusing on Kerala in comparison to other states, they found that increased development expenditures and government policies aimed at improving education greatly improved biological welfare and reduced inequality. Kerala moved from having the country's second-highest incidence of rural poverty in 1960 to one of the lowest incidences by the 1990s. These studies suggest that interstate differences exist and give clues as to their cause(s) but leave major questions about the same dynamics during the postcolonial period unanswered.

Figure 11.6, which aggregates the data into six regions, reveals that regional differences in women's stature existed in the postcolonial period. The stature of Western and Southern Indian women increased compared to the Northern, Northeastern, and Central states during the study period. Eastern states showed a moderate increase compared to the Northern and Central states. In general, women in the Northern states exhibit greater stature compared to the mean Indian value in the aggregated data. In contrast, women in Northeastern and Eastern states exhibit much lower stature. Interestingly, South India is the only region that experienced increasing stature even in the mid-1960s and early 1970s. The 1974 famine in Maharashtra and the 1973 oil crisis seem to have affected stature in the Western regions (Mumbai) more compared to other states. Compared to the colonial period (Guntupalli and Baten, 2006), Western Indian stature showed a positive trend whereas Eastern stature ranked lowest as in the 1915–1944 years.

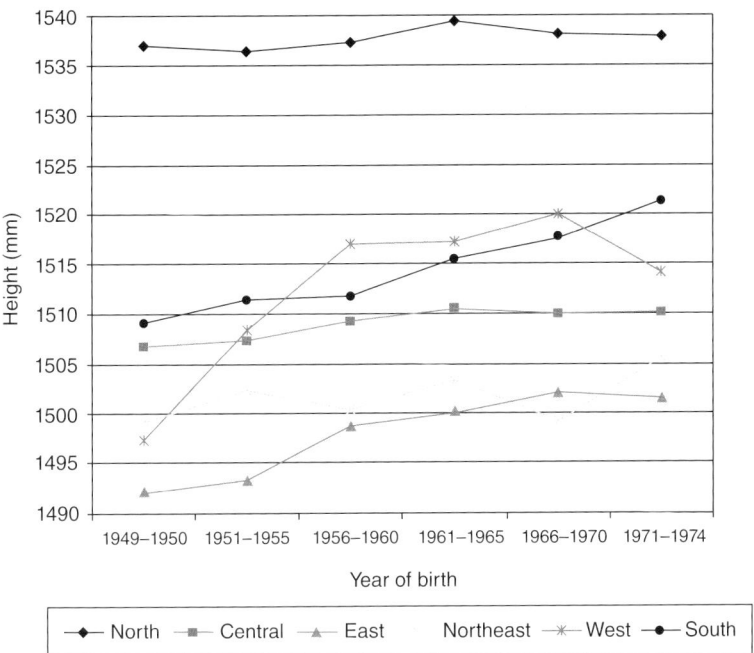

Figure 11.6. Stature of women from 26 states in India.

Interstate differences are also evident. In particular, the states of Punjab and Haryana had greater mean stature than the average for the nation. Perhaps unsurprisingly, agricultural productivity and protein supply in these regions are better than in other states. It is important to point that the interstate differences in postcolonial female stature do not corroborate with interstate differences in the male stature in precolonial India (Guntupalli and Baten, 2006). It is likely that regions and states that discriminate against females could have shorter females compared to the states that do not discriminate, leading to this contrasting picture. As datasets for comparing regional differences in stature between males and females in the postcolonial period do not exist, it is currently impossible to assess whether male stature would follow a similar pattern to female stature.

Religious Affiliation and Welfare

The 130 million Muslims in India form the second-largest Muslim population in the world following Indonesia. Unfortunately, however, limited economic data are available on the economic standing of Muslim and Hindu population from 1950 until recently, but stature data allows the investigation of whether these two groups are different in overall health and biological welfare. Several questions can be addressed. Did religious affiliation have an influence on stature between 1950 and 1974? Did the stature of Muslims differ from that of other religions like Hindus, Sikhs, and Jains?

Figure 11.7 demonstrates that Muslim women exhibited slightly higher stature and therefore greater health and biological well-being in comparison to Hindus. Sikh women, however, were the tallest. In general, they reside in Punjab and Haryana, where agricultural

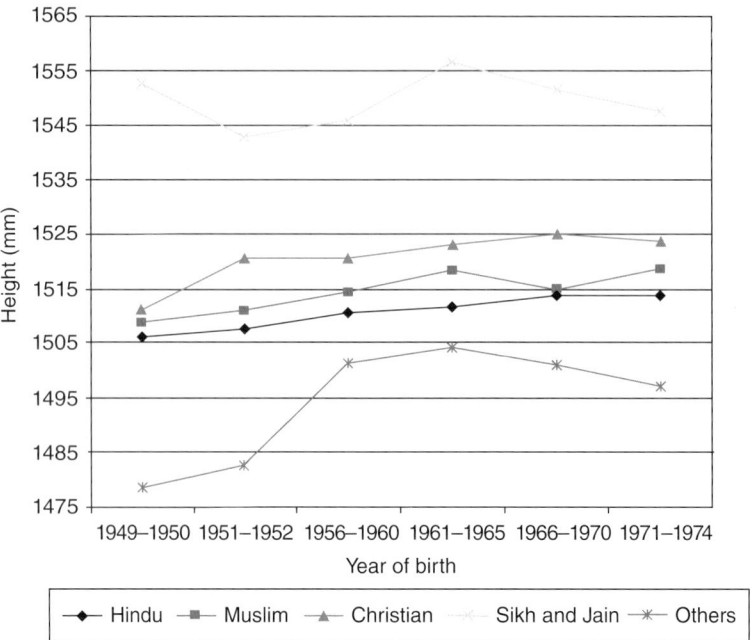

Figure 11.7. Stature in India by religion, 1949–1974.

productivity and access to protein was high during this period. Jains were mostly in trade occupations that yielded high incomes, which was reflected in their stature. Moreover, the diet of Jains includes milk, which is an important contributor to stature (see Koepke, this volume). Differences in stature in relation to religious affiliation may not be unique to women; previously published work on male stature between 1914 and 1944 showed that Sikh men were the tallest, followed by Muslim men (Guntupalli and Baten, 2006).

Welfare by Caste

In relation to caste, the most important question issue is whether overall health and biological welfare and therefore stature differed between various caste groups. Brennan et al. (1995, 1997, 2000, 2003) examined stature in relation to caste during the colonial period and found caste-based differences, which increased in magnitude after the mid-1930s. The dataset here allows investigation of the same, but for the postcolonial period, thus encompassing a period of substantial social change.

Figure 11.8 shows that female stature varied by caste. Perhaps unsurprisingly, stature was greatest among upper-caste women. Middle-caste women were the only group to have experienced increased stature and therefore improved welfare in the 1960s and 1970s. Stature also diverges between the scheduled and upper castes, with a greater gap (1–2 cm) between these two than any of the other groups. The same pattern of divergence exists between scheduled castes and middle castes; they possessed similar stature and trends in stature in 1950, but the gap between their respective mean statures had increased by nearly 1.2 cm by 1974. This suggests that the middle castes benefited most biologically during this period. Overall, these findings suggest that the trend of caste disparities in stature Brennan and colleagues identified persisted into the postcolonial period. The trends suggest that the

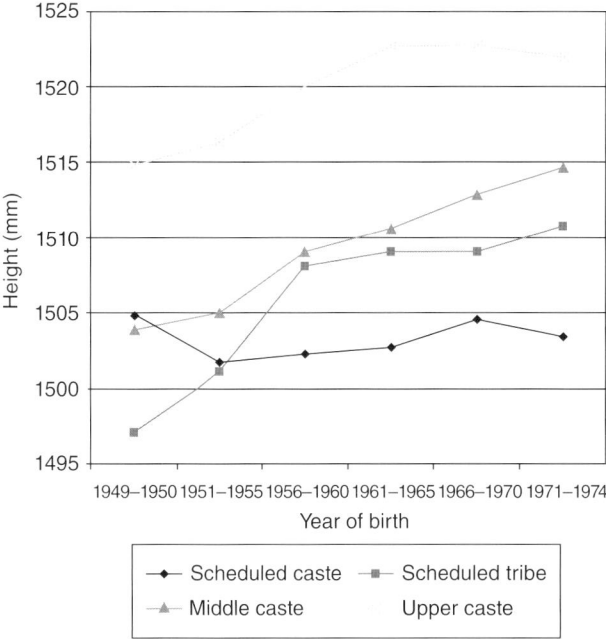

Figure 11.8. Stature in India by caste, 1949–1974.

middle castes benefited more from contemporary economic growth, especially in the non-agricultural sector, than did the scheduled castes. Perhaps unsurprisingly, given the generally higher SEP of upper castes, the stature series suggests that upper-caste women experienced higher-quality environments, better nutrition, and a lower disease burden than the other castes. However, they experienced declining stature from the 1960s on, suggesting declines in their welfare even as middle-caste women benefitted. This may be due to the benefits of the social welfare programs instituted during the drought years, whereas upper-caste women experienced costs associated with the declining GDP.

Rural–Urban Differences in Welfare

Numerous studies, including several in this volume, have found that urbanization has a profound effect on health. High population densities and poor environmental quality, among other factors, increase the risk of transmission of infectious diseases, and urban populations often also have poorer nutrition. However, studies of Indian populations suggest that rural children are more likely than their urban peers to suffer undernutrition and malnutrition (Lipton, 1977; Duncan and Howell, 1992), largely due to poor employment opportunities in rural areas. As agriculture relied exclusively on rainfall during the study period (irrigation was not widely practiced), rural populations were also highly vulnerable to drought. Information on childhood place of residence in the dataset was used to assess whether a rural–urban divide in stature existed between 1949 and 1974. As Figure 11.9 demonstrates, women from urban areas possessed greater stature than did those from rural areas. This difference magnifies from the 1960s onward, which may be linked to the drought years in the 1960s.

However, a convergence in stature did not occur in the 1970s, following environmental and economic stability, suggesting that economic and social policies in the 1970s

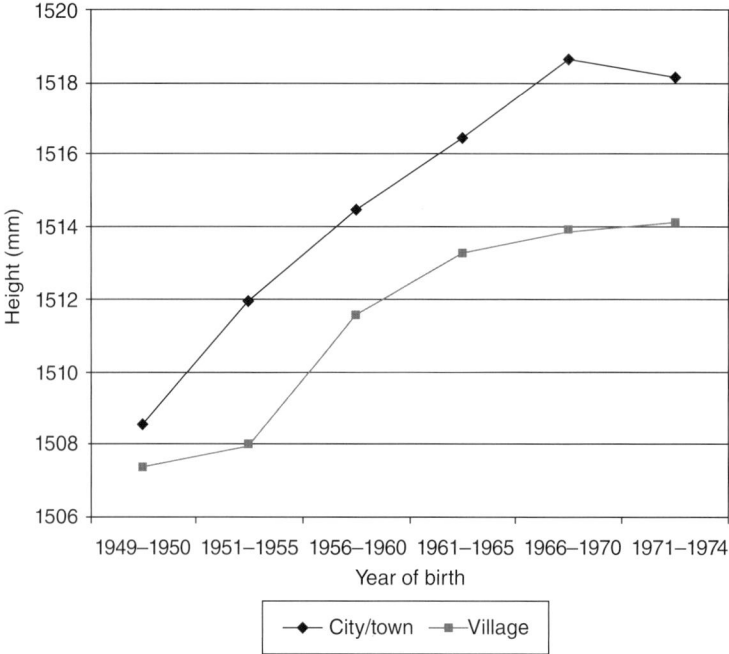

Figure 11.9. Stature in India by rural vs. urban childhood residence, 1949–1974.

that focused on building urban infrastructure and medical facilities—but not rural institutions—had a major impact on biological welfare of urban populations. Moreover, fertility was generally higher in rural areas than urban areas during the 1970s, creating the possibility that population pressure may have induced or exacerbated rural–urban welfare differentials in welfare.

Determinants of Welfare

QRs were performed for 0.10, 0.25, 0.50, 0.75, and 0.90 quintiles, respectively.[2] Simultaneous QR was conducted to run regressions for different quintiles at the same time. The analysis assessed state-level determinants of welfare, such as population growth, development expenditures, the head count index (% people that are below the poverty line in a state), real NDP per capita in the nonagricultural sector along with controls for the drought period 1964–1966, caste, religion, and states.[3]

Findings suggest that an increase in the share of development expenditures resulted in increased stature in almost all the quintiles (except for women in the uppermost quintile). Interestingly, the coefficients are higher in lower quintiles in comparison to the upper quintiles, suggesting that development expenditures have a greater impact on women with a low biological standard of living. Importantly, these findings can have a real-world impact by informing economic policy. They clearly suggest that investment in development programs directly and powerfully improve women's biological welfare. Given women's

---

[2] To control for heteroskedasticity, standard errors were estimated using the bootstrapping method.
[3] Haryana was not used in the regression due to multicollinearity.

role as caretakers and mothers, such programs would also have a positive economic and biological intergenerational effect for their children and other family members. In contrast, development programs targeting men may not have a trickledown effect to children and other family members.

In regards to the other variables, population growth did not affect all parts of the stature distribution, except for the 0.25 quartile (Table 11.5). The percentage of the population with no access to land was used as a proxy for agricultural land and home ownership, and

Table 11.5. Determinants of welfare in India, 1950–1975.

Variable	Q10	Q25	Q50	Q75	Q90
Real NDP per capita in the non-agricultural sector	6.89	3.85	8.38*	3.23	1.85
Head Count Index (Poverty)	−2.70	0.18	−2.40	1.89	0.7
Percentage of population with no land	−0.02	−3.23**	−0.43	−0.31	−1.58
Development expenditure share in GDP	7.95**	6.22**	5.15**	3.44*	5.03
Population Growth	−3.00	−3.76	−0.74	−0.5	0.84
Education in years	1.09**	1.12**	1.28**	1.44**	1.23**
Caste					
Low castes (reference)					
High caste	7.76**	8.37**	8.83**	8.41**	10.38**
Religion					
Hindu (reference Category)					
Muslim	−0.53	−0.03	2.17*	2.8*	2.2
Other religion	7.15**	5.48**	9.87**	8.72**	8.48**
Year 64–66	−0.71	−0.87	−0.60	−0.95	0.14
State					
Andhra Pradesh (reference)					
Assam	−11.32**	0.36	−13.46**	−17.98**	−18.78**
Bihar	−9.42**	−11.58**	−9.76**	−11.91**	−14.56**
Gujarat	1.28	7.66*	1.61	5.26	5.68
Karnataka	2.82	5.76*	7.11**	9.52**	9.27
Kerala	6.87*	12.14**	9.48**	12.65**	13.7
Madhya Pradesh	9.98**	13.35**	10.47**	10.79**	9.81
Maharashtra	−8.39	−2.74	−7.63*	−2.88	−4.27
Orissa	−8.59*	−5.26	−2.03	−1.44	−5.93
Punjab	19.53**	21.85**	16.61**	22.95**	20.58**
Rajasthan	23.02**	21.77**	23.22**	26.16**	23.16**
Tamil Nadu	−1.27	4.88	1.81	4.53	5.99
Uttar Pradesh	−7.73*	−10.78**	−8.44**	−6.91**	−11.21**
West Bengal	−14.88**	−13.67**	−15.34*	−10.24**	−13.62**
Constant	1470.62	1494.5	1530.45	1539.65	1580.13
Psuedo R^2	0.03	0.03	0.04	0.04	0.04
Number of observations	35552	35552	35552	35552	35552

*and **denote significance at the 5% and 10% confidence levels, respectively.

the negative impact of the lack of access to land on stature and therefore welfare is evident, but only for the 0.25 quartile. Years of education have a clear positive impact on stature for all parts of the distribution.

Caste also has a clear effect on welfare. Upper-caste women possessed greater stature than women of all other castes in all parts of the distribution. Religious affiliation does as well; irrespective of quintile, Sikh and Jain women consistently possess greater stature than do Hindus and Muslims. This is likely linked to dietary differences between the groups, specifically higher consumption of dairy products and therefore high-quality animal protein among the Sikh and Jain religious communities. However, in the 0.50 and 0.75 quartiles, Muslims have slightly greater stature in comparison to Hindus. This is likely linked to protein intake as well, specifically meat: some Hindus practice varying degrees of vegetarianism.

No significant relationships are evident between stature and head count index. In contrast, increased per capita NDP in the nonagricultural sector is significantly linked to greater stature in the 50th percentile of the distribution. The debate as to whether the economic and environmental shocks of 1964–1966 affected biological welfare, proxied by stature, was addressed by including control variables for these years. Results show that these years were not associated with decreased stature, suggesting that contrary to expectations, these events did not impact biological welfare. Analysis of interstate differences revealed several interesting patterns. Compared to Andhra Pradesh, Punjab, Kerala, Madhya Pradesh, and Rajasthan did well in all quintiles of the distribution. The states that performed poorly in all quintiles were West Bengal, Bihar, and Uttar Pradesh. Assam was poorer except at the 0.25 quartile, where it exhibited no significant difference in stature compared to Andhra Pradesh. Clarifying this issue, the findings suggest that interstate stature differences are similar to interstate poverty differences, as documented by Ravallion and Datt (1995). This is especially true in the case of Punjab, Assam, Bihar, Uttar Pradesh, and Kerala. The results clearly indicate that states that invested in and experienced improvements in education, health care, and infrastructure development underwent a transition in health earlier than those states that did not invest in these key areas.

Conclusion

How did the stature and therefore overall health and biological well-being of Indian women change between 1949 and 1974? It is evident that women's stature increased slightly over time, from 150.8 to 151.6 cm, which is in keeping with contemporary national-level trends of slow economic growth and an increase in life expectancy. Based on the stature series, welfare stagnated in the mid-1960s despite favorable economic and health conditions. This trend was not uniform across the country, though. Women from middle castes and southern Indian women experienced greater stature and therefore improved welfare in the 1960s and 1970s, pointing to regional and caste-based differences in living standards, such as nutrition, environmental quality, and disease burden, among other conditions during these decades.

Assessment of the determinants of welfare shed light into these dynamics. The stature series suggests than an improved per capita NDP in the nonagricultural sector improved the living standards of women in the central part of the distribution. This might be related to the fact that the nonagricultural sector improved economic conditions for the middle class by providing opportunities for semiskilled and skilled workers. This finding supports the arguments of Eswaran and Kotwal (1994) that the poor did not benefit from industrialization in India. In the 1950s, when India—on the national level—was moving into the second stage of epidemiologic transition due to an increase in life expectancy, economic development programs and investment in education seem to have played a vital role in improving living standards and biological welfare. Using stature as an indicator of net nutrition, inequalities in stature in relation to socioeconomic characteristics can be assessed. Findings here demonstrate that investment in public health writ large is positively associated with increased stature, highlighting the role of infrastructure development projects like water quality, sanitation, and even road quality in changing patterns of health and disease. The differences in stature and therefore in the biological welfare of women as well as in the determinants of welfare that produced them serve to emphasize that even within a single country, transitions in health are heterogeneous. Within region and interstate differences, as well as religious, caste-based, and rural–urban differences were all evident in the health of Indian women and emphasize that scholarship on health transitions must be attendant to the potential for such heterogeneous experiences.

Moreover, attention to these differences can have a practical, real-world, not just scholarly value. Debates continue as to the role of economic growth (i.e. the McKeown hypothesis) versus public health investment and infrastructure in improving health (Colgrove, 2002; Szreter, 2002). The finding that straightforward economic growth did not have the greatest impact on women's biological welfare, but instead that development expenditure and investment in education exerted the greatest benefit, can be used to inform current and future economic and social policy both in India and, potentially, in other countries and communities.

Acknowledgments

I am thankful to Joerg Baten and Deborah Rice for their comments on the previous version of this chapter.

References

Ahlburg DA. 1996. Population growth and poverty. In: Ahlburg DA, Kelley AC, Mason KO, editors. The impact of population growth on well-being in developing countries. New York: Springer-Verlag.
Armelagos GJ, Brown PJ, Turner B. 2005. Evolutionary, historical and political economic perspectives on health and disease. Soc Sci Med 61:755–765.
Baten J. 2000. Height and real wages: an international comparison. Jahrb für Wirtschaftsgesch 11:61–76.
Besley T, Burgess R. 2000. Land reform, poverty, and growth: evidence from India. Q J Econ 115:389–430.
Brennan L, McDonald J, Shlomowitz R. 1995. The variation in Indian height. Man India 75(4):327–337.
Brennan L, McDonald J, Shlomowitz R. 1997. Towards an anthropometric history of Indians under British rule. Res Econ Hist 17:185–246.
Brennan L, McDonald J, Shlomowitz R. 2000. Change in the stature of North Indians from British rule to early independence. Jahrb für Wirtschaftsgesch. 129–146.
Brennan L, McDonald J, Shlomowitz R. 2003. Long-term change in Indian health, in South Asia. J South Asian Stud 26(1):51–69.
Buchinsky M. 1998. Recent advances in quantile regression models: a practical guide for empirical research. J Hum Res 33(1):88–126.

Chandrasekhar S. 1968. How India is tackling her population problems? Demography 5(2):642–650.
Colgrove J. 2002. The McKeown thesis: a historical controversy and its enduring influence. Am J Public Health 925(5):725–730.
Datt G, Ravallion M. 1998. Why have some Indian states done better than others at reducing rural poverty? Economica 65(257):17–38.
Duncan A, Howell J, editors. 1992. Structural adjustment and the African farmer. Portsmouth: Heinemann Educational Books.
Eswaran M, Kotwal A. 1994. Why poverty persists in India? New Delhi: Oxford University Press.
Gage TB. 2005. Are modern environments really bad for us? Revisiting the demographic and epidemiologic transitions. Yearb Phys Anthropol 48:96–117.
Guntupalli AM, Baten J. 2006. The development and inequality of heights in North, West and East India, 1915–44. Explor Econ Hist 43(4):578–608.
Guntupalli AM, Moradi A. 2008. What does gender dimorphism in stature tell us about discrimination in rural India, 1930–1975? In: Pal M, Bharati P, Vasulu T, editors. Gender and discrimination: Health, nutritional status, and role of women in India. New Delhi: Oxford University Press.
International Institute for Population Sciences (IIPS) and ORC Macro. 1995. National family health survey (NFHS-1), 1992–93: India. Mumbai: IIPS.
International Institute for Population Sciences (IIPS) and ORC Macro. 2000. National family health survey (NFHS-2), 1998–99: India. Mumbai: IIPS.
International Institute for Population Sciences (IIPS) and ORC Macro. 2007. National family health survey (NFHS-3), 2005–06: India. Mumbai: IIPS.
Koenker R, Hallock KF. 2001. Quantile regression. J Econ Perspect 15:143–156.
Komlos J. 1996. Anomalies in economic history: towards a resolution of the Antebellum Puzzle. J Econ Hist 56:202–214.
Komlos J, Baten J, editors. 1998. The biological standard of living in comparative perspective. Stuttgart: Steiner Verlag.
Lipton M. 1977. Why poor people stay poor. Cambridge: Harvard University Press.
Mayer P. 1999. India's falling sex ratios. Popul Dev Rev 25(2):323–343.
McKeown R. 2009. The epidemiologic transition: changing patterns of mortality and population dynamics. Am J Lifestyle Med 3(S):19–26.
Office of the Registrar General. 2009. Report on causes of death in India, 2001–2003. New Delhi: Office of the Registrar General, India Ministry of Home Affairs.
Olshansky SJ, Ault AB. 1986. The fourth stage of the epidemiologic transition: the age of delayed degenerative diseases. Milbank Q 64(3):355–391.
Omran AR. 1971. The epidemiologic transition: a theory of the epidemiology of population change. Milbank Mem Fund Q 49(1):509–538.
Omran AR. 1982. International encyclopedia of population. New York: Free Press.
Özler B, Ravallion M, Datt G. 1996. A database on poverty and growth in India. Washington, DC: Policy Research Department, World Bank.
Popkin B. 2002. The shift in stages of the nutrition transition in the developing world differs from past experiences! Public Health Nutr 5(1A):205–214.
Ravallion M, Datt G. 1995. Growth and poverty in rural India. World Bank Policy Research Working Paper Vol. 1405. Washington, DC: World Bank.
Ravallion M, Datt G. 2002. Is India's economic growth leaving the poor behind? J Econ Perspect 16(3):89–108.
Reddy PH. 1989. Epidemiologic transition in India. In: Singh SN, Premi MK, Bhatia PS, Bose A, editors. Population Transitions in India. Vol. 1. Delhi: B.R. Publishing. p 281–290.
Rodgers G. 1989. Population growth and poverty in rural South Asia. New Delhi: Sage Publications.
Sahn D. 2003. Equality of what? Evidence from India. Paper presented at poverty, inequality, and development: A conference in honor of Erik Thorbecke, October 10–11, 2003, Ithaca.
Sivasubramonian S. 2004. The sources of economic growth in India. New Delhi: Oxford University Press.

Szreter S. 2002. Rethinking McKeown: the relationship between public health and social change. Am J Public Health 92(5): 722–725.

Varshney A. 2000. Is India becoming more democratic? J Asian Stud 59(1):3–25.

Virmani A. 2006. India's economic growth history: fluctuations, trends, break points and phases. Indian Econ Rev 41(1):81–103.

Waaler HT. 1984. Height, weight, and mortality: the Norwegian experience. Acta Med Scand 679:1–56.

Wilkinson S. 2000. Consociational theory and ethnic violence. Asian Surv 40(5): 767–791.

World Bank. 2012. Trends in maternal mortality: 1990–2010. Estimates developed by WHO, UNICEF, UNFPA and the World Bank. Accessed at: http://data.worldbank.org/indicator/SH.STA.MMRT (accessed on November 7, 2013).

Chapter 12
Tracking the Second Epidemiologic Transition Using Bioarchaeological Data on Infant Morbidity and Mortality

Megan A. Perry
Department of Anthropology, East Carolina University, Greenville, NC

Introduction

The second epidemiologic transition (Barrett et al., 1998) is a historically contextualized version of Omran's (1971) original model of the shift in mortality, fertility, and morbidity that purportedly occurred with industrialization. Omran used England and Wales to demonstrate the *Western* version of his model, wherein high morbidity and mortality from infectious diseases hindered population growth from antiquity until approximately the mid-19th century. Afterward, according to the model, mortality decreased significantly, driven by a sharp decline in infant and maternal mortality due to better prevention and treatment of epidemic infectious diseases and fewer complications with pregnancy and childbirth. Despite the model's oversimplification of population dynamics and disease ecology (Barrett et al., 1998) and the issue that many chronic and degenerative diseases have infectious causes, it provides a powerful summation of the impact of epidemic infectious disease on population dynamics.

Although Omran's model of decreased infant morbidity and mortality is supported by historical documents from the late 19th- to early 20th-century England (Woods et al., 1988, 1989), this demographic change has not been explored with skeletal remains from this period, particularly those of fetuses, neonates, and infants. Skeletal remains can help fill data gaps in lacunae-filled vital records and historical accounts and provide a counterbalance to historical documents. However, identifying the hallmarks of an epidemiologic transition from archaeological infant skeletal remains is also fraught with issues related to sampling bias and methodology. Despite these problems, careful assessment of health indicators during infancy could identify shifts in morbidity and mortality that may indicate an epidemiologic transition (Perry, 2006). First, I will discuss how we can identify individuals within the skeletal record who would have been considered infants within this historical context. Second, I will outline the trials and tribulations—and successes—in understanding mortality and morbidity in infants. Then skeletal data from medieval and industrial London on infants up to 1 year of age will be used to illuminate infant morbidity and mortality rates

Modern Environments and Human Health: Revisiting the Second Epidemiologic Transition, First Edition. Edited by Molly K. Zuckerman.
© 2014 John Wiley & Sons, Inc. Published 2014 by John Wiley & Sons, Inc.

up until the purported late-19th century epidemiologic shift to provide context for Omran's model. These data will show that Omran oversimplified the pretransition state of populations in England, who may have experienced lower-than-expected infant morbidity and mortality before the purported transition.

Background

Children in the Bioarchaeological Record

What Is an Infant?

Understanding patterns of infant mortality requires that we take a step back from our numbers and evaluate who would have been considered an infant in that particular time and space. Societal definitions such as *infant* and *child* come with certain cultural expectations and anticipated changes in behavior (Baxter, 2005). Regarding infant health, it is important to consider, for instance, when weaning occurred, as this process is linked with increased risk for disease and stress as the infant shifts from reliance upon their mother's to their own immune system. In addition, weaning often marks the end of maternal dependence and a more independent child in many societies around the globe, with acknowledgment of the stress the infant undergoes with this transition. In fact, the London Bills of Mortality identify infants and children who perished from *teeth* issues and, by implication, weaning, and physicians from the early modern period frequently mention *weaning illness* as a cause of mortality (Fildes, 1986: 390).

Thus in comparing patterns of infant health between populations, particularly neonatal vs. postneonatal health, one needs to consider at what age this period of increased risk for infection and malnutrition occurred. A newborn infant has to adapt to a new environment, and their success depends upon endogenous and exogenous factors related to maternal health and the external environment. Infant survival before, at, or near birth depends upon the uterine environment, maternal influences, and genetics, while after 41 gestational weeks (i.e., when an infant is approximately 1 month old), it is influenced by infectious disease, sanitation, immune system functioning, and living conditions (see Lewis and Gowland, 2007). Therefore, instead of relying strictly on comparing mortality at specific skeletal ages, mortality during different portions of the social life cycle should be compared.

Differential Burial

The first problem that archaeologists and bioarchaeologists encounter while investigating infant health is the sampling bias that can occur from differential burial practices due to age or varied preservation of infant skeletons. In medieval England, for example, infants and children seemed to have been clustered in one part of a cemetery away from the adults. The implication is that if an archaeologist did not select that area of the cemetery for excavation, these individuals would not be recovered and included within the archaeological skeletal sample. In the later medieval period, infants were considered inherently corrupted by the original sin of their conception and not included within community church cemeteries at all unless they were baptized before death (Lewis, 2007). This would mean that stillborn babies, infanticide cases, and infants who died suddenly may not be represented in typical samples. Furthermore, the widespread practice of middle- and upper-class families relying on wet-nurses, who would raise the infant in their own village until it was weaned (Fildes, 1988), would mean that infants who died while away in the wet-nurse's village would not be included in the birth family's community cemetery.

Careful excavation is also essential for generating a representative sample of infants. Infant bones are notoriously fragile and prone to deterioration in many burial environments (Gordon and Buikstra, 1981; Walker et al., 1988). In addition, incompletely developed bones and teeth may not be recognized as human by excavators. This potential underrepresentation of infants affects any attempt at estimating infant morbidity and mortality, in addition to fertility rates (Paine and Harpending, 1998). Age-related sampling biases in historical skeletal samples can be identified through calculating the relative number of infants and children vs. adults in the sample and comparing this to related mortality records (Walker et al., 1988; Saunders et al., 1995; Guy et al., 1997; Jones and Ubelaker, 2001). Others (Storey and Hirth, 1997) have used a combination of calculating the percentage of subadults under the age of 1 and comparing these with demographic models (e.g., Coale and Demeny, 1983) and looking at prevalence of subadults by mortuary practice to determine if burial context affects preservation, thus creating an underrepresentation of subadults. Thus studies have found that, in most cases, infants under 1 year are underrepresented in archaeological skeletal samples.

Birth or Death? What Influences Mortality Profiles?

Infant mortality rates provide a sensitive reflection of overall population fitness due to the susceptibility of infants to malnutrition and infectious diseases during growth and development (Saunders and Barrans, 1999). However, for several material and theoretical reasons (e.g., Wood et al., 1992), the mortality profile provided by a cemetery sample cannot directly indicate the overall mortality profile of the population and will not necessarily correspond with mortality bills, parish records, or other documents, especially in relation to the morbidity and mortality profiles of infants. For instance, historical demographers normally calculate infant mortality rates as cumulative exogenous mortality during the first year of life. However, Lewis (2007) points out that since cemeteries are used over a number of years, it is impossible to know the number of infant deaths by year. Skeletal age estimation techniques also do not allow the precision necessary for demographic analyses, such as knowing exact number of days since birth or distinguishing between a stillborn close-to-term fetus, a newborn, or an infant who lived only a few days after birth.

An even larger issue emerges in dealing with the reality that all populations are essentially demographically unstable—that is, they are experiencing growth through immigration and/or increased fertility or decline through emigration or decreased fertility. Paleodemographic studies have mathematically determined that small changes in fertility rates more dramatically alter the age-at-death profile of a sample than does even a significant change in mortality rates (Sattenspiel and Harpending, 1983; Wood et al., 1992; Paine, 1997). This means that skeletal samples that contain a relatively smaller number of deceased infants than does a comparative sample may indicate that the population is experiencing decreased fertility or decreased immigration (or even emigration) of families with young children or women of childbearing age, just as much as it may indicate a decrease in infant mortality.

Methodologically, a further problem stems from the bias resulting from the age estimation techniques applied to skeletal remains. Skeletal age estimation techniques are derived from samples of deceased individuals of known age-at-death. These *documented* or *reference* samples may have different age-at-death profiles than the actual profile of a given archaeological skeletal sample under investigation. Numerous studies have discovered that the resulting age-at-death profile of the target sample will *mimic* the age-at-death profile of the documented collection (Bocquet-Appel and Masset, 1982, 1996; Konigsberg and Frankenberg, 1992, 1994; Hoppa and Vaupel, 2002). To combat this bias, researchers have applied Bayesian statistics that assess the probability that a skeletal individual represents a certain age given their stage of skeletal growth or degeneration. This probability depends upon the prior probability of risk of death within a particular age category—that is, it must account for the fact that individuals at certain ages are more at risk of death than others. Different techniques can be used to generate this prior probability, but the most logical for archaeological skeletal sample involves using an unrelated but appropriate documented population to generate prior probabilities in terms of risk of death at certain ages (see Gowland and Chamberlain, 2002; Hoppa and Vaupel, 2002). These prior probabilities are combined with reference data providing age-at-death of an individual in relation to observations of the age assessment indicator in question (e.g., long bone length, pubic symphysis degeneration) to produce posterior probabilities that someone with a particular age indicator expression is of a certain age (Gowland and Chamberlain, 2002, 2005; Hoppa and Vaupel, 2002; Chamberlain, 2006; Samworth and Gowland, 2007). The Bayesian-produced age distribution is most powerful and least susceptible to bias if the reference samples used to estimate the conditional probabilities of age-indicator-given age, the prior probabilities, and the posterior probabilities have independent age-at-death distributions (Konigsberg and Frankenberg, 1992; Gowland and Chamberlain, 2002; Hoppa and Vaupel, 2002).

These age-at-death mortality profiles can then be used to evaluate the period at which an infant is most at risk of death in a sample. Mortality rates during the risky period—approximately 1 month surrounding birth—would stem from endogenous factors such as the health of the mother and childbirth practices. For infants older than 1 month, the largest risk for death would come from exogenous factors, particularly infectious pathogens the infant experiences during and immediately after weaning (Lewis and Gowland, 2007). In order to assess the relative impact of exogenous vs. endogenous factors on mortality, Saunders and colleagues (1995) and Lewis and Gowland (2007) compared the number of individuals who died between 28 and 40 weeks (neonatal deaths) vs. those who died between 41 and 48 weeks (postneonatal deaths). In their study of infant mortality in medieval and industrial rural and urban England, Lewis and Gowland found that the rural sites, regardless of the period, had relatively higher neonatal than postneonatal mortality, while the urban crypt of Spitalfields had fewer neonatal deaths but higher postneonatal deaths, which would have been due to exogenous factors. They attribute the decreased neonatal deaths to medical advances in childbirth and the relative wealth of the Spitalfields sample, and the retention of high postneonatal mortality to the relatively early weaning age of 7 months. On the other hand, Saunders and colleagues found an almost equal percentage of neonatal vs. postneonatal skeletons in the St. Thomas Cemetery in Belleview, Ontario, and consequently attributed this to the poor sanitary conditions likely experienced by these young babies, particularly after weaning, which occurred at approximately 5–7 months of age in this population.

In addition, the impact of weaning deaths on the mortality profile also should be assessed to understand the relative health of a population. Because the age of weaning shifted during the Industrial Revolution, the age of weaning-related mortality risk also changed. In pre-16th-century Europe, mothers generally started weaning their infants at 2 years of age, although some texts note that female infants were weaned anywhere from 6 to 12 months earlier than males (Fildes, 1986). This weaning age of 1–2 years has been confirmed by $\delta^{15}N$ studies of infants from the medieval site of Wharram Percy in England (Mays, 2010). However, as mothers headed off to work in factories in the 17th and 18th-century England, the weaning age of infants dropped to an average of 7.25 months (Fildes, 1986). Lewis (2002) also found that indicators of environmental stress possibly related to weaning are detectable in the Spitalfields skeletal sample from industrial London at approximately 7 months of age. This early weaning age actually may have preceded medical advice on the timing of weaning, which was shifting to fit with actual practice (Fildes, 1986). Therefore, in order to compare weaning-related deaths in medieval populations vs. industrial populations in this context, deaths occurring around 1–2 years of age in the medieval population should be compared with deaths from 6 months to 1 year of age in the industrial samples. Considering the age of weaning would also allow researchers to control for the effects of weaning on health as opposed to other conditions, if desired.

Identifying Morbidity Patterns and the Osteological Paradox

The epidemiologic transition model also postulates that infant morbidity and mortality declined specifically due to a decrease in the frequency of epidemic infectious disease. Critical analysis of the skeletal evidence may indeed indicate a corresponding decrease in infant mortality or morbidity, as noted earlier, but how do we know that decreased infectious disease was the culprit? Paleopathologists have traditionally quantified the extent of pathological lesions in a sample as an indicator of the quality of life of that population. That is, if more individuals were dying with signs of stress, infection, or malnutrition, then the inference can be made that that population had a greater frequency of these conditions than did a group with fewer skeletal lesions. On the other hand, because the individuals in a cemetery are there *for a reason* (i.e., they died), this introduces a mortality bias into the equation (Sattenspiel and Harpending, 1983; Wood et al., 1992). Two individuals in a population may have the same condition, but because of genetics or their immediate environment, one may be more at risk of dying than the other (hidden heterogeneity) and thus end up in the cemetery sample at an early age while the other individual survived and went on to live a healthy, long life. An added complication is that skeletal pathologies tend *not* to develop in acute infections, particularly those with high mortality such as bubonic plague, nor do all diseases result in bone pathology, such as measles. If an individual dies *before* a lesion can form, because the causal condition was acute or because they died—or recovered completely—before skeletal lesions began to manifest, their skeleton may appear to be *healthier* than that of someone who contracted the disease but survived into and beyond lesion formation.

So how can we interpret a skeletal sample to identify levels of infant morbidity due to epidemic infectious disease? As noted earlier, the data are limited to telling us only if an infant died while ill. If an infant perishes with an active (as opposed to healing or healed) pathology from an infectious condition, the presence of that condition may reflect their relative frailty in the population as opposed to another infant without that condition and, through careful interpretation, can reflect the overall disease burden of the population. The real problem stems from how to interpret morbidity from infants who perished with no signs of pathology. If they died while suffering from an infectious disease, they either suffered from one that did not result in skeletal pathologies or the disease was so virulent that they died before a pathological response could manifest. In the latter case, this means they actually may have been *frailer* than the infant who died with active skeletal pathology. In addition, they may not have had an infectious disease at all but may have instead died from violence or accidental trauma that did not impact the skeleton, as well as from nonskeletal congenital abnormalities, childbirth complications, or infanticide.

Unfortunately, we are left with inferring what happened to these infants through their age rather than through direct, primary evidence of cause of death. As noted earlier, once infants reach 27 days after birth, they are more likely to perish from exogenous factors rather than endogenous factors. At this point, the largest external force in terms of mortality for infants is infection and malnutrition. Therefore, we end up having to assume that these were the primary causes of death up to 1 year of age and thus infer that higher mortality during this age likely is due to greater infection and malnutrition. This does not mean discounting pathology frequencies altogether, but instead considering factors beyond visible infection that may have resulted in the creation of the sample.

Case Study: Infant Morbidity and Mortality during the Second Epidemiologic Transition

In order to explore changes in infant morbidity and mortality more deeply during the second epidemiologic transition, I turn to the context within which Omran first generated his model—England and Wales. The case study explored here is that of changes in infant health in London before and after the beginning of the 18th century, the moment that marks the shift from the Renaissance to industrial manufacturing and urbanization in England.

The Industrial Revolution generally refers to rapid changes in economic production and living conditions that occurred from the mid-1700s into the early 20th century. As Daunton (1995) points out, this includes increased production output of textiles, iron, glassmaking, and almost any manufacturing process that existed at this point. An increase in agricultural yields through mechanization meant that food was less expensive and more plentiful, although the associated decrease in the diversity of crops produced meant greater food insecurity and susceptibility to famine. These developments coincided with a growth of personal income at almost every level of the social hierarchy, although the actual value of the increased wages (i.e., real wages) decreased as costs increased. As a result, economic inequality increased not only within a locale but also between different regions of England. Fertility also increased, followed by mortality.

Increased agricultural yields opened up the possibility for urban migration, creating a new form of urbanism that required greater service infrastructure and complex administration. In the rush to house potential laborers, factory owners developed, for the most part, shoddy, crowded housing without proper sanitation. The growing link between poverty and disease, particularly with the publication of Chadwick's (1843) *The Sanitary Conditions of the Labouring Population*, led to the passing of multiple public health acts and the establishment of sanitary boards to push for cleanup of waste and trash that littered some urban neighborhoods. These organizations led to the construction of better sewage and drainage systems, which limited contamination of drinking water.

Numerous large cemeteries from both periods have been excavated in London as a result of late 20th- and 21st-century urban development, and the skeletal data diligently collected from these samples are freely available online. This case study also demonstrates the trials and benefits of working with published data, which in the end can tell us something about infant health up to the mid-1800s, the point when most of these excavated cemeteries went out of use. The remainder of this story about the epidemiologic transition relies on published studies of vital statistics dating through the turn of the 20th century AD.

Materials

This detailed examination of changes in infant morbidity and mortality at the beginning of the second transition utilized the Wellcome Osteological Research Database (WORD).[1] WORD contains data on a number of skeletal samples dating from AD 43 through the

---

[1] http://www.museumoflondon.org.uk/Collections-Research/LAARC/Centre-for-Human-Bioarchaeology/Database/

Table 12.1. Cemeteries in WORD used in this study.

Cemetery	Years in use (AD)	Total N	N Infants < 1 year (% of entire sample/% of subadults)
Pre-1700			
Guildhall Yard	1050–1230	68	1 (1/5)
East Smithfield Black Death	1348–1350	636	11 (2/6)
St. Mary Graces	1350–1538	389	4 (1/4)
Spital Square	1200–1500	124	1 (<1/2)
St. Thomas Hospital	1540–1714	193	6 (4/20)
Broadgate	1569–1714	137	29 (21/45)
Medieval St. Benet Sherehog	1250–1666	39	3 (8/2)
Total		**1586**	**55 (3/12)**
Post-1700			
Chelsea Old Church	1700–1850	198	9 (5/27)
St. Brides Lower	1770–1849	544	65 (12/38)
Cross Bones	1598–1853	148	87 (59/86)
Postmedieval St. Benet Sherehog	1670–1853	231	23 (10/36)
Total		**1121**	**184 (16/50)**

mid-19th century AD that are curated by the Museum of London. Ten samples from the database were selected for analysis based on sample size, preservation, and the representativeness of individuals under 1 year of age. These samples were divided into cemeteries dating prior to and after 1700 (see Table 12.1). The pre-1700 cemeteries include Guildhall Yard (1050–1230), the East Smithfield Black Death cemetery (1348–1350), St. Mary Graces (1350–1538), Spital Square (1200–1500), St. Thomas Hospital (1540–1714), Broadgate (1540–1714), and Medieval St. Benet Sherehog (1250–1666). The post-1700 cemeteries include Chelsea Old Church (1700–1850), St. Brides Lower (1770–1849), Cross Bones (1598–1853), and postmedieval St. Benet Sherehog (1670–1853). The records for 239 fetuses and infants less than 1 year old were investigated, $N = 55$ for the pre-1700 subset and $N = 184$ for the post-1700 subset. At present, WORD only groups together children over 1 year of age into the broader 1–5-year-old category, which unfortunately would conceal any variation in health and disease due to weaning. Thus I focus on infants under 1 year of age in this study.

The obvious difficulties in utilizing data from cemeteries in use over such a long period of time are the masking of short-term shifts in disease and mortality and the inability to exclude burials from a cemetery that includes both pre- and post-1700 burials due to imprecise dating of the graves (e.g., including potential 1598–1700 Cross Bones burials in the post-1700 group). Therefore, interpretation of the morbidity and mortality profiles will include the assumption that the samples represent the overall health and mortality of the era in question and that the pattern of individuals from the cemetery pre- or postdating the labeled era has a negligible impact on the overall profile.

In addition, the cemeteries contain individuals from a wide variety of social classes in London, which may skew health profiles of a particular era. Cross Bones, St. Brides Lower, St. Thomas Hospital, and the earlier burials at Broadgate represent poorer populations of medieval and industrial London, while Chelsea Old Church contains a more elite population (see DeWitte, this volume). Spital Square and St. Thomas contain individuals from hospital contexts and thus might show poorer health than other samples. The relative contributions that certain cemeteries make to the overall profile will be investigated in addition to interpreting the impact of industrialization on urban London. Finally, for individuals over the gestational age of 48 weeks, the age-at-death categories listed in the database were relied on to classify subadults by age.

Methods

WORD includes a detailed examination of pathologies of both a specific and nonspecific nature (Powers, 2012). Four pathological indicators of quality of life and health were focused up in this study: nonspecific bone infection or periosteal reactions, cribra orbitalia, porotic hyperostosis, and metabolic deficiencies (see Ortner, 2003). Nonspecific bone infections are associated with a number of conditions, including trauma and blood-borne infections, and do not have a pathological profile specific to a particular condition (see DeWitte, this volume). Cribra orbitalia and porotic hyperostosis are porous lesions in the upper eye orbits and the parietals and occipitals of the cranium, respectively, that are associated with several forms of anemia and vitamin deficiencies (see Walker et al., 2009). Metabolic disorders include skeletal reflections of insufficient amounts of vitamin C (i.e., scurvy) and vitamin D (i.e., rickets) in the body. The state of the condition (active vs. healed) was noted in the data analysis, if available in the database. For cribra orbitalia, infants with extensive diploic expansion (WORD code 4: foramina have linked into a trabecular structure, and WORD code 5: outgrowth in trabecular form from the outer table surface; see Powers, 2012) were considered separately from other infants with less-advanced stages of porosity. The statistical program JMP 9.0 was used for all statistical analyses. Overall frequencies of different pathological conditions were compared between the pre- and post-1700 groups using a Chi-square test. Age-related distributions of disease frequencies and overall age-at-death profiles were compared using the Kolmogorov–Smirnov test statistic.

The age distribution of fetuses and infants up to 48 gestational weeks was generated using regression formulae and Bayesian statistics based on femoral length. When observations of both left and right femora existed, the lengths of these two elements were averaged; otherwise, the existing left or right femur measurement was used. First, age-at-death estimates were calculated using Scheuer et al.'s (1980) regression method. Then, an alternate age-at-death distribution was generated using posterior probabilities of age based on femur length produced by Gowland and Chamberlain (2002). Gowland and Chamberlain developed prior probabilities of death at certain ages using a 1958 British perinatal mortality survey (Butler and Alberman, 1969) that tallies the gestational ages of 17,000 births over a period of time. From these, the posterior probabilities were developed from a sample of individuals of known age and femur length. This Bayesian-produced mortality distribution will theoretically eliminate bias of regression equations that mimic the mortality structure of the reference population (Konigsberg and Frankenberg, 1992; Hoppa and Vaupel, 2002). However, assessing associations by individual between age and another variable (e.g., the presence of a pathology) must rely upon the regression-produced age estimates of Scheuer and colleagues (1980).

Results

The age-at-death profiles for the pre-1700 (<1700) and post-1700 (>1700) combined samples showed rather significant differences, particularly the high frequencies of individuals below 1 month of age in the post-1700 sample and the inclusion of more individuals in

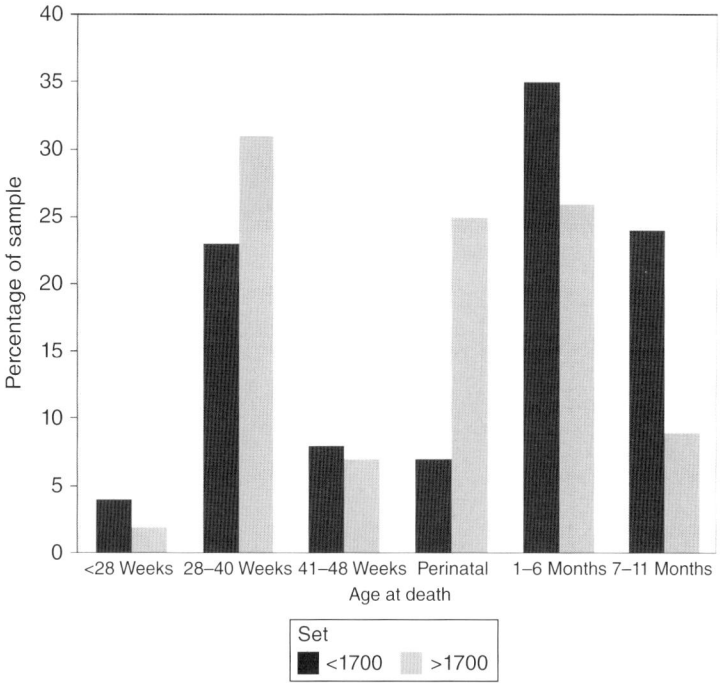

Figure 12.1. Age-at-death distributions of the pre- and post-1700 samples. *Weeks* refers to gestational weeks, *perinatal* includes all other fetuses/newborns for whom gestation age could not be estimated, 1–6 months and 7–11 months refer to time since birth.

the 7–11-month age category for the pre-1700 sample (Figure 12.1). The inclusion of subadults from a Black Death cemetery (East Smithfield), which was found to have a catastrophic, rather than attritional, age-at-death profile in terms of the adults in the sample (Gowland and Chamberlain, 2005), does not appear to greatly bias the overall pre-1700 age-at-death distribution of the subadults. The pre-1700 sample excluding the Black Death group has a similar mortality profile to the Black Death group alone.

Bayesian posterior probabilities were used to recalculate the pre- and post-1700 age-at-death distributions for individuals under 48 gestational weeks to compare the effects of the *in utero* and *ex utero* environments. These revised mortality profiles show similar peaks around 38 weeks, suggesting that stillborn babies, or babies dying shortly after birth, were common during both periods. The pre-1700 cemeteries contained more late second trimester infants (<28 gestational weeks) than the post-1700 cemeteries, and the post-1700 group had more infants dying during the third trimester and around the time of birth (28–40 weeks) than did the pre-1700 group. Deaths during the first 6 weeks of life (41–48 weeks) remained approximately the same for both groups (Figure 12.2).

The paleopathological data show that the post-1700 infants were dying with substantially more infections and metabolic conditions in all age categories than were those in the pre-1700 sample. The ages at which the post-1700 infants perished with signs of infection (active or healing) remains high with the exception of a small decrease from 1 to 6 months of age. The pre-1700 infants died with infection during the third trimester and from 1 to 6 months of age. Metabolic conditions in the post-1700 group remained high from the third trimester through 6 months of age, after which there is a substantial

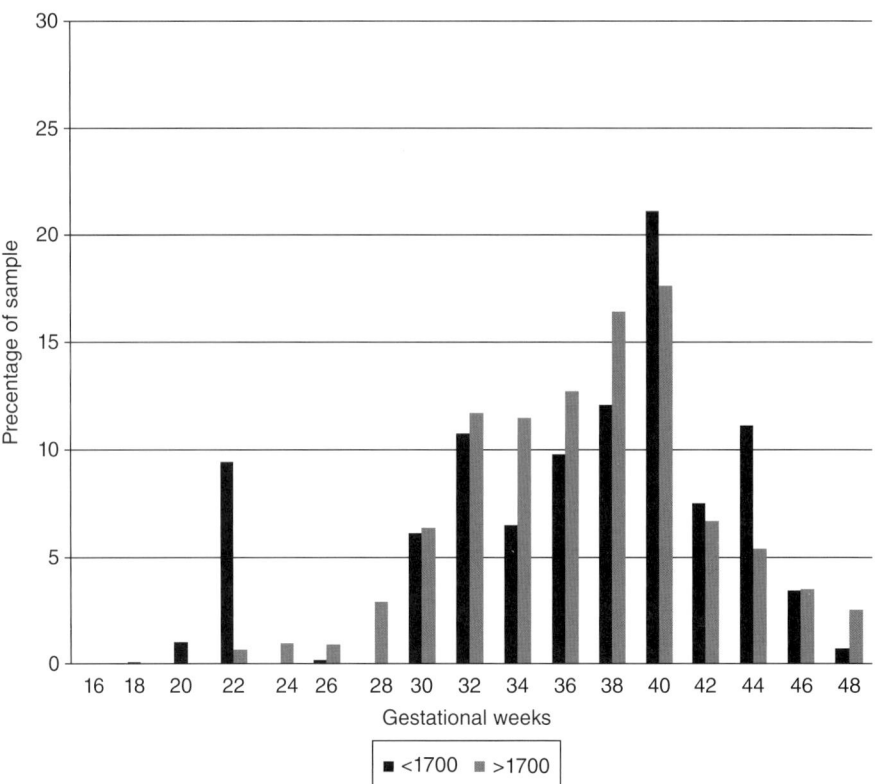

Figure 12.2. Bayesian-produced mortality profiles by gestational weeks.

decrease, while the pre-1700 group experienced these conditions in the third trimester and from 1 month to 1 year of age. Cribra orbitalia patterns also differed significantly between the two groups, with frequencies hovering between 40% and 50% from the late second trimester until 6 months of age in the post-1700 infants and then shooting up to 70% in the 7–11-month age group. The pre-1700 infants, on the other hand, experienced an incremental increase from the third trimester, peaking at 50% from 1 to 6 months of age, before seeing a slight decrease. However, if only those with advanced stages of cribra orbitalia (i.e., code 4 and code 5) are included in the analysis, only the post-1700 infants have examples of this advanced condition, particularly in the months before and around birth. Therefore, no pre-1700 infants had advanced stages of cribra orbitalia before the age of 1 year (Figure 12.3).

Breaking down these pathology frequencies by cemetery reveals that some intra-subset variation may be driving the overall differences between the groups. St. Thomas Hospital and the Smithfield Black Death cemeteries, which predate 1700, show the most infections in the subadult sample, particularly in the perinatal age category, while infants from the other cemeteries show no infections at all. Metabolic disorders are only seen in infants dying between 1 and 6 months of age at St. Thomas Hospital and infants from the Broadgate cemetery between 1 and 11 months of age. As with the overall sample, no divergent patterns in cribra orbitalia or porotic hyperostosis could be seen between the cemeteries.

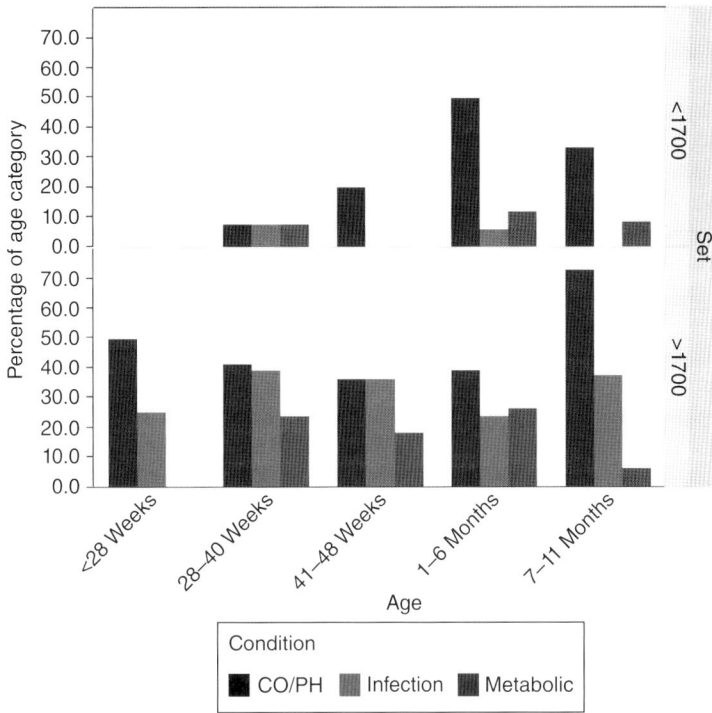

Figure 12.3. Frequency of pathological conditions by age-at-death for the pre- and post-1700 samples. *Weeks* refers to gestational weeks, and the other age categories are based on time since birth.

In the post-1700 samples, on the other hand, strong differences in the frequencies of different pathologies emerge between the cemetery samples. The Cross Bones cemetery clearly contains the most infants dying with infections, but metabolic disorders are prevalent in all cemeteries and only vary by age. The Cross Bones and St. Benet Sherehog cemetery samples exhibit high frequencies of these lesions in the perinatal and 1–6-month-old categories. Moderate-to-high frequencies also exist in the St. Brides Lower cemetery sample among infants between 1 and 11 months of age. In addition, very high frequencies (75–100%) of perinatal-age infants and 1–6-month-old infants manifest signs of cribra orbitalia and/or porotic hyperostosis in the Chelsea Old Church, St. Benet Sherehog, and St. Bride's Lower cemeteries. Intriguingly, these frequencies strongly decrease in these three samples in older infants, specifically those between 7 and 11 months of age.

While these intercemetery variations are not the focus of this study, it is possible that the influence of cemeteries containing more individuals of lower socioeconomic status in the post-1700 group (i.e., Cross Bones, St. Bridge's Lower) may explain some of the differences between the pre- and post-1700 periods, rather than an overall shift in health and disease in all segments of London's population during this period.

Discussion

Even this simple analysis demonstrates the difficulties in using skeletal data to understand shifts in population morbidity and mortality. Some of these issues are methodological. For instance, nonspecific infectious lesions and metabolic disorders were not consistently noted to be active, healing, or healed by the researchers who compiled data in the WORD database, meaning that it was impossible to determine which infants died while suffering from these conditions in contrast to those who had survived these insults only to perish from something else. However, the stage of progression of cribra orbitalia was coded, which allowed for some identification of individuals who likely had not experienced healing or cessation of the underlying condition before death. Others are more material. Bias is pervasive across the samples based on the nature of the population they derived from (e.g., St. Thomas' Hospital), how they ended up in the cemetery (e.g., the East Smithfield Black Death cemetery), and the potential underrepresentation of infants in the samples, particularly those predating 1700.

It is clear that the pre-1700 sample seems to have fewer infants than expected (3% of the entire pre-1700 sample, as compared to 16% of the entire post-1700 sample), thus any differences in pathological conditions in these age categories between the two eras may be affected strongly by sampling bias. In addition, poor preservation of subadult remains in the pre-1700 samples hindered age estimation efforts using femoral length, which would exclude these individuals from the Bayesian-generated age-at-death profile. However, it is unexpected that the pre-1700 sample actually had *more* preserved infants under 28 gestational weeks (i.e., with the least ossified skeletal tissue) than the post-1700 sample. This may suggest that factors other than preservation resulted in the lower-than-expected total number of infants in the pre-1700 sample. Differences in the two profiles, with more deaths occurring before 1 month of age in the post-1700 sample and a concentration of deaths in the pre-1700 sample after 1 month of age, suggest that endogenous factors played more of a role in the deaths of infants in the post-1700 group, while exogenous factors were more influential in the pre-1700 group. In addition, the decline of deaths of 7–11-month-old infants in the post-1700 group, which would presumably include infants that were being weaned, may suggest that weaning was not a strong exogamous factor in mortality. This supposition needs to be explored in more detail.

Not only did more infants perish before 1 month of age in the post-1700 sample than the pre-1700 sample, but many of them perished with signs of previous or current cribra orbitalia, nonspecific infection, and metabolic problems. Deaths during this period usually are attributed to endogenous factors of the mother and birth environment or congenital factors. The fact that a number of these late-term fetuses and neonates died with evidence of pathology further supports the presence of unhealthy uterine and peripartum environments during the industrial era. A number of infections of the mother can be transmitted vertically to the fetus, such as listeria, toxoplasmosis, syphilis, viral respiratory infections, and bacterial infections (Maclean et al., 2001). Many of these infections increase the morbidity and mortality risk of the neonate at birth; they can result in low birth weight, respiratory problems, organ malfunction, anemia, and death. Vitamin deficiencies, such as B_{12} deficiency, various anemias, and scurvy can also occur in neonates and infants (see Walker et al., 2009). These individuals are more likely to have low birth weight, premature delivery, poor health, and greater risk of death. In contrast, the pre-1700 group displayed proportionately fewer infant deaths before 1 month of age, and those who died had a lower frequency of cribra orbitalia, metabolic disorders, and nonspecific infectious conditions than the post-1700 group.

It seems that infants who survived to 1 month postpartum in the post-1700 group had lower chances of death after this period, although those who did die during the first month had significant frequencies of all the observed pathologies. The frequency of these conditions remains relatively high up to 1 year of age, with a slight drop from 1 to 6 months of age for nonspecific infections and cribra orbitalia, and a slight drop in metabolic disorders from 7 to 11 months of age. Cribra orbitalia remained a threat for the post-1700 group after

1 month of age and at a higher frequency than the post-1700 group for infants 1–6 months of age.

Despite the presence of sampling bias, it appears that the health of infants changed significantly before and after 1700. Infants had been exposed to disease, malnutrition, and vitamin deficiencies causing cribra orbitalia sometime before they perished more often in the post-1700 period than during the medieval era. Infants in industrial London began suffering from disease in the third trimester *in utero* and around the time of birth, implying that endogenous factors played a large role in perinatal mortality. It is possible that the increased mortality of infants before 1 month of age in the post-1700 group could reflect increased fertility during this period compared with the pre-1700 period; however, the fact that these infants were dying with signs of poor health and malnutrition may imply that a combination of environmental and demographic factors created this profile.

Unlike postepidemiologic transition societies, however, exogenous conditions during the first month of life for those living after 1700 were not salubrious for the infant. It is at this point that infants began dying with signs of nonspecific infection, vitamin deficiencies, anemia, and metabolic conditions, such as scurvy and rickets. In preindustrial London, the infants who perished did so without notable signs of pathology. It is difficult to hypothesize exactly what caused proportionately more mortality at this age, but one can presume that exogenous factors lead to acute conditions that quickly killed the infant before a pathological lesion could form.

The Age of Pestilence and Famine?

The changes in industry, family, and social structure, and migration patterns that marked the Industrial Revolution resulted in infants being weaned at earlier ages, increased family size, aggregation of different populations in urban centers, and an increase in environmental contamination. All of these factors should result in earlier sickness and death due to weaning, greater rates of respiratory and other conditions among younger individuals from the workplace environment, increased epidemic *crowd* diseases and other infections, and higher infant mortality in general than in previous eras.

Vital statistics confirming a general decline in health and higher rates of infant mortality have been identified in industrial urban populations in England (Woods et al. 1988, 1989; Williams and Galley, 1995). Omran (1971) labeled this period the *Age of Pestilence and Famine*, when infant mortality and fertility rates were high due to the preponderance of epidemic, acute infectious diseases. Although the consistency of this decline across England has been questioned (Lee, 1991), vital records indicate that a decrease in diarrheal diseases around 1900 that has been attributed to increased sanitation, and, among other factors, better infant feeding practices, led to healthier infants. The volatility of this change varied across cities with populations greater than 50,000 inhabitants, but the general trend was a decline, particularly from the 1890s to the early 1900s.

This decline in health with urbanization and industrialization is also mirrored in this and other studies of skeletal remains from cemeteries dated to pre- and post-1700 in England. However, similar to the vital statistics, these studies and the skeletal data analyzed here indicate that economic and geographic variability existed in the midst of this trend. As this study demonstrates, the urban context of London was clearly not conducive to the health of pregnant women and their fetuses, nor to newborns or infants under the age of 1. However, most of the industrial post-1700 cemeteries used here to generate conclusion represent the lower socioeconomic classes of the city. On the contrary, Lewis and Gowland's (2007) study of urban and rural populations from medieval, postmedieval, and industrial England revealed that endogenous factors did not strongly affect infant mortality. In addition, Lewis (2002) found that the skeletal stress indicators indicative of growth arrest, infection, and malnutrition, such as cribra orbitalia, dental enamel hypoplasias, nonspecific infectious lesions, maxillary sinusitis, and altered long bone growth, did not differ between the rural and urban samples. However, they did appear at a much younger age in urban and industrial populations, indicating the presence of maternal stress and infant stress during the first weeks of life. This stress continued into the postneonatal period in the samples from industrial London, though mortality in later infancy decreased in the other samples examined. As noted by Lewis and Gowland (2007), this difference likely stems not only from the lower endogenous deaths seen in rural environments, but also from the high social status of the urban sample from Spitalfields. In particular, Lewis and Gowland speculate that this latter factor may have translated into a greater ability to access medical advances, such as the use of forceps and male midwives in childbirth. Lewis (2002) also points out that upper-status infants often suffered disproportionately from metabolic disorders, which is attributed to their relatively early weaning and to the use of feeding formulas, such as pap, which had little to no nutritional value. It appears that the infants from other sectors of London, as shown in this study, were not shielded by the uterine environment, and many more lost their lives during the third trimester *in utero* or at birth than are is indicated for the Spitalfields sample.

In his model, Omran (1971) posits that morbidity and mortality from infectious diseases fluctuated in the centuries before 1650 but remained essentially high. On the other hand, in the case of London, according to the data presented here, infant morbidity and mortality increased dramatically around 1700, implying that the Age of Pestilence and Famine should be limited to the 200 years preceding the recorded transition in infectious disease morbidity and mortality rather than all of antiquity. This is not to say that pre-1700 populations did not suffer extensively from disease outbreaks and famine, but that these instances were limited and generally did not impact infant health in smaller communities from the 13th through 16th centuries in the area now covered by modern London anywhere near as much as they did during the industrial era.

The nature of the available skeletal data makes it impossible to track these trends past the 1850s. However, it is clear that the industrial revolution profoundly impacted the health of infants in London, with potentially higher infant mortality, higher disease loads, and greater stress from the third trimester *in utero* until 1 year of age. While increased fertility could explain the higher number of infant deaths post-1700, the increased frequency of these infants dying with pathological conditions compared to the pre-1700 group suggests that this is a true shift in morbidity and mortality.

According to epidemiologic transition theory, the second half of the 19th-century in England and Wales was marked by a decline in infant mortality and deaths from infectious disease, followed by a decrease in fertility. At this point, vital statistics serve to tell the rest of the story in terms of the transition. This narrowing of the evidence base is due to the lack of available skeletal samples from London after the 1850s, when burial within the city limits was outlawed. According to Omran's principles, the second transition would be marked by a decrease in diseases caused by nutritional deficiencies, an increase in public health measures, and higher neonatal mortality rather than postneonatal mortality. Consequent population growth should thus occur partly due to stabilization and the decrease in infant morbidity and mortality, particularly in the 1–4-year-olds (Omran, 1971). Mortality records report that London, and England as a whole, experienced a notable decline in infant mortality rates from 1860 to 1910, particularly after 1900 (Woods et al., 1988, 1989). Nonetheless, some internal variations in this decline have been noted based on social status, urban vs. rural environments, and economics (Woods et al., 1988, 1989; Williams and

Mooney, 1994; Haines, 1995; Williams and Galley, 1995). Reasons for this decline include local public health initiatives such as health visits to new mothers and regulating the sanitation of cow's milk and water (Williams and Mooney, 1994; Haines, 1995). In addition, the decade leading up to the 1890s was known for its abhorrent drought conditions, which compromised the water supply for many areas and led to increased diarrheal diseases in semilarge urban centers (Williams and Galley, 1995). In many areas, infant mortality peaked during this period, thus accentuating the decline experienced between 1890 and 1900.

While the role of decreased infant mortality rates in the epidemiologic transition seems to be well established, it appears that the impact of fertility is negligible (Watterson, 1988). Some demographers argue that the decline in infant and childhood mortality actually postdates the decline in fertility (Williams and Galley, 1995), contrary to Omran's model. This would put fertility's role in the transition on the back burner. On the other hand, other historians and demographers argue that increased fertility played a stronger role in population growth during the transition than did decreased mortality. Humphries (2010) suggests that population growth between 1750 and 1850 was driven by young women and men setting up independent households at earlier ages, extending the period of fecundity in women. In addition, Fildes (1986) points out that women had successively weaned their infants earlier and earlier in the 16th–19th centuries, which, in combination with earlier childbearing, would have been a strong impetus for population growth. Furthermore, having an increased number of children was not a financial problem, for real wages increased and younger and younger people experienced adequate adjusted earning power, reflecting what could be termed an *industrious* rather than *industrial* revolution (Humphries, 2010: 40). Based on the skeletal evidence up to 1850, which showed a sustained high infant morbidity and mortality, this may be the case, presuming that these trends do not reflect increased fertility in post-1750 urban populations.

Conclusion

The epidemiologic transition model appears on the surface to neatly package the effects of changes in morbidity, mortality, and fertility in industrializing or industrialized populations. However, other scholars, including Barrett and colleagues (1998), have noted that the model masks internal variation driven by differences in socioeconomic status, environmental conditions, and culture. Skeletal data and vital records clearly demonstrate the poor health of infants, and presumably their mothers, during the years leading up to the decline in epidemic infectious disease frequencies. Skeletal data from London dating prior to 1700 show that this high infant morbidity and mortality did not always exist at the levels seen during the industrial era, and in fact, there was a substantial increase *before* the transition-related decrease. The epidemiologic transition therefore could be viewed, as Barrett and colleagues do, as one of many shifts in diseases and mortality that have occurred for millennia with sociocultural, technological, subsistence, and economic changes.

It is clear that the problems faced by bioarchaeologists can only be solved through relying on datasets beyond skeletal remains, including vital records, environmental effects, and historical, social, and cultural views of infancy and childhood. Bioarchaeologists can provide valuable raw data on age-at-death and morbidity patterns during this period, but identifying the underlying factors of these patterns can only come through incorporation of contextual information from historical texts. This amalgamation must include critical assessment of both data sources—historical texts should not be privileged over other data and bioarchaeological data need to be understood within the context of the factors involved in creating the skeletal sample, such as the characteristics of the original living population, as reconstructed through historical texts. In addition, the interpretation of skeletal data has to incorporate a culturally informed understanding of infancy and childhood in order to understand what risk factors different individuals may be exposed to at different life stages, which can fluctuate based on relevant social and economic factors. This chapter seeks to achieve these goals and bridge these gaps for understanding the second epidemiologic transition.

References

Barrett R, Kuzawa CW, McDade T, Armelagos GJ. 1998. Emerging and re-emerging infectious diseases: the third epidemiologic transition. Annu Rev Anthropol 27:247–271.
Baxter JE. 2005. The archaeology of childhood. Walnut Creek: Alta Mira Press.
Bocquet-Appel J-P, Masset C. 1982. Farewell to palaeodemography. J Hum Evol 11:321–333.
Bocquet-Appel J-P, Masset C. 1996. Palaeodemography: expectancy and false hope. Am J Phys Anthropol 99:571–584.
Butler NR, Alberman ED. 1969. Perinatal problems. Edinburgh: Livingstone.
Chadwick E. 1843. Report on the sanitary condition of the labouring population of Great Britain: A supplementary report on the results of a spiecal [sic] inquiry into the practice of interment in towns. London: Clowes.
Chamberlain AT. 2006. Demography and archaeology. Cambridge: Cambridge University Press.
Coale AJ, Demeny P. 1983. Regional model life-tables and stable populations. New York: Academic Press.
Fildes V. 1986. Breasts, bottles, and babies: A history of infant feeding. Edinburgh: Edinburgh University Press.
Fildes V. 1988. Wet-nursing: A history from antiquity to present. New York: Basil Blackwell.
Gordon CC, Buikstra JE. 1981. Soil pH, bone preservation, and sampling bias at mortuary sites. Am Antiq 48:566–571.
Gowland RL, Chamberlain AT. 2002. A Bayesian approach to ageing perinatal skeletal remains: implications for the evidence of infanticide in Roman Britain. J Archaeol Sci 29:677–685.
Gowland RL, Chamberlain AT. 2005. Detecting plague: palaeodemographic characterization of a catastrophic death assemblage. Antiquity 79:146–157.
Guy H, Masset C, Baud C-A. 1997. Infant taphonomy. Int J Osteoarchaeol 7:221–229.
Haines MR. 1995. Socio-economic differentials in infant and child mortality during mortality decline England and Wales, 1890–1911. Popul Stud 49:297–315.
Hoppa RD, Vaupel JW. 2002. The Rockstock Manifesto for paleodemography: the way from age to stage. In: Hoppa RD, Vaupel JW, editors. Paleodemography: Age distributions from skeletal samples. Cambridge: Cambridge University Press. p 1–8.
Humphries J. 2010. Childhood and child labour during the British industrial revolution. Cambridge: Cambridge University Press.
Jones E, Ubelaker DH. 2001. Demographic analysis of the Voegtly cemetery sample, Pittsburgh, Pennsylvania. Am J Phys Anthropol Suppl 32:86.
Konigsberg LW, Frankenberg SR. 1992. Estimation of age structure in anthropological demography. Am J Phys Anthropol 89:235–256.
Konigsberg LW, Frankenberg SR. 1994. Palaeodemography: "not quite dead". Evol Anthropol 3:92–105.
Lee CH. 1991. Regional inequalities in infant mortality in Britain, 1861–1971: patterns and hypotheses. Popul Stud 45:55–65.
Lewis ME. 2002. The impact of industrialization: comparative study of child health in four sites from medieval and post-medieval England (AD 850–1859). Am J Phys Anthropol 119:211–223.
Lewis ME. 2007. Bioarchaeology of children: Perspectives from biological and forensic anthropology. Cambridge: Cambridge University Press.
Lewis ME, Gowland R. 2007. Brief and precarious lives: infant mortality in contrasting sites from Medieval and post-Medieval England (AD 850–1859). Am J Phys Anthropol 134:117–239.

Maclean AB, Regan L, Carrington D, editors. 2001. Infection and pregnancy, 3rd edn. London: RCOG Press.

Mays, S. 2010. The effects of infant feeding practices on infant and maternal health in a medieval community. Child Past 3:63–78.

Omran AR. 1971. The epidemiologic transition: a theory of the epidemiology of population change. Milbank Med Fund Q 49:509–538.

Ortner DJ. 2003. Identification of pathological conditions in human skeletal remains. New York: Academic Press.

Paine RR. 1997. The need for a multidisciplinary approach to prehistoric demography. In: Paine RR, editor. Integrating archaeological demography: Multidisciplinary approaches to prehistoric populations. Carbondale: Center for Archaeological Investigations, Southern Illinois University. p 1–18.

Paine R, Harpending HC. 1998. Effect of sample bias on paleodemographic fertility estimates. Am J Phys Anthropol 105:231–240.

Perry MA. 2006. Redefining childhood through bioarchaeology: toward an archaeological and biological understanding of children in antiquity. In: Baxter JE, editor. Children in action: Perspectives on the archaeology of childhood. Archeological Papers of the American Anthropological Association 15. Washington, DC: American Anthropological Association. p 89–111.

Powers N. 2012. Human osteology method statement. London: Museum of London.

Samworth R, Gowland R. 2007. Estimation of adult skeletal age-at-death: statistical assumptions and applications. Int J Osteoarchaeol 17:174–188.

Sattenspiel L, Harpending HC. 1983. Stable populations and skeletal age. Am Antiq 48:489–498.

Saunders SR, Barrans L. 1999. What can be done about the infant category in skeletal samples? In: Hoppa RD, Fitzgerald CM, editors. Human growth in the past: Studies from bones and teeth. Cambridge: Cambridge University Press. p 183–209.

Saunders SR, Herring DA, Boyce G. 1995. Can skeletal samples accurately represent the living population they come from? The St. Thomas' cemetery site, Belleville, Ontario. In: Grauer AL, editor. Bodies of evidence: Reconstructing history through skeletal analysis. New York: Wiley-Liss. p 69–89.

Scheuer L, Musgrave JH, Evans SP. 1980. The estimation of late fetal and perinatal age from limb bone length by linear and logarithmic regression. Ann Hum Biol 7:257–265.

Storey R, Hirth K. 1997. The El Cajon, Honduras skeletal population: Archaeological and paleodemographic approaches to understanding juvenile representation and low population fertility. In: Paine RR, editor. Integrating archaeological demography: Multidisciplinary approaches to prehistoric population. Carbondale: Center for Archaeological Investigations. p 131–149.

Walker PL, Johnson JR, Lambert PM. 1988. Age and sex biases in the preservation of human skeletal remains. Am J Phys Anthropol 76:183–188.

Walker P, Bathurst R, Richman R, Gjerdrum T, Andrushko V. 2009. The causes of porotic hyperostosis and cribra orbitalia: a reappraisal of the iron-deficiency-anemia hypothesis. Am J Phys Anthropol 139(2):109–125.

Watterson PA. 1988. Infant mortality by father's occupation from the 1911 census of England and Wales. Demography 25:289–306.

Williams N, Mooney G. 1994. Infant mortality in an 'Age of Great Cities': London and the English provincial cities compared, c. 1840–1910. Contin Chang 9:185–212.

Williams N, Galley C. 1995. Urban–rural differentials in infant mortality in Victorian England. Popul Stud 49:401–420.

Wood JW, Milner GR, Harpending HC, Weiss KM. 1992. The osteological paradox: problems in inferring prehistoric health from skeletal samples. Curr Anthropol 33:343–358.

Woods RI, Watterson PA, Woodward JH. 1988. The causes of rapid infant mortality decline in England and Wales, 1861–1921. Part I. Popul Stud 42:343–366.

Woods RI, Watterson PA, Woodward JH. 1989. The causes of rapid infant mortality decline in England and Wales, 1861–1921. Part II. Popul Stud 43:113–132.

Chapter 13
The Biological Effects of Urbanization and In-Migration on 19th-Century-Born African Americans and Euro-Americans of Low Socioeconomic Status: An Anthropological and Historical Approach

Carlina de la Cova
Department of Anthropology, University of South Carolina, Columbia, SC

Introduction

The 19th and early 20th centuries were turbulent times in American history, characterized by industrialization, the Civil War, the dissolution of slavery, financial depressions, in-migration, and increased competition in the labor market. The marginalized lower classes were especially susceptible to the periods of instability wrought by this turbulence, given their limited access to important resources such as stable work, health care, and sanitary housing. Importantly, understanding how individuals in these communities were biologically affected by these processes allows for a better comprehension of historic differences in health and mortality rates. Studies on past salubrity can also assist researchers in comprehending health disparities in contemporary American populations, as health disparities continue to track socioeconomic disparities. Central to this enterprise is discerning when the second epidemiologic transition occurred in the U.S. This transition is associated with economic change and industrialization in the 19th and early 20th centuries in Western Europe and the U.S., and is characterized by "a long-term shift in mortality and disease patterns whereby pandemics of infection are gradually displaced by degenerative and man-made disease as the chief form of morbidity and primary cause of death" (Omran, 1971: 516). However, evidence suggests that this transition was not uniformly experienced across populations within the U.S. In particular, historical and epidemiologic evidence suggests that minority and low socioeconomic status (SES) communities may have actually experienced an increase in various diseases and health disparities over the past two centuries (Riis, 1914; Giffin, 2005).

Skeletal remains can be studied to determine when the epidemiologic transition occurred and if all Americans, regardless of race and SES, experienced it. Bones can show evidence of poor nutrition and disease through the presence of pathologies. For instance, previous studies of skeletal health among the 19th- and early 20th-century-born African Americans and Euro-Americans indicate the presence of health disparities. Enslaved African Americans suffered from malnutrition, disease, growth disruption, enamel hypoplasia (EH), anemia, arthritis, trauma, and high mortality rates (Kelley and Angel, 1987; Martin et al.,

1987; Owsley et al., 1987; Rathbun, 1987; Rathbun and Scurry, 1991; Rankin-Hill, 1997; Blakey, 2001; Null et al., 2004; de la Cova, 2010). Emancipated African Americans born after 1866 also experienced similar issues, ranging from nutritional stress, growth disturbances, to higher rates of infectious diseases, such as tuberculosis (TB), suggesting a decline in health after liberation (Rose, 1985, 1989; Watkins, 2003). Euro-Americans born during these periods also experienced poor health, with skeletal remains exhibiting growth disruption, EH, poor oral health, porotic hyperostosis (PH), disease, periosteal reactions, and trauma (Lanphear, 1990; Rathbun and Scurry, 1991; Higgins et al., 2002; Rathbun and Steckel, 2002; Saunders et al., 2002; Sledzik and Sandberg, 2002; de la Cova, 2010).

These studies highlight health disparities, but few have examined a large sample of African Americans and Euro-Americans of low SES from similar environments for temporal transitions in salubrity associated with the second transition. This chapter analyzes the skeletal health of African Americans and Euro-Americans ($n = 651$) born during the Antebellum (1812–1860), Civil War (1861–1865), and Reconstruction (1866–1877) eras, to determine if ethnic and temporal differences in skeletal health were present. Since these individuals lived through the eras of industrialization and urbanization, their disease patterns were observed to determine if traces of the second transition could be detected. Furthermore, African Americans born during enslavement (1812–1865) and after emancipation (1866–1877) were also examined for dissimilarities in health. It is argued that the health disparities observed in this sample are the result of migration, urbanization, low SES, and discrimination.

Background

Historical Overview

In the U.S., the second transition was contemporaneous with the Industrial Revolution (Barrett et al., 1998). The U.S. began to industrialize during the Antebellum era (1812–1860). This process, mostly focused in the northern states, changed the face of the country through population increase, propagated wealth, and better transportation. Expansion of national borders in the 19th century, along with new machinery, resulted in an agricultural surplus that prompted population growth (Donald et al., 2001). The invention of the steam engine and steam-related transport also made travel faster and more efficient, which increased disease mobility. Denser populations and ease of travel also meant a rise in infectious disease and competition for access to food resources (Haines, 1998; Haines et al., 2003).

Despite rapid economic growth, the health of the country declined. U.S. height data from this time period shows a reduction in stature from the 1830s until the 1860s (Komlos, 1996, 1998; Haines, 1998; Haines et al., 2003). This height decrease, known as the "Antebellum Puzzle," is concurrent with the economic advances and the improvements in the standard of living associated with the Industrial Revolution. These factors usually result in a positive secular trend in stature, making the stature reduction an enigma. Many economists believe that Antebellum Puzzle was in fact caused by population explosion, increased mobility, and resource competition (Haines, 1998; Haines et al., 2003). Contending for food in a landscape experiencing a population explosion may have induced periodic dietary deficiencies and nutritional stress, which may have resulted in compromised immune systems and easier acquisition of infectious diseases, thus translating into reduced stature. The Antebellum Puzzle may also have been influenced by race and SES (Komlos, 1996; Lang and Sunder, 2003; Carson, 2009; Haines et al., 2011). Stature records of Euro-Americans of high SES show no evidence of the Antebellum Puzzle (Lang and Sunder, 2003; Sunder, 2004; Haines et al., 2011), likely because of the access to high-quality resources and good nutrition available to high SES Euro-Americans during growth and development throughout the U.S. Intriguingly, heights for enslaved African Americans also do not show significant declines during this period (Komlos and Coclanis, 1997; Carson, 2009; Haines et al., 2011). This contrasts with the decreasing stature observed among low SES Euro-Americans and free-born African Americans. It has been argued that the lack of a "puzzle" among slaves may have resulted from rationed food and the control slave owners exerted over nutritional and living environments (Haines et al., 2011).

The Civil War slowed the economic prosperity of the country and greatly altered its social structures. Many African Americans were involved in the conflict, but regardless of race, soldiers on both sides suffered from wounds, disease, and malnutrition, particularly in association with camp life. Rural recruits that had never been exposed to infectious childhood diseases such as measles or whooping cough sparked epidemics in regimental camps. Poor sanitation caused dysentery, typhoid fever, jaundice, and typhus. Warm weather and insect vectors exposed soldiers to unwanted fevers and malaria. Colds, coughs, pneumonia, bronchitis, and TB were also common. Ultimately, diseases claimed the most lives, resulting in two-thirds of the 600,000 war-related deaths (Bollet, 2002).

As the North mustered African American recruits after the Emancipation Proclamation in 1863, the Confederacy forced free and enslaved African Americans to build fortifications and work in munitions factories throughout the South (Nolen, 2001). However, slaves fled and sought refuge with the Union army in overwhelming numbers as the war continued. Officers initially placed the runways in contraband camps on the periphery of Federal encampments and assigned the men work as laborers, cooks, launderers, and servants (Nolen, 2001; Reid, 2002). However, in 1863, after President Lincoln approved the enlistment of African Americans, all able-bodied African American males were conscripted into the U.S. Army. They were denied the same pay, rations, and medical care as Euro-American soldiers and assigned to humiliating fatigue duty, which consisted of military manual labor (Higginson, 1984). Meanwhile, their elders, wives, and children remained behind in overcrowded contraband camps, where food and supplies were limited and diseases spread rapidly. By late 1862, these conditions forced the army to organize the refugees into designated settlements and provide some rations, clothing, and medicine. However, high mortality rates, crowding, and malnutrition still plagued the camps (Click, 2001; Reid, 2002). Many would perish from disease and nutritional deficiencies before attaining freedom (Reid, 2002).

The Civil War's end ushered in Reconstruction (1866–1877), the Gilded Age (1878–1900), and later the Progressive Era (1900–1917). The Southern economy stagnated during these eras while the industrializing North advanced economically. Many Southern African Americans chose to flee the oppressive discrimination, Jim Crow laws, sharecropping system, and Ku Klux Klan persecution associated with the South and relocate to industrialized cities like Cleveland, Ohio, and St. Louis, Missouri (Phillips, 1999; Giffin, 2005). This in-migration, from 1916 to 1930, of over two million Southern African Americans to Northern, Midwestern, and Western industrial cities in search of improved lives, better financial opportunities, and racial integration was known as the Great Migration (Marks, 1989).

Given the economic and social processes operating during these time periods, a decline in health would be expected, especially for African Americans and Euro-Americans of low SES. Overall, industrialization and disruption caused by the Civil War would have resulted in competition for resources and poor nourishment. Military camp conditions encouraged the rise of infectious diseases. Reconstruction and the years that followed introduced Southern-born African Americans to new urban environments, with poor housing and limited access to quality health care, which led to increased rates of infectious diseases. The 1880, 1890, and 1900 U.S. censuses support this, indicating that African Americans had higher rates of TB, pneumonia, scrofula, and venereal diseases.[1] However, few anthropological studies have addressed long-term shifts in health during these time periods using a large sample of African Americans and Euro-Americans (Rathbun and Steckel, 2002). Even fewer have examined the relationship between race, SES, and the second transition through the analysis of skeletal remains. By studying the skeletons of low SES African Americans and Euro-Americans who lived through industrialization and the second transition, a more complete picture of trends in salubrity, health disparities, and the interrelationship between industrialization, urbanization, race, and class can emerge.

Materials, Methods, and Urban Settings

Skeletal Data

A sample of Euro-American ($n = 290$) and African American males ($n = 361$) born from 1822 to 1877 were selected from the Hamann-Todd, Robert J. Terry, and William Montague Cobb anatomical collections (Table 13.1). For the purposes of this study, Euro-American is defined as an individual of European descent born in the U.S. Precautions were taken against the inclusion of Europeans by reviewing the U.S. census and morgue records to ensure subjects' places and years of birth. If discrepancies existed between census and morgue records with respect to birth year and age, then U.S. census data were relied upon.

All individuals were analyzed for evidence of biological stress, disease, and dietary deficiencies based on the disorders, pathologies, and methodologies cited in Table 13.2.

Skeletons were macroscopically examined for the presence of rickets, syphilis, osteomyelitis (Ortner, 2003), PH (Stuart-Macadam, 1989), and skeletal TB (Kelley and Micozzi, 1984; Roberts et al., 1994; Ortner, 2003). PH, which causes pitting and porosity of the cranial vault, is most often associated with biological stress as its etiology has been tied to various anemias and vitamin deficiencies (Walker et al., 2009). To ensure that syphilis and skeletal TB were accurately accessed, morgue records of individuals with these suspected diseases were examined to confirm that this was diagnosed at or before death. The presence of additional infectious diseases, such as polio, which causes atrophied limb bones, was also recorded.

Dental health, also an indicator of salubrity, hygiene, diet, and socioeconomic factors, was evaluated by edentulism, or complete loss of all teeth, and the presence of dental caries, abscesses, fillings, and EH (Sledzik and Moore-Jansen, 1991; Steckel et al., 2002). All available teeth were macroscopically examined for dental caries, which were recorded based on the presence of crown destruction. Abscesses were recorded based on alveolar bone destruction in the maxillae and mandible. If destruction was present, then the corresponding tooth was considered abscessed if it was carious. Caries and abscesses were recorded as present or absent by subject and tooth affected. Individuals were also examined for EH, which is a nonspecific indicator of biological stress associated with malnutrition, disease, weaning, and parasitism. All available anterior teeth undamaged by dental wear or postmortem breakage were macroscopically examined for EH and recorded by subject and tooth (Goodman et al., 1987; Goodman and Rose, 1990).

---

[1] John S. Billings, Report on the Mortality and Vital Statistics of the U.S. as Returned at the Tenth Census (June 1, 1880), Part I, Department of the Interior, Census Office, Washington, Government Printing Office, 1885. John S. Billings, Report on Vital and Social Statistics in the U.S. at the Eleventh Century: 1890, Part I, Analysis and Rate Tables, Department of the Interior, Census Office, Washington, D.C., Government Printing Office, 1896. William A. King, Census Reports, Volume III, Twelfth Census of the U.S., Taken in the Year 1900, Vital Statistics, Part I, Analysis and Ration Tables, Washington, U.S. Census Office, 1902.

Table 13.1. Sample size and birth cohorts examined in data analysis.

	Hamann-Todd	Terry	Cobb
African American	171	117	73
Euro-Americana[a]	19	258	13
Total	190	375	86

Birth cohorts

Birth cohorts (years of birth)		
Ancestry cohorts		
African American	Antebellum (1812–1860)	Pre-Reconstruction (1812–1865)
Euro-American	Civil War (1861–1865)	Reconstruction (1866–1877)
	Reconstruction (1866–1877)	

Collection (location)	Ancestry/birth cohorts	Birth status cohorts
Terry (St. Louis, Missouri)	Antebellum White	Enslaved Black (1812–1865)
Hamann-Todd (Cleveland, Ohio)	Antebellum Black	Pre-Reconstruction White (1812–1865)
Cobb (Washington, DC)	Civil War White	Liberated Black (1866–1877)
	Civil War Black	Reconstruction White (1866–1877)
	Reconstruction White	
	Reconstruction Black	

Table 13.2. Pathological conditions examined.

Dietary deficiencies	Infectious diseases	Dental defects
(present/absent)	(present/absent)	(present/absent)
Rickets—(Ortner, 2003)	Syphilis—(Ortner, 2003)	Enamel hypoplasia
Porotic hyperostosis (Stuart-Macadam, 1989)	Tuberculosis—(Kelley and Micozzi, 1984; Roberts et al., 1994; Ortner, 2003)	Caries
		Fillings
		Abscesses

Remains were also examined for the pathologies listed in Table 13.2. All were recorded as present or absent and statistically analyzed by the cohorts defined in Table 13.1 using chi-squared analysis. The exception to this was EH as due to the poor preservation of most teeth in the sample, EH could not be scored as absent. Therefore, only the prevalence of EH is reported. Individuals were placed in ancestry, birth, combined ancestry/birth, birth status, and collection cohorts to determine if ethnic and temporal differences, as well as shifts among enslaved and liberated African Americans, in disease rates, and in biological stress, existed.

Stature, which is considered a sensitive indicator of childhood health influenced by environment, SES, nutrition, and disease exposure, was also studied (Bogin, 1999). The maximum femoral lengths of 621 males from the Cobb, Hamann-Todd, and Terry collections were measured. If left bones were absent or pathological, their antimere was used. Stature was calculated using Trotter's (1970) formulae. Height estimates were examined using t-tests for the ancestry cohorts and ANOVA analyses for birth, combined ancestry/birth, birth status, and collection cohorts. Statistical significance was set at $p \leq 0.05$ for all tests.

Skeletal Collections

The skeletons examined in this study were from the Hamann-Todd, Robert J. Terry, and William Montague Cobb anatomical collections, which allowed for the construction of a large sample of African Americans and Euro-Americans who lived in similar industrialized cities (i.e., Cleveland, Ohio; St. Louis, Missouri; and Washington, DC). The Hamann-Todd Collection, housed at the Cleveland Museum of Natural History, contains the remains of about 3100 persons born between 1825 and 1910 that expired in Cleveland, Ohio. Hamann-Todd is mostly comprised of unclaimed bodies from public city hospitals and the Cuyahoga County Morgue. Each individual has been documented to ensure that the correct data on name, age, sex, race, and cause of death were recorded (Meindl et al., 1990; Jones-Kern and Latimer, 1999).

The Robert J. Terry Collection, from St. Louis, Missouri, contains 1728 individuals born from 1822 to 1943, whose age, sex, pathologies, cause of death, and place of death are known (Hunt and Albanese, 2005). Most of the collection is comprised of unclaimed bodies from the St. Louis City Hospital, City Sanitarium, morgue, Missouri state mental institutions, and other charity hospitals, but shifted to willed donations in 1955 (Trotter, 1981). Studies have demonstrated that willed individuals have longer femoral lengths than the unwilled (Ericksen, 1982); however, the unclaimed are more representative of the general population in regard to secular trends in femoral length, indicating that body donors had better economic standing.

The William Montague Cobb Collection, housed at Howard University in Washington, DC, is comprised of mostly unclaimed African Americans that died from 1932 to 1969 in local Washington, DC, almshouses and hospitals. It contains 987 skeletons, 700 of which are documented with demographic, anatomical, and medical information (Rankin-Hill and Blakey, 1994; Muller, 2004). Records indicate that many of the individuals are from local hospitals and geriatric care homes that acted as charitable organizations and provided health care to the city's African American and indigent population, particular the poor, unemployed, and disabled (Watkins, 2003).

Using these collections has advantages and disadvantages. Unlike archaeological skeletal samples, which vary in size, social status, and geographic location, anatomical collections can provide large samples of individuals from similar socioeconomic backgrounds and environments. However, anatomical collections are usually skewed toward older persons, males, and African Americans. Regardless of this, these samples can still provide insight into life and health in the past through the examination of their demography and pathologies. For this study, the greater numbers of African Americans permitted a large-scale analysis of health during enslavement and after liberation. Furthermore, the persons examined were of low SES and received their health care from public and charity institutions such as local poor houses, public hospitals, or city sanitariums. They perished in these organizations, and given the unclaimed nature of their remains, laws dictated that their bodies be delivered to medical schools for dissection. To ensure that SES remained consistent, morgue and personal records were examined to guarantee that only unclaimed individuals were analyzed. Willed bodies were excluded from this study. This allowed for a comparison of health among African Americans and Euro-Americans, as they were of a similar SES when they expired.

Cleveland, Washington, DC, and St. Louis as Urban Industrial Centers

The collections are associated with three major urban centers and industrialized cities. Washington, DC, home to the Cobb Collection, was and still remains a major metropolis. The cities associated with the Hamann-Todd and Terry collections, Cleveland and St. Louis, were major industrial towns in the 19th and 20th centuries, which specialized in steel, iron, and automobile production (Primm, 1998; Miller and Wheeler, 1990). Cleveland, home to the Hamann-Todd Collection, rose to prominence as an industrial and manufacturing center during the Civil War. By 1837, the city was a major hub for iron and steel production. The petroleum refining industry also played an important role in the local economy, spurring the development of the city's chemical, paint, and varnish businesses including Sherwin Williams and Glidden. Auto production was the last industry to emerge at the end of the 19th century through the creation of high-end luxury gas, steam, and electric-powered vehicles (Miller and Wheeler, 1990). By 1910, Cleveland lost its prominence in the motorcar industry but continued to play a major role in the production of steel, aircraft, and auto parts until the 1950s.

St. Louis, like Cleveland, was an industrial town located in the Midwest. In the 1880s, Anheuser-Busch made brewing the city's most popular and largest commerce, followed by iron and steel production, tobacco processing, flour milling, slaughtering, brickmaking, paint making, and machining (Primm, 1998; Hodes, 2009). From the 1920s to the 1930s, clothing and chemicals were manufactured in St. Louis, and the city contained numerous car assembly plants through the 1950s (Eyssell, 2006). American involvement in World War II shifted manufacturing to ammunitions, Curtiss-Wright aircrafts, chemicals, uniforms, and medicines (Primm, 1998).

Historical Research and Methodologies

Skeletal analyses were used in collaboration with historical research to comprehend observed patterns of disease and infection. Understanding the historical context of any sample is critical to comprehending the impact that environment, culture, and personal behavior may have on health (Rankin-Hill, 1997). Therefore, historical methods were employed by reading and analyzing primary sources such as census records, family papers, medical records, public health documents, newspapers, and personal letters for information that could be used to reconstruct the socioeconomic and cultural contexts of the shifting environments in which individuals included in the sample lived. These sources can also provide insight into living environments and social behaviors from the first-person perspective. Records from the National Archives, Library of Congress, Missouri Historical Society, Ohio Historical Society, and from contemporary newspapers (sourced using America's Historical Newspapers and 19th Century U.S. Newspapers databases) were researched to assess what the environment and daily life was like for African Americans, Euro-Americans, wage laborers, and persons of low SES living in 19th- to 20th-century Cleveland, St. Louis, and Washington, DC.

Basic demographic data on all persons came from morgue records accompanying the remains. Further information on the subjects, including birthplace, residence, and occupation, was acquired by investigating the U.S. census. Individuals were searched by city, name, and birth date. Additional data from morgue records, including the hospital in which the person expired, parent's birthplace, or occupation were utilized, if recorded, to ensure that the proper subjects were located in the census. Previous demographic studies done on the collections were also consulted (Cobb, 1935; Watkins, 2003; Hunt and Albanese, 2005).

Results

The average age at death in the sample was 65.51 years (Table 13.3). African Americans and Euro-Americans were of comparable ages for all time periods. A decline in age is present between the birth cohorts, with Reconstruction-born individuals being the youngest. Hamann-Todd also had younger individuals when compared to the other collections. These differences probably reflect the manner in which cadavers were amassed (de la Cova, 2010).

Chi-squared and frequency analyses of PH, rickets, syphilis, skeletal TB, and osteomyelitis are reported in Table 13.4. More than three-fourths (85.6%) of the sample suffered from PH. Euro-Americans, the Civil War cohort, and Pre-Reconstruction Euro-Americans had

Table 13.3. Average ages of cohorts.

Cohort	N	Mean
Entire sample	651	65.515
African American	360	65.019
Euro-American	291	64.127
Antebellum	175	75.029
Civil War	133	68.000
Reconstruction	343	59.697
Antebellum White	91	75.077
Antebellum Black	84	74.976
Civil War White	60	67.917
Civil War Black	73	68.068
Reconstruction White	140	59.543
Reconstruction Black	203	59.803
Terry	375	67.149
Hamann-Todd	189	58.545
Cobb	87	73.609

Table 13.4. Chi-squared analyses of ancestry, birth, and collection cohorts and pathologies

Skeletal pathologies

Cohorts	Porotic hyperostosis N/total (%), p	Rickets N/total (%), p	Treponematosis N/total (%), p	Tuberculosis N/total (%), p	Osteomyelitis N/total (%)[a]
Entire sample	495/578 (85.6)	26/644 (4.0)	30/644 (4.7)	28/644 (4.3)	9/605 (1.5)
African American	263/313 (84.0), 0.229	17/357 (4.8), 0.297	24/357 (4.2), 0.006	23/357 (6.4), 0.004	8/321 (2.5)
Euro-American	232/265 (87.5)	9/287 (3.1)	6/287 (2.1)	5/287 (1.7)	1/284 (0.4)
Antebellum	129/154 (83.8), 0.696	7/175 (4.0), 0.957	5/175 (2.9), 0.277	3/175 (1.7), 0.073	0/166 (0.0)
Civil War	103/118 (87.3)	6/135 (4.5)	9/134 (6.7)	5/134 (3.7)	2/126 (1.6)
Reconstruction	263/306 (85.9)	13/335 (3.9)	16/335 (4.8)	20/335 (6.0)	7/313 (2.2)
Antebellum White	69/81 (85.2), 0.802	3/91 (3.3), 0.943	2/91 (2.2), 0.048(b)	1/91 (1.1), 0.015(a)	0/90 (0.0)
Antebellum Black	60/73 (82.2)	4/84 (4.8)	3/84 (3.6)	2/84 (2.4)	0/76 (0.0)
Civil War White	52/58 (89.7)	2/60 (3.3)	2/60 (3.3)	0/60 (0.0)	1/60 (1.7)
Civil War Black	51/60 (85.0)	4/74 (5.4)	7/74 (9.5)	5/74 (6.8)	1/66 (1.5)
Reconstruction White	111/126 (88.1)	4/136 (2.9)	2/136 (1.5)	4/136 (2.9)	0/134 (0)
Reconstruction Black	152/180 (84.4)	9/199 (4.5)	14/199 (7.0)	16/199 (8.0)	7/179 (3.9)
Pre-Reconstruction (PR)	232/272 (85.3), 0.823	13/309 (4.2), 0.833	14/309 (4.5), 0.883	8/309 (2.6), 0.036	2/292 (0.7)
Reconstruction	263/306 (85.9)	13/335 (3.9)	16/335 (4.8)	20/335 (6.0)	7/313 (2.2)
Enslaved Black	111/133 (83.5), 0.667	8/158 (5.1), 0.758	10/158 (6.3), 0.046	7/158 (4.4), 0.007	1/142 (0.7)
FR White	121/139 (87.1)	5/151 (3.3)	4/151 (2.6)	1/151 (0.7)	1/150 (0.7)
Liberated Black	152/180 (84.4)	9/199 (4.5)	14/199 (7.0)	16/199 (8.0)	7/179 (3.9)
Reconstruction White	111/126 (88.1)	4/136 (2.9)	2/136 (1.5)	4/136 (2.9)	0/134 (0)
Terry	309/357 (86.6), 0.008	14/374 (3.7), 0.889	17/374 (4.5), 0.784	12/374 (3.2), 0.106	3/374 (0.8)
Hamann-Todd	156/178 (87.6)	8/185 (4.3)	10/185 (5.4)	13/185 (7.0)	2/185 (1.1)
Cobb	30/43 (69.8)	4/85 (4.7)	3/85 (3.5)	3/85 (3.5)	4/46 (8.7)

(Continued)

Table 13.4. (Continued)

Dental pathologies

Cohorts	Edentulism N/total (%), p	Abscesses N/total (%), p	Caries N/total (%), p	Dental work N/total (%), p	Enamel hypoplasia N/total (%), p
Entire sample	105/609 (17.2)	323/588 (54.9)	378/438 (86.3)	38/368 (10.3)	333 (51.8)
African American	50/328 (15.2) 0.159	184/315 (58.4), 0.068	211/248 (85.1), 0.396	**14/208 (6.7), 0.010**	177 (50.1)
Euro-American	55/281 (19.6)	139/273 (50.9)	167/190 (87.9)	**24/160 (15.0)**	156 (54.4)
Antebellum	**43/167 (25.7), 0.001**	92/163 (56.4) 0.899	85/98 (86.7) 0.676	6/74 (8.1), 0.763	67 (38.3)
Civil War	**22/127 (17.3)**	66/122 (54.1)	80/90 (88.9)	9/78 (11.5)	71 (53.8)
Reconstruction	**40/315 (12.7)**	165/303 (54.5)	213/250 (85.2)	23/216 (10.6)	195 (58.6)
Antebellum White	27/89 (30.3), 0.005	45/88 (51.1), 0.395	39/36 (84.8), 0.480	6/32 (18.8), 0.049 (b)	37 (40.7)
Antebellum Black	**16/78 (20.5)**	47/75 (62.7)	46/52 (88.5)	0/42 (0.0)	30 (35.7)
Civil War White	**9/60 (15.0)**	33/60 (55.0)	39/45 (86.7)	7/40 (17.5)	38 (63.3)
Civil War Black	**13/67 (19.4)**	33/62 (53.2)	41/45 (91.1)	2/38 (5.3)	33 (45.8)
Reconstruction White	**19/132 (14.4)**	61/125 (48.8)	89/99 (89.9)	11/88 (12.5)	81 (59.6)
Reconstruction Black	**21/183 (11.5)**	104/178 (58.4)	124/151 (82.1)	12/128 (9.4)	114 (57.9)
Pre-Reconstruction (PR)	**65/294 (22.1), 0.002**	158/285 (55.4), 0.811	165/188 (87.8), 0.439	15/152 (9.9), 0.809	138 (45.0)
Reconstruction	**40/315 (12.7)**	165/303 (54.5)	213/250 (85.2)	23/216 (10.6)	195 (58.6)
Enslaved Black	**29/145 (20.0), 0.013**	80/137 (58.4), 0.291	87/97 (89.7) 0.232	**2/80 (2.5), 0.015**	63 (40.4)
PR White	**36/149 (24.2)**	78/148 (52.7)	78/91 (85.7)	**13/72 (18.1)**	75 (49.7)
Liberated Black	**21/183 (11.5)**	104/178 (58.4)	124/151 (82.1)	**12/128 (9.4)**	114 (57.9)
Reconstruction White	**19/132 (14.4)**	61/125 (48.8)	89/99 (89.9)	**11/88 (12.5)**	81 (59.6)
Terry	**68/368 (18.5), 0.000**	**188/366 (51.4), 0.033**	224/256 (87.5), 0.562	25/208 (12.0), 0.326	204 (54.7)
Hamann-Todd	**11/181 (6.1)**	**114/181 (63.0)**	138/162 (85.2)	13/150 (8.7)	116 (63.0)
Cobb	**26/60 (43.3)**	**21/41 (51.2)**	16/20 (80.0)	0/10 (0.0)	13 (15.1)

Bolded sections denote statistical significance at the $p = 0.05$ level. [a]Prevalence was too small for statistical analyses. (i) 1 cell (14.7%) has an expected count of less than 5; (ii) 4 cells (33.3%) have expected counts of less than 5.

the highest rates, but these results were not significant. The Hamann-Todd Collection had significantly more cases of PH when compared to other collections. Rickets was present in 4% of the individuals examined, with African Americans, the Civil War–born cohorts, and Enslaved African Americans having the most cases. However, none of the groups exhibited statistical differences.

Syphilis was prevalent in 4.7% of the sample (Table 13.4), but significantly more African Americans suffered from the disease than Euro-Americans ($p = 0.004$). Birth status cohort analyses indicated that liberated African Americans born during Reconstruction had significantly higher rates of syphilis when compared to other groups ($p = 0.046$). Findings on TB mirrored these results with 4.4% of the sample suffering from the disease. African Americans (4.5%) had significantly higher rates ($p = 0.004$). TB also increased thorough time, with Liberated African Americans having the highest prevalence ($p = 0.007$). Collection analyses for both syphilis and TB were not significantly different, although Hamann-Todd had the most cases.

Nine individuals, or 1.5% of the sample, had osteomyelitis (Table 13.4). Due to the low prevalence, only frequencies are reported. African Americans, especially those born liberated during Reconstruction, suffered from higher rates when compared to Euro-Americans. Additionally, four African Americans, three of whom were from the Hamann-Todd Collection, had polio.

Dental results are also reported in Table 13.4. Analyses revealed that, of 609 individuals, 17.2% ($n = 105$) were edentulous. African Americans and Euro-Americans were not significantly different in regard to edentulism, but the Antebellum era ($p = 0.001$) and Pre-Reconstruction groups ($p = 0.002$) had the highest rates. The combined ancestry/birth and birth status cohorts told a different story though: significantly more Antebellum-born Euro-Americans ($p = 0.005$) and Pre-Reconstruction Euro-Americans ($p = 0.013$) were edentulous when compared to their counterparts. Collection analyses revealed that the Cobb sample had significantly higher frequencies of edentulism ($p = 0.000$). Caries rates were also higher, with 86.3% of the sample affected. None of the cohorts were statistically significant, but Euro-Americans, the Civil War, Pre-Reconstruction, Civil War African American, Reconstruction Euro-American, and the Terry Collection cohorts had the highest rates of cavities. Abscesses were also present in 54.9% of the sample. African Americans, the Antebellum cohort, the Pre-Reconstruction group, Antebellum-born African Americans, and African Americans born enslaved and liberated had the largest frequencies of abscesses, but these findings were not significant. Hamann-Todd had the most individuals with abscesses when compared to the other collections ($p = 0.033$).

EH frequencies indicated that 51.8% of the persons studied had enamel defects (Table 13.4). Euro-Americans, the Reconstruction era, and Hamann-Todd had the highest frequencies. The combined ancestry/birth cohorts indicated that Civil War–born Euro-Americans were the most afflicted with EH. When the data were examined by birth status cohorts, Reconstruction Euro-Americans had the highest rates of EH.

Approximately 10.3% of the sample had dental work, with Euro-Americans having significantly ($p = 0.010$) more instances (Table 13.4). Individuals possessed either gold, silver, or amalgam fillings, gold crowns, or a combination of fillings and crowns. Birth cohorts were not statistically significant, but the Civil War era had the highest rates of dental repair. The combined ancestry/birth cohorts echoed these findings with Civil War Euro-Americans having significantly larger frequencies of dental work ($p = 0.049$), but these results may be erroneous due to low expected cell counts. However, birth status cohorts indicated that significantly more Pre-Reconstruction Euro-Americans had dental restorations when compared to the other groups ($p = 0.015$). The Terry Collection also had the highest rates of individuals with dental work.

Table 13.5. Significant results of stature and cohorts analyses.

	T-test analysis of stature and ancestry cohort			
Cohort	N	Mean	t	Sig.
African American	347	169.161	−2.551	0.011
Euro-American	274	170.315		
	ANOVA and birth status cohorts			
Cohort	N	Mean	F	Sig.
Enslaved Black[b]	154	168.944	4.609	0.003
Pre-Reconstruction White	149	169.509		
Liberated Black[b]	193	169.334		
Reconstruction White[b]	125	171.277		
	ANOVA and collections cohorts			
Cohort	N	Means	F	Sig.
Terry[c]	367	170.143	3.259	0.039
Hamann-Todd[c]	180	168.907		
Cobb	74	169.182		

[b]Bonferroni *post hoc* tests indicated that the Reconstruction White cohort was 2.332 cm taller than the Enslaved Black ($p=0.003$) and 1.942 cm taller than the Liberated Black cohorts ($p=0.015$).
[c]Bonferroni *post hoc* tests indicated that the Terry Collection cohort was 1.234 cm taller than Hamann-Todd Collection cohort ($p=0.047$).

Stature analyses revealed significant differences among the ancestry, enslaved/liberated, ancestry/birth, and collection cohorts (Table 13.5). African Americans were significantly shorter than Euro-Americans ($p=0.011$) by 1.15 cm. Bonferroni *post hoc* tests indicated that Reconstruction-born Euro-Americans were significantly taller than Enslaved ($p=0.003$) and Liberated African Americans ($p=0.015$) and those born during the Antebellum ($p=0.004$) and Reconstruction eras ($p=0.038$). An analysis of the collection cohorts indicated that Hamann-Todd was significantly shorter ($p=0.030$) than Terry.

Discussion

Omran's epidemiologic transition model hypothesizes that epidemic infectious diseases will decline as nations modernize and be replaced with an increase in *man-made* and degenerative disorders. In regard to the current study, Omran's model implies that skeletal evidence of infectious diseases in this sample should temporally wane and decline to their lowest frequencies in the Reconstruction and liberated cohorts. However, the results contradicted this expectation. Higher rates of infectious diseases prevailed and differed according to race and birth period. Euro-Americans born during the Civil War and Pre-Reconstruction eras were more affected with PH and had significantly higher rates of dental work and edentulism. Reconstruction-born Euro-Americans were also significantly taller than their African American counterparts and had the largest frequencies of EH. In contrast, African Americans, especially those born free during Reconstruction, had higher rates of osteomyelitis and suffered significantly more from syphilis, TB, and dental abscesses. They were also significantly shorter than Euro-Americans, contradicting previous research about enslaved African Americans, who were believed to be similar in height to contemporary Euro-Americans and taller than West Africans and Europeans from the same time period (Fogel and Engermann, 1974; Steckel, 1979, 1986). This difference may be an artifact of the sample analyzed, which encompasses the birth periods of slavery and Reconstruction, and is biased toward persons of low SES. However, poor nutrition or chronic disease exposure during growth and development also results in reduced stature. In regard to the collections, Hamann-Todd had significantly shorter individuals and the highest frequencies of syphilis, TB, abscesses, PH, and EH.

These findings indicate that liberated African Americans, especially those in the Hamann-Todd Collection, suffered from significantly higher rates of infectious diseases and experienced more biological stress during their lifetimes than did Euro-Americans. Furthermore, TB and syphilis did not decrease through time, but increased, reaching their highest frequencies in the Reconstruction and liberated African American cohorts. Osteomyelitis, although not statistically significant, also shared this pattern. Reconstruction-born individuals, who spent their youth in the postindustrialization period, were expected, based on the epidemiologic transition model, to have lower rates of infectious diseases. These results disagree with Omran's model and imply that the transition was not a universal event that happened simultaneously within the American population, but varied based on race and SES.

The cities associated with the skeletal samples studied, as mentioned earlier, were major urban and industrialized centers in the 19th and 20th centuries, which provided job opportunities that attracted immigrants from throughout the U.S. Advancements in transportation also facilitated such movements. This was especially true for southern African Americans, who were desperate to flee the oppressive south in search of better lives. St. Louis served as a major Union command center during the Civil War, which attracted numerous runaway slaves seeking freedom, and many of these individuals may have stayed for the long term. The city also saw a large influx of African Americans passing through its borders at the end of the 19th century. For instance, in January 1886, the St. Louis *Globe-Democrat* reported that "another large party" of Southern African Americans, "numbering several hundred," had passed through St. Louis in route to Arkansas. Exorbitant rent, "bad crops, and the defective tenant system" were cited as reasons for the migration.[2]

This was the precursor to the Great Migration (Marks, 1989). The rise of the Ku Klux Klan, racial violence, lynchings, Jim Crow laws, natural disasters, and the institution of sharecropping in the Post-Civil War and Reconstruction South served as major push factors for this exodus of African Americans to Northern and Midwestern states (Tolnay and Beck, 1990; Phillips, 1999). Many southern African Americans relocated to cities like St. Louis and Cleveland after 1910, when Northern industrial corporations began to employ African Americans (Phillips, 1999; Giffin, 2005; Wilkerson, 2010). In 1917, war-time shortages of unskilled Euro-American workers in steel mills, foundries, and railroads resulted in the creation of new jobs that were previously unavailable to African Americans and companies traveled to southern states to actively recruit African American males (Tolnay and Beck, 1990; Phillips, 1999). In addition, the increased demand for African American laborers in the North provided attractive opportunities that served as 'pull factors" for African American southerners to migrate and escape the racial persecution prevalent in the South (Mandle, 1978). In 1919, the Cleveland *Advocate*, a historically African American newspaper, reported

---

[2]"Negro Exodus from the South," St. Louis *Globe-Democrat*, January 8, 1886, 7.

that many were migrating in large numbers from three states in particular: "60,000 Southern Blacks had left Alabama, 22,000 Tennessee, and 12,000 Florida."[3] Thousands of African American males found employment in the city's rolling mills, foundries, railroad yards, blasting furnaces, cast and wrought iron factories, and iron works (Phillips, 1999). The Cleveland branches of the Baltimore & Union Railroad and the New York Central Railroad (NYCR) also recruited southern African American males to load and unload freight in their depots, promising housing and decent wages (Phillips, 1999).

While the Great Migration impacted the population and demographics of Cleveland, it also affected the composition of the Cobb, Hamann-Todd, and Terry collections. Studies of the collections and census research indicated that most of the African Americans examined were southern-born (Cobb, 1935; Watkins, 2003, 2007; Hunt and Albanese, 2005; de la Cova, 2010). They came from 27 states, but were mostly born in Georgia, Alabama, and South Carolina, as well as Tennessee, Kentucky, Virginia, Mississippi, North Carolina, and Arkansas. Cobb (1935) observed that African American remains in the Hamann-Todd Collection began to increase in 1915 and, by 1930, had surpassed the total sum of Euro-Americans, which he attributed to the Great Migration. Census data support this, indicating that from 1910 to 1930, Cleveland's African American population surged from 8448 to 72,120 (Cobb, 1935). The Terry Collection shared this demographic pattern, with most of its African Americans being southern-born. Research by Watkins (2003) on the Cobb Collection also revealed that most of the individuals were native to Washington, DC, Maryland, and Virginia. In contrast, Euro-Americans were from the cities and regions associated with the anatomical samples or from neighboring states. The demographics of the Hamann-Todd Collection indicated that many of the Euro-Americans were first-generation Americans, with the majority born in Ohio, New York, and Pennsylvania (Cobb, 1935).

The southern roots of African Americans and the Midwestern and Northern origins of the Euro-Americans examined in this study may explain their health disparities. Findings in Table 13.4 show that both African Americans and Euro-Americans suffered from PH (85.6%), rickets (4%), abscesses (54.9%), caries (86.3%), and EH (51.8%). The 85.6% affliction rate of PH, especially among Euro-Americans, the Civil War and Pre-Reconstruction-born cohorts, and Enslaved African Americans suggests high levels of malnutrition, synergistic disease, and parasitism. Hookworm, roundworm, and pinworm were endemic in the rural South and Midwest in the 19th and early 20th centuries, including the states associated with the anatomical collections (Savitt, 1978; Coelho and McGuire, 2007). For instance, the U.S. Sanitary Commission reported that between 1909 and 1914 over 43% of 17.5 million southerners were infected with hookworm. Persons afflicted with parasites suffer from appetite loss, chronic diarrhea, and vomiting (Coelho and McGuire, 2007), which leads to poor nutrient absorption, and can produce dietary deficiencies, immunosuppression, and increased susceptibility to disease. All of these could in turn manifest on the skeleton as PH, rickets, and EH.

Individuals who were born during and lived through the late Antebellum and Civil War periods also experienced environmental stressors associated with the war, which may have further contributed to the observed health disparities. Historical accounts indicate deleterious food shortages among soldiers, slaves, and civilians, particularly in southern states, from conditions in military camps to economic disruptions among civilian populations caused by the conflict.[4] The war was particularly difficult for African Americans, as evidenced by the documented hardships many endured in the transition from slavery to freedom. Numerous accounts describe the deplorable conditions of Union contraband camps. Sanitary

---

[3] "Here's Figures on the Exodus," *Cleveland Advocate*, December 2, 1916, 1.
[4] T. T. Treadway to Cornelius T. Chase, August 5, 1865, Cornelius Chase Family Papers, Box 3, Folder 8, Cornelius Thurston Chase (son) Correspondence, 1865–1867, Library of Congress (LOC).

Commission agent Maria Mann described the abuse and neglect she witnessed toward African American refugees while working, in February 1863, at the "sickly, pestilential, [and] crowded" government-administered St. Helena Contraband Camp in Helena, Arkansas. Food was lacking, beef was scarce, and army rations and pork formed the dietary staple. The quarters provided for the refugees lacked "comfort or decency" and the hospital was a "wretched hovel" where many went to die.[5] The contrabands had "narrowed [living] habitations, half of them with ground floors, without [a] window or closet," and mud floors. They had been destitute for 4 months with no relief except the males, who, through the military, were "comfortably clothed."[6] Three months later, T. A. Goodwin wrote that "able-bodied" male contrabands "fair well enough, as laborers or as servants or soldiers, but the thousands of women and children … huddled in the filthy quarters appeal to our sympathy."[7] In December 1863, an editorial in the Washington, DC, *Daily Constitutional Union* revealed that the poor conditions of the contrabands continued. Fit males were "forced into the army, either for labor or military service," but the elderly, women, and children were "provided with insufficient shelter and abandoned to the chances of hunger, nudity, and disease." The government failed to support the "miserable wretches" who were "deluded" into military lines "by the promise of freedom."[8] Lt. Col. James T. Call, of the First New York Volunteer Engineers, also expressed similar concerns for runaways entering his camp in the South Carolina Low Country "in a condition of extreme destitution." He wrote that "mortality, especially among the children is very great" and protested that he lacked sufficient quarters to care for the increasing numbers of refugees.[9] These food shortages, the lack of resources, and poor living conditions experienced by contrabands may have induced the biological stress observed in this study among Antebellum- and Civil War-era enslaved African Americans. The lack of food, poor sanitary conditions, and dirt floors would have also exposed children to hookworm infection and made them susceptible to various deficiencies.

However, health and environmental conditions did not improve with emancipation. Results here demonstrate that liberated African Americans had significantly higher rates of TB and syphilis and were of shorter stature than their Euro-American counterparts, who suffered more from PH, EH, and caries. This may be attributable to the fact that census data and demographic studies indicate that the African Americans examined in this study were part of the Great Migration. These emigrants may have spent their childhoods in less populated environments with lower levels of infectious disease than were present in the densely populated urban centers that they emigrated to, namely, St. Louis, Cleveland, and Washington, DC. Emigration exposed them to new climates, social networks, and living conditions, which may have heightened exposure and susceptibility to various infectious diseases. This may be especially true for those in the Hamann-Todd Collection, who had the largest frequencies of PH, EH, syphilis, TB, osteomyelitis, and abscesses (Table 13.4).

Upon relocating, many African Americans toiled as unskilled laborers, receiving wages lower than their Euro-American counterparts, and living in substandard tenement housing (Phillips, 1999). Some worked longer hours for less pay, which made them attractive hires for Northern companies (Kusmer, 1978; Phillips, 1999), but poor Euro-Americans in Cleveland, St. Louis, and other industrial cities did not eagerly welcome the new competitors for

---

[5] Maria R. Mann to Elisa, February 10, 1863, The Papers of Mary T. P. Mann, 1863, LOC.
[6] Maria R. Mann to Rev. William L. Ropes, April 13, 1863, *Ibid.*
[7] T. A. Goodwin to Robert Dale Owen, May 1, 1863, The Papers of James Morrison MacKaye, LOC.
[8] "The Contrabands in the South," Washington *Daily Constitutional Union*, December 19, 1863, 2.
[9] James Call to W. L. M. Burger, October 25, 1864, Order Book: Companies A to K, Vol 5, Book Records of Volunteer Union Organizations, 33rd U.S. Colored Infantry, Records of the Adjutant General's Office, RG 94, NARA.

housing, jobs, and resources, particularly given existing competition with European immigrants (Bonacich, 1972, 1975). Matters worsened when companies began to use southern African American in-migrants as strikebreakers to fill labor losses caused by striking Euro-American workers protesting increased hours associated with war-time production, the growing presence of African Americans in the work force, and the lack of recognition of organized labor unions (Phillips, 1998). These sentiments erupted in violence during the summer of 1917 and the Red Summer of 1919 when race riots flared up throughout the U.S., instigated by job competition, housing shortages, and racial tension (Bonacich, 1972, 1975; Wilson, 1978). The 1917 East St. Louis Riots, the bloodiest of all the unrests, resulted in the recorded deaths of 39 African Americans and 9 Euro-Americans.[10] This escalation of violence may have resulted in the high rates of gunshot wounds observed among Reconstruction-born African American males in this study, especially since many were African American southerners that had migrated to Cleveland and St. Louis.

In addition to job competition and violence, finding shelter was another obstacle faced by African American emigrants. The large influx resulted in housing shortages, tenement overcrowding, and homelessness in many cities; in Cleveland and St. Louis, for instance, emigrants faced limited options and unequal access to housing due to racial discrimination (Kusmer, 1978; Phillips, 1999; Giffin, 2005). Housing scarcities prompted landlords to increase rent, often targeting African Americans, who paid 50–75% more than their Euro-American counterparts (Donald, 1921; Giffin, 2005). In addition to high leases, African American in-migrants were marginalized to specific neighborhoods and tenements, which were often substandard (Phillips, 1999). In 1919, the *Advocate* condemned the lodgings and sanitation in several of these neighborhoods in Cleveland, where "hundreds of respectable Colored citizens are forced to live because of confounded color prejudice." The dwellings were described as an "unsightly, unsanitary, disease-breeding plague," a "veritable miasma," and a "menace to the health and moral civic progress of the Colored people who are forced to inhabit them."[11] Public health officials in many places, such as Ohio, recognized that these conditions were a direct threat to the health of local African American populations (Giffin, 2005). In Cleveland, for instance, 50% of industrial workers living in tenements occupied dwellings with less than one room per person, and African American dwellings were among the most crowded.[12] Even company lodgings, such as those provided by the NYCR, were overcrowded and had "inadequate light and ventilation, old, soiled stained bed coverings" and "filthy floors."[13] These unsanitary living conditions would have induced infectious illnesses, especially respiratory diseases, and may explain why the liberated African Americans in this study were significantly more afflicted with TB when compared to Euro-Americans.

Climate may have also contributed to the high rates of TB observed among the emancipated African Americans examined in this study. Many of them migrated from the south with inadequate clothing, unprepared for the cold weather (Donald, 1921), which was particularly an issue in Cleveland. The combined effects of poor housing, inadequate clothing, and a lack of acclimation or adaptation to the local climate likely resulted in high frequencies of respiratory conditions, such as pneumonia, bronchitis, and TB; in Cincinnati, Ohio, for instance, pneumonia rates among African Americans were three times higher than among Euro-Americans (Donald, 1921). Mortality rates in Cincinnati also reflected these disparities.

---

[10] Carlos F. Hurd, "E. St. Louis Riot," *St. Louis Post-Dispatch*, July 3, 1917.
[11] "Central Avenue Needs Attention," *Cleveland Advocate*, September 27, 1919, 8.
[12] *A Popular Summary of the Cleveland Hospital and Health Summary*, Cleveland Hospital Council, Cleveland, 1920, p. 7; *ibid.*, p. 53; *ibid.*, p. 353.
[13] *Ibid.*, p. 53.

The overall death rate for African Americans was double that of Euro-Americans. Three times as many African American children died before their first birthday when compared to their Euro-American counterparts (Donald, 1921).

Ill African Americans would have also needed medical treatment. Morgue and census records indicate that the persons examined in this study received their medical care at municipal poorhouses, public hospitals, and charity institutions, where medical fees were minimal or gratuitous and care was abysmal. These establishments, despite their benevolent motives, may also have served as further vectors for disease and malnutrition. In Cleveland, political conflicts, poor funding, and shortages of full-time doctors and nurses impeded local public health care, especially at the town's two indigent health-care providers, the Cleveland City and Cleveland State hospitals. The 1920 Cleveland Council Hospital Survey, for instance, noted that the Cleveland City Hospital was highly overcrowded, was poorly headed, lacked artificial light, and had inadequate sanitary services.[14] Many persons in the Terry Collection both received medical care and perished at the St. Louis City Hospital. This institution lacked trained nurses and bedside care until the Civil War,[15] and in 1875, the St. Louis Board of Health review described the hospital as "unfit for the accommodation and treatment of the numerous patients crowded into the institution." Six years later, conditions were still badly crowded. Contemporaries also noted that the meals provided to patients were inadequate in terms of nutrition and quantity, and were centered on small quantities of milk, bread, soup, potatoes, and rice, with only small quantities of fruit and meat.[16] These conditions likely exacerbated or even created nutritional deficiencies among the patients and increased their susceptibility to disease.

These conditions seem to have taken a biological toll on contemporary African Americans, as evidenced by the high frequencies of indicators and poor health in the skeletal samples, particularly in the Hamann-Todd collection. These findings are substantiated by historical data and current work on health disparities. The 1880, 1890, and 1900 U.S. censuses consistently recorded that African Americans had higher rates of TB, pneumonia, scrofula, and venereal diseases.[17] Even at the turn of the 20th century, TB still remained a major killer in the African American community, with infection rates three times that of Euro-Americans (Farley, 1970; Kiple and King, 1981). In Cleveland, for instance, TB- and pneumonia-related death rates in Cleveland in 1920 were more than twice that of Euro-Americans (Giffin, 2005). This is consistent with the much higher rates of TB in African Americans found in this study, especially through the Reconstruction period. These patterns of higher rates of infection from TB and other diseases, such HIV/AIDS and syphilis, remain persistent in urban African American communities into the present (Stead et al., 1990; Gaylin and Kates, 1997), reflecting the continuing influences of racism, social inequality, and socioeconomic disparities on patterns of health and disease in minority and low SES populations in the U.S.

The Euro-Americans examined in this study were from Cleveland, St. Louis, and neighboring Northern states. Most of their lives were probably spent in more densely populated and

---

[14] *Ibid.*, pp. 9, 15–17.
[15] "Patients in the First City Hospital Slept on the Floor," *St. Louis Globe-Democrat*, October 29, 1944.
[16] "City Hospital: Visit of Inspection by the Board of Health," *St. Louis Globe-Democrat*, June 28, 1875; "The City's Sick: People Who Are Carted Away to the City Hospital to Languish and Die," *St. Louis Globe-Democrat*, August 28, 1881; p. 9.
[17] John S. Billings, Report on the Mortality and Vital Statistics of the U.S. as Returned at the Tenth Census (June 1, 1880), Part I, Department of the Interior, Census Office, Washington, Government Printing Office, 1885. John S. Billings, Report on Vital and Social Statistics in the U.S. at the Eleventh Century: 1890, Part I.-Analysis and Rate Tables, Department of the Interior, Census Office, Washington, D.C., Government Printing Office, 1896. William A. King, Census Reports, Volume III, Twelfth Census of the U.S., Taken in the Year 1900, Vital Statistics, Part I, Analysis and Ration Tables, Washington, U.S. Census Office, 1902.

urbanized settings. This metropolitan lifestyle would have exposed low SES Euro-Americans to poor sanitation, substandard housing, and competition for resources throughout their lives, particularly during the vulnerable stages of growth and development. In turn, these conditions may have resulted in the higher rates of PH and EH observed in Euro-Americans. However, the presence of dental work suggests that many of these individuals were not poor their entire lives, as dental restoration was only available to those who could purchase it in the 19th and early 20th centuries. However, the high rates of edentulism and caries suggest an inability to afford long-term oral care and a decline in access to resources with age that we do not see among the African Americans.

Conclusion

Evidence for the second epidemiologic transition was absent among the African American individuals examined in this sample as infectious disease rates rose through time, reaching their peak during the Reconstruction. This increase among particular subpopulations suggests that the epidemiologic transition did not occur simultaneously among all communities in the U.S. Omran did acknowledge this possibility to a certain extent, noting that the transition's onset varied by race, gender, and SES. In particular, the transition seems to have initiated earlier among the Euro-American upper classes, due to their higher levels of education and greater access to resources, including nutrition and health care (Omran, 1977). As this study, among others, suggests, non-Euro-Americans who lacked these resources progressed through the transition later and at a slower rate. Omran (1977) has argued that, as time passes, these differences eventually converge, with infectious disease rates declining and longevity increasing among non-Euro-Americans and non-elites. However, this pattern was not observed here. Therefore, these results suggest that the model cannot explain the nonconvergence evident in this study nor does it consider the differences in ancestry, gender, SES, or migration that may exist between various subgroups within a population. By neglecting these important factors, Omran's model generalizes the transition at the population and global levels, overlooking relationships that may exist between health and mortality disparities in regard to race, class, and environment (Mackenback, 1994; Gaylin and Kates, 1997). However, knowing how these factors affect salubrity and understanding important social and cultural differences among population subgroups are essential to comprehending differential rates of mortality and disease (Kunitz, 1990; Gaylin and Kates, 1997). Gaylin and Kates (1997: 609) indicate that these dissimilarities should be emphasized to "create a more complete and accurate representation of population morbidity and mortality." If this is done, then the portrait drawn by the epidemiologic theory would significantly develop "to reflect numerous epidemiologic transitions intricately connected to socioeconomic, sex, and racial differences" (Gaylin and Kates, 1997: 609).

Nonetheless, the model is far from useless or irrelevant. The results of this study demonstrate that Omran's theory can serve as a foundation for understanding large-scale changes in patterns of health over time across and within populations in response to economic and environmental change, but only if modified to include ancestry and SES. By doing so, addressing issues related to ancestry, SES, and culture, this study demonstrates the deleterious effects urbanization, in-migration, and industrialization had on southern-born African Americans that migrated to the North in search of better lives. African American in-migrants experienced the hardships of discrimination, tenement housing, and poor health care, which resulted in significantly higher rates of infectious diseases. Sadly, they died in poverty, holding on to a dream that was not realized in their lifetimes.

References

Barrett R, Kuzawa CW, McDade T, Armelagos GJ. 1998. Emerging and re-emerging infectious diseases: the third epidemiologic transition. Annu Rev Anthropol 27:247–271.
Blakey ML. 2001. Bioarchaeology of the African Diaspora in the Americas: its origins and scope. Annu Rev Anthropol 30:387–422.
Bogin B. 1999. Patterns of human growth. 2nd ed. Cambridge: Cambridge University Press.
Bollet AJ. 2002. Civil War medicine: Challenges and triumphs. Tucson: Galen Press.
Bonacich E. 1972. A theory of ethnic antagonism: the split labor market. Am Soc Rev 37:547–559.
Bonacich E. 1975. Abolition, the extension of slavery, and the position of free African Americans: a study of split labor markets in the U.S., 1830–1863. Am J Soc 81:601–628.
Carson SA. 2009. African-American and Euro-American inequality in the nineteenth century American south: a biological comparison. J Popul Econ 22:739–755.
Click PC. 2001. Time full of trial: The Roanoke Island Freedmen's Colony, 1862–1867. Chapel Hill: University of North Carolina Press.
Cobb WM. 1935. Municipal history from anatomical records. Sci Mon 40:157–162.
Coelho PRP, McGuire RA. 2007. Racial differences in disease susceptibilities: intestinal worm infections in the early twentieth-century American south. Soc Hist Med 19:461–482.
de la Cova CM. 2010. Cultural patterns of trauma among 19th-century-born males in cadaver collections. Am Anthropol 112:589–606.
Donald HH. 1921. The Negro migration of 1916–1918. Washington, DC: Assoc Study Negro Life Hist.
Donald DH, Baker JH, Holt MF. 2001. The civil war and reconstruction. New York: W. W. Norton & Company, Inc.
Ericksen MF. 1982. How "representative" is the terry collection? Evidence from the proximal femur. Am J Phys Anthropol 59:345–350.
Eyssell TH. 2006. St. Louis and the automobile. In: Rosenfeld R, editor. Hidden assets: Connecting the past to the future of St. Louis. St. Louis: Missouri History Museum Press. p 13–42.
Farley R. 1970. Growth of the African American population: A study of demographic trends. Chicago: Markham Publishing Company.
Fogel RW, Engerman SL. 1974. Time on the cross: The economics of American Negro slavery. New York: W. W. Norton & Company, Inc.
Gaylin DS, Kates J. 1997. Refocusing the lens: Epidemiologic transition theory, mortality differentials, and the AIDS pandemic. Soc Sci Med 44:609–621.
Giffin WW. 2005. African Americans and the color line in Ohio, 1915–1930. Columbus: Ohio State University Press.
Goodman AH, Rose JC. 1990. Assessment of systemic physiological perturbations from dental enamel hypoplasias and associated histological structures. Yearb Phys Anthropol 33:59–110.
Goodman AH, Allen LH, Hernandez GP et al. 1987. Prevalence and age at development of enamel hypoplasias in Mexican children. Am J Phys Anthropol 72:7–19.
Haines MR. 1998. Health, height, nutrition, and mortality: evidence on the "Antebellum Puzzle" from Union Army recruits in the middle of the nineteenth century. NBER Historical Working Papers 0107. In: Komlos J, Baten J, editors. The biological standard of living in comparative perspective: Contributions to the conference held in Munich, January 18–22, 1997, for the XIIth International Economic History Association. Berlin: Franz Steiner Verlag. p 154–181.
Haines MR, Craig LA, Weiss T. 2003. The short and the dead: nutrition, mortality, and the 'Antebellum Puzzle' in the U.S. J Econ Hist 63:385–416.
Haines MR, Craig LA, Weiss T. 2011. Did African Americans experience the 'Antebellum Puzzle'? Evidence from the U.S. colored troops during the Civil War. Econ Hum Biol 9:45–55.
Higgins RL, Haines MR, Walsh L, Sirianni JE. 2002. The poor in the mid-nineteenth century northeastern U.S.: evidence from the Monroe County poorhouse, Rochester, New York. In: Steckel RH, Rose JC, editors. The backbone of history: Health and nutrition in the Western Hemisphere. New York: Cambridge University Press. p 162–183.
Higginson TW. 1984. Army life in a black regiment. New York: Putnam.

Hodes FA. 2009. Rising on the river: St. Louis 1822 to 1850, explosive growth from town to city. Tooele: The Patrice Press.

Hunt DR, Albanese J. 2005. History and demographic composition of the Robert J. Terry Anatomical Collection. Am J Phys Anthropol 127:406–417.

Jones-Kern K, Latimer B. 1999. History of the Hamann-Todd osteological collection: Skeletons out of the closet. Avaliable at: http://www.cmnh.org/collections/physanth/documents/HamannTodd_Osteological_Collection.html (accessed November 6, 2013).

Kelley MA, Micozzi MS. 1984. Rib lesions in chronic pulmonary tuberculosis. Am J Phys Anthropol 65:381–386.

Kelley JO, Angel JL. 1987. Life stresses of slavery. Am J Phys Anthropol 74:199–211.

Kiple KF, King VH. 1981. Another dimension to the African American diaspora: diet, disease and racism. Cambridge: Cambridge University Press.

Komlos J. 1996. Anomalies in economic history: toward a resolution of the "Antebellum Puzzle." J Econ Hist 56:202–214.

Komlos J. 1998. Shrinking in a growing economy? The mystery of physical stature during the industrial revolution. J Econ Hist 58:779–802.

Komlos J, Coclanis P. 1997. On the 'Puzzling' Antebellum cycle of the biological standard of living: the case of Georgia. Explor Econo Hist 34:433–459.

Kunitz SJ. 1990. The value of particularism in the study of the cultural, social and behavioral determinants of mortality. In: Caldwell JC, Findley F, Caldwell P, et al., editors. What we know about health transition: The cultural, social, and behavioral determinants of health. Canberra: Australian National University. p 92–109.

Kusmer KL. 1978. A ghetto takes shape: African American Cleveland, 1870–1930. Urbana: University of Illinois Press.

Lang S, Sunder M. 2003. Nonparametric regression with Bayes X: a flexible estimation of trends in human stature in nineteenth century America. Econ Hum Biol 1:77–89.

Lanphear KM. 1990. Frequency and distribution of enamel hypoplasias in a historic skeletal sample. Am J Phys Anthropol 81:35–43.

Mackenback JP. 1994. The second epidemiologic transition. J Epidemiol Community Health 48:329–332.

Mandle JR. 1978. The roots of African American poverty: The southern plantation economy after the Civil War. Durham: Duke University Press.

Marks C. 1989. Farewell—We're good and gone: The Great African American migration. Bloomington: Indiana University Press.

Martin DL, Magennis AL, Rose JC. 1987. Cortical bone maintenance in an historic Afro-American cemetery sample from Cedar Grove, Arkansas. Am J Phys Anthropol 74:255–264.

Meindl RS, Russell KF, Lovejoy CO. 1990. Reliability of age at death in the Hamann-Todd collection: validity of subselection procedures used in blind tests of the summary age technique. Am J Phys Anthropol 83:349–357.

Miller CP, Wheeler R. 1990. Cleveland: A concise history, 1796–1990. Bloomington: Indiana University Press.

Muller J. 2004. Evidence of interpersonal violence in the W. Montague Cobb skeletal collection. Am J Phys Anthropol 123(S38):150.

Nolen CH. 2001. African American Southerners in Slavery, Civil War, and reconstruction. Jefferson: McFarland & Co.

Null CC, Blakey ML, Shujaa KJ, Rankin-Hill LM, Carrington SHH. 2004. Osteological indicators of infectious disease and nutritional inadequacy. In: Blakey ML, Rankin-Hill LM, editors. The New York African Burial Ground, skeletal biology: Final report. Vol. 1. Washington, DC: The African Burial Ground Project, Howard University, for the U.S. General Services Administration Northeast and Caribbean Region. p 351–402.

Omran AR. 1971. The epidemiologic transition: a theory of the epidemiology of population change. Milbank Mem Fund Q 49:509–538.

Omran AR. 1977. A century of epidemiologic transition in the United States. Prev Med 6(1):30–51.

Ortner DJ. 2003. Identification of pathological conditions in human skeletal remains. San Diego: Academic Press.

Owsley DW, Orser, CE, Jr., Mann RW, Moore-Jansen PH, Montgomery RL. 1987. Demography and pathology of an urban slave population from New Orleans. Am J Phys Anthropol 74:185–197.

Phillips KL. 1999. Alabama North: African-American migrants, community, and working-class activism in Cleveland, 1915–1945. Urbana: University of Illinois Press.

Primm JN. 1998. Lion in the valley: A history of St. Louis, Missouri, 1764–1980. St. Louis: Missouri History Museum Press.

Rankin-Hill L. 1997. A biohistory of 19th-century Afro-Americans: The burial remains of a Philadelphia cemetery. Westport: Bergin & Garvey.

Rankin-Hill L, Blakey ML. 1994. W. Montague Cobb (1904–1990): physical anthropologist, anatomist, and activist. Am Anthropol 96:74–96.

Rathbun TA. 1987. Health and disease at a South Carolina plantation: 1840–1870. Am J Phys Anthropol 74:239–253.

Rathbun TA, Scurry JD. 1991. Status and health in colonial South Carolina: Belleview plantation, 1738–1756. In: Powell ML, Bridges PS, Mires AMW, editors. What mean these bones? Studies in southeastern bioarchaeology. Tuscaloosa: University of Alabama Press. p 148–164.

Rathbun TA, Steckel RH. 2002. The health of slaves and free African Americans in the East. In: Steckel RH, Rose JC, editors. The backbone of history: Health and nutrition in the Western Hemisphere. New York: Cambridge University Press. p 208–225.

Reid RM. 2002. Government policy, prejudice, and the experience of African American Civil War soldiers and their families. J Fam Hist 27:374–398.

Riis JA. 1914. How the other half lives studies among the tenements of New York. New York: Charles Scribner's Sons.

Roberts C, Lucy D, Manchester K. 1994. Inflammatory lesions of ribs: an analysis of the Terry Collection. Am J Phys Anthropol 95:169–182.

Rose JC. 1985. Gone to a better land: A biohistory of a rural African American Cemetery in the post-Reconstruction South. Arkansas Archaeological Survey Research Series, No. 25. Fayetteville: Arkansas Archaeological Survey.

Rose JC. 1989. Biological consequences of segregation and economic deprivation: a post-slavery population from southwest Arkansas. J Econ Hist 49:351–360.

Saunders S, Herring A, Sawchuk L, Boyce G, Hoppa R, Klepp S. 2002. The health of the middle class: The St. Thomas Anglican Church cemetery project. In: Steckel RH, Rose JC, editors. The backbone of history: Health and nutrition in the Western Hemisphere. New York: Cambridge University Press. p 130–161.

Savitt TL. 1978. Medicine and slavery: The diseases and health care among African Americans in Antebellum Virginia. Urbana: University of Illinois.

Sledzik PS, Moore-Jansen P. 1991. Dental disease in nineteenth century military skeletal samples. In: Kelly M, Larsen C, editors. Advances in dental anthropology. New York: Wiley-Liss. p 215–224.

Sledzik PS, Sandberg LG. 2002. The effects of 19th century military service on health. In: Steckel RH, Rose JC, editors. The backbone of history: Health and nutrition in the Western Hemisphere. New York: Cambridge University Press.

Stead WW, Senner JW, Reddick WT, Lofgren JP. 1990. Racial differences in susceptibility to infection by Mycobacterium tuberculosis. N Engl J Med 322:432–437.

Steckel RH. 1979. Slave mortality: analysis of evidence from plantation records. Soc Sci Hist 3:86–114.

Steckel RH. 1986. Birth weights and infant mortality among American slaves. Explor Econ Hist 23:173–198.

Steckel RH, Rose JC, Sciulli PW. 2002. A health index from skeletal remains. In: Steckel RH, Rose JC, editors. The backbone of history: Health and nutrition in the Western Hemisphere. New York: Cambridge University Press. p 61–93.

Stuart-Macadam P. 1989. Porotic hyperostosis: relationship between orbital and vault lesions. Am J Phys Anthropol 80:187–193.

Sunder M. 2004. The height of Tennessee convicts: another piece of the 'Antebellum Puzzle.' Econ Hum Biol 2:75–86.

Tolnay SE, Beck EM. 1990. African American flight: lethal violence and the Great Migration, 1900 to 1930. Soc Sci Hist 14:347–370.

Trotter M. 1970. Estimation of stature from intact long limb bones. In: Stewart TD, editor. Personal identification in mass disasters. Washington, DC: Smithsonian Institution Press. p 71–83.

Trotter M. 1981. Robert J. Terry, 1871–1966. Am J Phys Anthropol 56:503–508.

Walker PL, Bathurst RR, Richman R, Gjerdrum T, Andrushko VA. 2009. The causes of porotic hyperostosis and cribra orbitalia: a reappraisal of the iron-deficiency-anemia hypothesis. Am J Phys Anthropol 139:109–125.

Watkins RJ. 2003. To know the brethren: A biocultural analysis of the W. Montague Cobb skeletal collection. Unpublished Ph.D. dissertation, Department of Anthropology. Chapel Hill: University of North Carolina at Chapel Hill.

Watkins RJ. 2007. Knowledge from the margins: W. Montague Cobb's pioneering research in biocultural anthropology. Am Anthropol 109:186–196.

Wilkerson I. 2010. The warmth of other suns: The epic story of America's Great Migration. New York: Random House.

Wilson WJ. 1978. The declining significance of race. Chicago: University of Chicago Press.

Part 5
The Environment and the Second Epidemiologic Transition

Chapter 14
Reassessing the Good and Bad of Modern Environments: Developing a More Comprehensive Approach to Health Trend Assessment

Lawrence M. Schell
Department of Epidemiology and Biostatistics and Department of Anthropology, University of Albany, State University of New York, Albany, NY

> *Life is never fair,...And perhaps it is a good thing for most of us that it is not.*
> —spoken by Lord Goring in *An Ideal Husband* by Oscar Wilde

Introduction

The transition from hunting and gathering to village and then urban settlements based on the intensification of agriculture is clearly the basis for the first epidemiologic transition (Barrett et al., 1998). However, the transition is marked not only by a change in subsistence and settlement patterns and health, but also by a far-reaching social change, the development of social stratification (Flannery and Marcus, 2012). This *invention* has characterized and affected human societies ever since. It was well established in the countries experiencing the second epidemiologic transition around the turn of the 19th century and continues today to structure new changes in health profiles within and among societies.

Social stratification is evident to varying degrees in every type of social organization except for hunter-gatherer bands (Service, 1971; Flannery and Marcus, 2012) and predominates overwhelmingly among contemporary societies. The common definition of social stratification is the categorization of persons in a society into groups based on any features of individuals, though usually economic ones, and valuing those divisions unequally. Many have viewed the rise of social stratification in association with agriculture as a mechanism for organizing labor and a form of resource allocation. In addition to this, though, from a health ecology point of view, social stratification is also the chief means of assigning all manner of physical risks within a society. These include the risk of injury, the risk of poor diet, the risk of violence, and the risk of death. Culture, which has been called *Man's Adaptive Dimension* (Montagu, 1968), may allocate resources and provide adaptive solutions to challenges at the population level (Rappaport, 1967), but it is also a system for allocating various forms of risk within a population (Schell 1992, 1997).

Acknowledging the overwhelming effect of social stratification also reveals the fallacy of considering health changes over the course of human existence solely in terms of large demographic units such as nations. Recognizing the importance of social stratification raises the possibility of there being radically different mortality characteristics of different social strata (or groups) within communities and societies (Gaylin and Kates, 1997). If epidemiologic transitions are defined by changes in common causes of death, then different transition stages may exist within a society that are not observable when mortality data are assembled on the level of a society or nation. Furthermore, relying on mortality to define changes in health over time necessarily and understandably leads to a particular focus in studying health changes over time. Mortality measures are based on counts of deaths that are dichotomous as persons are either dead or alive. The dead or alive distinction is unambiguous and contributes to the certainty of the count. However, there is a price. Not everyone who is alive is equally healthy so this focus does not adequately consider the very real variations in the quality of health among the living. If it is based on national data, it also does not consider within-society variation in the quality of health.

In medical anthropology, health is conceived of as a continuum wherein death is only the worst form of health, and there are many gradations of better health. This view of health is usually applied in medical anthropology to counts of morbidity, as it is too obvious a critique of mortality counts. Counts of morbidity are dichotomous also and suffer because there are many gradations of health that do not cross over into the case definition of a disease. For example, people may be overweight before they meet the case definition of obesity; they may have precancerous lesions before the diagnosis of cancer can be applied; they are often infected with a disease-causing agent before being symptomatic and symptomatic before exhibiting the disease syndrome. In morbidity counts typical of epidemiology, the preobese and the precancerous and the presyndromic are placed in the same category of healthy as they do not have a disease diagnosis. A more sensitive analysis needs not only to distinguish the dead and living, the diseased and not diseased, but also those less healthy from those more healthy. Developing a measurement scale of health that assesses the variation in health in a society is a requisite beginning point for a more sensitive analysis of health trends.

To some extent, this has been avoided by practitioners in epidemiology because of the difficulty of assessing different gradations of health. It is akin to assessing different degrees of adaptation, a notoriously difficult and contentious subject. The difficulty in measuring health reflects the development of the field itself. Typically, in science, the earliest attempts to measure the significant entities begin with the crudest scales. But with an increase in knowledge gained from that crude scale, more gradations are seen and more sensitive measuring scales are developed. Epidemiologic transition theory has been based on changes in common causes of death and as such assesses part of the health pyramid, but if it is to address changes in health sensitively and more comprehensively, the development of more fine-grained measuring techniques would facilitate theory development.

Background

The Allocation of Risk

The allocation of risk among different socially defined groups is painfully clear in many societies today. Persons in the higher strata of many industrial societies have preferential opportunities for occupations and housing that carry low risks of health consequences while people in other strata must take jobs that entail more risk of injury and disease and are able to find housing predominantly in environments that have greater health risks from crime, pollution, traffic, etc. (Schell et al., 2003).

The differential distribution of risk in stratified societies may be partially exhibited in different death rates. According to the most recent analysis by the U.S. Centers for Disease Control (CDC), for example, in 2007, the infant death rate among U.S. Black Americans was 13.31, 2.4 times the rate for non-Hispanic Whites.[1] Further, while the two most common causes of death are diseases of the heart and malignant neoplasms for both non-Hispanic Blacks and Whites, the percent of deaths from the other causes varies considerably. When specific age and gender categories are compared, the differences can be large. Comparing men 45–54 years of age, suicide accounts for 6.4% of deaths among Whites but is not among the 10 common causes of death for same-aged U.S. Black Americans. On the other hand, assault accounts for 3.1% of Black American deaths but is not among the 10 common causes of death for same-aged White men. The death rate due to HIV infection in this age group is nearly five times higher among Black Americans than Whites. The health profile of Native Americans is equally grim. For instance, Alaskan Natives have a mortality rate that is 1.6 times the rate for US Whites (Day and Lanier, 2003). Similarly, American Indian populations have about twice the mortality rate of the general US population (Kunitz, 2008).

Such differences in mortality arise through multiple pathways. Some pathways lead to increased disease incidence and others to less care, poorer health, and poorer survival. In the same way, morbidity is multicausal. Even infectious disease occurrence depends on qualities of the host (e.g., exposure, susceptibility), the pathogenic agent, and their environment. The noninfectious diseases and other debilitating conditions are particularly multicausal. Disentangling these many contributions occupies a good deal of health researchers' efforts (see Kelsey et al., 1996; Merril, 2008).

Pollutant Exposures as Pathways of Risk: Lead

Lead exposure and its effects illustrate many of these points. Lead reaches humans from along several pathways (Figure 14.1). Some lead enters the environment through simple crustal weathering, but the largest contribution comes through human-directed, anthropogenic processes. For instance, industrial processes and the addition of lead to products such as paint and gasoline lead to human exposure. Lead components of water system structures (i.e., pipes and solder) can contribute lead as can lead components of food storage items. Most of these sources have been eliminated or severely reduced in many countries, and lead is often considered a legacy pollutant as current exposure comes from past uses. The most common exposures for children today are from lead in dust from lead-painted buildings that have deteriorated, flaked, and lost bits of their leaded paint into rooms and adjacent outdoor areas, and from objects painted with leaded paint. The latter can be a source because young children frequently mouth such objects.

Lead exposure contributes to morbidity and mortality in various ways. At lower levels, it contributes to disability and indirectly to underlying causes of morbidity and mortality, while at higher levels, it contributes more directly by producing clinical syndromes and death. The CDC has described the range of these effects in terms of blood lead levels (Centers for Disease Control and Prevention, 1991). Among adults, levels of 150 μg/dL were considered deadly, and those just below, permanently injurious to the central nervous system. Exposures like these have been linked most often to specific occupations wherein lead is used in manufacturing and to unsafe practices such as *ad hoc* reclamation of lead storage batteries, but may also be produced by children's ingestion of paint chips containing

---

[1] NCHS Data Brief, #74; http://www.cdc.gov/nchs/data/databriefs/db74.pdf.

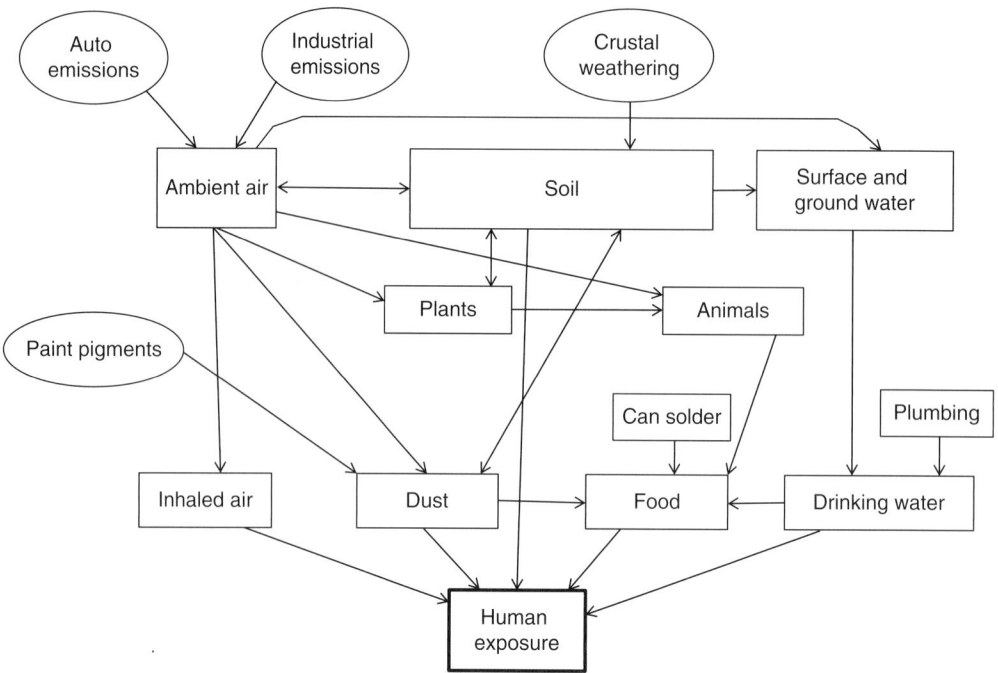

Figure 14.1. Pathways of human exposure to lead. Adapted from Agency for Toxic Substances and Disease Registry (1988).

lead. Levels of 70 µg/dL produce peripheral neuropathies and reduce hemoglobin synthesis. Levels as low as 20 µg/dL slow nerve conduction. Levels of 10 µg/dL produce developmental toxicity, such as reductions in cognitive performance (i.e., IQ), growth, and hearing. Today, levels above 10 µg/dL in children are considered injurious and levels above 5 µg/dL in need of follow-up, although there is considerable evidence for effects at still lower levels (Centers for Disease Control and Prevention, 2005).

In general, exposure to pollutants is unequally distributed in highly industrialized countries, and its distribution in less industrialized ones is becoming more and more unequal as well (Schell and Czerwinski, 1998; Rollin et al., 2009). In general, males exhibit higher levels than females and older persons higher levels than younger adults. More industrialized urbanized communities have higher levels than ones that are less industrialized and less urban (Schell et al., 2004).

Culturally influenced behavior patterns can explain a good deal of these differences in exposure. In the US, according to the National Health and Nutrition Survey (NHANES) for the period of 1999–2002, significant differences in lead levels are present among socially defined groups. In the U.S., Black Americans of every age category, for instance, have higher average lead levels than same-aged Whites. More importantly, the frequency of persons with elevated lead levels (elevated was considered to be 10 µg/dL until mid-2012, when the level was reset to 5 µg/dL by the U.S. CDC) is substantially greater among non-Hispanic Blacks than Whites. Comparing children of 5 years of age or less, the rate is nearly three times higher for non-Hispanic Blacks than non-Hispanic Whites. Problematically, lead is also passed from mother to child across the placenta (correlations between mother's blood lead and newborns are high) and by lactation. As a result, these differences are not static and confined to a single generation; one generation's lead exposure is not erased but is passed to the next.

In 1982, the cutoff point for elevated lead requiring clinical intervention of some kind was 30 μg/dL of blood, not 10. At that time, the U.S. National Center for Health Statistics (NCHS) found that the rate of elevated lead levels was more than four times greater in young African American children from poor families than in poor White children of comparable age. Among African American children between 6 months and 5 years of age, 18% had an elevated lead level of 30 μg/dL of blood or above (Annest et al., 1982).

More recent NCHS surveys show reduced lead levels nationally but do find pockets of higher lead levels. For instance, using data from the 1991 to 1994 NHANES survey, it can be seen that the rate of elevated lead levels in Black non-Hispanic children less than 5 years of age living in housing built before 1942 was 10 times higher than among same-aged Whites living in housing built after 1973. Housing age, location, and condition are strong influences on exposure to lead, and due to the history of residential segregation in the U.S., these index both social category and exposure risk.

The effects of lead directly contribute to morbidity and, when exceptionally high, to mortality as well, but the indirect effects that are evident through a holistic, biocultural approach show the far-reaching and devastating effects of exposure to this toxicant. At the low exposures that are most common, lead produces cognitive deficits. These have myriad effects that contribute to morbidities by contributing to poverty, downward social mobility, and even criminality (Carpenter and Nevin, 2010). In real terms apparent to educators, consider a hypothetical yet typical kindergarten classroom in 1987 of perhaps 28 African American children from relatively poor, inner-city backgrounds. Based on the frequency of elevated lead in this demographic group, five (18%) will have elevated lead levels (30 μg/dL at the then current standard), which would affect their central nervous system and their ability to learn. These children may demand extra attention from the teacher, thereby reducing the teacher's attention to others in the classroom with lower lead levels. Overall, the delivery of education will suffer. Consider also that the action level for children's blood lead has been reduced from 30 to 10 μg/dL on the basis of the existence of proven cognitive deficits associated with lead at lower and lower levels. We now know that a classroom of the 1980s serving a Black inner-city community might contain, in addition to the five children with lead levels above 30, many children with lead levels between 10 and 29 whose cognitive performance and other behaviors would have also been affected. Between the children with elevated lead and those with enough lead to affect learning, the classroom for the children as a whole would have been a highly challenging environment for learning. As educational attainment is one of many significant predictors of morbidity and mortality (Gakidou et al., 2010; Montez et al., 2012), this dynamic alone could translate into and contribute to lasting, pervasive, negative economic and health consequences for urban U.S. Black communities.

The effects of lead can be seen early in life, but lead exposure and its unseen effects begin with the previous generation; lead is a multigenerational problem. Children raised in an older urban area with higher environmental lead levels are more likely to absorb lead and experience the problems associated with elevated blood lead levels, including cognitive effects. These contribute to poor school performance, poor acquisition of life skills (including nutrition knowledge), and job skills. This translates into early school-leaving and often to early family-starting as well as poor preparation for employment. All of these factors contribute to poor diet and to likely residence in poor areas of the cities where lead levels are high. When a woman raised in these conditions becomes a mother, her child will be born with a lead burden transferred across the placenta. The child will be exposed to lead from its environment in postnatal life as well, and its diet may be deficient in micronutrients that afford some protection from lead absorption and lead effects (Schell et al., 1992). It repeats the cycle of exposure and the effects that its mother experienced as a child. Two important questions that arise from this dynamic are: How to break the cycle? How low must lead levels be in order to insure healthy development?

Work done in Albany, New York, sought to determine if very low levels of lead affected infant development, and what factors might influence the uptake of lead and provide a window for intervention (Schell et al., 1997, 2000). Mothers included in this study were largely undereducated, with half not having finished high school, and nearly all were single. Their nutritional intakes were high in protein, but more than 50% were deficient in intakes of kilocalories, zinc, calcium, iron, and vitamin D. Maternal nutrition is related to the lead level of their newborn; women who gain more weight and add more arm circumference (reflecting fat gain) during pregnancy have newborns with lower blood lead levels. In addition, mothers with diets poor in vitamin D, calcium, and iron have newborns with higher lead levels, while those with diets higher in these nutrients have newborns with lower levels (Schell et al., 2003). This association has also been confirmed by more recent findings from experimental work: calcium supplementation during pregnancy can partially blunt transmission of lead from mother to fetus as measured in the newborn, although other nutrients modify transmission as well (Gulson et al., 2004).

The Albany infants' own dietary intakes were extremely high in protein, and most other nutrients were consumed at adequate levels, though consumption below the recommended dietary allowance most often occurred for iron, vitamin D, calcium, and zinc, especially as the infant entered its second year of life (Nolan et al., 2002). Importantly, the zinc, iron, and calcium intakes of infants at 6 months of age were negatively associated with their lead levels at the same age (Schell et al., 2004). Overall, maternal nutrient intake during pregnancy and the infant's own intake in the first year of life are critical dietary factors in the acquisition of a child's lead burden before starting school. Lead levels typically rise in proportion to the lead in the environment in the preschool years, and blocking exposure is critical. Lead levels also affect child growth and development. Reduced growth is a common feature of lead-exposed children (Habercam et al., 1974; Schwartz et al., 1986; Little et al., 1990; Greene and Ernhart, 1991; Hamilton and O'Flaherty, 1995; Denham et al., 2005). In keeping with this, the infants followed in the Albany study exhibited reduced growth, especially in head circumference, in proportion to their mother's lead level in mid-pregnancy (Schell et al., 2009).

The example of lead demonstrates that environmental exposures that are detrimental to development and health are unevenly distributed within a population. Importantly, they also show that the uneven distribution exists along socially created categories. More to the point, the effects of these exposures are unevenly distributed as well. Furthermore, these exposures and their effects can circulate within families from one generation to the next, reproducing deficits and poverty for some while others who are unexposed enjoy advantages. As the cycle is repeated, health and socioeconomic gradients become established, become normal, and become accepted as a feature of society.

Pollutant Exposures as Pathways of Risk: Organic Compounds

The example of lead exposure also shows that local environments contribute to health disparities within a society by creating immediate sources of health deficits. However, although lead exposures typically reflect the local environment, not all detrimental exposures have local or immediate sources. Urban environments typically have distant footprints, as they export wastes and depend on distant sources for energy generation, food, and water, and because city-serving industries are located far away on less valuable land. The result is that in modern industrialized and industrializing nations, industrial processes often negatively affect rural areas. This pattern contrasts with the urbanism typical of England, Western Europe, and the

U.S. in the 19th century. In early industrializing cities, the disposal of pollutants from industrial production and organic waste was largely local (Tarr, 1996). Although services were in place to transport human waste out of cities, waste contributed powerfully to the contamination of urban water supplies. Today's cities are far more efficient at displacing pollution of all kinds away from urban areas into more rural areas, although clearly air pollution is an exception. In modern industrialized and industrializing nations, waste disposal sites are unequally distributed in society. A preponderance of these sites occur near poorer and or ethnic minority communities (Berry, 1977; Bullard, 1990) and polluting urban facilities (i.e. bus garages and parking lots, incinerators) are often situated in poorer neighborhoods (Mott, 1995).

Another system for distributing pollutants is through the food chain. Thus, industrial wastes may be absorbed by animals in one place and consumed by animals elsewhere that, in turn, are consumed by people in another location. One of the earliest documented instances of this was in Minamata, Japan (Smith and Smith, 1975). Industrial waste that contained inorganic mercury was deposited in a nearby ocean bay where it was absorbed by tiny ocean-bed-dwelling organisms. These organisms converted the inorganic mercury to organic mercury, or methylmercury, which is highly toxic. These organisms were eaten by larger ones and eventually by fish. When these fish were in turn caught and eaten by humans, the exposure to methylmercury produced devastating effects, including loss of physical coordination, difficulty with speech, hearing impairment, blindness, and death. Children who had been exposed *in utero* through their mothers' ingestion were also affected with a range of symptoms including motor difficulties, sensory problems, and mental retardation. Other similar examples of food-chain-dispersed pollutants include organic pollutants. Among the populations most highly exposed to organic pollutants from specific manufacturing processes are the Inuit of Canada (Dewailly et al., 1996). Following traditional subsistence strategies, many Inuit diets are based on consumption of large fish and sea mammals that are at the top of the food chain. However, in modern environments, these apex predators in turn have consumed numerous smaller fish, each one contributing small amounts of organic toxicants. Thus, at each trophic level, any toxicant that is not metabolized is retained in prey organisms, and the toxicants become biologically accumulated up the food chain from prey to predators.

Some American Indian groups also suffer from both the polluting effects of industrialization located in rural areas as well as contamination of the local food chain from both local and more distant processes. For instance, the Akwesasne Mohawk people, a sovereign tribal nation bordering the U.S. and Canada and bridging the St. Lawrence River, have been exposed to polychlorinated biphenyls (PCBs) from nearby industrial plants on the St. Lawrence River as well as to pesticides that communities are more commonly exposed to across the U.S., such as DDT. In addition, the Akwesasne Mohawk reservation is just downstream from a U.S. federal Superfund site. This is a designation that requires substantial proof of contamination and serves as testimony to the real contamination of the local river ecology. The fact that there are two New York state superfund sites along the St. Lawrence River nearby is further evidence of the problematic local environment (Gallo et al., 2005).

Evidence that exposure to PCBs is detrimental comes from studies of both Akwesasne Mohawk youth and adults. For instance, Akwesasne youth (10- to 17-year-olds) have higher thyroid stimulating hormone (TSH) levels and lower free thyroxin levels in proportion to their PCB burden. Elevated TSH is a classic symptom of hypothyroidism, which contributes to several morbidities, including overweight. This in turn is a contributing factor to numerous other morbidities, especially chronic and degenerative conditions such as diabetes. Accordingly, the same youth also display high rates of overweight and obesity (Gallo et al., 2005). There is also evidence of more rapid attainment of menarche, a valuable marker of sexual maturation, in relation to PCB level in young women (Denham et al., 2005). Later as young adults, both males and females with elevated antithyroid peroxidase antibody (TPOAb) also exhibit higher levels of PCBs. TPOAb is a marker of autoimmune disease, which is consistent with reports by local health authorities of elevated frequencies of such diseases (e.g., polycystic ovary syndrome, type 1 diabetes, Hashimoto's thyroiditis, Grave's disease, systemic lupus erythematosus, and others). Deleterious effects on cognitive performance are also found to be present (Newman et al., 2006, 2009), and these are consistent with other studies showing adverse cognitive effects from acute exposure to PCBs (Chen et al., 1992).

Among adults, Akwesasne males display lower levels of testosterone in proportion to their PCB burden (Goncharov et al., 2009), more diabetes (Codru et al., 2007), and higher levels of serum lipids, which are associated with heart disease (Goncharov et al., 2008). Additionally, studies of other populations exposed to PCBs show that these pollutants are positively related to blood pressure in adults (Goncharov et al., 2010). Several studies among different populations have further found that simply being in residence near a hazardous waste site is associated with negative health outcomes, such as metabolic syndrome and asthma, even after controlling for the most common risk factors (Carpenter et al., 2008; Sergeev and Carpenter, 2010). Such residential associations are a substantial part of the basis for the environmental justice movement in the U.S. The crux of this movement is the improbably frequent coincidence of hazardous waste sites and other types of environmental hazards located near to minority communities.

A Brief History of Social Stratification, Pollution, Toxicants, and Human Health

The contribution of pollution hazards to human health has a long history reaching well into prehistory, starting with the implementation of fire several hundred thousand years ago. Indeed, some scholars argue that our long history of exposure to the byproducts of combustion, particularly aromatic hydrocarbons, has preadapted us to deal harmlessly with exposure to many modern synthetic chemicals that have similar structures (Silkworth and Brown, 1996). However, there is much evidence to the contrary. First, the intensity and quantity of many exposures to pollutants have changed substantially throughout time and have greatly intensified within modern environments. For instance, ice cores reveal much higher levels of industrial pollutants in the atmosphere of the northern hemisphere than were present in the preindustrial era (Committee on Measuring Lead in Critical Populations, 1993). Furthermore, if adaptation to smoke has occurred throughout human populations, smoke-related diseases should not be common, but they are. Occupational studies clearly show that workers routinely exposed to smoke do not adapt to the exposure but exhibit higher rates of smoke-related diseases. Indeed, one of the earliest cancers linked to a particular occupation was scrotal cancer among chimney sweeps. In 18th-century London, the surgeon Percivall Pott (better known to anthropologists for identifying a tubercular vertebral deformity) linked exposure to soot, which typically contains high concentrations of aromatic hydrocarbons, to high frequencies of cancer among the sweeps, noting that soot lodged in scrotal crevices where cancer later developed (Waldron, 1983).

Industrialization in Western Europe introduced myriad occupational hazards to contemporary populations through exposure to various chemicals, dust, and other materials. Ice core data suggest that atmospheric pollution was likely a major component of these new hazards. Atmospheric pollution has many effects, including detrimental changes in respiratory structures and functions as well as producing systemic effects through neurological pathways (Waldbott, 1978). Like lead, it is a contributing factor to many diseases, sometimes in seemingly indirect ways. For instance, in industrial cities of 19th-century England, where clothing styles, dawn-to-dusk factory work hours, and tall buildings coupled with narrow streets limited people's exposure to sunlight, the addition of air pollution reduced ultraviolet exposure so severely that rickets became common among industrial workers (Loomis, 1970). The example of rickets, usually thought of as a nutritional disease, illustrates how pollutants can indirectly but significantly contribute to disease occurrence in modern human populations.

High socioeconomic status communities and individuals avoid exposure to pollutants as such exposure is linked in the popular mind, if not always in the scientific literature, to disease and to poorer health. Social position and the power it entails allow these groups to avoid as much as they wish any perceived risk to health. (High-status activities, such as boat or car racing and mountain climbing, that carry high risks are the exception.) The ability to afford safe homes in safe neighborhoods and low-risk occupations is part of this larger picture. Similarly, they are frequently able to blunt other, more global threats such as air pollution by such expensive devices as air-filtering systems for homes and in automobiles that filter air and add oxygen for the daily commute. Such amenities, and the reduced exposure to risk that they carry, are generally out of reach for lower-status communities, with concordant, persistent, and even intergenerational impacts upon their health.

Conclusion

Today, more than one half of the world's population is urban, and urban growth is increasing (Department of Economic and Social Affairs, 2000). Because social stratification functions as a potent means of displacing risk from the more powerful in society to the less powerful, it produces differences in morbidity and mortality profiles across socioeconomic strata. Seen in the context of epidemiologic transitions, such differences temper the conclusiveness of models of such transitions and analyses of the transition in different nations that are based on whole-society mortality data. Different stages might be seen to occur among the different strata if data are analyzed in finer units than whole societies.

Human industry in premodern cities produced pollution that was both organic (e.g., butchery) and synthetic (e.g., lead, soot) in nature. Modern production of synthetic compounds in the 20th and 21st centuries poses an entirely additional set of different hazards. Studies of contemporary populations demonstrate that exposure to the pollutants characteristic of the modern era as well as to those, such as lead or mercury, which would have been typical in premodern cities, is associated with alterations in development and functioning that are almost always indicative of poorer health. Measures of growth, development, and overall functioning provide sensitive yardsticks of health. Tanner (1978: 219) wrote that "growth emerges as the prime measure of a child's physical and mental health." This equation of one child's growth and health can be further translated to the population level: "It [growth] is equally powerful for studying the effect of political organization upon the relative welfare of the various social, cultural and ethnic groups which make up a modern state." This, for instance, is the basis for the study of economic history through analysis of changes in stature (Fogel, 1986). In short, growth data and other measures of development can provide additional sensitivity to the more typical and conventional measurements of health that are provided by mortality statistics.

Shortly after the discovery in 1978 of Love Canal, when the impact of toxic waste dumpsites was being hotly debated, advocates of what became known as the precautionary principle argued that we could not wait for dead bodies to accumulate from toxic exposures before taking action. This point of view has gained many adherents and has led to a far greater emphasis on the evaluation of the danger from hazardous waste sites on functional measures (e.g., cognitive performance) as well as prenatal and postnatal growth and development. Mortality data may be a traditional and strong basis for conclusions about health, but they are not sufficient if health is construed as more than the absence of disease and death. Measures of function and development can add a more fine-grained dimension of health and provide a more sensitive scale to access changes in health over time. A synthetic approach to epidemiologic transition studies would include such measures as much as possible along with the more traditional sources. In this way, it may be possible to distinguish health trends within societies and over time with more sensitivity.

James Tanner examined the record of height of the French Count PhilippeGuéneau de Montbeillard's son, who was born in 1759. At this time, French society was racked with inequality and frequent food shortages that led to social upheaval, and most French children grew quite poorly in comparison to today's growth references. However, the son of the well-off count grew very well and his height tracked at or above the 50th centile of modern references; he was as tall as you or I and must have towered over his lower-born countrymen. Oscar Wilde's Lord Goring was quite correct, in that "Life isn't fair," and we who are interested in the history of health know that it is not. We can incorporate that realization it into our investigations. Ironically, one should note that de Montbeillard's son died by an instrument of social upheaval, the guillotine.

References

Agency for Toxic Substances and Disease Registry. 1988. The nature and extent of lead poisoning in children in the United States: a report to congress. U.S. Atlanta: Department of Health and Human Services.
Annest JL, Mahaffey KR, Cox DH, Roberts J. 1982. Blood lead levels for persons 6 months–74 years of age: United States, 1976–80. Adv Data 79(79):1–23.
Barrett R, Kuzawa CW, McDade TW, Armelagos GJ. 1998. Emerging and re-emerging infectious diseases: the third epidemiologic transition. Annu Rev Anthropol 27:247–271.
Berry JLB. 1977. The social burdens of environmental pollution: A comparative metropolitan data source. Cambridge: Ballinger Publishing Company.
Bullard RD. 1990. Dumping in dixie: Race, class, and environmental quality. San Francisco: Westview Press.
Carpenter DO, Nevin R. 2010. Environmental causes of violence. Phys Behav 99(2):260–268.
Carpenter DO, Ma J, Lessner L. 2008. Asthma and infectious respiratory disease in relation to residence near hazardous waste sites. Ann N Y Acad Sci 1140(1):201–208.
Centers for Disease Control and Prevention. 1991. Preventing lead poisoning in young children. Atlanta: CDC.
Centers for Disease Control and Prevention. 2005. Preventing lead poisoning in young children. Atlanta: CDC.
Chen YC, Guo YL, Hsu CC, Rogan WJ. 1992. Cognitive development of Yu-Cheng ("oil disease") children prenatally exposed to heat-degraded PCBs. JAMA 268(22):3213–3218.
Codru N, Schymura MJ, Negoita S, Rej R, Carpenter DO, Akwesasne Task Force on the Environment. 2007. Diabetes in relation to serum levels of polychlorinated biphenyls and chlorinated pesticides in adult Native Americans. Environ Health Perspect 115(10):1442–1447.
Committee on Measuring Lead in Critical Populations. 1993. Measuring lead exposure in infants, children, and other sensitive populations. Washington, DC: National Academy Press.
Day GE, Lanier AP. 2003. Alaska native mortality, 1979–1998. Public Health Rep 118(6):518–530.
Denham M, Schell LM, Deane G, et al. 2005. Relationship of lead, mercury, mirex, dichlorodiphenyl-dichloroethylene, hexachlorobenzene, and polychlorinated biphenyls to timing of menarche among Akwesasne Mohawk girls. Pediatrics 115(2):e127–e134.
Department of Economic and Social Affairs, P.D. 2000. World population monitoring 1999. New York: United Nations.

Dewailly E, Ayotte P, Laliberte C, Weber JP, Gingras S, Nantel AJ. 1996. Polychlorinated biphenyl (PCB) and dichlorodiphenyl dichloroethylene (DDE) concentrations in the breast milk of women in Quebec. Am J Public Health 86(9):1241–1246.

Flannery KV, Marcus J. 2012. The creation of inequality: How our prehistoric ancestors set the stage for monarchy, slavery, and empire. Cambridge: Harvard University Press.

Fogel RW. 1986. Physical growth as a measure of the economic well-being of populations: the eighteenth and nineteenth centuries. In: Falkner F, Tanner JM, editors. Human growth: A comprehensive treatise. New York: Plenum. p 263–281.

Gakidou E, Cowling K, Lozano R, Murray CJ. 2010. Increased educational attainment and its effect on child mortality in 175 countries between 1970 and 2009: a systematic analysis. Lancet 376(9745):959–974.

Gallo MV, Schell LM, Akwesasne Task Force on the Environment. 2005. Height, weight and body mass index among Akwesasne Mohawk youth. Am J Hum Biol 17:269–279.

Gaylin DS, Kates J. 1997. Refocusing the lens: epidemiologic transition theory, mortality differentials, and the AIDS pandemic. Soc Sci Med 44(5):609–621.

Goncharov A, Haase RF, Santiago-Rivera A, et al. 2008. High serum PCBs are associated with elevation of serum lipids and cardiovascular disease in a Native American population. Environ Res 106(2):226–239.

Goncharov A, Rej R, Negoita S, et al. 2009. Lower serum testosterone associated with elevated polychlorinated biphenyl concentrations in Native American men. Environ Health Perspect 117(9):1454–1460.

Goncharov A, Pavuk M, Foushee HR, Carpenter DO, Anniston Environmental Health Research Consortium. 2010. Blood pressure in relation to concentrations of PCB congeners and chlorinated pesticides. Environ Health Perspect 119(3):319–325.

Greene T, Ernhart CB. 1991. Prenatal and preschool age lead exposure: relationship with size. Neurotoxicol Teratol 13(4):417–427.

Gulson BL, Mizon KJ, Palmer JM, Korsch MJ, Taylor AJ, Mahaffey KR. 2004. Blood lead changes during pregnancy and postpartum with calcium supplementation. Environ Health Perspect 112(15):1499–1507.

Habercam JW, Keil JE, Reigart JR, Croft HW. 1974. Lead content of human blood, hair, and deciduous teeth: correlation with environmental factors and growth. J Dent Res 53(5):1160–1163.

Hamilton JD, O'Flaherty EJ. 1995. Influence of lead on mineralization during bone growth. Fundam Appl Toxicol 26(2):265–271.

Kelsey J, Whittemore AS, Evans AS, Thompson W. 1996. Methods in observational epidemiology. New York: Oxford University Press.

Kunitz SJ. 2008. Ethics in public health research: changing patterns of mortality among American Indians. Am J Public Health 98(3):404–411.

Little BB, Snell LM, Johnston WL, Knoll KA, Buschang PH. 1990. Blood lead levels and growth status of children. Am J Hum Biol 2(3):265–269.

Loomis WF. 1970. Rickets. *Scientific American* 223(6): 76–91.

Merril RM. 2008. Environmental Epidemiology. Sudbury: Jones & Barlett.

Montagu MFA. 1968. Culture: Man's adaptive dimension. New York: Oxford University Press.

Montez JK, Hummer RA, Hayward MD. 2012. Educational attainment and adult mortality in the United States: a systematic analysis of functional form. Demography 49(1):315–336.

Mott L. 1995. The disproportionate impact of environmental health threats on children of color. Environ Health Perspect 103(S6):33–35.

Newman J, Aucompaugh AG, Schell LM, et al. 2006. PCBs and cognitive functioning of Mohawk adolescents. Neurotoxicol Teratol 28(4):439–445.

Newman J, Gallo MV, Schell LM, et al. 2009. Analysis of PCB congeners related to cognitive functioning in adolescents. Neurotoxicology 30(4):686–696.

Nolan K, Schell LM, Stark AD, Gomez MI. 2002. Longitudinal study of energy and nutrient intakes for infants from low-income, urban families. Public Health Nutr 5(3):405–412.

Rappaport RA. 1967. Ritual regulation of environmental relations among a New Guinea people. Ethnology 6(1):17–30.

Rollin HB, Sandanger TM, Hansen L, Channa K, Odland JO. 2009. Concentration of selected persistent organic pollutants in blood from delivering women in South Africa. Sci Total Environ 408(1):146–152.

Schell LM. 1992. Risk focusing: an example of biocultural interaction. In: Huss-Ashmore R, Schall J, Hediger ML, editors. Health and lifestyle change. Philadelphia: Masca, University of Pennsylvania. p 137–144.

Schell LM. 1997. Culture as a stressor: a revised model of biocultural interaction. Am J Phys Anthropol 102:67–77.

Schell LM, Czerwinski SA. 1998. Environmental health, social inequality and biologic difference. In: Strickland S, Shetty P, editors. Human biology of social inequality. Cambridge: Cambridge University Press. p 114–131.

Schell LM, Madan M, Davidson GK. 1992. Auxological epidemiology and methods for the study of effects of pollution. Acta Medica Auxologica 24(3):181–188.

Schell LM, Stark AD, Gomez MI, Grattan WA. 1997. Blood lead level by year and season among poor pregnant women. Arch Environ Health 52(4):286–291.

Schell LM, Czerwinski S, Stark AD, Parsons PJ, Gomez M, Samelson R. 2000. Variation in blood lead and hematocrit levels during pregnancy in a socioeconomically disadvantaged population. Arch Environ Health 55(2):134–140.

Schell LM, Denham M, Stark AD, et al. 2003. Maternal blood lead concentration, diet during pregnancy, and anthropometry predict neonatal blood lead in a socioeconomically disadvantaged population. Environ Health Perspect 111(2):195–200.

Schell LM, Denham M, Stark AD, Ravenscroft J, Parsons P, Schulte E. 2004. Relationship between blood lead concentration and dietary intakes of infants from 3 to 12 months of age. Environ Res 96(3):264–273.

Schell LM, Denham M, Stark AD, Parsons PJ, Schulte EE. 2009. Growth of infants' length, weight, head and arm circumferences in relation to low levels of blood lead measured serially. Am J Hum Biol 21(2):180–187.

Schwartz J, Angle C, Pitcher H. 1986. Relationship between childhood blood lead levels and stature. Pediatrics 77(3):281–288.

Sergeev AV, Carpenter DO. 2010. Residential proximity to environmental sources of persistent organic pollutants and first-time hospitalizations for myocardial infarction with comorbid diabetes mellitus: a 12-year population-based study. Int J Occup Med Environ Health 23(1):5–13.

Service ER. 1971. Primitive social organization: An evolutionary perspective. New York: Random House.

Silkworth JB, Brown JF, Jr. 1996. Evaluating the impact of exposure to environmental contaminants on human health [published erratum appears in Clin Chem 1997;43(2):410]. Clin Chem 42(8): 1345–1349.

Smith WE, Smith AM. 1975. Minamata. New York: Holt, Rinehart and Winston.

Tanner JM. 1978. Foetus into man: Physical growth from conception to maturity. Cambridge: Harvard University Press.

Tarr JA. 1996. The search for the ultimate sink: Urban pollution in historical perspective. Akron: The University of Akron Press.

Waldbott GL. 1978. Health effects of environmental pollutants. St. Louis: Mosby.

Waldron HA. 1983. A brief history of scrotal cancer. Br J Ind Med 40(4):390–401.

Chapter 15
Childhood Lead Exposure in the British Isles during the Industrial Revolution

Andrew Millard[1], Janet Montgomery[1], Mark Trickett[2], Julia Beaumont[3], Jane Evans[4], and Simon Chenery[5]

[1] Department of Archaeology, Durham University, Durham, UK
[2] Department of Archaeology, Thomas Jefferson's Monticello, Charlottesville, VA
[3] Division of Archaeological, Geographical and Environmental Sciences, University of Bradford, Bradford, UK
[4] NERC Isotope Geoscience Laboratory, Keyworth, UK
[5] British Geological Survey, Keyworth, UK

Introduction

The second epidemiologic transition, as originally characterized by Omran (1971), includes a shift in patterns of morbidity and mortality wherein pandemics of acute infectious disease are gradually displaced by degenerative and *man-made* diseases as the chief form of morbidity and primary cause of death. Omran's characterization has been widely accepted as describing a shift in causes of mortality from approximately 1840 to 1920 in Britain and other industrializing nations (Barrett et al., 1998; Mooney, 2007; Noymer and Jarosz, 2008; McKeown, 2009). Although historians have considered and debated the claim of a concomitant shift in morbidity and its relation to *man-made diseases* (Riley, 2001), paleopathologists working with human remains have paid less attention to it. In this chapter, we approach the question of one anthropogenic cause of morbidity, but not by direct examination of the morbidity. Instead, we do so by measuring exposure to a toxin, lead. We investigate the concentration of lead in the tooth enamel of 18th- and 19th-century individuals, using this as a proxy for lead exposure during childhood, and attempt to assess its impact on morbidity.

It has been shown using measurements on ice cores that atmospheric lead pollution in the northern hemisphere has increased about 400 times between prehistory and the mid-1960s, with a major increase from about 500 BC, a decline in the second half of the first millennium AD, and then a continuous increase during the second millennium AD (Hong et al., 1994). However, global atmospheric pollution is not the main route of exposure of humans and is not directly reflected in human remains, which record local exposure to lead, although the peaks of exposure recorded in tooth enamel reflect the same trends in the intensity of lead exploitation from the Neolithic to the early 16th century AD (Budd et al., 2004; Montgomery et al., 2010). Environmental and historical evidence indicates that there were further increases in lead usage in the postmedieval period. In the 16th century, the increase seems to have been rapid, as Harrison (1587: 188) recorded that within living memory, there had been "exchange of vessell, as of treene (i.e., wooden) platters into pewter." Previous investigations of lead in human remains have not covered this postmedieval increase in exposure to lead.

Modern Environments and Human Health: Revisiting the Second Epidemiologic Transition, First Edition. Edited by Molly K. Zuckerman.
© 2014 John Wiley & Sons, Inc. Published 2014 by John Wiley & Sons, Inc.

Background

Health Effects of Lead Exposure

Lead poisoning, or plumbism, causes a range of symptoms though the effects vary depending on the amount and duration of exposure, as well as the age of the individual exposed, with children having the greatest susceptibility to lead poisoning (Putnam, 1986; Garrettson, 1990). Acute poisoning is rare and usually related to the ingestion of soluble lead salts; it causes death within 1 or 2 days if untreated. Chronic long-term exposure to lead may cause gastrointestinal irritation (*dry belly-ache*), reduced libido and infertility, anorexia, anemia, blue-lines on the gum-margins, peripheral neuropathy (wrist-drop), convulsions, and encephalopathy (i.e., neural dysfunction), including impairment of visuospatial/visual motor functioning, short-term memory loss, and confusion and fatigue (Weisskopf et al., 2004; Patrick, 2006). It has also been associated with gout due to impairment of the renal system (Matte et al., 1992). In addition, plumbism causes anemia by interfering with the action of various enzymes in the biosynthesis of heme and through other effects that reduce production or increase destruction of red blood cells (Patrick, 2006). It is thus primarily an iron-deficiency anemia, but it also has aspects of hemolytic anemia. The level of lead in the blood is correlated directly with the severity of anemia (WHO, 2010: 24). The mechanisms by which plumbism causes infertility differ in males and females. In males, libido and sperm count are reduced, probably via the disruptive effect of lead on the production of reproductive hormones including testosterone, and with no safe lower limit of exposure (Vigeh et al., 2010). In females, the main effect is through an increase in spontaneous abortion even at low exposures (Hertz-Picciotto, 2000).

Subclinical poisoning is most common in childhood. Previously, this has been associated with elevated levels of lead in the blood, but recent work has suggested that the lower limit for effects is below the lowest level of 1 μgdL^{-1} that has been studied, and there may be no safe lower limit (WHO, 2010: 25). Subclinical poisoning may result in developmental and behavioral changes, including reduced mental acuity (Lanphear et al., 2000; Dietrich et al., 2001; Needleman, 2004) and a tendency toward behavioral abnormality (Dietrich et al., 2001).

Childhood exposure starts with exposure in the womb, as fetal blood lead levels are very close to maternal levels. Exposure to maternal lead will reduce with breast-feeding as little lead is transmitted in breast milk, but this is offset by a higher rate of absorption in the immature gut. The hand-to-mouth behavior of young children can also lead to oral exposure to lead-bearing objects (WHO, 2010: 22).

Lead Pollution in the 18th and 19th Centuries

The industrialization that occurred in England during the 18th and 19th centuries has been characterized as the transition from an *organic* to *mineral* economy, with concomitant increase in coal use (Wrigley, 1988; Daunton, 1995). It is the smoke and fog resulting from coal use that dominates popular images of the period (thanks to the descriptions of contemporary authors such as Dickens), but other sources of pollution, including lead pollution, were also increasing. The transition from an *organic* to *mineral* economy involved not only fuel, but also the replacement of wooden vessels with inorganic ones, including lead-glazed ceramics and lead and pewter vessels. Acidic materials stored in lead containers would rapidly leach lead into the food or beverage (Richards, 1999). Probate inventories, which are biased toward middle-class households and mostly cover the period between 1550 and 1740, suggest that pewter items were common, particularly in middle-class households until the late 18th century. By the early 19th century, they had declined in popularity in these households and were being replaced by earthenware, particularly *Staffordshire creamware*, which had a lead glaze. Instead, pewter became particularly common in poorer households throughout the 19th century (De Vries, 2008). Although by the 1750s, there was some understanding of the dangers associated with acidic materials in lead-glazed or-bearing vessels (e.g., Lind, 1754; Hardy, 1778), this only slowly reduced the frequency of lead poisoning caused by the exposure of alcoholic drinks to lead or lead-glazed vessels (Handler et al., 1986). Deliberate contamination of food with lead, either as a sweetener or a colorant, was a well-known health issue in the mid-19th century (Drummond and Wilbraham, 1939: 292). Likewise, the utilization of lead to sweeten wine continued into the 19th century (Eisinger, 1982).

The burning of coal was a possible source of lead exposure as well. Farmer et al. (1999) estimated that in 1830, between 9% and 33% of atmospheric lead released in the UK was from coal, with the rest from lead smelting, increasing to between 11% and 38% by 1855. Away from lead production areas, therefore, the atmospheric releases from coal are likely to have been the majority of atmospheric lead releases. Today, however, direct exposure by inhalation is considered to be a minor source of lead poisoning because the particles produced by domestic use of coal are usually too large to be inhaled (WHO, 2010: 18). Whether coal burning could have been a contributing factor to past lead exposure is therefore unclear.

Occupational lead exposure was also acknowledged for deleterious health effects, if not specifically plumbism, from the 17th-century onward (Weeden, 1984). By the 19th century, those professions that were most susceptible to lead poisoning were listed by Thackrah (1831), including miners, ironworkers, founders, potters, brass workers, and solderers. However, this chapter's examination of tooth enamel confines the lead exposure that can be investigated to childhood.

Lead exposure was likely therefore widespread in all ranks of society. However, due to the cost of items such as pewter, lead crystal, and wine, elites and upper socioeconomic status individuals were more likely to experience long-term lead exposure than lower status or poor individuals—except those with specific occupational exposure. Lead was also more widely used in urban than rural contexts, putting urbanites at greater risk. Indeed, those living on a subsistence diet in rural areas may have escaped exposure to the major sources of anthropogenic lead. This is in contrast to the present day, where lead poisoning is predominantly a disease of the poor (WHO, 2010: 35).

Materials and Methods

Study Sites

All the study sites were selected within projects designed primarily to investigate migration rather than lead exposure *per se*. The sample is therefore not optimized to relate lead exposure to social factors. Their locations are shown in Figure 15.1.

Coventry

In 1801, Coventry was a regional city of some 16,000 inhabitants. The population had grown rapidly over the preceding century from about 6700 in 1694 and more than doubled again by 1851 to over 36,000, with migration a major factor in this expansion (Lancaster and Tomlinson, 1969: 5). The primary industries of the city were ribbon weaving (though this declined and virtually disappeared between 1840 and 1860) and watchmaking (Lancaster, 1969).

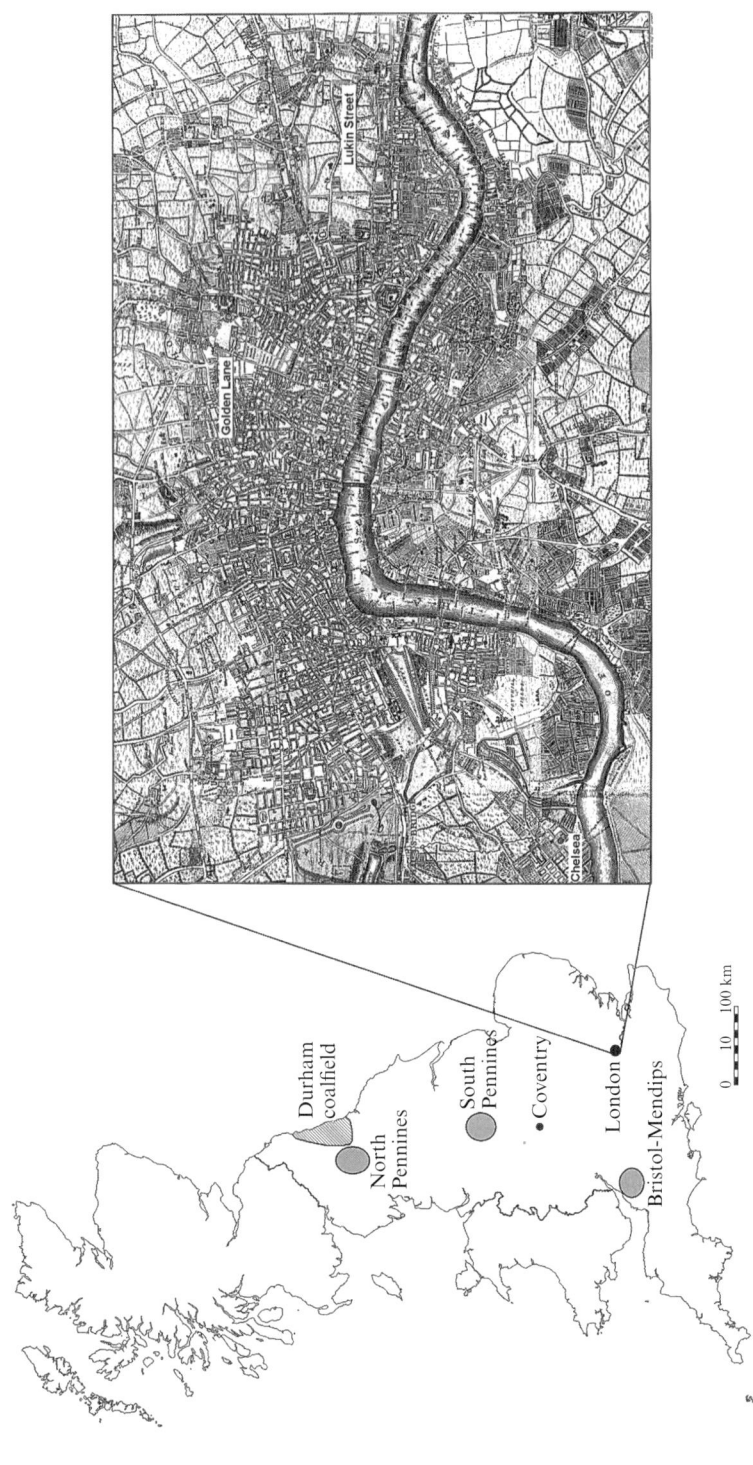

Figure 15.1. Locations of the sites and of British lead ore-fields. Outline reproduced from Ordnance Survey map data by permission of the Ordnance Survey © Crown copyright 2010. The approximate locations of the London sites are shown overlain on Rocque's map of 1741–1745, depicting London at the very start of the period considered. The locations of major lead ore-fields are shown approximately, based on Rohl (1996).

The cemetery investigated was the churchyard of the parish of Holy Trinity, Coventry, with remains recovered from an area representing an extension in 1776 of that cemetery over the demolished nave and aisles of the former St. Mary's cathedral. The cemetery was formally closed in 1856, though burials had become increasingly rare in the preceding years, with only 38 since 1849. Some additional burials within family tombs continued until as late as 1890. Excavations ahead of redevelopment in 1999 entailed the removal of 1706 articulated human skeletons (Rylatt and Mason, 2003). The majority of remains were reburied, but some were retained for further study. The 10 individuals sampled represent retained skeletons that had associated coffin plates giving partial or complete identification; their dates of death range from 1825 to 1847 and their dates of birth $c.1780$ to $c.1823$. Although Coventry was not a very wealthy city, the selection of skeletons with coffin plates will have biased the sample toward higher socioeconomic status groups.

Chelsea

In the early 18th century, Chelsea was a village on the river Thames and the main road west from the City of London, about 3 miles from Westminster, and not within the urban area of London. In 1674, it is recorded as having only 172 houses, but by 1795, this had increased to 1350 households. With a population of less than 12,000 in 1801, it was smaller than Coventry, but by 1851, this had increased to almost 54,000, and it was included within the expanding metropolitan area of London (Rudé, 1971; Croot, 2004). Chelsea had a reputation as the residence of the upper classes—Defoe (1724) described it as a *town of palaces*—but the 18th century saw the development of a service and pleasure industry due to those upper-class residents and Chelsea's favorable location close to the Cities of Westminster and London (Rudé, 1971; Cathcart-Borer, 1973; Insley and Croot, 2004). The epitome of this was the Chelsea Bun House, a famous bakery favored by the Hanoverian royalty, which was run by the Hand family. The economy of Chelsea was diverse, with significant areas of market garden persisting into the 19th century and supplying the needs of London (Bryan, 1869). In the 18th century, there were porcelain manufactories, and until approximately 1825, foundries for bell and weapon manufacture (Croot and Insley, 2004).

Excavations in 2000 investigated part of the graveyard of Chelsea Old Church. A total of 285 skeletons were recovered, mostly from the 18th and 19th centuries. The majority were interred in wooden coffins, though some were in lead coffins, including five of those considered here (Cowie et al., 2008). Twenty-four individuals were sampled with a preference for those identifiable from coffin plates, so that nine identified individuals are considered here.

Lukin Street, Whitechapel

Since the 17th century, Whitechapel and London's docklands have been home to successive waves of immigrants seeking unskilled work and cheap accommodation (White, 2007). With a long-established Irish community, the population here was increased in the 1840s first by slum clearances in St. Giles (*Little Ireland*) to build New Oxford Street and then by the mass migration of the very poor escaping the Great Irish Famine of 1847–1851. The population density in Whitechapel was 15 times that of Chelsea in 1841, and overcrowding and poor sanitation coupled with extreme poverty led to this area having a lower life expectancy than other areas of London (Graham, 1843).

Excavations at the cemetery of the Catholic Mission of St. Mary and St. Michael, Whitechapel (LUK04), provided a sample of 705 individuals (268 adults and 437 subadults) buried between 1843 and 1854. Epigraphic evidence suggests this burial ground served a population chiefly of Irish descent, some of whom came to England during the Great Famine of 1847–1848 from the poorest rural areas of Ireland (Powers, 2008). This is corroborated by documentary evidence that the Catholic Mission served a first-generation Irish community. In order to identify the first-generation migrants, 120 individuals have been studied using a suite of stable isotope ratios from hair, bone, and teeth, of whom 45 were also sampled for enamel lead concentrations.

Golden Lane, London

Golden Lane, on the northern side of the City of London, was described by a police sergeant in mid-19th century London as a *bad, ruffianly, thievish place* (Mayhew, 1861: 237). A poor area with a large Irish community, it was particularly noted for its large number of taverns. Even as late as 1874, Golden Lane was described as "the 'slummiest' of slums" (Greenwood, 1874: 19). The surrounding area, however, was the focus for nonconformists such as the Quakers, and John Wesley's Chapel was established nearby in 1778 (Connell and Miles, 2010).

The City Bunhill Burial Ground was on the site of a former brewery. This dissenters' cemetery was in use from October 1833 to August 1853, and in just over 20 years, more than 18,000 burials appear in the registers. A sample of addresses suggests that more than half were not from the local area (i.e., within 500 m) and that people were brought to the cemetery for burial from a wide area of London. During excavations in 2006, the Museum of London Archaeology excavated the remains of 239 individuals. Osteological analysis of these individuals revealed high infant mortality and disease patterns consistent with other low socioeconomic groups in London. The 13 legible coffin plates from the excavation do not contain any Irish surnames, although as a nonparochial cemetery, it is likely that Catholics will have been buried here (Connell and Miles, 2010).

Skeletal Material

The data used in this study were collected within two separate PhD projects at different universities, using slightly different protocols, but the methods yield commensurate results. Of the skeletal materials available, enamel was selected as the tissue for analysis due to its resistance to diagenesis and its retention of biogenic concentrations and isotope ratios for lead (and for parallel studies of strontium), in contrast to bone and dentine, which often exhibit diagenetic changes due to the burial environment (Budd et al., 2000; Trickett et al., 2003; Millard, 2006). For the majority of skeletons, we selected second premolar or second molar teeth, though the lack of preserved teeth meant that other teeth had to be selected for a minority of individuals. The enamel of the second premolar and the second molar is formed between approximately 2.5 and 6.5 years of age (Moorrees et al., 1963), so, in the main, our results relate to exposure to lead in this period of childhood.

Lukin Street and Golden Lane

For each tooth, the outer surface of the enamel was removed to a depth of more than 100 μm with a tungsten carbide burr. Enamel samples were cut from the tooth with a flexible dental saw, and all saw-cut surfaces, and any adhering dentine was then rigorously cleaned with a tungsten carbide bur. After cleaning with dilute acid, a sample of 5–10 mg of enamel was rinsed with ultrapure water, dried, and weighed. The enamel was then dissolved in acid, and lead concentrations were measured by inductively coupled plasma mass spectrometry (ICP-MS) on an Agilent 7500cx quadrupole mass spectrometer MS at the British Geological Survey, Nottingham. Full details of the protocols are given in Montgomery et al. (2010). The reproducibility of the lead concentration data was ±10% (2σ).

Chelsea and Coventry

Each tooth was half-sectioned, and the dentine removed with a dental burr. The outermost 200 to 300 μm of the internal, external, and cut surfaces of the enamel cap were removed to eliminate any surface contamination. After cleaning in ultrapure water and acetone, approximately 50 to 100 mg of enamel was dissolved in acid and split into two aliquots, one of which was spiked with ^{208}Pb. Lead was purified using ion exchange chromatography. Concentrations were determined by isotope dilution using Thermal Ionization Mass spectrometry on a Finnigan MAT262 multicollector mass spectrometer and the lead isotope compositions were determined using a VG Elemental Axiom MC-ICP-MS. The data were normalized to the reported values of NBS 981 (Thirlwall, 2002). Full details of these protocols are given by Trickett (2006). Lead concentrations were measured with a precision better than ±10%.

As neither concentrations nor isotope ratios of lead are expected to be normally distributed, nonparametric statistical tests were used. In all of the comparisons, Levene's test, which is based on medians, was used to compare variances of the groups. If no significant difference was found, a Mann–Whitney test (for two groups) or Kruskal–Wallis test (for more than two groups) was used to compare central tendency. Where significant differences were found in comparing multiple groups, pairwise Mann–Whitney tests (or in the case of unequal variances, Kolmogorov–Smirnov tests) were used to establish which groups differed. All calculations were performed using the Palaeontological Statistics Package (Hammer et al., 2001).

Results

The results of this study are shown in Table 15.1 and Table 15.2.

Identifying Geogenic and Anthropogenic Lead Exposure

Before evaluating the extent of anthropogenic exposure to lead, it is necessary to establish what range of lead concentrations in tooth enamel can occur naturally. Figure 15.2 shows the dataset compiled by Montgomery et al. (2010) together with our data from Chelsea and Coventry, plotted as concentration versus ^{207}Pb/^{206}Pb. The spread of lead isotope ratios clearly decreases with increasing lead concentration. This phenomenon is attributed to high levels of exposure occurring with lead-rich anthropogenic sources with a limited isotopic range and low levels of exposure deriving from isotopically diverse natural sources. This inverse correlation of isotopic ratio with concentration has therefore been termed *cultural focussing* (Montgomery et al., 2005). On the basis of a visual inspection of a similar figure, Montgomery et al. (2010) suggested that lead concentrations in excess of about 0.5 ppm should be regarded as having an anthropogenic component, as this is the point at which cultural focussing becomes apparent. Here we extend the analysis with a more formal investigation into defining a suitable cut-off point above which we regard the lead content of human tooth enamel to have an undoubted anthropogenic contribution. If the data of Figure 15.2 are partitioned into two groups on the basis of their lead concentration, and the variance of the higher concentration group is plotted as a function of the cut-off between the two groups (Figure 15.3), it is clear that there are major changes in variance at 0.87 and 30 ppm. We therefore take values at and above 0.87 ppm as falling in a *cultural* category where the anthropogenic component of lead exposure dominates. Conversely, values at and below 0.68 ppm are considered primarily natural, though some anthropogenic component cannot be ruled out. For instance, all Neolithic and Bronze Age humans

Table 15.1. Sample details and lead concentrations.

Site code[a]	Skeleton	Tooth[b]	Sex	Cribra orbitalia	Year of Death	Age at death[c]	Title	Forename	Surname	Coffin plate	Offspring?	Pb ppm
COV	50	14	F	N		43*		…	…ein…	Y		29.92
COV	76	25	F	N	1827	30*		Eliza(beth)	Burton	Y		9.51
COV	417	44	F	Y	1847	60*		Sarah	Green	Y		7.71
COV	433/4	35	F		1825	22*		Hannah	Denney	Y		18.40
COV	516	37	M	N	1845	65*		…	Cooper	Y		
COV	672	47	F		1844	30*		Eliza	Sparkes	Y		32.04
COV	808	44	M		1846	48*		William	Wagstaffe	Y		26.86
COV	866	47	F	N	1846	25*		Harriet	Parsons	Y		21.38
COV	978	44	M		1846	61*		James	Brown	Y		16.96
COV	1248	37	M	Y	1842	19*		John	Chattaway	Y		9.18
GDA06	522											31.14
GDA06	744											7.46
GDA06	757(H)											26.55
GDA06	837(H)											66.10
GDA06	991(H)											32.27
LUK04	13	37	F	N		Early middle adult						7.88
LUK04	47	17	U	Y		18–25						92.22
LUK04	288	45	M	N		Early middle adult						0.87
LUK04	413	16	M	Y	1850	12–18		John	Crawley	Y		26.85
LUK04	419	46	F	Y		12–18			y	Y		16.21
LUK04	557	35	F	N		Late middle adult			…ona…oe	Y		6.27
LUK04	597	17	F	N	1852	Late middle adult		Bridget	McNally	Y		21.93
LUK04	633	27	M	N		Late middle adult						1.79

LUK04	755	27	M	N		Late middle adult			6.37
LUK04	813	27	F	N		Early middle adult			59.96
LUK04	833	27	M	N	1847	Late middle adult		John	Berry	Y	2.01
LUK04	840	17	F	N		Old adult				2.46
LUK04	848	17	M	Y		Early middle adult		Timothy	Sullivan	Y	2.12
LUK04	873	17	F	N		Late middle adult	Mrs.	…	….acklin	Y	0.47
LUK04	881	27	M	N		Late middle adult				0.77
LUK04	903	45	M	Y	1851	Late middle adult		Michael	Ryan		2.84
LUK04	913	35	M	Y		Late middle adult				3.38
LUK04	948	35	M	N		Early middle adult		Keon		Y	5.00
LUK04	1012	27	M	N		Late middle adult				37.07
LUK04	1014	17	M	Y	1845	Early middle adult		Alexander Henry	Creamer	Y	26.35
LUK04	1031	17	F	N	1847	Late middle adult		Bridgett	Muldary	Y	2.77
LUK04	1041	17	F	N		Early middle adult				2.72
LUK04	1081	37	M	N		Old adult				10.45
LUK04	1113	27	M	Y		Late middle adult				0.63
LUK04	1129	17	F	N	1851	Early middle adult				Y	54.13
LUK04	1142	27	F	N		Early middle adult	Mrs.	Jane	Su……	Y	2.34
LUK04	1162	27	F	Y	1844	Late middle adult				Y	1.84
LUK04	1174	17	F	N		18–25	Miss			Y	0.91
LUK04	1210	27	M	Y		Late middle adult					1.27
LUK04	1220	45	F	N		Old adult					6.84
LUK04	1222	27	M	N		Early middle adult		P	Sullivan	Y	23.11
LUK04	1282	33	F	N	1847	Old adult				Y	7.32
LUK04	1290	17	M	N	184–	Late middle adult				Y	71.17

(Continued)

Table 15.1. (Continued)

Site code[a]	Skeleton	Tooth[b]	Sex	Cribra orbitalia	Year of Death	Age at death[c]	Title	Forename	Surname	Coffin plate	Offspring?	Pb ppm
LUK04	1312	43	F	N	1845	Late middle adult		Georgiana	Neale	Y		30.00
LUK04	1314	27	M	Y		Old adult						1.40
LUK04	1337	17	M	N	1848	Old adult		John	Regan	Y		4.77
LUK04	1348	35	M	N	1846	Early middle adult		Miguel	Penethera	Y		9.72
LUK04	1363	17	M	N		Old adult						1.63
LUK04	1396	17	M	Y		Late middle adult	Mr.	Jam..	...oll	Y		39.29
LUK04	1404	17	F	N		Early middle adult		Jul...		Y		18.33
LUK04	1430	27	M	Y	1845	Late middle adult	Mr.	C	Hart	Y		2.02
LUK04	1459	27	M	N	1846	18–25				Y		18.06
LUK04	1476	35	F	N		Late middle adult		Bridgett	Hi	Y		2.51
LUK04	1495	27	F	Y		12–18	Miss	So...	Fl....ery	Y		33.02
LUK04	1567	17	F	N		Late middle adult						4.14
OCU00	18	35	F	Y		36–45						82.42
OCU00	19	35	F	N		>45						9.42
OCU00	31	16	F	N		36–45						4.65
OCU00	35	45	M	N	1821	60*	Mr.	Gideon Richard	Hand	Y	N	75.04
OCU00	39	35	F			36–45						44.75
OCU00	104	37	F	N		36–45						50.84
OCU00	147	44	M	N	1808	67*	Mr.	Thomas	Langfield	Y	N	14.03
OCU00	161	35	F	Y		18–25						70.25
OCU00	285	37	M	N		36–45						12.49
OCU00	392	35	F	Y		18–25						42.69
OCU00	552	17	F	N		>45						19.39
OCU00	654	17	M	N	1827	78*	Mr.	Thomas	Long	Y	Y	40.25

OCU00	697	15	F	N		>45				5.01
OCU00	713	18	M	N	1822	68*	Esq.	John	Long	83.44
OCU00	744	15/25	M	N	1793	70*	Mr.	John	Long	Y	N	12.96
OCU00	750	15/25	F	N		>45				Y	Y	28.95
OCU00	792	25	F	N	1807	68*	Mrs.	Milborough	Maxwell	Y	Y	83.42
OCU00	841	37	F	Y		>45				27.88
OCU00	856	25	M	N		26–35				37.06
OCU00	918	37	F	N		>45				27.46
OCU00	980	35	F	N	1806	54*		Sarah	Adams	Y	Y	6.04
OCU00	990	37	F	N	1781	32*		Charity	Adams	Y	Y	13.29
OCU00	994	35	M	N		36–45				19.80
OCU00	1051	36/46	U		179	1–5			Collon/Collum/Collins?	Y		23.60

Blank cells indicate a lack of data.

[a] Site codes: COV, Coventry St. Mary; GDA, Golden Lane; LUK04, Lukin Street; OCU00, Chelsea Old Church (in Trickett (2006) the code CHE was used for this site).

[b] Using the FDI system.

[c] Age at death from coffin plate is indicated by *; other ages are osteological estimates.

? indicates that the year is uncertain.

Table 15.2. Lead isotope ratios for Chelsea and Coventry individuals.

Site code	Skeleton	$^{206}Pb/^{204}Pb$	%RSD	$^{207}Pb/^{204}Pb$	%RSD	$^{208}Pb/^{204}Pb$	%RSD	$^{207}Pb/^{206}Pb$	%RSD	$^{208}Pb/^{206}Pb$	%RSD
COV	50	18.4495	0.003	15.6286	0.005	38.4143	0.007	0.84710	0.002	2.08215	0.005
COV	76	18.4218	0.004	15.6309	0.006	38.3971	0.008	0.84849	0.002	2.08447	0.005
COV	417	18.4014	0.003	15.6083	0.003	38.3301	0.004	0.84820	0.001	2.08300	0.002
COV	433/4	18.4373	0.002	15.6239	0.003	38.4015	0.003	0.84739	0.001	2.08279	0.001
COV	516	18.4456	0.004	15.6254	0.006	38.4142	0.007	0.84714	0.002	2.08267	0.005
COV	672	18.4314	0.003	15.6193	0.003	38.3846	0.004	0.84744	0.001	2.08253	0.002
COV	808	18.4360	0.002	15.6253	0.002	38.3944	0.003	0.84756	0.001	2.08264	0.001
COV	866	18.4415	0.007	15.6298	0.008	38.4233	0.009	0.84752	0.002	2.08352	0.003
COV	978	18.4438	0.002	15.6222	0.002	38.3967	0.003	0.84702	0.001	2.08182	0.001
COV	1248	18.4183	0.002	15.6131	0.003	38.3719	0.004	0.84771	0.001	2.08336	0.002
OCU00	18	18.4232	0.006	15.6197	0.006	38.3972	0.007	0.84782	0.002	2.08411	0.003
OCU00	19	18.4286	0.008	15.6142	0.009	38.3759	0.012	0.84727	0.002	2.08232	0.005
OCU00	31	18.3766	0.009	15.6168	0.011	38.3471	0.014	0.84987	0.002	2.08686	0.005
OCU00	35	18.4460	0.007	15.6179	0.009	38.3936	0.012	0.84667	0.001	2.08138	0.004
OCU00	39	18.4206	0.008	15.6118	0.009	38.3731	0.012	0.84751	0.001	2.08307	0.005
OCU00	104	18.4398	0.005	15.6153	0.006	38.3829	0.009	0.84683	0.001	2.08145	0.004
OCU00	147	18.3745	0.005	15.6163	0.006	38.3465	0.008	0.84991	0.002	2.08694	0.003
OCU00	161	18.4188	0.005	15.6173	0.006	38.4069	0.008	0.84790	0.001	2.08517	0.003
OCU00	285	18.4396	0.005	15.6168	0.006	38.3743	0.009	0.84691	0.001	2.08105	0.003
OCU00	392	18.4369	0.005	15.6264	0.006	38.4388	0.009	0.84758	0.002	2.08496	0.005
OCU00	552	18.4445	0.005	15.6271	0.006	38.4101	0.010	0.84730	0.002	2.08262	0.005
OCU00	654	18.4520	0.005	15.6277	0.006	38.4290	0.009	0.84692	0.002	2.08267	0.005

OCU00	697	18.4357	0.007	15.6286	0.008	38.4212	0.010	0.84774	0.002	2.08420	0.006
OCU00	713	18.4523	0.004	15.6302	0.004	38.4388	0.006	0.84707	0.002	2.08316	0.004
OCU00	744	18.4441	0.006	15.6291	0.006	38.4221	0.006	0.84739	0.002	2.08314	0.002
OCU00	750	18.4402	0.006	15.6285	0.006	38.4174	0.006	0.84752	0.002	2.08332	0.002
OCU00	792	18.4332	0.006	15.6324	0.006	38.4403	0.006	0.84806	0.002	2.08536	0.002
OCU00	841	18.4502	0.006	15.6274	0.006	38.4238	0.006	0.84700	0.002	2.08261	0.002
OCU00	856	18.4524	0.003	15.6286	0.004	38.4157	0.005	0.84696	0.002	2.08192	0.003
OCU00	918	18.4541	0.004	15.6289	0.004	38.4173	0.005	0.84695	0.002	2.08192	0.003
OCU00	980	18.4058	0.003	15.6293	0.004	38.3932	0.006	0.84914	0.002	2.08594	0.004
OCU00	990	18.4069	0.004	15.6172	0.003	38.3735	0.006	0.84844	0.002	2.08475	0.004
OCU00	994	18.4495	0.007	15.6260	0.008	38.4163	0.010	0.84699	0.002	2.08230	0.004
OCU00	1051	18.4364	0.004	15.6312	0.005	38.4203	0.007	0.84785	0.002	2.08397	0.005

%RSD, % relative standard deviation.

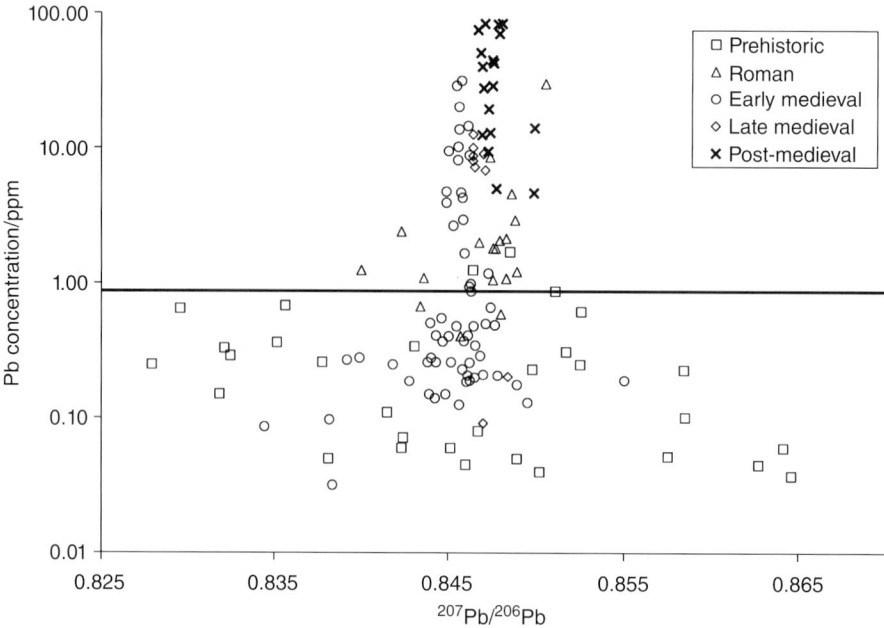

Figure 15.2. Lead concentration compared with ^{207}Pb/^{206}Pb ratio. Note the logarithmic scale for concentration. The solid horizontal line is our lower limit of 0.87 ppm for undoubted *cultural* lead exposure.

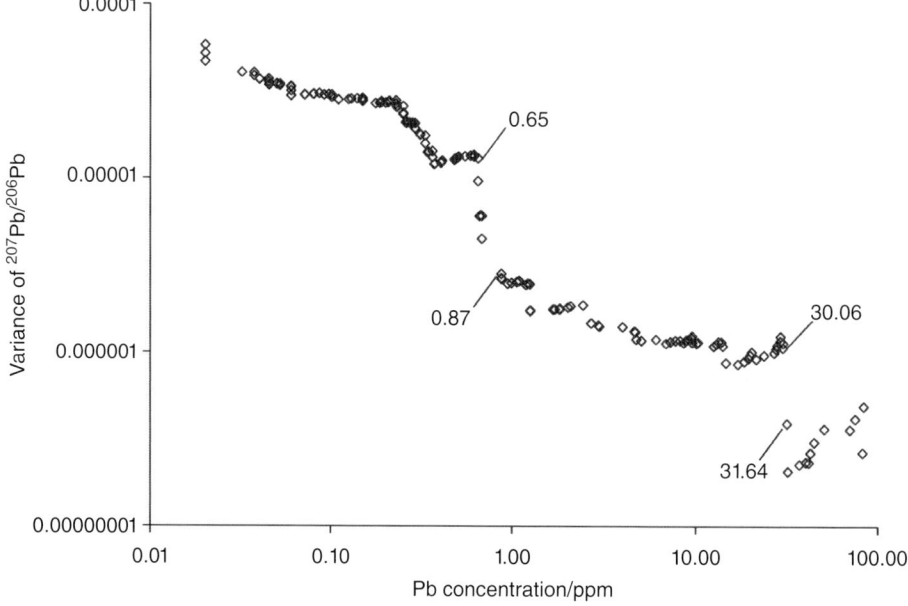

Figure 15.3. Variance of ^{207}Pb/^{206}Pb for samples above a given Pb concentration. Note that both scales are logarithmic. Labels show concentrations at points between which changes in variance occur.

from the dataset compiled by Montgomery et al. (2010) exhibit concentrations that fall within the *natural* category. We have no isotopic data for concentrations between 0.68 and 0.87 ppm to better define the upper limit of natural exposure. The change in variance at 30 ppm is discussed later.

Statistical Comparisons

Within our dataset, Lukin Street is the only site where people display *natural* levels of lead, with two individuals below 0.68 ppm and one ambiguous at 0.77 ppm. A Kruskal–Wallis test comparing the four sites shows that there is a very significant difference between sites in lead concentrations ($H = 19.34$, $p = 0.0002$), and pairwise Mann–Whitney tests indicate that the difference is between Lukin Street and the other sites: Chelsea, Coventry, and Golden Lane do not differ significantly from one another.

Comparison of lead concentrations by sex reveals no statistically significant difference (Kruskal–Wallis test, $H = 14.24$, $p = 0.23$). Subdividing by sex and site reveals a significant difference (Kruskal–Wallis test, $H = 19.2$, $p = 0.0018$), which pairwise Mann–Whitney tests indicate arises because males and females from Lukin Street differ from males and females from Chelsea.

The effect of socioeconomic status on lead exposure may be investigated by comparisons between sites or between individuals. However, at the individual level, the only indicator of status that we have is the presence or absence of coffin plates. For the whole dataset, there is no difference between those with and without coffin plates (Mann–Whitney $U = 816$, $p = 0.70$).

Iron-deficiency anemia has been identified in the paleopathological literature as a cause of cribra orbitalia, a pathology that usually occurs in childhood and is observed in skeletal material as a pitting of the bone of the orbits, which still remains visible in adults (Stuart-Macadam, 1989; but see Walker et al., 2009). Individuals from all the sites have been scored in a consistent manner for the extent of cribra orbitalia using the system of Stuart-Macadam (1991). If lead exposure was sufficient to cause significant iron-deficiency anemia, we may expect a correlation of the extent of cribra orbitalia with lead concentration. However, comparison of lead concentrations between individuals showing cribra orbitalia and those without cribra orbitalia showed no significant difference (Mann–Whitney $U = 536$, $p = 0.67$), nor was there any correlation between the severity score for cribra orbitalia and lead concentration (Spearman's rho $= 0.01$, $p = 0.91$).

Comparison of our isotope data for Chelsea and Coventry with the dataset compiled by Montgomery et al. (2010) shows that the ^{207}Pb/^{206}Pb ratio for humans with *cultural* lead concentrations differs in variance between periods (Levene's test, $p = 0.0018$, see also Figure 15.2), and comparisons using pairwise Kolmogorov–Smirnov tests show that postmedieval individuals' ratios differ significantly from those of early and late medieval people, but there is no significant difference when comparing postmedieval and Roman individuals.

Discussion

Lead Exposure

Reported lead concentrations in human tooth enamel from Britain, from this and previous studies, vary by a factor of approximately 30,000, from 3 ppb to over 90 ppm (Figure 15.2). Variability resulting from natural exposure may be up to 300 times, but at least a further 100-fold variation in concentration arises from cultural exposure to lead. Within that cultural variation, the postmedieval populations presented in this chapter show a threefold increase in maximum lead concentrations compared to the maxima reported by Montgomery et al. (2010) for Roman (30.1 ppm) and Viking (31.6 ppm) populations. With 24% (20/83) of our postmedieval individuals exceeding the maximum for previous periods, and only 4% (3/83) having natural levels of exposure, the consistent and high exposure of postmedieval populations is clear. The individuals above 30 ppm come from all four of our sites, but a higher proportion of individuals from Chelsea (10/23) and Golden Lane (3/5) occur in this group than from Coventry (1/10) and Lukin Street (7/45) (χ^2–test, p = 0.01), concordant with the expectation that higher-status populations had greater lead exposure.

Gulson (1996) reported enamel lead concentrations up to 30 ppm in children who had ingested dust from lead paint, but these were extreme, and even those who had high exposure from living in a lead-mining community typically had values of 2 to 10 ppm. The values exhibited by our postmedieval populations are therefore high to extremely high compared to modern populations in which the effects of lead are well documented.

Relating this high lead exposure to morbidity is much more difficult than simply identifying it. Sampling tooth enamel means that the measurement relates to the average lead exposure during the period of formation of the enamel during childhood, which is at least 2 to 3 years. Any acute exposure to lead will therefore be averaged out in our samples, and the total variability in exposure to lead must be much greater than we observe. The literature on clinical manifestations of lead poisoning is based around levels of lead in blood, and there are very few studies relating blood lead levels to enamel levels. The findings of those studies that do make this extrapolation are also very difficult to interpret, as the samples taken must necessarily relate to different periods of life.

Direct inference of morbidity from enamel lead concentrations is therefore not possible, and we must therefore seek other lines of evidence. Corroboration of many of the effects of lead poisoning using the skeletal or historical records is not easy, but two of them may be detectable: infertility and/or reduced libido, and anemia. Of the named adults in our dataset, we currently only have information on the reproductive success of eight (Table 15.1). While within that eight the three with the lowest concentrations did have children, the dataset is really too small to establish any effect, and the difference in lead concentrations for those with and without children is not significant (Mann–Whitney U = 3, p = 0.23). More historical data is needed to investigate this, and it would be preferable to be able to separately consider male and female reproductive success. In addition, it would be necessary to compare the historical data on individuals to more general data on the period, which suggests that the percentage of adults who never married was 9% to 12% in the 18th to early 19th centuries (Schofield, 1985) and primary sterility occurred in 7% of marriages (Wrigley et al., 1997).

Although lead exposure is known to cause iron-deficiency anemia and cribra orbitalia has been associated with anemia (Stuart-Macadam, 1989), we have not observed any association between childhood lead concentrations as recorded in enamel and the occurrence or severity of cribra orbitalia. This may be because of small sample sizes, but is also highly consistent with recent work that has proposed other etiologies for cribra orbitalia (Walker et al., 2009).

As anticipated from the historical record, there appear to be differences by social class, with the clearest comparison being between Lukin Street in the deprived area of Whitechapel and well-to-do Chelsea. However, this is not a simple comparison as the Chelsea sample is slightly earlier chronologically and the Lukin Street sample probably includes a high proportion of immigrants from rural Ireland. Therefore, this finding is more likely to be due to rural–urban differences. This suggestion is supported by the small sample from Golden Lane who were probably poor but not Irish immigrants and who exhibit higher lead levels than the Lukin Street population. It is additionally supported by the fact that the intermediate social status group from Coventry also has higher lead levels than the

Lukin Street sample. The economy of rural Ireland was at a subsistence level and would have involved less use of lead and the burning of peat rather than coal, so that the exposure of the population to lead was as low as in prehistoric periods. The individuals from Lukin Street exhibiting high lead levels seem less likely to be the first-generation immigrants from rural areas.

Sources of Lead Exposure

Figure 15.4 compares the isotope ratios of lead from the individuals from Coventry and Chelsea with the ratios reported in the literature for major lead ore sources and coal from England and Wales. In comparison to the ratios from coal, the humans plot in an unlikely position for an average of coal sources and at and beyond the upper end of the distribution of values from the Durham coalfield, which was the major source of coal for London at this time (Daunton, 1995). It seems unlikely therefore that coal smoke played a major part in the exposure of children to lead in these populations.

The group with high lead concentrations also has a narrower range of lead isotope ratios than in previous periods, which might indicate either a reduction in the range of ore sources exploited or further cultural focussing of lead isotope ratios by recycling of lead. As the 18th and early 19th centuries were a time of expansion in lead production and in the number of mines, a reduction in the range of sources seems unlikely. When compared to the lead isotope ratios of lead ores from the three major ore-fields, most of the tooth enamel values cluster over the southern Pennines values and between the Bristol/Mendips and north Pennines values. During the period for which there are good production figures,

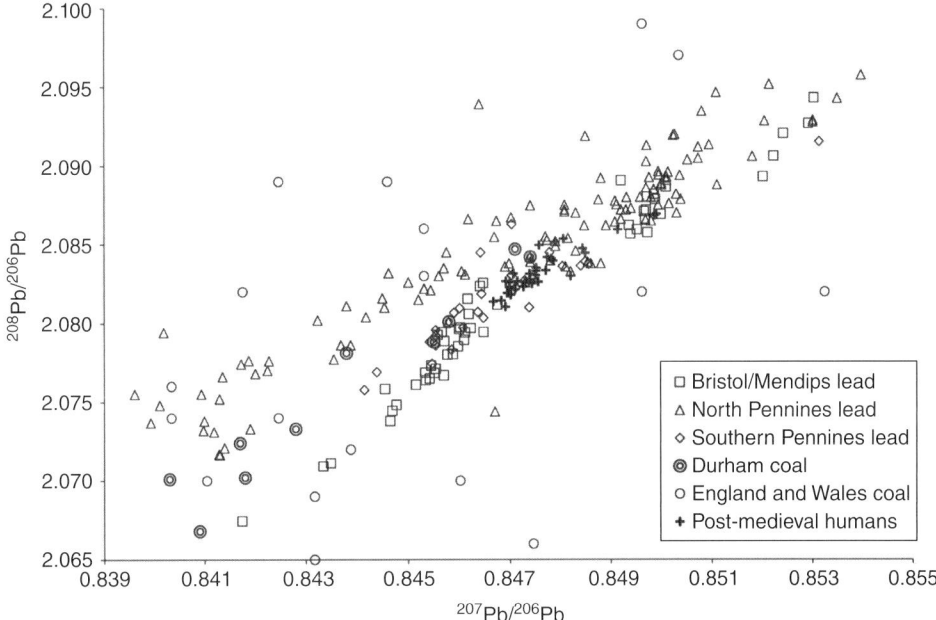

Figure 15.4. ^{208}Pb/^{206}Pb versus ^{207}Pb/^{206}Pb for postmedieval human tooth enamel compared to the major lead ore-fields and coal from England and Wales. Data sources for mineral isotope compositions: Bristol/Mendips lead (Rohl, 1996), North Pennines lead (Rohl, 1996; Scaife et al., 2001; Shepherd et al., 2009), Southern Pennines lead (Rohl, 1996), Durham coal (Farmer et al., 1999; Shepherd et al., 2009), and other England and Wales coal (Farmer et al., 1999).

the North Pennines ore-field dominated production, with 35% to 40% of all UK production in the period 1851 to 1881 (computed from Burt, 1984: 194). Although this is slightly later than the period of exposure of the individuals studied here, the relative productivity is unlikely to have changed much. It therefore seems unlikely that the south Pennines orefield was the major source of lead to which people were exposed, even though this orefield is not as well characterized as the other two. Instead, the tight grouping of values implies that humans were being exposed to lead in such a way that the isotope ratio is an average of English lead ores. This could have been via exposure to a variety of items containing lead derived from different ores, or the use of recycled lead, which would average the isotope ratios of lead in circulation and yield a consistent isotope ratio for all lead exposure.

Three individuals stand out as having distinctly higher lead isotope ratios, and two of these are very close together and fall among a group of lead ore samples that derive from the Mendips. One of these individuals is Thomas Langfield (CHE147), who is known to have spent his childhood in Somerset, where one might expect the primary source of anthropogenic lead to be the Mendip mines. The averaging seen in the majority of the sample investigated might therefore not be characteristic of the country as a whole, but reflect the nature of urban exposure to lead.

The immediate source of exposure cannot easily be identified without isotopic analysis of contemporary lead, pewter, and lead glaze. This would allow us to identify whether there are multiple isotopically distinct lead sources averaged in the human body, or whether the recycling of lead has averaged the ore-source compositions before exposure. Nevertheless, the fact that early and late medieval people differ in isotopic composition from Roman and postmedieval people indicates that the balance of ore sources being exploited differs between time periods. The postmedieval period saw a return to the exploitation of sources, or to a balance of sources, which was similar to that of the Roman period, when lead use and exposure reached its preindustrial peak.

Conclusion

Our results show clearly that the postmedieval populations examined here include individuals who exhibit higher levels of childhood lead exposure than any others reported in the literature. Although it is difficult to correlate this exposure with morbidity, we do not doubt that there must have been effects from this lead exposure, with the most likely effects being anemia, reduced mental acuity, and, if the body burden of lead continued into adulthood, reduced fertility. The source of the exposure was primarily lead derived from lead ores rather than lead released by the burning of coal, despite the contribution that the latter made to atmospheric releases of lead. Rural and poor populations were less exposed than rich and urban ones, but there are no differences by sex, which suggests that boys and girls were exposed to similar degrees.

In the context of the second epidemiologic transition, we can clearly corroborate the claim that exposure to environmental pollution (in this case lead) increased with industrialization and that exposure reached levels unheard of before or since that time. From skeletal material and historical records, it is difficult to establish what specific morbidities arose from that exposure, though comparisons with studies in the late 20th century suggest that they must have occurred. However, contrary to those studies, we find that wealth and urban dwelling were major drivers of exposure during the 18th and early 19th centuries, which supports the assertion that the change in morbidity of second epidemiologic transition was driven by industrialization.

Acknowledgments

This work was funded by the UK Natural Environment Research Council and Economic and Social Research Council (NERC-ESRC PhD studentship to MAT), the Arts and Humanities research Council (studentship to JB), the NERC Isotope Geoscience Facility Steering Committee (grants to ARM and JM), and the Royal Irish Academy (grant to JM). Access to skeletal materials and information about the sites were provided by Dr. Jenny Wakely and Dr. Richard Thomas at Leicester University (Coventry), Iain Soden of Northamptonshire Archaeology (Coventry), Adrian Miles and the late Dr. Bill White at Museum of London Archaeology and the Museum of London (Chelsea), and Natasha Powers at Museum of London Archaeology (Chelsea, Lukin Street. and Golden Lane).

References

Barrett R, Kuzawa CW, McDade T, Armelagos GJ. 1998. Emerging and re-emerging infectious diseases: the third epidemiologic transition. Annu Rev Anthropol 27:247–271.

Bryan G. 1869. Chelsea in the olden & present times. London: George Bryan.

Budd P, Montgomery J, Barreiro B, Thomas RG. 2000. Differential diagenesis of strontium in archaeological human dental tissues. Appl Geochem 15:687–694.

Budd P, Montgomery J, Evans J, Trickett MA. 2004. Human lead exposure in England from approximately 5500BP to the 16th century AD. Sci Total Environ 318:45–58.

Burt R. 1984. The British lead mining industry. Redruth: Dyllansow Truran.

Cathcart-Borer M. 1973. Two villages: The story of Chelsea and Kensington. London: W.H. Allen.

Connell B, Miles A. 2010. The city bunhill burial ground, Golden Lane, London. London: Museum of London Archaeology.

Cowie R, Bekvalac J, Kausmally T, Museum of London, Archaeology Service. 2008. Late 17th- to 19th-century burial and earlier occupation at all saints, Chelsea Old Church, Royal Borough of Kensington and Chelsea. London: Museum of London Archaeology Service.

Croot PEC. 2004. Population. In: Croot PEC, editor. A history of the county of Middlesex, Vol. XII. Chelsea. Oxford: Oxford University Press. p 13.

Croot PEC, Insley R. 2004. Economic history. In: Croot PEC, editor. A history of the county of Middlesex, Vol. XII. Chelsea. Oxford: Oxford University Press. p 146–165.

Daunton MJ. 1995. Progress and poverty: An economic and social history of Britain 1700–1850. Oxford: Oxford University Press.

De Vries J. 2008. The industrious revolution: Consumer behaviour and the household economy, 1650 to the present. Cambridge: Cambridge University Press.

Defoe D. 1724. A tour thro' the whole island of Great Britain. Divided into circuits or journies. London: G. Strahan.

Dietrich KN, Ris MD, Succup PA, Berger OG, Bornschein RL. 2001. Early exposure to lead and juvenile delinquency. Neurotoxicol Teratol 23:511–518.

Drummond JC, Wilbraham A. 1939. The Englishman's food: Five centuries of English diet. London: Pimlico.

Eisinger J. 1982. Leaded wine: Eberhard Gockel and the colica Pictonum. Med Hist 26:279–302.

Farmer JG, Eades LJ, Graham MC. 1999. The lead content and isotopic composition of British coals and their implications for past and present releases of lead to the UK environment. Environ Geochem Health 21:257–272.

Garrettson LK. 1990. Lead. In: Haddad LM, Winchester JF, editors. Clinical management of poisoning and drug over-dose. Philadelphia: W. Saunders Co. p 1017–1023.

Graham G. 1843. Fifth annual report of the registrar general of births, deaths and marriages in England. London: Her Majesty's Stationery Office.

Greenwood J. 1874. In strange company, being the experiences of a roving correspondent. London: Henry S. King & Co.

Gulson BL. 1996. Tooth analyses of sources and intensity of lead exposure in children. Environ Health Perspect 104:306–312.

Hammer Ø, Harper DAT, Ryan PD. 2001. PAST: palaeontological statistics software for education and data analysis. Palaeontol Electron 4:article 4.

Handler JS, Aufderheide AC, Corruccini RS, Brandon EM, Wittmers LE, Jr. 1986. Lead contact and poisoning in Barbados slaves: historical, chemical, and biological evidence. Soc Sci Hist 10:399–425.

Hardy J. 1778. A candid examination of what has been advanced as the colic of poitou and devonshire, with remarks on the most probable and experiments intended to ascertain the true causes of gout. Devonshire: J. Hardy.

Harrison W. 1587. The description of England. In: Holinshed R, editor. The chronicles of England, Scotland and Ireland. London: Henry Denham.

Hertz-Picciotto I. 2000. The evidence that lead increases the risk for spontaneous abortion. Am J Ind Med 38:300–309.

Hong S, Candelone J-P, Patterson C, Boutron C. 1994. Greenland ice evidence of hemispheric lead pollution two millennia ago by Greek and Roman civilizations. Science 265:1841–1843.

Insley R, Croot PEC. 2004. Social and cultural activities. In: Croot PEC, editor. A history of the county of Middlesex, Vol. XII: Chelsea. Oxford: Oxford University Press. p 166–176.

Lancaster JC. 1969. The city of Coventry: Crafts and industries. In: Stephens WB, editor. The history of the county of Warwick, Vol. VIII: The city of Coventry and borough of Warwick. London: Oxford University Press. p 151–189.

Lancaster JC, Tomlinson M. 1969. The city of Coventry: Introduction. In: Stephens WB, editor. The history of the county of Warwick, Vol. VIII: The city of Coventry and borough of Warwick. London: Oxford University Press. p 1–23.

Lanphear BP, Dietrich K, Auinger P, Cox C. 2000. Cognitive deficits associated with blood lead concentrations <10 µg/dL in US children and adolescents. Public Health Rep 115:521–529.

Lind J. 1754. To the author of the Scots Magazine. Scots Magazine, 18 May 1754.

Matte TD, Landrigan PJ, Baker EL. 1992. Occupational lead exposure. In: Needleman HL, editor. Human lead exposure. Boca Raton: CRC Press. p 155–168.

Mayhew H. 1861. London labour and the London poor (extra volume). London: Griffin, Bohn & Co.

McKeown RE. 2009. The epidemiologic transition: changing patterns of mortality and population dynamics. Am J Lifestyle Med 3:19S–26S.

Millard A. 2006. Comment on Martinez-Garcia et al. "heavy metals in human bones in different historical epochs". Sci Total Environ 354:295–297.

Montgomery J, Evans JA, Powlesland D, Roberts CA. 2005. Continuity or colonization in Anglo-Saxon England? Isotope evidence for mobility, subsistence practice, and status at West Heslerton. Am J Phy Anthropol 126:123–138.

Montgomery J, Evans J, Chenery S, Pashley V, Killgrove K. 2010. "Gleaming, white and deadly": using lead to track human exposure and geographic origins in the Roman period in Britain. J Roman Archaeol Suppl 78:199–226.

Mooney G. 2007. Infectious diseases and epidemiologic transition in Victorian Britain? Definitely. Soc Hist Med 20:595–606.

Moorrees CFA, Fanning EA, Hunt EE. 1963. Age variation of formation stages for ten permanent teeth. J Dent Res 42:1490–1502.

Needleman HL. 2004. Low level lead exposure and the development of children. Southeast Asian J Trop Med Public Health 35:252–254.

Noymer A, Jarosz B. 2008. Causes of death in nineteenth-century New England: the dominance of infectious disease. Soc Hist Med 21:573–578.

Omran A. 1971. The epidemiologic transition. Milbank Mem Fund Q 49:509–538.

Patrick L. 2006. Lead toxicity, a review of the literature. Part 1: exposure, evaluation, and treatment. Altern Med Rev 11:2–22.

Powers N. 2008. "All the Outward Tinsel Which Distinguishes Man from Man Will Have Then Vanished…" an assessment of the value of post-Medieval human remains to migration studies. In: Brickley M, Smith M, editors. Proceedings of the eighth annual conference of the British association for biological anthropology and osteoarchaeology. Oxford: Archaeopress. p 41–49.

Putnam RD. 1986. Review of toxicology of inorganic lead. Am Ind Hyg Assoc 47:700–703.
Richards S. 1999. Eighteenth-century ceramics: products for a civilised society. Manchester: Manchester University Press.
Riley JC. 2001. Rising life expectancy: A global history. Cambridge: Cambridge University Press.
Rohl BM. 1996. Lead isotope data from the Isotrace Laboratory, Oxford: Archaeometry Data Base 2, galena from Britain and Ireland. Archaeometry 38:165–180.
Rudé GFE. 1971. Hanoverian London, 1714–1808. London: Secker & Warburg.
Rylatt M, Mason P. 2003. The archaeology of the medieval cathedral and priory of St. Mary, Coventry. Coventry: Coventry City Council.
Scaife B, Barreiro BA, McDonnell JG, Pollard AM. 2001. Lead isotope ratios of 36 galenas from the Northern Pennines. Available at: http://brettscaife.net/lead/npennine/npennine.html (accessed November 6, 2013).
Schofield R. 1985. English marriage patterns revisited. J Fam Hist 10:2–20.
Shepherd TJ, Chenery SRN, Pashley V, et al. 2009. Regional lead isotope study of a polluted river catchment: River Wear, Northern England, UK. Sci Total Environ 407:4882–4893.
Stuart-Macadam PL. 1989. Nutritional deficiency diseases: a survey of scurvy, rickets, and iron-deficiency anemia. In: Iscan MY, Kennedy KAR, editors. Reconstruction of life from the skeleton. New York: Liss. p 201–222.
Stuart-Macadam PL. 1991. Anemia in Roman Britain: Poundbury Camp. In: Bush H, Zvelebil M, editors. Health in past societies: Biocultural Interpretations of human skeletal remains in archaeological contexts. Oxford: BAR. p 101–113.
Thackrah CT. 1831. The effects of the principle arts, trade and professions, and of civic states and habits of living, on health and longevity. Philadelphia: L. Johnson.
Thirlwall MF. 2002. Multicollector ICP-MS analysis of Pb isotopes using ^{207}Pb/^{204}Pb double spike demonstrates up to 400 ppm/amu systematic errors in Tl-normalisation. Chem Geol 184:255–279.
Trickett MA. 2006. A tale of two cities: diet health and migration in post-medieval Coventry and Chelsea through biographical reconstruction, osteoarchaeology and isotope biogeochemistry. Ph.D. thesis. Durham: Durham University.
Trickett MA, Budd P, Montgomery J, Evans J. 2003. An assessment of solubility profiling as a decontamination procedure for the ^{87}Sr/^{86}Sr analysis of archaeological skeletal tissue. Appl Geochem 18:653–658.
Vigeh M, Smith DR, Hsu P-C. 2010. How does lead induce male infertility? Iran J Reprod Med 9:1–8.
Walker PL, Bathurst RR, Richman R, Gjerdrum T, Andrushko VA. 2009. The causes of porotic hyperostosis and cribra orbitalia: a reappraisal of the iron-deficiency-anemia hypothesis. Am J Phys Anthropol 139:109–125.
Weeden RP. 1984. Poison in the pot: The legacy of lead. Carbondale: Southern Illinois University Press.
Weisskopf MG, Howard H, Mulfern RV, et al. 2004. Cognitive deficits and magnetic resonance spectroscopy in adult monozygotic twins with lead poisoning. Environ Health Perspect 112:620–625.
White J. 2007. London in the nineteenth century "A human awful wonder of god". London: Jonathan Cape.
WHO. 2010. Childhood lead poisoning. Geneva: World Health Organization.
Wrigley EA. 1988. Continuity, chance and change: The character of the industrial revolution in England. Cambridge: Cambridge University Press.
Wrigley EA, Davies RS, Oeppen J, Schofield RS. 1997. English population history from family reconstitution, 1580–1837. Cambridge: Cambridge University Press.

Chapter 16

The Hygiene Hypothesis and the Second Epidemiologic Transition

Molly K. Zuckerman[1] and George J. Armelagos[2]
[1] *Department of Anthropology and Middle Eastern Cultures, Mississippi State University, Starkville, MS*
[2] *Department of Anthropology, Emory University, Atlanta, GA*

Introduction

The hygiene hypothesis (i.e., microbial deprivation hypothesis) claims that in developed nations, the lack of exposure to certain environmental microorganisms early in life increases susceptibility to chronic inflammatory disorders (CIDs), including allergic and autoimmune diseases in adulthood (Strachan, 1989, 2000; Björkstén, 2009). These microorganisms include helminthic parasites, numerous chronic viruses and bacterial infections, environmental saprophytes (Rook, 2010), and most likely, gut microbiota. All of these are ubiquitous outside of westernized, urbanized, and hygienic environments and have been throughout mammalian evolutionary history. Therefore, the hygiene hypothesis posits that not only are we immunologically tolerant of them, but that we exist in a state of immunological *evolved dependence* with them. More specifically, it proposes that because of this evolved state, exposure to them, particularly during childhood, is critical for successful immunological functioning, specifically the development of an immunomodulatory response that maintains tolerance of self-antigens and abrogates autoimmune diseases. However, in developed, industrialized nations where these microorganisms are scarcer, children experience diminished or delayed exposure. In turn, the hygiene hypothesis proposes that this diminishing exposure produces an immunoregulatory failure, leading to increased incidences of some CIDs, such as allergic and autoimmune diseases, ranging from asthma and allergic rhinitis to inflammatory bowel disease (IBD), multiple sclerosis (MS), and insulin-dependent diabetes mellitus (T1D) (Yazdanbakhsh et al., 2002; Yazdanbakhsh and Matricardi, 2004). Here, the hygiene hypothesis is examined in light of the second transition, in which a high burden of mortality and morbidity from infectious disease was replaced with a high burden of degenerative conditions in industrializing, modernizing nations. The same practices that limited contact with infectious agents in these populations, and thus precipitated the transition, seem to have also reduced their exposure to more beneficial organisms and therefore may have impaired their immunological development.

Modern Environments and Human Health: Revisiting the Second Epidemiologic Transition, First Edition. Edited by Molly K. Zuckerman.
© 2014 John Wiley & Sons, Inc. Published 2014 by John Wiley & Sons, Inc.

The objective of this chapter is to evaluate epidemiologic transitions in human populations through time, from the Paleolithic to the Neolithic to the present day, in light of how they have played into changing human–environment interactions and contacts between humans and generally nonpathogenic microorganisms. While the hygiene hypothesis has been examined in relation to epidemiologic transitions before (Armelagos, 2009; Rook, 2010), the treatment here focuses on the second epidemiologic transition and prioritizes a consideration of how relationships between humans and these microorganisms have shifted over the course of the long-term changes in urbanization, modernization, and sanitization that frame the first and second epidemiologic transitions. The intent is to conceptualize the hygiene hypothesis within an evolutionary and epidemiologic perspective and bring it into consideration as a key component of the fundamental changes in human–environment and human–microorganism interactions that occurred during the second transition. Importantly, as the hygiene hypothesis has yet to be included in any empirical investigations of the second transition, we also conclude with a series of propositions as to how the hygiene hypothesis could be embedded into interdisciplinary investigations of the second epidemiologic transition.

Background

Epidemiologic Transitions

The Paleolithic experienced the first epidemiologic transition (Barrett et al., 1998). The first transition marked the shift to increased morbidity and mortality from acute, epidemic infectious diseases, like smallpox and cholera, which accompanied the intensification of agriculture and the rise of increasingly sedentary human populations during the Neolithic, around 10,000–12,000 years ago. This pattern characterized the next 10 millennia, until industrialization began, bringing with it the rise of modernized, westernized, and urbanized environments in the 19th and 20th centuries. These changes reduced exposure to infectious pathogens and risk of death from infectious disease in developed nations, and precipitated a transition—the second epidemiologic transition—to greater longevity and a greater proportion of mortality and morbidity due to noncommunicable disease (NCD) and chronic disease, such as cardiovascular disease and cancer (Gage, 2005). In addition to the increase in noncommunicable and chronic diseases, the second transition also involved an apparent increase in allergies and diseases of the immune system in some segments of these populations. Starting in the last half of the 20th century, numerous researchers have noted increased rates of these CIDs in many high-income industrialized nations over the past four decades (Munoz-Lopez, 2006).

The Rising Incidence of Allergic and Autoimmune Diseases

The hygiene hypothesis was formulated to explain this trend of increasing incidence of CIDs, namely, allergic and autoimmune diseases. The allergic conditions that have increased in incidence include asthma, rhinitis, and atopic dermatitis; the autoimmune diseases include MS, T1D, and IBD, among others. The pattern has been noted in many different developed nations and for many different conditions. The incidence of these disorders apparently began to increase in the 1950s and continues to do so today (Bach, 2002). For instance, in 1998, approximately one in five children in industrialized nations had allergic diseases, including asthma, allergic rhinitis, and atopic dermatitis (ISAAC, 1998). This proportion has increased during the intervening decade, particularly with asthma, which is now considered to be nearly epidemic. In developed nations, such as the United Kingdom and Australia, more than

15% of children suffer from asthma. Children in developing nations are also increasingly affected; Peru, Costa Rica, and Brazil, for example, report rates above 8% (Eder et al., 2006). The prevalence of atopic dermatitis has also doubled or tripled in developed countries over the past three decades, now affecting 15–30% of children and 2–10% of adults (Beiber, 2008). Autoimmune diseases are also on the rise. T1D, for example, has become a major public health problem in the US and some European countries, and is occurring much earlier in life than previously (Harjutsalo et al., 2008). Incidences of IBD, such as Crohn's disease, are also increasing (Bach, 2002; Rautiainen et al., 2007). Part of the increased incidence of these conditions may be because of improved diagnosis and better access to medical care in developing and developed nations. However, this cannot explain the dramatic rise in the prevalence of allergic and immunological diseases that has occurred in such a short period of time in these nations, particularly for conditions that are conspicuous and easy to diagnose, such as MS and T1D (Okada et al., 2010).

Infectious Diseases in Developed Nations

Instead, many epidemiologists have noticed that the rise in incidences of CIDs is concomitant with an obvious decrease in the incidence of many infectious diseases in developed countries, namely, the reduction in the burden of infectious disease that occurred throughout the developing, industrializing world in the 19th and 20th centuries as part of the second epidemiologic transition. These declines were primarily ushered in through various public health measures, such as decontamination of water supplies, pasteurization and sterilization of milk and other food products, vaccination against common childhood infections, and the broad use of antibiotics, all of which occurred in the 19th and early 20th centuries in high-income nations. Improved hygiene and improved socioeconomic conditions also likely played a role. The declines are particularly apparent for diseases like hepatitis A, childhood diarrhea, and in the 20th and 21st centuries, parasitic diseases such as filariasis, onchocercosis, and schistosomiasis (Zaccone et al., 2006) but also include more common childhood or crowd infections, like measles and mumps. Working from this observation, the hygiene hypothesis proposes that there is a link between these declines and the increased incidence of CIDs; the lauded improvements in health care, sanitation, hygiene, and public health, and their celebrated consequences of reduced infectious disease mortality, may not have been without cost. Instead, they may have produced a substantial trade-off in health and quality of life, with developed, high-income nations trading a high burden of infectious disease for one of CIDs.

Heirloom and Souvenir Microorganisms and the Hygiene Hypothesis

The hygiene hypothesis recognizes several groups of microorganisms with which humans have developed a state of evolved dependence. The majority of these are heirloom diseases. Studies of disease ecology suggest that the infectious diseases affecting humans can generally be grouped into two broad categories: heirlooms, or those with a long-standing, millennial relationship with humans and their ancestors, and souvenirs, which have been more recently acquired. Sprent (1969a,b) was the first to make this distinction, proposing that both classes of microorganisms affected Paleolithic foragers. Heirloom species are pathogens that originated in our anthropoid ancestors and continued to infect hominins and eventually humans during the Paleolithic. In contrast, souvenirs are newer evolutionary acquisitions that are typically picked up via exposure to areas with existing zoonotic reservoirs or vectors (Kilks, 1990). These diseases are generally zoonoses or infections that can be transmitted from animals to humans. Zoonoses can be contracted through a number of routes: from sympatric reservoir host species through cross-transfer, through insect or animal bites, or through the preparation and consumption of contaminated animal flesh. Some zoonotic species are capable of temporary host-switches without substantial levels of adaptation, as long as the novel host is fairly similar to the natural host in terms of resources available to the pathogen (Kellog, 1896, in Brooks and Ferrao, 2005: 1292). In this case, the microorganism will remain specialized to its natural host, but continue to infect humans and other novel hosts across a wide ecological niche. However, other souvenir species enter human populations and assume permanent habitation, adapting to the novel host and becoming established; this type of souvenir includes several of the major diseases that affect human societies, such as malaria (*Plasmodium falciparum*) and HIV/AIDS.

According to Rook (2010), there are three general groups of microorganisms that are relevant to the hygiene hypothesis. The first are the heirlooms, including a variety of helminthic parasites, which were ubiquitous in Paleolithic and Neolithic environments but have been nearly eliminated in hygienic modern environments in high income, developed nations (see Reinhard and Pucu de Araújo, this volume). They also include various chronic viral infections, such as herpesviruses and papovaviruses (papilloma), and bacteria, such as *Helicobacter pylori*, Salmonella, and Staphylococcus, which humans and their ancestors have also been affected by for millennia (Rook, 2010). The second are what Rook designates as psuedo-commensals, such as environmental saprophytes and lactobacilli, which are harmless microorganisms associated with mud, untreated water, and fermenting vegetable matter. These had been ubiquitous in food and water throughout mammalian and human evolutionary history up until the rise of westernized, modernized urban environments in the late 19th and 20th centuries. Others have also included pseudo-commensal gut flora as a third important category, as several studies have shown that gut microbiota changes contribute to the modulation of immune disorders, especially IBD (see Okada et al., 2010). As Bach (2002) notes, exposure to all of these microorganisms has declined with modernization. For instance, the age at which colonization of the intestinal microbiota occurs differs among nations, with colonization by gram-negative bacteria occurring later in developed than less developed and developing nations, both quantitatively and qualitatively (Adlerberth et al., 1991; Adlerberth et al., 1998). The high prevalence of parasitic infections, especially plasmodia and schistosoma in southern nations, also sharply contrasts with their absence in developed nations. Additionally, the frequency of infection by minor parasites, like pinworms, over the past decade has greatly decreased in developed nations (Vermund and MacLeod, 1988; Gale, 2002). Intriguingly, in the less developed, low-income nations where these microorganisms remain endemic, the prevalence of CID, such as allergy, also remains low. However, as these nations eradicate these common infections, they have witnessed a rise in CIDs and are following the trajectory of dramatically increased incidences of these conditions now seen in developed nations (Okada et al., 2010).

The Geographic Distribution of Allergic and Autoimmune Diseases

Allergic and autoimmune diseases are not uniformly distributed across nations, regions, or even communities and ethnic groups. Indeed, examination of the distribution reveals several interrelated phenomena. For instance, an overall North–South gradient exists for immune disorders in North America (Wallin et al., 2004), Europe (Bach, 2002), and China (Yang et al., 1998). A West–East one exists in Europe, with incidence of some conditions, such as T1D, much higher in Eastern than Western Europe (Green and Patterson, 2001). These differences cannot be fully explained by genetic differences within or between the affected populations (Okada et al., 2010). For instance, migration studies have demonstrated that the children of migrants from countries with low incidence acquire the same incidence as the host country, and as quickly as the first generation for some conditions, such as T1D

(Bodansky et al., 1992) and MS (Leibowitz et al., 1973; Hammond et al., 2000). As Okada et al. emphasize, these data also do not exclude the importance of genetic factors for these immunological disorders, such as the high concordance of asthma and T1D in monozygotic twins or differences in some genetic factors according to ethnicity, like Human Leukocyte Antigen differences between Caucasians and Asians. However, these issues likely play a minor role in the geographic distribution of these conditions. Likewise, while deficiencies in medical resources could lead to underdiagnosis of these diseases in low-incidence nations, the differences in frequency also involve southern countries with more than adequate medical resources, like Greece and Spain (Bach, 2002). Instead, studies have now convincingly demonstrated that this uneven geographical distribution is a mirror image of the geographical distribution of various infectious diseases, including hepatitis A virus (HAV), gastrointestinal infections, and parasitic infections (Okada et al., 2010).

This linkage between infectious and allergic and autoimmune diseases was not always recognized, however. Originally, the earliest formulations of the hygiene hypothesis attributed this distribution and the differential incidence of CIDs to exposure to childhood infections. Strachan (1989), seeking to explain the postindustrial rise in allergic rhinitis (e.g., hay fever) in the United Kingdom, proposed that a direct link existed between childhood infections and susceptibility to this allergy. He examined longitudinal data from a national sample of nearly 20,000 children born in 1958 (the National Child Development Study) in the UK and found a strong link between family size during childhood, sibling order, and presence of hay fever in young adults. Strachan proposed that changes in family size, improvements in household amenities, and higher standards of personal cleanliness during the mid-20th century had reduced opportunities for cross-infection between siblings, later leading to a higher susceptibility to hay fever during adulthood. Strachan tentatively linked cross-infection to unhygienic contact with older siblings or prenatal acquisition from a mother infected by contact with her older children; he proposed that later infection or reinfection by younger siblings might confer additional protection against allergies such as hay fever. Other studies have found similar protective effects between the presence of older siblings and development of MS (Ponsonby et al., 2005) and T1D (Cardwell et al., 2008).

However, since Strachan's earliest formulation, evidence has consistently accumulated that contradicts relationships between childhood infections and allergic and autoimmune disease, leading epidemiologists to further clarify and extend the hypothesis. For instance, Bremner and colleagues (2008) demonstrated that exposure to childhood infections does not exert a protective effect against allergies later in life. They investigated relationships between 30 acute infectious conditions, both more traditionally researched ones, such as upper respiratory tract infections and diarrhea, and more rare ones, such as measles, during infancy and early childhood and the later development of hay fever. The study employed a large data set of medical records from over 100,000 children in two birth cohorts from the 1980s and 1990s in Britain. However, their results yielded no strong or consistent associations between any of these infections and hay fever.

Instead, numerous other lines of evidence have come together over the past three decades to convincingly suggest that environmental factors, namely, decreased incidence of infectious diseases and, with them, diminished or delayed exposure to the heirloom and pseudocommensal microorganisms, are foremost in driving the rising incidence of CIDs with development and modernization. These data come from studies of geographic distribution of CIDs and infectious disease, animal models of autoimmune and allergic diseases, and, to a lesser degree, clinical intervention studies. For the most part, these show a direct, inverse relationship between incidence of infectious disease and that of CID, but not always a straightforward one; clinical intervention studies in particular have revealed diverse and occasionally ambiguous interactions between these variables (Okada et al., 2010). For instance, animal models, specifically mice, have found a direct relationship between pathogen exposure and sanitary conditions and the incidence of T1D, with a low infectious burden translating into a high disease incidence (Like et al., 1991; Bach, 2002). Other studies have shown that infecting nonobese diabetic (NOD) mice with a wide variety of bacteria, viruses, and parasites protects completely (clean NOD mice) from diabetes (Bach, 2002). Clinical intervention studies focusing on infections have found, for example, that intentional helminthic infection with the swine-derived parasite *Trichuris suis* improved symptoms in patients with active Crohn's disease as well as ulcerative colitis (Summers et al., 2005a,b). Others have shown that helminth eradication can increase atopic skin sensitization (Lynch et al., 1993), though studies of some of the same populations found that the same treatment improved asthma symptoms (Lynch et al., 1997), a contradictory finding that might be related to asthma's complex pathophysiology. Studies of the effects of probiotics have produced ambiguous results. For instance, some have found that treatment of expectant mothers with *Lactobacillus GG* greatly reduced incidence of atopic dermatitis among their children, while others have found no effect (see Okada et al., 2010). While the evidence on gut microbiota is ambiguous, several studies have shown that the microbiota diversity in pediatric patients with allergy is different from those without (Björkstén et al., 1999) and, more conclusively, that patients with Crohn's disease have a greatly diminished diversity of commensal bacteria (Manichanh et al., 2006). Murine models have also demonstrated that supplementing the microbiota diversity with specific bacterium, such as *Bacteroides fragilis*, exerts a protective effect against experimental ulcerative colitis (Mazmanian et al., 2008). However, it is difficult to extrapolate between IBD and other autoimmune disorders (see Okada et al., 2010).

The result of this increasing body of literature is the modified—and now standard—version of the hygiene hypothesis, known as the "old friends hypothesis" or the microbial deprivation hypothesis (Björkstén, 2009) (and referred to in this chapter simply as the hygiene hypothesis). As mentioned earlier, this proposes that the underlying mechanism between lifestyle changes and the incidence of CID is that the lack of exposure to heirloom and pseudo-commensal microorganisms during childhood prevents the development of immunoregulatory pathways that inhibit CID. Exposure to these microorganisms, due to the long evolutionary history of contact, does not provoke an aggressive immune response; they are the "old friends" of humans and other terrestrial mammals. Instead, daily life-long contact with these microorganisms over the millennia seems to have created a state of evolved dependence, wherein rather than just being tolerated, they are critical to normal immune functioning (Rook and Brunet, 2005b; Rook, 2010). However, exactly how this mechanism occurs remains unclear (Pulendran and Artis, 2012). Both helminthic infections and inflammatory allergic conditions involve allergic type 2 immune responses; type 2 immunes responses are induced by and confer protection against helminthes, but also promote acute and chronic inflammatory responses to pathogens (Paul and Zhu, 2010; Pulendran et al., 2010; Zhu et al., 2010; Lambrecht and Hammad, 2012). Type 2 responses are characterized by the induction of CD^+4T helper cells (T_H2). In the immune system, pathogen recognition receptors, which are expressed on the surface of a variety of cell types, have evolved to sense a variety of stimuli. When triggered, they activate dendritic cells (DCs), which stimulate antigen-specific T cells and tune the type of T_H response generated. DCs are usually highlighted in studies of this process, but recent work has shown that pathways such as the enzymatic activities of allergens or recognition of tissue damage and metabolic changes caused by allergens or helminths may also trigger a type 2 response. Rook (2007, 2010) has proposed that when exposure to old friend organisms occurs, DCs mature into regulatory DCs, which drive regulatory T-cell (T_{reg}) responses to the antigens of the organisms. T_{reg} cells are a subpopulation of T_H cells that modulate immune responses, including suppression of T_H1 and T_H2 responses (Zhu et al., 2010), maintain tolerance to self antigens, and abrogate autoimmune disease, effectively operating as a self check to prevent excessive reactions. According to Rook, this activation of T_{reg} cells leads to two mechanisms that control inappropriate inflammation. In the first, continuing exposure to the organisms in gut flora, food, or resident as parasites leads to continuous background release of regulatory cytokines from the T_{reg}, which exerts bystander suppression of other responses. Second, the increased numbers of DC_{reg} lead to increased processing by DCs of self-antigens, gut content antigens, and allergens. Consequently, the numbers of T_{reg} triggered by these antigens is also increased, which downregulates autoimmunity, IBDs, and allergies, respectively (Rook, 2007: 1073). In effect, this teaches the immune system to be less sensitive and not to react to every invading pathogen or protein and thus may lower the incidence of allergy and autoimmune disease. In turn, when environmental exposure to old friends does not occur, T_{reg} are not triggered, specific and bystander regulation are defective, pathologic type 2 responses to allergens occur, and on a population level, the incidence of allergy and autoimmune disease increases (Rook and Brunet, 2005a). However, as Pulendran and Artis (2012) emphasize, type 2 immune responses are highly diverse, and many underlying aspects remain unknown and unclear. There is great—and still poorly understood—diversity in the range of stimuli that trigger prototypic type 2 responses, the mechanisms by which the innate immune system senses such stimuli and the cellular and molecular pathways that orchestrate the responses. It also remains unclear as to whether there are distinct kinds of type 2 responses, some protective and some pathogenic. Therefore, the exact immunomodulatory mechanism(s) underlying the hygiene hypothesis remains unclear (Okada et al., 2010).

Despite continuing ambiguity on the underlying mechanisms, the bulk of evidence suggests that there is a strong causal relationship between lifestyle changes, gut flora, exposure to heirlooms and pseudo-commensal microorganisms, and the incidence of CID (Okada et al., 2010). Given this pattern, and the continuing trend of urbanization and modernization in many developed and developing nations, it is perhaps not surprising that allergies are an epidemic phenomenon in modern, westernized environments (Wills-Karp et al., 2001). The increase in allergies is pushed by affluence, smaller family size, stable intestinal flora, heavy antibiotic use, the absence of a helminthic burden, and a low orofecal burden due to improved sanitation. Since these changes occurred in such a short time, the genetic structure of the populations was likely to have remained stable, leaving environmental factors associated with modern environments to blame.

Intriguingly, the hygiene hypothesis has also been pulled forth from the scientific literature and interjected into the popular media over the past two decades, though in a profoundly incorrect form. Popular interpretations of the hygiene hypothesis have focused on domestic cleanliness practices such as bathing, use of soap, detergents, and clean cutting boards as the agents of reduced exposure to our old friends. In other words, it is the artificial, bleach-powered cleanliness of our proximate environments, such as the bathroom, and the reduced incidences of childhood infections that follow that are responsible for high incidences of CID. However, Bloomfield and coworkers (2006) reported that increases in allergic disorders were not correlated with changes in domestic hygiene or decreased incidences of common childhood infections. They instead suggested that fundamental changes in lifestyle led to decreased exposure to commensal microbes and helminthic parasites, and declines in gut microbiota diversity, which affected the immunological response to usually harmless allergens.

The Hygiene Hypothesis and Epidemiologic Transitions

This chapter evaluates the hygiene hypothesis in relation to the second epidemiologic transition, employing a model of epidemiologic transitions modified from Omran's (1971) original formulation. Epidemiologic transition theory provides a means of understanding the changing relationship between humans, pathogens, and other disease insults from the

Paleolithic period to the present (Barrett et al., 1998); here, it foregrounds discussion of the hygiene hypothesis within an evolutionary context, beginning with the Paleolithic (Armelagos, 2009). Whereas previous studies (i.e., Armelagos and Harper, 2010) have employed the Paleolithic as a baseline to better understand the epidemiologic and environmental shifts that occurred with the first transition during the Neolithic, this study is aimed at a better understanding of the second, and how it precipitated the current epidemic of CIDs affecting westernized, industrialized nations and, increasingly, those in the developing world. Therefore, our lens here is refocused on shifts in the prevalence of helminthic parasites, commensal bacterial, and gut microbiota from the Paleolithic to the present.

The Paleolithic Baseline

In order to reconstruct the dramatic shift in environmental quality and disease-scapes that occurred with the second transition, it is necessary to reconstruct the Paleolithic pattern as a baseline (Harper and Armelagos, 2010). Various lines of evidence suggest that the disease ecology of human populations during the Paleolithic was rich with heirloom and pseudo-commensal microorganisms, but with likely low levels of other forms of pathogen exposure. Paleolithic environments, represented by the time period stretching from approximately 2.6 mya up until 10,000 years BP, are reconstructed using ethnographic data from modern foraging populations, archaeological evidence, as well as genomic analysis of humans, domesticated plants and animals, and various pathogens (Armelagos and Harper, 2005a,b). During the Paleolithic, small population sizes, and to a lesser extent, low population density would have generally limited the range of heirloom species affecting hominins and humans and prevented the sustained transmission of many viruses and bacteria (Dunn et al., 2010: 2590). In particular, ethnographic evidence suggests that population sizes were likely too small to support continuous transmission of many acute, epidemic crowd conditions, like measles and smallpox (Burnet, 1962). Nonetheless, some species of pathogens, including various helminthes, and bacteria seem to have thrived in this setting. Characteristically, heirlooms are able to persist in small, dispersed populations, produce incomplete or short-lived host immunity, or produce chronic infection, enabling prolonged transmission to new host individuals or communities. Several potential heirloom species that would have infected Paleolithic foragers have been identified, including several macroparasites, such as head and body lice (*Pediculus humanus*) (Reed et al., 2004) and helminthes, such as pinworms (*Enterobius vermicularis*) (Hugot et al., 1999) and tapeworms (Hoberg et al., 2000; Hoberg et al., 2001), and most of the internal protozoa found in modern humans. They also include bacteria such as *H. pylori*, Pneumocystis, Salmonella, and Staphylococcus (Cockburn, 1967, 1971), and several chronic viral infections, such as herpesviruses, papovaviruses (papilloma), adenoviruses, parvoviruses, picornaviruses, HAV, and perhaps hepatitis B (Rook, 2010). Other possible heirlooms include the causative agents of yaws (*Treponema pallidum* subsp. *pertenue*) (Harper et al., 2008), typhoid (*Salmonella typhi*) (Roumagnac et al., 2006), and herpes (HSV-1, Epstein-Barr) (Van Regenmortel et al., 2000).

Ethnographic evidence also suggests that Paleolithic foragers were likely exposed to a great range and quantity of pseudo-commensal microorganisms. This would have been due to intensive and life-long contact with soil, mud, decaying organic matter and nonsanitized water sources. For instance, Bengmark (2000: 612) comments:

> The most dramatic difference between Paleolithic and modern Western food is that the diet of our forefathers contained at least a billion times more non-pathogenic health-promoting bacteria, mainly of the *Lactobacillus* species. Our ancestors' diet was rich with these bacteria because they consumed unprocessed natural food and they stored most of their food in the soil, where it became rich in fiber-fermenting lactobacilli.

While archaeological and ethnographic evidence is ambiguous on how much food Paleolithic foragers might have stored underground, *Lactobacilli* would certainly have been encountered through exposure to decaying vegetable material. Fermentation has also been widely used as a form of food processing up until recently, which would have also supplemented past diets with *Lactobacillus*. Bengmark also notes that there are substantial differences in gut microbiota in modern populations that can be tied to lifestyle and levels of environmental hygiene. In developed, high-income nations, humans typically carry a gut microbiota that weighs less than 1.3 kg. However, in low-income rural and developing nations, humans' microbiota typically weighs approximately 2 kg (Bengmark, 2000). Given the increasingly recognized role of gut microbiota in maintaining immunological function (Maslowski and Mackay, 2011; Clemente et al., 2012; Leslie, 2012), extrapolating from this figure to Paleolithic populations suggests that individuals in these environments may have experienced high levels of immunological regulation. This further suggests that epidemiologic transitions involve shifts in both health-promoting and health-destroying microorganisms (Harper and Armelagos, 2010). Viewed in the context of the hygiene hypothesis, both extensive and intensive levels of life-long exposure to these old friends—heirloom and pseudo-commensal microorganisms—suggest that Paleolithic foragers may represent an immunological baseline, possessing immunoregulatory pathways that were finely tuned toward inhibiting CID, in great contrast to the disease ecology and immunology of industrialized societies.

The First Epidemiologic Transition: Disease in Agropastoral Populations

Human environments and disease ecology changed radically with the first transition, but in ways that likely only increased exposure to heirloom and pseudo-commensal microorganisms, while at least maintaining gut microbial diversity. However, the domestication of animals increased human exposure to a wider range of zoonoses. Plant domestication encouraged sedentism, which would have increased orofecal transmission and exposure to water-borne pathogens, and the arrival of consistent yields and food surpluses gradually produced larger populations and higher population density. Agricultural practices like irrigation and the use of human and animal feces as fertilizer would have increased exposure to various pathogens, such as the one responsible for schistosomiasis (Cockburn, 1971). Cultivation practices, such as loosening sod and planting, would have also exposed early farmers to chiggers, which carry the causal agent for scrub typhus, *Orientia tsutsugamushi* (Audy, 1961), while also further exposing them to saprophytic microorganisms. Many major souvenir zoonotic infections emerged, such as caliciviruses, rotaviruses, coronaviruses, orthomyxoviruses (i.e., influenza B and C), and paramyxoviruses (e.g., measles, mumps, parainfluenza, smallpox) (Rook, 2010), as well as infections acquired from peri-domestic animals, like rodents, who assumed habitation in and near human dwellings, such as bubonic plague. Some of these, including the caliciviruses and rotaviruses, may already have infected hominins or were at least encountered sporadically or as zoonotic infections, but the current human strains show great similarity to strains of bovid and suid origin, suggesting that they further evolved during the Neolithic (Van Blerkom, 2003). In low-income developing nations, nearly all children are seropositive for rotavirus, suggesting that it might play a role in the hygiene hypothesis as well (Rook, 2010). In human communities, larger population sizes, reduced mobility, and increased population density allowed many of these diseases, such as influenza (B and C), smallpox, mumps, and measles to become acute, epidemic crowd diseases in human populations, but not until communities reached the level of several hundred thousand residents, which did not occur until the presence of cities approximately 2000–3000 years ago. Once human populations achieved this milestone through urbanization, however, a novel epidemiologic regime was ushered in, consisting of increased morbidity and mortality from infectious disease, particularly among infants and children.

As Rook (2010) notes, since large urban communities represent a fairly recent development, and up until recently, only a small proportion of humanity lived in large urban centers, it is unlikely that the strong selective pressure required to develop evolved dependence emerged between humans and these more sporadically encountered pathogens.

Changes in nutrition and food handling would have also altered both disease risk and exposure to various commensal bacteria. The shift to primary food production narrowed the dietary range, increasing human susceptibility to various dietary deficiencies and metabolic diseases, which are evident on contemporary skeletal samples from many parts of the world where plant and animal domestication occurred. In addition to producing diseases on their own, these deficiencies would have also altered host immunocompetence, making them more susceptible to diverse pathogens and parasites (Chandra, 1999). Storing food surpluses underground and the increasing use of fermentation would have also increased human exposure to *Lactobacillus*, as well as increased exposure to the wide range of yeasts, molds, and bacteria associated with food spoilage and food poisoning.

Overall, these dramatic shifts introduced many human populations to very different epidemiologic regimes than those they had experienced during the Paleolithic. However, these same trends likely only intensified humans' exposure to heirloom and pseudo-commensal microorganisms, increasing the frequency of contact and the range of encountered microorganisms. This situation likely persisted until the 19th century for many communities, as up until very recently, the great majority of humans have lived in rural areas, maintaining contact with their old friends. In short, the first transition may have had the effect of reducing longevity and increasing mortality from acute, epidemic infectious diseases, initiating the epidemiologic trends that came to an end only with the second transition (Cohen and Armelagos, 1984; but see Gage, 2005; Gage and DeWitte, 2009). However, it may also have further improved human immunological health by fine-tuning and developing the immunological pathways primed during the Paleolithic and further protecting these communities from CID.

The Second Epidemiologic Transition: The Rise of Degenerative Disease

The second transition marked a progressive reversal in these environmental conditions and, with them, concomitant decreases in the incidence and associated mortality from infectious diseases, increases in longevity, and increased mortality from noncommunicable and degenerative diseases. Over the last 200 years, developed countries have experienced medical advances and improvements in public health, urban infrastructure, pasteurization, and sanitization of water supplies and sewage systems and, with them, a higher standard of living, reduced exposure to pathogens, and more hygienic modernized environments (Gage, 2005). According to the hygiene hypothesis, however, the second transition also drastically altered human relationships with our old friends, pulling populations in developed nations away from lives filled with dirt (Ashenburg, 2007) and toward those replete with cleanliness (Smith, 2007).

The Unequal Burdens of Infectious Disease and CID

Importantly, this effect has been variable, not only between developing nations and developed ones, but also within them. As discussed throughout this volume, the second transition was experienced differently in timing and scope between different nations as well as within them. This is true of high-income nations that completed the transition in the 19th and 20th centuries, as well as low-income nations experiencing an incomplete form now, like Tanzania, which shoulders a high burden of infectious disease alongside a rising burden of degenerative conditions (WHO, 2006). Much of this difference between and within nations can be attributed to inequality and its impacts upon environmental quality and population health. While inequality originated with the Neolithic (Flannery, 2012) and has been generating health disparities since the first epidemiologic transition (Wilkinson, 2001; Armelagos et al., 2005), due to rapid and uneven development, degrees of income divergence and wealth sequestration between different nations have become greatly exaggerated (Saez, 2008). For instance, prior to the Industrial Revolution in Western Europe, per capita incomes there were only about 30% higher than China's and India's, but by 1870, the world's richest nation was 900% richer than the poorest; by 1990, the difference was 4500% (Baldwin et al., 1998). Inequality is significant on more than an interpopulation level, however. For instance, it has been estimated that within recent years, within-nation income inequality has accounted for more than 80% of global inequality figures (Korzeniewicz and Moran, 1997; Melchior, 2001). This translates into differences in public health infrastructure, environmental quality, access to medical resources, and lifestyle not only between high-income, developed and low-income, developing nations, but also within them. While this produces the regional and national-level patterning in the epidemiologic regimes of infectious vs. degenerative diseases and NCDs traditionally associated with epidemiologic transitions, it seems to also translate into the differences in the CID that have become a key epidemiologic feature of modern, western environments. While socioeconomic status exerts a buffering effect against infectious disease and poor environmental quality, it seems to deprive privileged members of both high-income and low-income nations of their old friends, thus leaving them susceptible to CID. As Okada et al. note, several studies have found a positive correlation between gross national product and the incidence of asthma, T1D, and MS in Europe (Bach, 2002). This has been demonstrated on the national level, but more intriguingly, it also exists at the regional level. For instance, in Northern Ireland, a low incidence of T1D is correlated with a low-average socioeconomic status (Patterson et al., 1996). Similar findings have also been uncovered for Crohn's Disease in Manitoba, Canada. Even more evocatively, positive correlations between socioeconomic status and CID have also been documented at the household-level; incidence of atopic dermatitis, for instance, is correlated directly with family income (Blanchard et al., 2001). While these findings do not pinpoint which factor within socioeconomic indices is directly responsible for the disorder, several studies have found a positive correlation between sanitary conditions and T1D (Patterson et al., 1996) and MS (Leibowitz et al., 1966), further implicating our old friends.

Placed within an evolutionary context, these findings suggest that modernization has generated substantial trade-offs in human health. Developed nations are experiencing historically low levels of mortality from both infectious and noncommunicable and degenerative causes and levels in many developing nations are on the decline (Gage, 2005). However, reductions in infectious disease specifically may have come at the cost of increased morbidity from CID for some members of the population and, with them, a decreased quality of life. Many critiques of the concept of epidemiologic transitions have argued that because it focuses solely on trends in mortality, the model is inadequate for addressing the ramifications of increased longevity for quality of life and well-being (Riley and Alter, 1989; Johansson, 1992; Riley, 1992). The fact that the hygiene hypothesis has not yet been incorporated into studies of the second transition falls under the scope of this critique. Incorporating it into epidemiologic transition theory, and into models of the second transition specifically, would reform these models to employ a more holistic definition of health, to extend quality of life issues past those associated with longevity and end of life concerns, and to include patterns of CID, with their early life origins and life-long impacts on health and quality of life. The latter would also further usher studies of the second transition into relevance with current scholarship in epidemiology, much of which is focused on the developmental origins of health outcomes in later life (see also Hallman and Gagnon, this volume). It would also provide yet another critical avenue for investigating relationships between environmental quality and change and human health—specifically interactions with generally nonpathogenic microorganisms—into considerations of the second epidemiologic transition and epidemiologic transition theory.

Conclusion: Incorporating the Hygiene Hypothesis into Investigations of the Second Epidemiologic Transition

Despite its relevance to the second epidemiologic transition and studies of the human–environment interactions associated with modernization, the hygiene hypothesis has yet to be integrated into empirical investigations of the second transition. Importantly, despite being placed within an evolutionary framework by Rook and Armelagos, it also has yet to be explicitly tested and investigated using historical and archaeological evidence of human exposure to helminthic parasites, commensal bacteria, and pseudo-commensal gut bacteria. Given the importance of understanding the dynamic between lifestyle, environmental quality, and CIDs for comprehending the legacy of the second transition and its continuing impacts upon high-income developed and low-income developing nations, in conclusion, we propose here a number of venues for correcting this scarcity. Our hope is that these recommendations might guide and encourage scholarship on the subject in current environments as well as past ones and generate further insights into the dynamic interactions between our old friends and our current health. We propose here that approaches using direct evidence, such as paleoparasitological analysis of the remains of past bacteria and parasites, as well as more indirect evidence, such as phylogenetic and genomic analyses of humans and their bacteria and parasites, paleoimmunological studies of past human–parasite interactions, and historical material on sanitation and hygiene throughout time, could be employed in these investigations. All of these approaches are currently being used in various disciplines to better understand diverse aspects of human–environment interactions from prehistory to the present, but they have yet to be applied to understanding the hygiene hypothesis and the major trade-offs in human health created by this dynamic. Importantly, as we propose here, they could be used to generate empirical investigations of relationships between humans, their environments, and their old friends instead of the current reliance upon hypothetical scenarios and ethnographic evidence from modern analogues, such as foragers and non-industrial agriculturalists.

Direct Approaches

The most direct evidence for interactions between humans, their environments, and helminthic parasites, commensal bacteria, and pseudo-commensal gut bacteria in the past can be generated from paleoparasitological and paleoecological studies that employ the physical remains of these organisms. Directly detecting bacteria in archaeological deposits relies upon cultures of species that can survive in a dormant state for long periods of time. These can be recovered from environmental contexts, such as in dust and occupational debris preserved in sediment strata at archaeological sites (Reitz and Shackley, 2012). Detecting direct evidence of parasites is more straightforward, in part because of the greater size of most helminthic parasites in comparison to bacteria as well as because the microorganism's capacity for dormancy is not a limiting factor. Paleoparasite evidence consists mostly of eggs and only very rarely of the larvae (developmental stages) of intestinal parasites, or of the chitinous shells of ectoparasites, such as lice, mites, and fleas (Araújo et al., 2000; Dittmar, 2000). While the conditions conducive to preservation vary by region throughout the globe (Reinhard, 1992), different forms of paleoparasite evidence can be recovered from many different microenvironments recovered from the archaeological record, such as latrine soils and related sediments, sediments from the pelvic cavity of skeletons, coprolites (i.e., preserved feces), preserved hair, and mummified remains (Bouchet et al., 2003). There are several issues involved in interpretation, such as difficulties with identifying different species as well as estimating the epidemiologic consequences of infection. With the latter, due to a number of diagenetic factors that can influence parasite remains, it is difficult to accurately assess the parasitic load of a given individual or environment, and thus estimate prevalence (Mendonça de Souza et al., 2003).

Nonetheless, with these issues taken into consideration, paleoparasite evidence can be used to directly reconstruct relationships between human behavior, cultural development, and environmental change, as reflected in conditions of hygiene and sanitation (Bouchet et al., 2003). Within the context of investigating the hygiene hypothesis specifically, these data could then be used to investigate how human–parasite interactions changed with the first epidemiologic transition and the associated development of agriculture, increased sedentism, animal and plant domestication, urbanization, and increased population density; industrialization and increased urbanization; and the improvements in sanitation, hygiene, and water and food safety that precipitated the second transition, the associated reductions in infectious disease morbidity and mortality, and the rise of sanitized modern environments. Importantly, because paleoparasite evidence preserves within specific microenvironments, various resolutions of analysis are possible—from individual households to neighborhoods, villages, towns, and, through aggregating data from multiple sites, whole cities and even large geographic regions—that are rendered largely impossible in conventional investigations of sanitation and hygiene that use historical material, which is not consistently available at these scales.

While no published studies employing paleoparasite evidence have tackled the hygiene hypothesis directly, several demonstrate how such research could be focused and conducted. For instance, though infection rates are considered to have been fairly low among prehistoric hunter–gatherers, and little paleoparasite evidence predating agriculture has been recovered from the archaeological record (Sianto et al., 2009), Bathurst (2005) employed innovative, noninvasive methods to recover surprisingly substantial evidence of helminthic parasites, such as roundworms and tapeworms, from multiple several-thousand-year-old hunter–gatherer habitation sites. Such an approach could be used to provide empirical evidence for better understanding the Paleolithic baseline. Focusing on preindustrial urban sites, Fernandes and colleagues (2005) analyzed sediments from suspected latrines excavated from a 16th-century Belgian village; identification of numerous species of parasite eggs and counts of these eggs suggest a high prevalence of infection and, in turn, very poor levels of sanitation. Approaches such as this, focusing on urban sites, could be useful for directly tracking how levels of sanitation, hygiene, and, consequently, interactions with our old friends changed following the second transition and the rise of urbanization. Others (e.g., Mrozowski, 2006) have reconstructed class-based differences in sanitation through analysis of latrine sediments in historic urban environments. Such an approach would enable direct analysis of the role of socioeconomic status in levels of environmental quality, sanitation, and interactions with different microorganisms in various contexts over time. Fisher et al.'s (2007) analysis of paleoparasite data from historical Albany, New York, featured an investigation of the dynamics between rapid population growth, urban expansion, and parasite numbers; quantitative estimates of the effectiveness of different medical treatments and sanitary innovations, such as new techniques of privy construction and water sanitization, on parasite numbers; and on a household-level, how these innovations and treatments were used. In this volume, Reinhard and Pucu de Araújo demonstrate how both site-level and aggregate paleoparasite data collected from numerous sites across time and from both the New and Old worlds can be used to detect exactly when and where the changes in sanitation associated with the first and second epidemiologic transitions took place, as well as the effectiveness of different sanitation and water sanitization techniques on parasite levels, and how this varied by household and neighborhood socioeconomic status, as well as by community and region. Overall, if applied to the hygiene hypothesis, such approaches could generate direct, empirical evidence for fluctuating levels of sanitation and interactions between humans and our old friends across different geographic, cultural, and chronological contexts.

Indirect Approaches

As we propose here, more indirect approaches, such as those employed in archaeogenetics, archaeogenomics, phylogenetics, and paleoimmunology, could also be fruitfully applied to empirical investigations of the hygiene hypothesis. The presence of helminthic parasites and other microorganisms can be indirectly inferred from the presence of vectors or hosts in the archaeological record, such as the shells of the various species of gastropod freshwater snails that act as intermediate hosts for schistosomiasis, as well as from the pathological lesions they may potentially leave on human remains (Reitz and Shackley, 2012). Their presence can also be inferred from archaeogenetic analysis of the DNA these microorganisms may deposit in and on preserved human tissues. For instance, several studies have now recovered both calcified and noncalcified reactive bacterial aDNA of several species from human dental calculus using transmission electron microscopy (TEM) and gold-labeled antibody TEM as well as PCR (Preus et al., 2011; De La Fuente et al., 2013). Due to the frequent, long-term preservation of dental calculus in the archaeological record from a number of sites, this approach opens the possibility of identifying specific bacterial species that infected past populations, reconstructing and comparing oral and gastric microbiomes in past populations, and reconstructing genetic covariation between bacteria and humans in numerous contexts stretching back several thousand years (De La Fuente et al., 2013). Such approaches and their insights could not only be used to reconstruct levels of sanitation in the past and infer the presence of specific helminthic parasites in past environments, but also to reconstruct how human microbiomes of commensal bacteria have varied in different contexts. By extrapolation, these data could in turn potentially be used to infer the immunological processes these types of exposures may have elicited in past human populations and how these relate to the immunomodulatory processes implicated in the hygiene hypotheses.

Numerous researchers have also used gut contents recovered from mummified human remains to reconstruct the microbiota of pseudo-commensal gut bacteria present in past populations. While mummified remains are rare and do not exist for many regions of the globe, analysis of aDNA from the gut contents of numerous bog bodies, frozen remains, and intentionally (culturally) and unintentionally desiccated remains has been used to reconstruct the genetic libraries of gut microbiota in numerous historic and prehistoric contents, largely through PCR (e.g., Ubaldi et al., 1998; Cano et al., 2000; Rollo et al., 2007). Importantly, for investigations of the hygiene hypothesis, many of these studies have found that especially when compared to microbial libraries of the gut microbiota from modern populations, there is remarkable constancy in species composition and proportion of gut microbiota across populations through time (Wilson, 2005). Among other issues, researchers must be cautious that some bacterial species, such as clostridia, can be introduced through taphonomic processes rather than being indigenous (Rollo et al., 2007). However, when linked with data from the growing body of literature on the immunomodulatory effects of pseudo-commensal gut bacteria and placed within the context of the hygiene hypothesis, these studies have the potential to grant tremendous insight into how gut microbiota have changed through time and in different environments and epidemiologic regimes and, in turn, tie this to estimations of the immunological functioning of past populations.

Lastly, archaeogenomic, phylogenetic, and paleoimmunological techniques could also be used to generate key insights into the human–environment–microorganism interactions that are fundamental to the hygiene hypothesis. Phylogenetic and archaeogenomic analyses have been helpful for clarifying the timing and nature of when initial interactions between various parasites, pathogenic and nonpathogenic bacteria, and human hosts settled into more established parasitic relationships. Biogeographic evidence has also been used to complement genomic data, meaning that the general geographic region for these interactions can also be estimated. Such analyses have been conducted for a variety of pathogens as well as helminthic parasites, and therefore can be used to provide empirical though indirect evidence of the longevity of human relationships with parasites, and the environments and contexts in which these interactions arose. For instance, phylogenetic analysis of human taenid tapeworms (i.e., *Taenia saginata*, *Taenia asiatica*, *Taenia solium*) suggests that contrary to the long-standing hypothesis that they represent a zoonotic infection acquired from domesticated cattle and pigs, they instead predate the evolution of humans and likely evolved from species affecting felid and hyaenid hosts. Furthermore, molecular evidence indicates that divergence between the human-infecting sister species, *T. saginata* and *T. asiatica*, occurred between 780,000 and 1.7 million years ago, far before animal domestication. Together, these findings suggest that human tapeworms arose from hominins scavenging from bovid carcasses left by felid and hyaenid predators somewhere in Africa (Hoberg et al., 2000; Hoberg et al., 2001).

Paleoimmunological studies could also be useful for investigating the longevity and context of human–parasite interactions in the context of the hygiene hypothesis. As the immunomodulatory processes underlying the hygiene hypothesis are still unclear, we do not propose here that paleoimmunological data could currently be used to reconstruct evidence for these dynamics in the past. However, through detecting the immunological responses elicited by helminthic infection, paleoimmunological approaches could be used to indirectly infer infection in the past, as well as its epidemiologic consequences. In contrast to the limitations for estimating prevalence posed by paleoparasite evidence, immunologic examinations of preserved human tissues, such as that from mummified remains, can be used to reconstruct various epidemiologic parameters for helminthic infection in past populations. For example, Campbell Hibbs et al. (2011) conducted *ELISA* on preserved human tissues associated with several agricultural communities in ancient Nubia to estimate age- and sex-based profiles for prevalence, as well as associations between prevalence and different agricultural techniques. This approach allowed them to reconstruct not only differential behavioral patterns for water contact in relation to age and gender, but also the specific impacts of human alteration of the environment on the transmission of schistosomiasis. Such approaches, if applied to questions related to the hygiene hypothesis, could shed key insights into which patterns of human behavior and environmental modification have produced different levels of interactions with many of our old friends throughout time, and what these patterns may entail for present and future interactions.

The many insights into present-day medical dilemmas granted by evolutionary medicine over the past several decades clearly demonstrate the value of a long-sighted, evolutionary perspective for investigating human disease and human–pathogen interactions. The hygiene hypothesis may add the origins of CID, such as allergy and autoimmune conditions, to this list. While clinical intervention studies and analyses of animal models for type 2 immune responses continue at a furious pace (Pulendran and Artis, 2012), an evolutionary perspective that draws upon diverse lines of direct and indirect evidence from the archaeological record, as well as phylogenetic approaches, can be used to generate novel, empirical, and otherwise inaccessible insights into the history and complexity of long-standing relationships between humans and their old friends. Furthermore, while it is currently impossible to detect the diverse type 2 responses implicated in the hygiene hypothesis in past populations, researchers can cautiously extrapolate backward by using data from modern populations experiencing these exposures to infer what immunological processes humans may have experienced in the past. Such investigations would be fueled by the confidence of knowing that when parasitism and other more symbiotic relationships are directly related to specific environmental conditions, present-day relationships among parasites, other microbes, and their hosts can be extended and, by analogy, used to infer similar relationships in the past (Reitz and Shackley, 2012). It is our hope that working from these assurances, researchers will use diverse, interdisciplinary approaches to extend our understanding of the hygiene hypothesis back through the millennia and use these empirical insights to improve the well-being and lessen the chronic disease burden of modern populations.

References

Adlerberth I, Carlsson B, De Man P, et al. 1991. Intestinal colonization with Enterobacteriaceae in Pakistani and Swedish hospital-delivered infants. Acta Paediatr 80(6–7):602–610.

Adlerberth I, Jalil F, Carlsson B, et al. 1998. High turnover rate of *Escherichia coli* strains in the intestinal flora of infants in Pakistan. Epidemiol Infect 121(3):587–598.

Araújo A, Ferreira L, Guidon N, Freire N, Reinhard K, Dittmar K. 2000. Ten thousand years of head lice infection. Parasitol Today 7:269.

Armelagos G. 2009. The paleolithic disease-scape, the hygiene hypothesis, and the second epidemiological transition. In: Rook G, editor. The hygiene hypothesis and Darwinian medicine. Basel: Birkhauser Publishing. p 29–43.

Armelagos GJ, Harper KN. 2005a. Genomics at the origins of agriculture, part one. Evol Anthropol 14(2):68–77.

Armelagos GJ, Harper KN. 2005b. Genomics at the origins of agriculture, part two. Evol Anthropol 14(3):109–121.

Armelagos G, Harper K. 2010. Emerging infectious diseases, urbanization and globalization in the time of global warming. In: Cockerham WC, editor. The new Blackwell companion to medical sociology. Hoboken: Wiley Publishing. p 291–311.

Armelagos G, Brown P, Turner B. 2005. Evolutionary, historical and political economic perspectives on health and disease. Soc Sci Med 61:755–765.

Ashenburg K. 2007. The dirt on clean: An unsanitized history. New York: North Point Press.

Audy JR. 1961. The ecology of scrub typhus. In: May JM, editor. Studies in disease ecology. New York: Hafner Publishing. p 389–432.

Bach J-F. 2002. The effect of infections on susceptibility to autoimmune and allergic diseases. N Engl J Med 347:911–920.

Baldwin R, Martin P, Ottaviano G. 1998. Global income divergence, trade and industrialization: the geography of growth take-offs. Working Paper 6458. Cambridge: National Bureau of Economic Research.

Barrett R, Kuzawa CW, McDade T, Armelagos GJ. 1998. Emerging infectious disease and the third epidemiological transition. In: Durham W, editor. Annual Review of Anthropology. Palo Alto: Annual Reviews Inc. p 247–271.

Bathurst R. 2005. Health and settlement implications of parasites from Pacific Northwest coast archaeological sites. PhD Dissertation. Hamilton: McMaster University.

Beiber T. 2008. Atopic dermatitis. N Engl J Med 358:1483–1494.

Bengmark S. 2000. Bacteria for optimal health. Nutrition 16(7–8):611–615.

Björkstén B. 2009. The hygiene hypothesis: do we still believe in it? In: Brandtzaeg P, Isolauri E, Prescott S, editors. Microbial–host interaction: Tolerance versus allergy. Basel: Karger. p 11–22.

Björkstén B, Naaber P, Sepp E, Mikelsaar M. 1999. The intestinal microflora in allergic Estonian and Swedish 2-year-old children. Clin Exp Allergy 29:342–346.

Blanchard J, Bernstein C, Wajda A, Rawsthorne P. 2001. Small-area variations and sociodemographic correlates for the incidence of Crohn's disease and ulcerative colitis. Am J Epidemiol 154:328–335.

Bloomfield SF, Stanwell-Smith R, Crevel RWR, Pickup J. 2006. Too clean, or not too clean: the hygiene hypothesis and home hygiene. Clin Exp Allergy 36:402–425.

Bodansky H, Staines A, Stephenson C, Haigh D, Cartwright R. 1992. Evidence for an environmental effect in the aetiology of insulin dependent diabetes in a transmigratory population. BMJ 304: 1020–1022.

Bouchet F, Guidon N, Dittmar K, et al. 2003. Parasite remains in archaeological sites. Mem Inst Oswaldo Cruz 98(S1):47–52.

Bremner SA, Carey IM, DeWilde S, et al. 2008. Infections presenting for clinical care in early life and later risk of hay fever in two UK birth cohorts. Allergy 63(3):274–283.

Brooks D, Ferrao A. 2005. The historical biogeography of co-evolution: emerging infectious diseases are evolutionary accidents waiting to happen. J Biogeogr 32:1291–1299.

Burnet FM. 1962. Natural history of infectious disease. Cambridge: Cambridge University Press.

Cano R, Tiefenbrunner F, Ubaldi M, et al. 2000. Sequence analysis of bacterial DNA in the colon and stomach of the Tyrolean Iceman. Am J Phys Anthropol 3(1):297–309.

Cardwell C, Carson D, Yarnell J, Shields M, Patterson C. 2008. Atopy, home environment and the risk of childhood-onset type 1 diabetes: a population-based case–control study. Pediatr Diabetes 9:191–196.

Chandra R. 1999. Nutrition and immunology: from the clinic to cellular biology and back again. Proc Nutr Soc 58:681–683.

Clemente JC, Ursell LK, Parfrey LW, Knight R. 2012. The impact of the gut microbiota on human health: an integrative view. Cell 148(6):1258–1270.

Cockburn T. 1967. Infections of the order primates. In: Cockburn T, editor. Infectious diseases: their evolution and eradication. Springfield: CC Thomas.

Cockburn T. 1971. Infectious disease in ancient populations. Curr Anthropol 12(1):45–62.

Cohen MN, Armelagos GJ, editors. 1984. Paleopathology at the origins of agriculture. Orlando: Academic Press.

De La Fuente C, Flores S, Moraga M. 2013. DNA from human ancient bacteria: a novel source of genetic evidence from archaeological dental calculus. Archaeometry 55(4):767–778.

Dittmar K. 2000. Ectoparasites on guinea pig mummies of El Yaral, Chungara. Review of the Anthropology of Chile 32:123–127.

Dunn R, Davies T, Harris N, Gavin M. 2010. Global drivers of human pathogen richness and prevalence. Proc R Soc B 277:2587–2595.

Eder W, Ege M, von Mutius E. 2006. The asthma epidemic. N Engl J Med 355:2226–2235.

Fernandes A, Ferreira LF, Gonçalves ML, et al. 2005. Intestinal parasite analysis in organic sediments collected from a 16th-century Belgian archeological site. Cad Saúde Pública 21(1):329–332.

Fisher C, Reinhard K, Kirk M, DiVirgilio, J. 2007. Privies and parasites: the archaeology of health conditions in Albany, New York. Hist Archaeol 41(4):172–197.

Flannery K. 2012. The creation of inequality: How our prehistoric ancestors set the stage for monarchy, slavery, and empire. Cambridge: Harvard University Press.

Gage T. 2005. Are modern environments really bad for us? Revisiting the demographic and epidemiologic transitions. Yrbk Phys Anthropol 48:96–117.

Gage T, DeWitte S. 2009. What do we know about the agricultural demographic transition? Curr Anthropol 50(5):649–655.

Gale EA. 2002. A missing link in the hygiene hypothesis? [see comment]. Diabetologia 45(4):588–594.

Green A, Patterson C. 2001. Trends in the incidence of childhood-onset diabetes in Europe 1989–1998. Diabetologia 44(S3):B3–B8.

Hammond S, English D, McLeod J. 2000. The age-range of risk of developing multiple sclerosis: evidence from a migrant population in Australia. Brain 123:968–974.

Harjutsalo V, Sjoberg L, Tuomilehto J. 2008. Time trends in the incidence of type 1 diabetes in Finnish children: a cohort study. Lancet 371:1777–1782.

Harper K, Armelagos G. 2010. The changing disease-scape in the third epidemiological transition. Int J Environ Res Public Health 7(2):675–697.

Harper K, Ocampo P, Steiner B, et al. 2008. On the origin of the treponematoses: a phylogenetic approach. PLoS Negl Trop Dis 2(1):e148.

Hibbs AC, Secor W, Van Gerven D, Armelagos G. 2011. Irrigation and infection: the immunoepidemiology of schistosomiasis in ancient Nubia. Am J Phys Anthropol 145:290–298.

Hoberg EP, Jones A, Rausch RL, Eom KS, Gardner SL. 2000. A phylogenetic hypothesis for species of the genus Taenia (Eucestoda: Taeniidae). J Parasitol 86(1):89–98.

Hoberg EP, Alkire NL, de Queiroz A, Jones A. 2001. Out of Africa: origins of the Taenia tapeworms in humans. Proc R Soc Lond B 268(1469):781–787.

Hugot J, Reinhard K, Gardner S, Morand S. 1999. Human enterobiasis in evolution: origin, specificity and transmission. Parasite 6(3):201–208.

ISAAC. 1998. Worldwide variation in prevalence of symptoms of asthma, allergic rhinoconjunctivitis, and atopic eczema: ISAAC. The International Study of Asthma and Allergies in Childhood (ISAAC) Steering Committee. Lancet 351:1225–1232.

Johansson SR. 1992. Measuring the cultural inflation of morbidity during the decline in mortality. Health Transit Rev 2(1):78–89.

Kilks M. 1990. Helminths as heirlooms and souvenirs: a review of New World paleoparasitology. Parasitol Today 6(4):93–100.

Korzeniewicz R, Moran T. 1997. World-economic trends in the distribution of income, 1965–1992. Am J Sociol 102(4):1000–1039.

Lambrecht B, Hammad H. 2012. Lung dendritic cells in respiratory viral infection and asthma: from protection to immunopathology. Annu Rev Immunol 30:243–270.

Leibowitz U, Antonovsky A, Medalie J, Smith H, Halpern L, Alter M. 1966. Epidemiological study of multiple sclerosis in Israel. II. Multiple sclerosis and level of sanitation. J Neurol Neurosurg Psychiatry 29:60–68.

Leibowitz U, Kahana E, Alter M. 1973. The changing frequency of multiple sclerosis in Israel. Arch Neurol 29:107–110.

Leslie M. 2012. Gut microbes keep rare immune cells in line. Science 335(6075):1428.

Like A, Guberski D, Butler L. 1991. Influence of environmental viral agents on frequency and tempo of diabetes mellitus in BB/Wor rats. Diabetes 40:259–262.

Lynch N, Hagel I, Perez M, Di Prisco M, Lopez R, Alvarez N. 1993. Effect of anthelmintic treatment on the allergic reactivity of children in a tropical slum. J Allergy Clin Immunol 92:404–411.

Lynch N, Palenque M, Hagel I, DiPrisco M. 1997. Clinical improvement of asthma after anthelminthic treatment in a tropical situation. Am J Respir Crit Care Med 156:50–54.

Manichanh C, Rigottier-Gois L, Bonnaud E, et al. 2006. Reduced diversity of faecal microbiota in Crohn's disease revealed by a metagenomic approach. Gut 55:205–211.

Maslowski KM, Mackay CR. 2011. Diet, gut microbiota and immune responses. Nat Immunol 12(1):5–9.

Mazmanian S, Round J, Kasper D. 2008. A microbial symbiosis factor prevents intestinal inflammatory disease. Nature 453:620–625.

Melchior A. 2001. Global income inequality: beliefs, facts, and unresolved issues. World Economics 2(3):87–108.

Mendonça de Souza SM, Maul de Carvalho D, Lessa A. 2003. Paleoepidemiology: is there a case to answer? Mem Inst Oswaldo Cruz 98(S1):21–27.

Mrozowski S. 2006. The archaeology of class in urban America. Cambridge: Cambridge University Press.

Munoz-Lopez F. 2006. Validity of the hygiene hypothesis [comment]. Allergol Immunopathol 34(4):129–130.

Okada H, Kuhn C, Feillet H, Bach JF. 2010. The 'hygiene hypothesis' for autoimmune and allergic diseases: an update. Clin Exp Immunol 160(1):1–9.

Omran AR. 1971. The epidemiologic transition: a theory of the epidemiology of population change. Milbank Mem Fund Q 49(4):509–538.

Patterson C, Carson D, Hadden D. 1996. Epidemiology of childhood IDDM in Northern Ireland 1989–1994: low incidence in areas with highest population density and most household crowding. Northern Ireland Diabetes Study Group. Diabetologia 39:1063–1069.

Paul W, Zhu J. 2010. How are TH2-type immune responses initiated and amplified? Nat Rev Immunol 10:225–235.
Ponsonby A-L, van der Mei I, Dwyer T, et al. 2005. Exposure to infant siblings during early life and risk of multiple sclerosis. JAMA 293(4):463–469.
Preus H, Marvik O, Selvig N, Bennike P. 2011. Ancient bacterial DNA (aDNA) in dental calculus from archaeological human remains. J Archaeol Sci 38(8):1827–1831.
Pulendran B, Artis D. 2012. New paradigms in type 2 immunity. Science 337(6093):431–435.
Pulendran B, Tang H, Manicassamy S. 2010. Programming dendritic cells to induce TH2 and tolerogenic responses. Nat Rev Immunol 11(8):647–655.
Rautiainen H, Salomaa V, Niemelä S, et al. 2007. Prevalence and incidence of primary biliary cirrhosis are increasing in Finland. Scand J Gastroenterol 42:1347–1353.
Reed D, Smith V, Hammond S, Rogers A, Clayton D. 2004. Genetic analysis of lice supports direct contact between modern and archaic humans. PLoS Biol 2(11):e340.
Reinhard K. 1992. Parasitology as a tool for the archaeologist. Am Antiq 57:231–245.
Reitz E, Shackley M. 2012. Viruses, bacteria, archaea, protists, and fungi. In: Reitz E, Shackley M, editors. Environmental Archaeology. New York: Springer. p 161–189.
Riley J. 1992. From a high mortality regime to a high morbidity regime: is culture everything in sickness? Health Transit Rev 2(1):71–78.
Riley J, Alter G. 1989. The epidemiologic transition and morbidity. Ann Demogr Hist 199–213.
Rollo F, Luciani S, Marota I, Olivieri C, Ermini L. 2007. Persistence and decay of the intestinal microbiota's DNA in glacier mummies from the Alps. J Archaeol Sci 34(8):1294–1305.
Rook GA. 2007. The hygiene hypothesis and the increasing prevalence of CID. Trans R Soc Trop Med Hyg 101(11):1072–1074.
Rook GAW. 2010. 99th Dahlem conference on infection, inflammation and CID: Darwinian medicine and the 'hygiene' or 'old friends' hypothesis. Clin Exp Immunol 160(1):70–79.
Rook GA, Brunet LR. 2005a. Microbes, immunoregulation, and the gut. Gut 54(3):317–320.
Rook GA, Brunet LR. 2005b. Old friends for breakfast. Clin Exp Allergy 35(7):841–842.
Roumagnac P, Weill F, Dolecek C, et al. 2006. Evolutionary history of *Salmonella typhi*. Science 314:1301–1304.
Saez E. 2008. Striking it richer: the evolution of top incomes in the United States (Update using 2006 preliminary estimates). Working Paper. Berkeley: University of California, Department of Economics.
Sianto L, Chame M, Silva C, et al. 2009. Animal helminths in human archaeological remains: a review of zoonoses in the past. Rev Inst Med Trop São Paulo 51(3):119–130.
Smith VS. 2007. Clean: A history of personal hygiene and purity. New York: Oxford University Press.
Sprent J. 1969a. Evolutionary aspects of immunity of zooparasitic infections. In: Jackson G, editor. Immunity to parasitic animals. New York: Appleton. p 3–64.
Sprent J. 1969b. Helminth "zoonoses": an analysis. Helminthol Abst 38:333–351.
Strachan DP. 1989. Hay fever, hygiene, and household size. Br J Med Psychol 299:1259–1260.
Strachan DP. 2000. Family size, infection and atopy: the first decade of the 'hygiene hypothesis'. Thorax 55(S1):S2–S10.
Summers R, Elliott D, Urban J, Jr., Thompson R, Weinstock J. 2005a. Trichuris suis therapy for active ulcerative colitis: a randomized controlled trial. Gastroenterology 128:825–832.
Summers R, Elliott D, Urban J, Jr., Thompson R, Weinstock J. 2005b. Trichuris suis therapy in Crohn's disease. Gut 54:87–90.
Ubaldi M, Luciani S, Marota I, Fornaciari G, Cano R, Rollo F. 1998. Sequence analysis of bacterial DNA in the colon of an Andean mummy. Am J Phys Anthropol 107(3):285–295.
Van Blerkom LM. 2003. Role of viruses in human evolution. Am J Phys Anthropol 122(S37):14–46.
Van Regenmortel MH, Fauquet CM, Bishop DHL, et al. 2000. Virus taxonomy: classification and nomenclature of viruses. Seventh Report of the International Committee on Taxonomy of Viruses. San Diego: Academic Press.
Vermund S, MacLeod S. 1988. Is pinworm a vanishing infection? Laboratory surveillance in a New York City medical center from 1971 to 1986. Am J Dis Child 142:566–568.
Wallin M, Page W, Kurtzke J. 2004. Multiple sclerosis in US veterans of the Vietnam era and later military service: race, sex, and geography. Ann Neurol 55:65–71.

WHO. 2006. WHO Africa region: Tanzania: Country health system fact sheet, 2006, United Republic of Tanzania. Brazzaville: World Health Organization-Regional Office for Africa.
Wilkinson R. 2001. Mind the gap: Hierarchies, health and human evolution. New Haven: Yale University Press.
Wills-Karp M, Santeliz J, Karp CL. 2001. The germless theory of allergic disease: revisiting the hygiene hypothesis. Nat Rev Immunol 1:69–75.
Wilson M. 2005. Microbial inhabitants of humans. Cambridge: Cambridge University Press.
Yang Z, Wang K, Li T, et al. 1998. Childhood diabetes in China. Enormous variation by place and ethnic group. Diabetes Care 21(4):525–529.
Yazdanbakhsh M, Matricardi PM. 2004. Parasites and the hygiene hypothesis: regulating the immune system? Clin Rev Allergy Immunol 26(1):15–24.
Yazdanbakhsh M, Kremsner PG, Ree RV. 2002. Allergy, parasites, and the hygiene hypothesis. Science 296:490–494.
Zaccone P, Fehervari Z, Phillips J, Dunne D, Cooke A. 2006. Parasitic worms and inflammatory diseases. Parasite Immunol 28:515–523.
Zhu J, Yamane H, Paul W. 2010. Differentiation of effector CD4 T cell populations. Ann Rev Immunol 28:445–489.

Chapter 17
Comparative Parasitological Perspectives on Epidemiologic Transitions: The Americas and Europe

Karl J. Reinhard[1] and Elisa Pucu de Araújo[2]
[1] School of Natural Resources, University of Nebraska at Lincoln, Lincoln, NE
[2] Manter Laboratory of Parasitology, University of Nebraska at Lincoln, Lincoln, NE

Introduction

To address the second epidemiologic transition from the perspective of parasitology, it is necessary to define the first epidemiologic transition based on archaeoparasitology and pathoecology. With regard to parasitism (by helminthic parasites), Barrett et al. (1998) presented a paleopathological perspective on the development of parasitic disease. They defined three ancient parasitological states separated by two transitions. Originally, there was the Paleolithic state of low levels of parasitism in small scattered hunter–gatherer groups. The first epidemiologic transition coincides with the Neolithic Transition and the intensification of agriculture and rise of sedentism. This transition marks the rise of mortality from acute, epidemic infectious disease and, in regard to parasites, an increased prevalence of these organisms, especially those transmitted from human to human. The second epidemiologic transition, marked by the decline of mortality from acute, epidemic infectious disease, coincided with the Industrial Revolution and the rise of modern, urban environments. In turn, it is also associated with a reduction in infectious parasitic diseases.

Archaeoparasitology is devoted to the recovery and analysis of parasites from historic and prehistoric sites. Specifically, it entails the recovery and contextualization of parasites within a cultural perspective. Past infections are related to behavior, environment, subsistence, trade, social complexity, and other details of past life reconstructed from archaeological and historical sources. This field depends on the fusion of archaeological method and theory with biological understandings of parasite life cycles and ecology. Archaeoparasitology has much to offer in understanding changes in patterns of parasitic disease associated with the development of complex societies, environmental change, and alterations in subsistence practices.

Pathoecology is the reconstruction of the ecology of disease based on evidence recovered from archaeological sites. Infectious diseases are the product of ecological interactions between hosts and pathogenic organisms. The material remains of the can sometimes be recovered from archaeological sites. This is especially true of multicellular parasites, specifically macroparasites such as helminths. Helminth parasites can be recovered from coprolites, mummies, latrines or privies, ceramic vessels, hair, and in sediments from trash, yards, streets, drains, and activity surfaces. Additionally, organic sediments recovered from stratigraphic layers in archaeological sites can also be sources for recovering evidence of parasites and demonstrating changes in parasite diversity and abundance through time (Gonçalves et al., 2003; Rocha et al., 2006). The regions that are best studied include Europe and the Americas. For these regions, there is sufficient time depth to assess thousands of years of human–parasite interaction. Throughout the past five decades of research, parasitological data from the Americas and Europe have shown stark contrasts between these regions as well as between time periods. The contrasts are so dramatic that one can ask whether a first epidemiologic transition—from a parasitic perspective—did occur, and therefore whether the second occurred as well. In turn, it is necessary to ask whether the concept of epidemiologic transitions is suitable for modeling parasitic infections and human–parasite interactions in the past. This study examines these issues and investigates what archaeoparasitological and pathoecological evidence can tell us about environmental change and human–environment interactions in the context of epidemiologic transitions in Europe and the Americas. This is relevant for understanding how trends in parasitism fit into epidemiologic transition models. It is also, as Zuckerman and Armelagos (this volume) discuss, increasingly important for generating insights into how our evolutionary interactions with parasites have influenced other aspects of human health and how levels of hygiene have changed over time in light of current research on the important relationship between helminth parasites and maturation of the immune system (i.e., the hygiene hypothesis). This study discusses the current state of knowledge on helminthic parasites, specifically whipworm and maw-worm in Europe and the Americas, and presents a case study on changes in parasitism over time in Albany, New York.

The First Epidemiologic Transition in Europe and the Americas

Evidence of the First Epidemiologic Transition in Europe: Filth and Filth-Borne Parasites

Filth, and specifically the sort of filth derived from feces, defined the pathoecology of parasitism in Europe from the Neolithic Transition onward. The archaeoparasitological record of Europe is characterized by the ubiquity of fecal-borne geohelminths, or soil-transmitted helminthic worms, particularly maw-worm (*Ascaris lumbricoides*) and whipworm (*Trichuris trichiura*). Their physical presence in the archaeological record is complemented by mention of the parasites in ancient medical texts.

These two parasites have similar life cycles. The eggs are passed in the feces of infected human hosts. They then need a period of time to embryonate to the infective stage in soil. Once ingested in soil or soil-contaminated materials, the eggs hatch and mature to sexually reproductive adults. However, once they reach infective stage, the eggs stay viable for many years. The resistance of the eggs to environmental conditions is astounding: 6 years for whipworm and 10 years for maw-worm. Both species, but especially maw-worm, are resistant to conditions including a wide temperature range, desiccation, and even being formalin immersion. Maw-worm eggs are covered with a sticky coat that spontaneously attaches to anything that comes into contact. Whether the eggs adhere to laboratory glassware or the succulent shoots of cultivated plants in a farm field fertilized with human feces, the adherent nature of the eggs leads to a complex web of contamination from the fecal source. The eggs are as stunning in their evolutionary perfection with regard to achieving infection.

Despite these similarities in life cycle and infection, the adult worms themselves are quite different. Maw-worm is also called the *giant intestinal roundworm*. This alternative common name hints at the fact that the adult worms are large, about a foot long, and a quarter inch in diameter. They are vigorous squirmers and maintain their place in the intestinal lumen by actively swimming against the current caused by peristalsis. When passed, either due to expiration or active squirming from the anus, the worms are noticeable to their hosts. The maw-worm spends its adult life wriggling, snakelike, to resist the intestine's wave-inducing contraction and relaxation of muscles, which push the contents toward excretion. While maintaining their position in the intestine, the parasites manage reproduction. The numbers of eggs produced is remarkable at 200,000 per female per day.

In contrast, whipworms are subtle parasites. Their subtlety contributed to the fact that for most of post-Roman history, they were not recognized as a human parasite. The adult worms are nearly microscopic. Morphologically, they look like a whip, with a handle-like posterior portion and a long filament-like anterior portion; the former scientific name, *Trichocephalus*, more accurately describes the whip-like morphology of the latter portion. Whipworms embed themselves in the lining of the intestinal mucosa with their broad anterior firmly entrenched. So anchored, they are mostly protected from the chemistry and mechanics of the intestinal lumen, with only their posterior portion projecting beyond the intestinal lining. Up to 10,000 eggs per day per female are laid. The human host usually faces no pathology from the infection and is unaware of the worms (Roberts and Janovy, 2000).

The archaeoparasitological record of Europe represents all time periods and main cultural developmental stages, from Paleolithic hunter–gatherers to the industrial period. Almost all investigations have revealed evidence of maw-worms or whipworms—or both parasites—at archaeological sites ranging from the Paleolithic to the beginning of the modern period (Table 17.1 and Table 17.2). From Roman times onward, these parasites accompanied their human hosts wherever Europeans settled; evidence of both geohelminths can be found in Europe dating from the Roman and medieval periods (Table 17.1 and Table 17.2; See Fernandes et al., 2005; Ferreira et al., 2011; Gonçalves et al., 2003; Rocha et al., 2006). Maw-worm in particular is extremely abundant in European archaeological sites and acts as an indicator of environmental conditions, denoting the presence of filth and fecal pollution. Importantly, the number and abundance of eggs allows the study of different degrees of contamination, between privies, stables, and other structures (Roberts & Janovy, 2000; Cox, 2002; Ferreira et al., 2011).

The eggs of these parasites were ubiquitous in the medieval environment, as shown by the analysis of sediments from a variety of archaeological sites. Jones (1985) first showed that hundreds of eggs of whipworm per milliliter of sediment recovered from a given medieval archaeological site indicates that whipworm was a normal part of the background fauna. This phenomenon has since been documented for all countries in Europe where studies have been conducted. Historical records suggest that uncontained disposal of excrement was responsible for this problem. During the medieval and early modern periods, feces were deposited in refuse pits or yards, in front of houses or in streets. And personal hygiene that could have limited infection was primitive. Nonelites usually had few changes of clothes and therefore could not clean filth from their clothing. Many people wore the same clothes during the day and at night. Moreover, while some records are ambiguous about this, it is likely that overall, people rarely washed, in part due to beliefs stemming from the principles of humoralism, which held that in some states water could be harmful and weaken the human body (Bartošová et al., 2011).

The composition of waste in medieval European urban areas was almost exclusively organic. Most wastes consisted of biodegradable food scraps, human waste, animal waste, and offal. Human excrement and garbage were treated similarly in urban areas, and in rural settlements, waste—both animal and human—was applied directly to the land as a fertilizer. Thus, the cultivated land was contaminated with adherent geohelminth eggs and became

Table 17.1. Maw-worm (*Ascaris lumbricoides*) finds, locality, country, and date from ancient remains in Europe.

Archaeological site/mummy	Country	Date
Grand Grotte, Arcy-sur-Cure, Yonne	France	30,160 ± 140–24,660 ± 330 BP
Clairvaux, Jura	France	3600 BC
Arbon, Thurgau	Swiss	3384–3370 BC
Chalain, Jura	France	2700–2440 BC
Somerset	England	4100–2600 BC
Drobintz girl	Prussia	600 BC
Hallstat	Austria	2300 years
Hallein, Salzburg	Austria	2000 years
Bremerhaven	Germany	100 BC–AD 500
Valkenburg on Rhine	The Netherlands	AD 42–100
Winchester	England	Roman age
Lindow Man	England	2nd century AD
Bobigny	France	2nd century AD
York	England	2nd–3rd century AD
Karwinden Man	Prussia	AD 500
Ribe	Denmark	AD 750–800
York	England	9th–10th century AD
Winchester	England	1000 years
Winchester	England	AD 1100
Utrecht	The Netherlands	13th–14th century AD
Southampton	England	13th–14th century AD
Amsterdam	The Netherlands	AD 1370–1425
Paris	France	14th–15th century AD
York	England	14th–16th century AD
Worcester	England	15th century AD
Oslo	Norway	15th century
Lübeck	Germany	15th century AD
Montbeliard	France	15th–16th century
Schleswig	Germany	Medieval age
Berlin	Germany	Medieval age
Braunschweig	Germany	Medieval age
Hameln	Germany	Medieval age
Höxter	Germany	Medieval age
Göttingen	Germany	Medieval age
Marburg	Germany	Medieval age
Freiburg	Germany	Medieval age
Breisach	Germany	Medieval age
Regensburg	Germany	Medieval age
Landshut	Germany	Medieval age
Marly-le-Roy, Yveline b	France	17th–18th century
Namur	Belgium	AD 18th century

Every context analyzed after the Neolithic Revolution is positive for this parasite. References can be found in Gonçalves et al. (2003) and Leles et al. (2010).

Table 17.2. Whipworm (*Trichuris trichiura*) finds, locality, country, and date from ancient remains in Europe.

Archaeological site/mummy	Country	Date
Clairvaux, Jura	France	3600–3500 BC
Swifterbant	The Netherlands	5400 ± 40–5230 ± 40 BP
Arbon, Thurgau	Swiss	3384–3370 BC
Otzal, Tyrol	Austria	3300–3200 BC
Chalain, Jura	France	32nd–25th century BC
Somerset	England	4100–2600 BP
Hulin, Central Moravia	Czech Republic	1600–1500 BC
Drobintz girl	Prussia	600 BC
Hallstatt	Austria	2300 years
Tollund Man, Central Jutland	Denmark	210 BC
Vilshofen	Germany	150–140 BC
Bremerhaven	Germany	100 BC–500 AD
Hallein, Salzburg	Austria	2000 years
Valkenburg on Rhine	The Netherlands	42–100 AD
Winchester	England	Roman Age
Lindow Man, Manchester	England	2nd century AD
Bobigny	France	2nd century AD
Nahal-Mishmar Valley	Israel	160 AD
York	England	2nd–3rd century AD
Grauballe Man, Silkeborg	Denmark	3rd–4th century AD
Karwinden Man	Prussia	500 AD
Ribe	Denmark	750–800 AD
York	England	9th–11th century AD
Winchester	England	11th century AD
Winchester	England	1100 AD
Acre	Israel	13th century AD
Southampton	England	13th–14th century AD
Utrecht	The Netherlands	13th–14th century AD
Paris	France	4th–15th century AD
York	England	14th–16th century AD
Amsterdam	The Netherlands	1370–1425 AD
Worcester	England	15th century AD
Lübeck	Germany	15th century AD
Oslo	Norway	15th century AD
Schleswig	Germany	Medieval age
Lübeck	Germany	Medieval age
Berlin	Germany	Medieval age
Braunschweig	Germany	Medieval age
Hameln	Germany	Medieval age
Höxter	Germany	Medieval age
Göttingen	Germany	Medieval age

(*Continued*)

Table 17.2. (Continued)

Archaeological site/mummy	Country	Date
Marburg	Germany	Medieval age
Freiburg	Germany	Medieval age
Breisach	Germany	Medieval age
Regensburg	Germany	Medieval age
Landshut	Germany	Medieval age
Montbeliard	France	15th–16th century AD
Marly-le-Roy, Yveline	France	17th–18th century AD
Namur	Belgium	18th century AD

Every context analyzed after the Neolithic Revolution is positive for this parasite. References can be found in Gonçalves et al. (2003) and Leles et al. (2010).

reservoirs of infection. In urban areas, waste was also dumped onto the unpaved streets and was consequently eaten by pigs or other animals that lived in the city or was soaked up into urban mud. This mud was then used as fertilizer (Sterner, 2008).

Medieval agriculture relied on all the waste, and urban areas were dependent on the agricultural production. Additionally, most medieval residences had a vegetable garden or larger green space behind the house, which would be fertilized from the same sources. However, when after AD 1300 population densities began to increase, a crisis in sanitation ensued, as urban areas became overwhelmed with an excess of human and, to a lesser extent, animal wastes. Additionally, medieval innovations in agricultural technology and practice, such as three-field rotation, new crops, windmills, and watermills, produced food surpluses, further exacerbating population pressures. Further, when streets in some urban areas began to be paved in the late medieval period, wastes could no longer be absorbed into the soil, which added to the deterioration of environmental conditions in already stressed urban areas, and contributed to the emergence of infectious diseases, both acute, epidemic conditions and parasitic ones (Gottfried, 1983; Sterner, 2008).

Conditions in urban areas during the medieval period contributed to the mass infection of their populations with maw-worm and whipworm. According to Roberts and Janovy (2000), there are two requirements for these two worms to become a serious health problem: poor standards of sanitation (e.g., feces deposited into the soil) and combinations of physical conditions that allow the worms' survival and development, such as warm climate, high levels of precipitation and humidity, moisture-retaining soil, and dense shade. There are also additional sources for the transmission of maw-worm; cockroaches can carry the eggs and transmit them, and even windborne dust can carry eggs when the conditions permit (Roberts and Janovy, 2000).

The use of human and animal wastes in agricultural fields predisposed whole urban areas to geohelminthic infections through exposure to contaminated agricultural products. For instance, uncooked and unwashed vegetables function as vectors of the eggs. However, washing vegetables in contaminated water is as risky as not washing them at all. There is historical evidence that indicates that the later medieval diet was mostly composed of cereals, especially wheat, with varying amounts of meat and fish being consumed by different social groups. Meat and fish were the main food expenditure in elite households: the aristocracy, the upper clergy, and also wealthy urbanites. Most lower-status households complemented their cereals with boiled pottage and seasonal pulses of fresh vegetables. The main source of protein was milk, cheese, and eggs (Müldner and Richards, 2005).

Based on the geohelminth's life cycle, reinfection was therefore inevitable: since the main food source was vegetables, there was a constant reexposure to parasites. However, because of the high prevalence of parasitic disease during that time, there was likely a great degree of tolerance to the disease. Hosts have defenses against parasites, which can be physiological or immunological responses (Poulin, 2002). The parasite's environment is considered to be the host, and therefore the habitat of that host, such as where the host lives, can influence the parasite's occurrence and abundance. As the host can vary in age, sex, and immune condition, parasites are dependent on those heterogeneities of its *environment* and also with the actual physical, cultural, and built environment outside of the host (Thomas et al., 2002). For instance, some studies show a difference in parasitic abundance and occurrence in relation to physical environmental conditions, such as the humidity of the ecosystem (Fenton and Hudson, 2002; Monello and Gompper, 2009).

Parasites also regulate themselves to survive within their hosts. They can be either pathogenic or nonpathogenic and can show different patterns of adaptations in relation to their respective hosts. They can also manipulate their hosts to increase the transmission rate, increasing susceptibility to predation, for instance. Additionally, they can increase persistence time by resource diversion and resource defense (Brown, 1999). This adaptive dynamic between the intricacies of host and parasite biology creates an opportunity for high-resolution research into human–environment interactions and environmental change in the face of urbanization prior to the industrial revolution (see also Koepke; Zuckerman and Armelagos, both this volume); deeper studies of medieval cities and their conditions should be performed to correlate behavior with infection (Sterner, 2008).

The extreme levels of parasitism evidenced by the archaeoparasitological record for medieval Europe did not go unnoticed by the people represented by that evidence nor by others affected by parasites in antiquity. Parasitism was recognized throughout written history. The first descriptions of parasites in the Old World are from Egyptian medical records dating to between 3000 and 400 BC. Other descriptions of various diseases that may or may not have been caused by parasites, such as fevers recorded by Greek physicians between 800 and 300 BC, are also evident in early medical records. More accurate descriptions of parasitic diseases were recorded in later periods, with accurate data and medical works from Islamic physicians (Cox, 2002). In Europe, maw-worm and other parasitic worms were recognized by Greeks and Romans (Sandison, 1967).

In medieval Europe, medical progress was impeded by religious and superstitious beliefs. Any progress toward recognizing and controlling parasitism that might have been achieved during Roman times was lost in the medieval period, a part of a wholesale loss of classical medical knowledge. Medieval literature on parasitism was therefore limited, but there are references to geohelminthic worms, which in some cases were recognized as causes of disease (Cox, 2002). It is safe to say that medieval communities had no concept of how maw-worm and whipworm originated and spread.

Evidence of the First Epidemiologic Transition in the Americas: Did It Happen?

There is a near-absence of maw-worm in the prehistoric Americas. Whipworms were present, but rare. For many years, archaeoparasitologists in the Americas worked under the impression that this paucity of evidence could be due to sample size or geographical bias in the data sources. Also, the source of data differs between the Americas and Europe, which was thought to have potentially introduced a bias; coprolites and mummies are the main data sources for the Americas as opposed to mummies, privy sediments, and urban sediments for Europe. However, by 2010, archaeoparasitologists became convinced that the data set from the Americas was sufficiently robust: the low presence of geohelminths evident in the archaeoparasitological record was the reality of prehistoric life in the Americas. When data from burials and mummies were compared between the Americas and Europe, the difference is obvious. In mummy and burial contexts, maw-worm and whipworm are nearly ubiquitous in Europe and nearly absent in the Americas. Leles and colleagues (2010) have explored the reasons for this low to nonexistent American infection with maw-worm and whipworm. They suggest that several aspects of prehistoric American life limited opportunities for infection and kept prevalence low: high environmental quality with low amounts of fecal contamination, the presence of anthelminthic therapies, and lower population densities. Diet also made a profound difference.

In contrast to Europe, prehistoric Americans did not totally subsist on cultivated food in all areas of the Americas. Coprolite and dental calculus analyses show that in the Southwest, Mexico, northwest Brazil, and coastal Brazil, for instance, a large proportion of the plant foods consumed, and a majority of the animal protein, came from wild sources (Fry, 1977; Reinhard, 1992a; Boyadjian et al., 2007; Reinhard and Bryant, 2008; Wesolowski et al., 2010). The area that contrasts with this tendency is the Andes, where the majority of plant and animal foods have a domestic origin. In general, the reliance on farms and fecal fertilization seen in Europe was not an aspect of prehistoric American subsistence.

Population density in the Americas, comparatively and generally speaking, was also less than that of Europe, especially in most of North America and in South America east of the Andes (Ubelaker, 1976a,b; Verano and Ubelaker, 1992). Therefore, prehistoric Americans did not suffer the crisis in sanitation that medieval communities experienced.

Prehistoric Americans also had a large pharmacopoeia that included anthelminthic plant-based therapies. Unlike Europeans, who looked for spiritual causation of disease and used ineffectual treatments based on humoralism, the archaeological record shows that Native Americans employed effective treatments for their specific infections, whether hookworms (Chaves and Reinhard, 2006), acanthocephalans (Fugassa et al., 2011), or pinworms (Reinhard et al., 1985).

The data regarding maw-worm and whipworm would suggest that an epidemiologic transition related to filth and fecal contamination did not occur in the Americas. Indeed, only one study hints at the emergence of disease transmitted through a fecal–oral route in the Americas. Santoro et al. (2003) found evidence that whipworm emerged in the Atacama Desert after the establishment of the Inca Empire. But the ubiquity of filth-borne parasites evident in Europe was never the norm in the prehistoric Americas.

However, another *crowd parasite* can be used as an indicator of an evidently unique epidemiologic transition in the Americas. This parasite is *Enterobius vermicularis*, the pinworm. Pinworm eggs are among the most fragile paleopathological structures and are highly susceptible to damage from poor environmental conditions. For this reason, they are found only in conditions highly favorable to preservation, such as within the bodies of mummies or in coprolites. Because of this—and the comparative scarcity of mummies and coprolites in the European archaeological record—they are nearly absent in Europe. The thin-shelled eggs rapidly decompose in other environments, and they are rarely found in archaeologically excavated privies, the source of most European archaeoparasitological data.

Pinworm prevalence is a good proxy gauge of the general level of infectious disease in a given community because the pinworm has evolved multiple routes of infection including anal–oral, hand-to-hand, and airborne routes (Reinhard, 1992b, 2008). Pinworms are particularly interesting among human parasites because the female worm exits the anus at night to disperse eggs, doing so by two different means. Some of the eggs are laid on the perianal folds accompanied by an irritating excretion. The resulting itching and nocturnal scratching then transfer the eggs to the host's fingers. Second, other eggs are distributed by aerosol when the female's body dries outside of the host's anus and bursts. When this happens, thousands of light eggs, adapted for air dispersal, contaminate the environment.

Several modes of infection can result from the female's nocturnal tours. Retroinfection occurs when the eggs hatch and the larvae reenter the body of the host. Hand-to-hand transfer of the eggs occurs when the host interacts with others upon waking and transfers the eggs to novel potential hosts. Autoinfection occurs when humans eat food contaminated with the eggs from their own hands or place their fingers in their mouths. Airborne infection occurs when humans inhale the eggs. Finally, the airborne dissemination of eggs results in the contamination of food and water. Therefore, infection can occur through many routes.

Pinworm is the penultimate crowd parasite. In compact communities, it flourishes. However, very few infected individuals pass eggs in their feces. Since the eggs are largely deposited external to the body, only 5% of infected people are positive for eggs in their feces (Jiménez et al., 2012). This is an important value to remember as the archaeological data are reviewed later.

Reinhard and Bryant (2008) recently completed an analysis of coprolites from 10,000 years of time from multiple sites in arid regions throughout the Americas: in Coahuila and Durango, Mexico; Texas, New Mexico, Arizona, Utah, Colorado, Nevada, and Oregon. The findings demonstrate that in archaic hunter–gatherer populations, pinworm was a rare human parasite and, if found, did not exceed 1% prevalence in coprolites. After horticulture was introduced, the prevalence ranged from 2% to 38% of coprolites (Jiménez et al., 2012; Reinhard, 1992a, 2005). Hugot and colleagues (1999) showed that lower prevalence was evident in small, open sites in the Americas, while higher prevalence was evident in large, multistory stone villages. In particular, the highest prevalence was found in large multistory stone villages built within caves. Therefore, for the arid regions (where coprolites have preserved best) of North America, pinworm was a crowd parasite that was associated with horticulture and large villages and towns, especially when large communities were packed into small cave spaces. Santoro and colleagues (2003) found a similar pattern in the Lluta Valley of northern Chile with the advent of the Inca Empire. When people lived in small, scattered farmsteads, pinworm was absent. When the Inca Empire resettled farmers into large towns, pinworm emerged and became established at a high prevalence.

Therefore, if the first epidemiologic transition is evident at all in the Americas, it is evidenced in pinworm prevalence related to crowding. In contrast, the pattern of filth and associated maw-worm and whipworm infection is essentially absent in the New World in the prehistoric period.

Colonization, the Spread of Filth to the Americas, and the Second Epidemiologic Transition

The absence of filth-borne parasitism in the Americas ended with European colonization. The archaeoparasitological data show that the European-style first epidemiologic transition was imposed upon the Americas soon after the arrival of European colonists in the late 15th and 16th centuries. The second epidemiologic transition, which included control of parasitic disease, followed around the turn of the 20th century.

Case Study: The First and Second Parasitic Epidemiologic Transitions in Albany, New York

Archaeoparasitological analysis of historic sites in the Americas demonstrates these trends. In particular, the record from Albany, New York, demonstrates this best (Fisher et al., 2007). The following synopsis is derived from Fisher et al. (2007). The Dutch West India Company established a colony in Albany in 1614, which was fortified in 1624. In the 1700s, it became the military headquarters for combat expeditions. As a military center, new features were constructed, including a city wall, guardhouses, barracks, hospital, stables, magazines, and storehouses. Soldiers and officers were bivouacked in civilian homes as the population grew. By 1756, Albany had 335 households and 4000 people in a tight 75 acres. The population continued to increase after the American Revolution; the 1790 census reveals a population of 3500 people, growing to over 50,000 by 1850, and 113,000 by 1860.

Despite the increase in population, Albany did not begin to construct and utilize a large-scale sanitary sewer system until the 1850s; it was complete and in widespread use by 1880. Prior to this, sewers were limited to drainage ditches. In 1773, the city was so foul that city ordinance mandated the emptying of tubs or chamber pots of human waste only in the river, not the ditches. The ditches had become clogged with trash and feces, much like European towns and cities. Urban-dwelling cows and pigs added to the foul, muddy streets of Albany. The city's early water systems were equally primitive. At the earliest period in the 17th century, water arrived in the town through a spring. Conduits from the spring were included in a 1698 map of Albany. These conduits were identified and exposed during excavations in the 20th century, which revealed that they were wooden, board-covered troughs. In the 1790s, the city began enclosing the streams that flowed through the city into the Hudson River. Some were channeled in culverts beneath streets, which helped to prevent accumulation of trash in the stream and its ravine. Streams and ravines were used for trash disposal in the early 1700s, as indicated by archaeological work. The culverts helped prevent accumulation of trash in the streams and ravines. Brick-lined and stone-capped drains were built in the late 1700s and early 1800s beneath one of the streets of affluent families.

Archaeological evidence shows that in the 1700s, residents ignored the prohibition regarding dumping sewage anywhere but in the river. Privies were built and maintenance was required to control the spread of filth from them as the surrounding soil became saturated with sewage. An 1829 law restricted the hours when the privies could be cleaned to between 11 at night and 3 in the morning. The business of cleaning out privies continued into the 1870s after privies were connected to the sewers, especially because of problems created when trash was dumped into the privies, which ultimately clogged the sewers. Discarding trash in privies was prohibited by 1872.

In the early 1790s, new methods of providing clean water were created. These included gravity-fed systems of wooden pipes and small reservoirs that channeled water into the city and distributed it to residents. However, this was done to provide water for fire fighting, not for public health. Systems to provide potable drinking water were installed in 1795, when the city built a system of bored pitch pine logs that brought water to residents from springs about 4 miles away. By 1850, the city had constructed additional reservoirs, but even this was insufficient to meet the demands of the expanding population. Instead, by 1875, the city had constructed a pumping station to take water directly from the Hudson River. With regard to parasitism, at this time, the cities of Troy, West Troy, and Cohoes were dumping their waste into the river north of Albany.

Archaeoparasitological evidence of the emergence and control of filth-borne parasites in Albany was based on the analysis of dozens of samples from the city. These included drains, privies, streets, and yards (Table 17.3). Samples from trash and yards dating to the 1600s and 1700s show that whipworm and maw-worm emerged at the time of colonization. A second set of samples came from privies (Table 17.4). The privy samples averaged 36,211 maw-worm eggs and 1,275 whipworm eggs per milliliter (ml) of soil. A milliliter is approximately two-tenths of a teaspoon. From the 1700s to 1900, samples from privies show a general increase followed by a decreased in numbers of eggs per milliliter. This reflects the introduction of fecal-borne parasites, followed by an explosive increase in their prevalence and

Table 17.3. Samples from nonlatrine contexts with the calculated egg concentrations in terms of numbers of eggs per milliliter.

Context	Date (AD)	A. lumbricoides	T. trichiura
Surface refuse near house	1640–1700	60	0
Surface refuse near house	1640–1700	10	1
Surface refuse near house	1640–1700	0	0
Surface sample, house, yard, trash	1650–1670	0	0
Surface, Brickyard	1600s	0	0
Surface, Brickyard	1600s	0	0
Surface Sample House Yard	1650s	37	0
Distillery, Below Vat	1750–1800	0	0
Distillery, Vat Sediment	1750–1800	0	0
Distillery, Vat Sediment	1750–1800	0	0
Street outside the north gate of early stockade wall	1740–1760	12,153	6145
Surface near stockade	1750s	113	0
Surface associated with British Army	1750s	2019	
Dog Burial, Pelvic Area	1760	1065	355
Surface near stockade	1760	2019	621
Surface near stockade	1760s	1249	0
Drain, cleaned	1780s	75	0
Surface along wharf	1780s	219	0
Surface sample	1780	0	0
Surface sample	1780	0	0
Stone culvert	1790s	70	0
Surface sample	1790s	0	0
Stone culvert	1800	0	0
Stone culvert	1805	14,918	1243
Drain	1820s	5	0
Surface sample	1820	155	0
Drain	1830	124	62

The counts show that eggs were common contaminants of the environment in Albany.

burden, concluding with a marked decline associated with the control of parasitism. The evidence for this relates not only to the presence or absence of eggs in *night soil*, levels of sediment from privies that are composed entirely of compressed feces, but also to the numbers of eggs in theses samples. The average concentrations of maw-worm egg per milliliter of night soil from privies were 51,460, 7890, and 3979 eggs for the time periods of 1740–1800, 1805–1850, and 1855–1900, respectively. The average concentration of whipworm eggs per ml of night soil were 2492, 399, and 185 eggs for the time periods of 1740–1800, 1805–1850, and 1855–1900, respectively.

Aside from the night soil, sediments from other contexts also yielded eggs. Overall, the results suggest that maw-worms were present in large numbers in the environment in Albany and had infected the inhabitants of the city since the 1600s. The presence of parasite eggs, especially maw-worms, in the soils surrounding various nonprivy structures from the 1600s and 1700s indicates that waste was discarded on the ground surface or in planting beds

Table 17.4. These data represent parasite egg concentrations from latrines in Albany, New York, associated with laborers.

Date	Maw-worm	Whipworm	Description
1800s	4972	104	From a communal privy probably associated with boarding houses. This privy had a stone drain that carried waste into the street.
1810s	25,199	678	From a boarding house and house owned by Uriah Benedict's widow.
1830s	35,277	797	Iron-banded barrel reused as a privy had a capacity of about 84 gallons.
1830s	2275	300	Boarding house latrine.
1830s	11,311	1119	Boarding house latrine.
1850s	26,281	711	Boarding house latrine.
1850s	60,202	2090	From a privy associated with laborer Thomas Higgins.
1850s	13,851	888	Boarding house latrine.
1860s	448	0	Associated with laborers who resided at the Buffalo House.
1860s	2260	0	Associated with several adjacent properties, housing warehouses, storefronts, and a boarding house.
1860s	228	0	Associated with workers and transients occupying the neighboring American House.
1880s	179	0	Associated with a warehouse owned by the New York Central and Hudson River Railroad most likely used by a number of laborers.

Like the other data, they show a decline of parasitism in the 1860s. This shows that parasitism was controlled in all economic classes of the city.

around buildings during this period. The samples from the earliest sediments, recovered from a brick maker's house dating from approximately 1640 to 1700, evidence the earliest infection. The greatest number of maw-worm eggs per sample relate to those obtained from privies associated with the period of the French and Indian War in the mid-1700s. This result was expected due to the increased population size and density within the walled city during this period of conflict.

The archaeologically documented presence of privies in the late 1700s appears to correlate with the decreasing number of parasite eggs in samples from the ground surfaces and from privies. The increased use of privies was related to increased prohibitions on open air dumping in Albany. The privies appear to have contained the spread of parasite eggs to within the privy shafts. Parasite eggs remained present in large numbers within privy contexts until about the 1830s, when the number of maw-worm eggs declined to less than a quarter of the number present in the 1780s. By 1860, there were few parasite eggs of any kind observed in the samples.

The initial decline in the number of parasite eggs in the 1830s corresponds to a general shift in the method of privy construction. The earlier wooden barrels set in the ground and used throughout the 18th century were replaced with wooden box vaults in the early 19th century, which were augmented later by the use of stone-lined vaults. This trend reflects the effort to control seepage from the privies into the surrounding soil. This finding is key in relation to the second epidemiologic transition; minor technological changes in sanitation systems, even on a household level, and small-scale city ordinances intended to improve environmental quality, produced a major, citywide decline in the prevalence of infectious disease over a span of just a few decades.

Social status also played a key role in defining parasitism. The records available for 17th- to 20th-century Albany link privies to specific households and to the general socioeconomic characteristics of these households. This provides a rare opportunity to evaluate parasitism in relation to socioeconomic status and therefore examine a critical aspect of variability in the mechanics of the second epidemiologic transition; how does status relate to hygiene, environmental quality, sanitary practices, and therefore to infectious disease? For Albany and surrounding communities, we can compare elite households to middle-status households to lower-status communities, specifically working-class households and itinerant workers living in boarding houses. In both, we can see evidence of progress in the process of controlling filth-born disease through two centuries of the city's history as sanitation systems were introduced and developed.

The highest egg concentrations were found in elite households. For example, the DeFreest household included successful tradesmen, merchants, and farmers who had large agricultural holdings outside the city. The concentrations of eggs in the DeFreest privy from 1760s sediments were 62,710 eggs per ml for maw-worm and 2508 eggs per ml for whipworm. The concentrations were even higher for this family from sediments dating to the 1780s with 223,248 eggs of maw-worm and 3763 of whipworm. John Bogart was a sailor and soldier during the Revolutionary War and later became a merchant. His household included his wife, four children, and four slaves. In sediments dating to the 1790s, his privy provided concentration values of 89,675 eggs per ml for maw-worm and 3763 for whipworm. Stewart Dean's privy sediments, which dated to around 1800, were also analyzed. He was a merchant-captain whose household included a number of children and three slaves. Analyses revealed maw-worm concentrations of 310,999 eggs per ml and whipworm concentrations of 12,181 eggs per ml.

Three later households dating from 1890 to 1910 from the nearby town of Troy were also studied and represent an emerging middle class. One was a couple with two sons, which revealed only traces of maw-worm eggs, and one was a couple with two daughters; analysis of sediments from their privy revealed and absence of parasite eggs. The third couple had no children living with them. Only traces of whipworm were found in their privy sediments. The diminishing traces of eggs from these privies indicate that fecal-borne parasites were under control in this community by 1890.

Night soil from three boarding houses was also sampled. One sample from one boarding house location revealed concentrations of 4972 maw-worm eggs and 104 whipworm eggs per ml of sediment. Counts from a similar boarding house dating to the 1830s were 2275 for maw-worm and 300 for whipworm. Another 1830s' boarding house had counts of 11,311 for maw-worm and 119 for whipworm. The lower concentrations of eggs in boarding house privies presumably relates to the fact that they were probably used primarily if not exclusively by adult males, who were likely to be resistant to infection. The higher concentrations of eggs in elite households are likely due to the fact that they were used by children of the house owners, who did not possess such resistance, and their slaves.

The data from working-class households are extensive and are presented in Table 17.5. Data from these households shows a sharp decline in parasite egg concentrations after 1860, similar to that found in the samples from the middle-status households. The data from the elite households shows a sharp difference between the parasitism of households dating to the late 1700s and early 1800s compared to the late 1800s and early 1900s. The exact decade of control is represented in the data from the boarding houses; the control of filth-borne parasites occurred in 1860, around the time that the sewer system was constructed in Albany and was linked to privies. Improved sanitary systems on a citywide level, but tied indirectly to household-level systems, are directly responsible for controlling parasitic infectious disease in this urban environment.

Table 17.5. Egg concentrations from working-class family latrines in Albany, New York.

Date	Maw-worm concentrations	Whipworm concentrations	Shaft construction/context
1810s	25,199	678	Stone/from a boarding house and house owned by Uriah Benedict's widow.
1830s	35,277	797	Iron-banded barrel reused as a privy/The privy was owned by Hugh Bradford, a builder.
1850s	26,281	711	Stone/from a boarding house (?) privy 5.
1850s	60,202	2090	Wood/associated with Thomas Higgins, a laborer.
1850s	13,851	888	Wood/from boarding house (?) privy.
1860s	448	0	Brick and stone/associated with laborers who resided at the Buffalo House.
1860s	2260	0	Wood/associated with several adjacent properties, housing warehouses, storefronts, and a boarding house.
1860s	228	0	Wood/associated with workers and transients occupied the neighboring American House.
1880s	179	0	Wood/associated with a warehouse owned by the New York Central and Hudson River Railroad at that time and, most likely, used by a number of laborers.
1830s	251	0	Wood/associated with a dwelling that had a number of occupants.

The data indicate that parasitism was controlled after 1860.

Conclusion

This archaeoparasitological comparison of Europe and America shows that the first epidemiologic transition from the Paleolithic state to the Neolithic state was quite different in these two settings. The first parasitic epidemiologic transition occurred initially in Europe as filth-borne parasitism was established with the Neolithic lifestyle and became very common by the Iron Age. In the medieval period, infection with fecal-associated parasites was seemingly unavoidable do the fouled conditions of towns and agricultural production and increases in population size and density. In contrast, fecal-borne parasites were certainly introduced into the Americas in prehistory, but a demonstrably more pristine lifestyle prevented these parasites from becoming ubiquitous. Maw-worm is nearly nonexistent, and whipworm is rare in the prehistoric Americas. In most regions, filth-borne parasites were rare, even when crowd parasites such as pinworm proliferated. Therefore, in regard to the concept of epidemiologic transitions, Europe presents evidence for a clear-cut first epidemiologic transition, specifically in regard to parasitic disease. In contrast, the first epidemiologic transition was greatly delayed in the Americas; parasites, and the poor hygiene, poor sanitation, and low environmental quality that their presence indicates, did not arrive until relatively late, and not until the arrival of Europeans and with them, European environments. The data from the Americas, specifically Albany, also demonstrate how, at least in one urban community, the sanitary aspect of the second epidemiologic transition played out. There, filth-borne parasites were ubiquitous in all classes of people until sewage systems were built that included the whole community, from the level of the city down to that of the household, finally initiating their control.

The behavioral factors in the Americas that show long-term control of the most common infectious parasites are not fully known. More research must be devoted to analyzing the pathoecology of Native American sites, with a focus on elucidating the development of anti-helminthic therapies, the establishment of sanitation, the perception of infection causality, agricultural practices that limited pathogen transmission, the role of diet in the spontaneous clearance of helminths, and other as yet unknown topics. Thus, there is a rich future for pathoecology in the New World. Future analyzes will lead to an understanding of the more healthy choices in the Americas that put off the first epidemiologic transition until it was imposed by colonization.

At the same time, more research is also needed into the conditions that spurred the second epidemiologic transition in relation to parasitism in the Americas—as well as in Europe—and in other areas. The case study from Albany demonstrates how parasitism varied over time and in relation to socioeconomic status. Therefore, it gives key insights into how variation in hygiene, the effectiveness of sanitary systems, and environmental quality. At the same time, it also elucidates how, in one urban context, city-level and household-level sanitary systems and citywide legal prohibitions on sanitation brought about the control of parasitism and, by inference, improved hygiene, sanitation, and environmental quality. This study begs for additional archaeoparasitologic work in other contexts and demonstrates the critical potential importance of archaeoparasitology and pathoecology to studies of the second epidemiologic transition.

References

Barrett R, Kuzawa CW, McDade T, Armelagos GJ. 1998. Emerging and re-emerging infectious diseases: the third epidemiologic transition. Annu Rev Anthropol 27:247–271.

Bartošová L, Ditrich O, Beneš J, Frolík J, Musil J. 2011. Paleoparasitological findings in medieval and early modern archaeological deposits from Hradební Street, Chrudim, Czech Republic. Interdiscip Archaeol II(1):27–38.

Boyadjian CHC, Eggers S, Reinhard K. 2007. Dental wash: a problematic method for extracting microfossils from teeth. J Archaeol Sci 34:1622–1628.

Brown SP. 1999. Cooperation and conflict in host-manipulating parasites. Proc R Soc Lond B 266:1899–1904.

Chaves SAM, Reinhard KJ. 2006. Critical analysis of prehistoric evidence of medicinal plant use, Piauí, Brazil. J Palaeogeogr, Palaeoclimatol Palaeoecol 237:110–118.

Cox FEG. 2002. History of human parasitology. Clin Microbiol Rev 15(4):595–612.

Fenton A, Hudson PJ. 2002. Optimal infection strategies: should macroparasites hedge their bets? Oikos 96(1):92–101.

Fernandes A, Ferreira LF, Gonçalves MLC, et al. 2005. Intestinal parasite analysis in organic sediments collected from a 16th century Belgian archaeological site. Cad Saude Publica 21(1):329–332.

Ferreira LF, Reinhard KJ, Araujo A. 2011. Fundamentos da paleoparasitologia. Rio de Janeiro: Editora FIOCRUZ, p 377–404.

Fisher CL, Reinhard KJ, Kirk M, DiVirgilio J. 2007. Privies and parasites: the archaeology of health conditions in Albany, New York. Hist Archaeol 41:172–197.

Fry GF. 1977. Analysis of prehistoric coprolites from Utah. Anthropological Papers 97. Salt Lake City: University of Utah.

Fugassa MH, Reinhard KJ, Johnson KL, Vieira M, Araújo A. 2011. Parasitism of prehistoric humans and companion animals from Antelope Cave Mojave County, Arizona. J Parasitol 97:862–867.

Gonçalves ML, Araújo A, Ferreira LF. 2003. Intestinal parasites in the past: new findings and a review. Mem Inst Oswaldo Cruz 98:103–118.

Gottfried RS. 1983. The black death: Natural and human disaster in medieval Europe. New York City: The Free Press.

Hugot JP, Reinhard KJ, Gardner SL, Morand S. 1999. Human enterobiasis in evolution: origin, specificity and transmission. Parasite 6:201–208.

Jiménez FA, Gardner SL, Araújo A, et al. 2012. Zoonotic and human parasites of inhabitants of Cueva de Los Muertos Chiquitos, Rio Zape Valley, Durango, México. J Parasitol 98(32):304–309.

Jones AKG. 1985. Trichurid ova in archaeological deposits: their value as indicators of ancient feces. In: Fieller NJR, Gilbertson DD, Ralph NGA, editors. Paleobiological investigations: Research design, methods and data analysis. BAR International Series 266. Oxford: British Archaeological Reports. p 105–114.

Leles D, Reinhard KJ, Fugassa M, Ferreira LF, Iniguez AM, Araújo A. 2010. A parasitological paradox: why is ascarid infection so rare in the prehistoric Americas? J Archaeol Sci 37:1510–1520.

Monello RJ, Gompper ME. 2009. Relative importance of demographics, locale, and seasonality underlying louse and flea parasitism of raccoons (*Procyon lotor*). J Parasitol 95(1):56–62.

Müldner G, Richards MP. 2005. Fast or feast: reconstructing diet in later medieval England by stable isotope analysis. J Archaeol Sci 32:39–48.

Poulin R. 2002. Parasite manipulation of host behaviour. In: Lewis EE, Campbell JF, Sukhdeo MVK, editors. The behavioural ecology of parasites. Wallingford/Oxon/New York: CABI Publications. p 243–257.

Reinhard KH. 1992a. Patterns of diet, parasitism, and anemia in prehistoric west North America. In: Stuart-Macadam P, Kent S, editors. Diet, demography, and disease: Changing perspectives on anemia. New York: Aldine de Gruyter. p 219–258.

Reinhard KJ. 1992b. Parasitology as an interpretive tool in archaeology. Am Antiq 57:231–245.

Reinhard KJ. 2008. Parasite pathoecology of Chacoan Great Houses: the healthiest and wormiest ancestral Puebloans. In: Reed PF, editor. Chaco's Northern Prodigies: Salmon, Aztec, and the Ascendancy of the Middle San Juan Region After AD 1100. Salt Lake City: University of Utah Press. p 86–95.

Reinhard KJ, Bryant VM. 2008. Pathoecology and the future of coprolite studies. In: Stodder AWM, editor. Reanalysis and reinterpretation in southwestern bioarchaeology. Tempe: Arizona State University Press. p 199–216.

Reinhard KJ, Ambler JR, McGuffie M. 1985. Diet and parasitism at Dust Devil Cave. Am Antiq 50:819–824.

Roberts LS, Janovy JJ. 2000. Foundations of parasitology. New York City: McGraw-Hill.

Rocha G, Harter-Lailheugue S, Le Bailly M, et al. 2006. Paleoparasitological remains revealed by seven historic contexts from íPlace díArmesî, Namur, Belgium. Mem Inst Oswaldo Cruz 101(S2):43–52.

Sandison AT. 1967. Sir Marc Armand Ruffer (1859–1917) pioneer of palaeopathology. Med Hist 11(2):150–156.

Santoro C, Vinton SD, Reinhard KJ. 2003. Inca expansion and parasitism in the Lluta Valley: preliminary data. Mem Inst Oswaldo Cruz 98(Suppl 1):161–163.

Sterner CS. 2008. Waste and city form: reconsidering the medieval strategy. Journal of Green Building 3(3):69–78.

Thomas F, Brown SP, Sukhdeo M, Renaud F. 2002. Understanding parasite strategies: the need for a state-dependent approach? Trends Parasitol 18(9):387–390.

Ubelaker DH. 1976a. The aboriginal population of America north of Mexico: a new appraisal. Am J Phys Anthropol 44:212–213.

Ubelaker DH. 1976b. Prehistoric New World population size: historical review and current appraisal of North American estimates. Am J Phys Anthropol 45:661–665.

Verano JW, Ubelaker DH. 1992. Disease and demography in the Americas. Washington, DC: Smithsonian Institution Press.

Wesolowski V, de Souza SMM, Reinhard KJ, Ceccantini G. 2010. Evaluating microfossil content of dental calculus from Brazilian Sambaquis. J Archaeol Sci 37:1326–1338.

Part 6
Epilogue

Chapter 18
The Second Epidemiologic Transition, Adaptation, and the Evolutionary Paradigm

George J. Armelagos
Department of Anthropology, Emory University, Atlanta, GA

Introduction

Epidemiologic transition models provide a means to evaluate the evolution of disease and the adaptive response of human populations. This is in contrast to traditional epidemiologic studies of disease, as epidemiologists have been reluctant to incorporate evolutionary models into their methodology. This reluctance creates an interesting distinction between epidemiologic transition studies and epidemiology as a science (Harper and Armelagos, 2010). Epidemiologic studies of disease conceptualize them as individual entities attributed to specific proximate causes, rather than as part of a suite of insults that are due to common, ultimate causes (Turshen, 1977). In contrast, epidemiologic transition models focus on a disease-scape, and broad, population-level trends in health rather than on individual diseases that affect health (see McKeown et al., this volume). The contributions to this volume provide numerous examples of the richness of problems that can be attacked and the variety of methods and perspectives that can be employed in studies of populations undergoing the second epidemiologic transition (SET) when population-level, broad, and evolutionary frameworks are employed.

When Zuckerman framed the 2011 conference and the subsequent volume, she envisioned the theme of an *interdisciplinary approach* to studies of the SET with a few chapters that would use an evolutionary perspective. In keeping with this, my contribution (Zuckerman and Armelagos, this volume) on the hygiene hypothesis, Hallman and Gagnon (this volume), and Perry (this volume) adopt an evolutionary perspective. In their study, Hallman and Gagnon demonstrated that early life exposure to influenza was a risk factor for mortality in subsequent epidemics. Perry employed a large skeletal series to test how trends in infant mortality played into the SET. In our study, Zuckerman and Armelagos examined how the rise of modern, more hygienic environments and consequent reductions in exposure to commensal bacteria and helminthic parasites may have contributed to immune dysregulation and a rise in allergy and autoimmune disorders (i.e., the hygiene hypothesis). Additionally, while many of the other chapters do not make an evolutionary perspective explicit, they feature an adaptive approach and complex interactions that are easily placed into an evolutionary framework.

Epidemiologic transition theory explains major population-level trends in the profile of human diseases and provides insight into their ultimate causes. Importantly, understanding the ultimate cause of various diseases and categories of disease helps to explain their emergence, the risk of human exposure to them, and spread; defining the ultimate cause of a disease or category of disease represents the first step toward containing, controlling, and preventing it (Ezzati and Riboli, 2012). In contrast, epidemiologists are usually concerned with single diseases and identifying their proximate causes. Within epidemiology, the identification of a pathogen or characterization of an emerging infectious disease requires immediate attention to risk factors, the life cycle of the pathogen, the ecological setting that produces the disease interaction, and the vectors that carry the insult, among other issues. For noncommunicable diseases (NCDs), determining the risk factor(s) is the key objective of disease prevention (Smith, 2012). Epidemiologic approaches are very effective for identifying immediate risks and causes, but less so for tackling the large-scale and long-term causes, which tend to be political economic in nature, very deeply rooted in a given context, and often much more intractable (Turshen, 1977; Farmer, 1996).

Though it is certainly important to recognize and describe the particulars of a given pathogen or other disease risks, I argue here that synthesizing such data using epidemiologic transition theory is likely to prove useful to epidemiologists as well. Situating a disease within a particular context using transition theory may provide clues to both possible proximate as well as ultimate causes, as well as prevention strategies and predictions regarding future trends.

This preface sets up the objective of this epilogue; I want to look at the examples of the SET presented by contributors to this volume. The chapters within this volume represent excellent examples of the use of adaptive approaches to examining human health and disease within varied settings, and draw upon a noteworthy array of data sources for illuminating the SET.

The Second Epidemiologic Transition

The SET represents a dramatic shift in the disease profiles of human populations, but is only one of three major changes in human disease trends (Barrett et al., 1998; see Zuckerman and Armalegos, this volume). The first epidemiologic transition (FET) began approximately 10,000 years ago with the Neolithic transition and agricultural intensification, sedentism. It marked an area of higher mortality from acute epidemic infectious disease, such as influenza and smallpox (Armelagos and Harper, 2010). When these infections were brought under control, beginning in the late 1800s in many high-income developed countries, and declined as a source of mortality, degenerative conditions, such as cardiovascular disease and cancer, replaced them as sources of mortality. In these nations, the SET first emerged following industrialization in the 19th and early 20th centuries and occurred alongside improved nutrition, public health practices that improved sanitation, and medical advances such as the germ theory of disease. Unfortunately, however, high levels of social stratification in many of these nations meant that not all of their citizens benefited from these improvements and the beneficial change in mortality trends that they accompanied. The third epidemiologic transition (TET), which is ongoing, involves a rise in mortality from emerging, reemerging, and antibiotic-resistant diseases. While high-income developed countries are experiencing the TET, many low- and middle-income countries (LMICs) are experiencing a simultaneous SET and TET. This can create a syndemic, which consists of a synergistic interaction of two or more coexistent diseases and resultant excess disease load (Singer and Clair, 2003; Singer, 2009), and inflict a double disease burden on a population (Bygbjerg, 2012; Barrett and Armelagos, 2013).

Data Sources and Methods in Studies of the Second Epidemiologic Transition

The contributions to this volume all capture one or more aspects of the SET and the major adaptive changes in human populations and human health that industrialization has elicited. Those aspects are dependent on the type and quality of available data, and all of the contributors extensively discuss the intricacies of data collection and application involved in these analyses. A uniting theme in studies of the SET is the interpretive complications created by the varying degrees of quality, completeness, and bias found in varied lines of evidence for the SET. For example, Beemer (this volume) demonstrates the many problems involved with using vital statistics, namely, death registries, to examine map-out temporal changes in cause of death. Looking at registries from Massachusetts, Beemer found that modifications to cause of death nomenclature can result in the decline of some infectious diseases and rise of others. While his study focused on tuberculosis, this phenomenon—an artifact of improved diagnosis and changing medical terminology—is not unique and should inspire caution in researchers using older death records.

The sources of data used to examine the SET in different contexts are varied. For instance, data from headstones from two Columbia, Missouri, cemeteries dating to before and during the SET provide an excellent comparative example of the effects of the transition, with a focus on small, nonmetropolitan communities (Sattenspiel and Shattuck, this volume). Archaeological skeletal remains presented another data source. Many of these studies demonstrate the great advantages of skeletal data for studies of the SET, particularly that of very large skeletal samples, or aggregates from multiple sites and large databases. Perry's (this volume) analysis of several thousand subadult individuals from preindustrial and industrial London took advantage of the scope provided by a large database of skeletal remains, the Wellcome Osteological Research Data Base, to investigate patterns of fertility and mortality among women, infants, and children that are otherwise obscured in historical materials and smaller skeletal samples. Using the same database to access a large sample size that, like Perry's sample, dates to the preindustrial and industrial period in London, DeWitte examined changes in age patterns of mortality and frailty, risks of mortality associated with physiological stress, at the beginning of the SET. Use of a large sample, and one oriented toward low-socioeconomic-status individuals, allowed DeWitte to examine how health and longevity changed among children, and male and female adults at the onset of industrialization, intensive urbanization, and modernization among those most affected by the heart of the English industrial revolution. Importantly, DeWitte also discusses some of the limitations of her study, which are universal in all studies that employ skeletal data; researchers, especially those from other fields or disciplines, interested in using skeletal data to assess the SET must also attend to these issues. First, traditional age estimation methods for adult skeletons yield interval estimates with a terminal age category of 50+ years, which makes it difficult to examine changes in longevity adult skeletons. DeWitte recommends more widespread application of age estimation methods that yield point-estimates of age, such as transition analysis (Boldsen et al., 2002). In addition to allowing for the estimation of point-estimates of age at all possible adult ages, this method uses statistical methods to avoid the problem of age mimicry associated with traditional methods (age mimicry refers to estimated ages that are biased toward the age distribution of a known-age reference sample) (Bocquet-Appel and Masset, 1982; Boldsen et al., 2002). A second limitation is its focus on lower-status people, which generates questions about the experience of the early stages of the SET by higher-status groups. Future studies might examine whether status exerted a buffering effect against the hazards of urban, modern, industrial living, as other studies (i.e., Guntupalli, this volume) suggest.

Skeletal data also allow investigation of other dynamics that are generally inaccessible in historical materials. The remains of 651 African Americans and Euro-American individuals from the Cobb, Todd, and Hamann–Todd collections, for instance, provided the means to measure the biological impact of urbanization on health among blacks and whites in multiple urban contexts (de la Cova, this volume). In another study, Koepke (this volume) used stature data from 18,500 skeletons from throughout prehistoric and historic Europe (dating from the 8th century BC to the 18th century AD) to reconstruct long-ranging trends in health and nutrition that served as the epidemiologic background to the SET.

Other materials less commonly employed in studies of the SET allowed Millard and colleagues (this volume) and Schell (this volume) to examine the role of toxicants and industrial pollutants in the SET. As is discussed later, industrial pollutants are a key component of industrialization and disproportionate mortality and morbidity from degenerative disease in modern LMICs. Presumably, they played the same role during the 19th- and 20th-century SET. However, they are neglected in studies of the SET as they are difficult to empirically assess in historical material. As I discuss in the following text, Millard and colleagues circumvented this limitation through the use of lead data from human teeth, while Schell drew connections between modern lead exposure, measured through blood lead levels, and pollution experienced by people in the past.

Documents have always been a mainstay of measures of health in the U.S. (Greven, 1972) and England (Wrigley and Schofield, 1989; Wrigley, 1990). For example, Swedlund's (2010) tour-de-force, *Shadows in the Valley: A Cultural History of Illness, Death, and Loss in New England, 1840–1916*, incorporated death registers, diaries, letters, and court records to triangulate a synthetic, interdisciplinary study of health in the Connecticut Valley. In this volume, death records (Beemer), bills of mortality (Anroman), cemetery burial records (Sattenspiel and Shattuck), letters and diaries (Anroman), parish records (Kitson), and National Health Surveys (Guntupalli) are all compiled to assess the diverse causes and impacts of the SET in different settings.

Levels of Variation in the Second Epidemiologic Transition

Contributors also assessed a diverse array of diseases and health conditions that, viewed in aggregate, holistically demonstrate the dynamics of the SET. These include acute, epidemic infectious diseases (Anroman), degenerative conditions, mortality, longevity, and fertility (Perry; Kitson; Sattenspiel and Shattuck; Budnik; DeWitte), growth and stature in living populations (Guntupalli) and archaeological samples (Koepke), specific diseases such as tuberculosis (Beemer) and influenza (Hallman and Gagnon; Orbann et al.), parasitic infections (Reinhard and Pucu de Araújo; Zuckerman and Armelagos), and lead contamination (Millard et al.), and other industrial pollutants (Schell).

The conclusions reached in the various contributions generate a rich source of evidence on the impacts of the SET. For instance, Guntupalli used stature to measure the well-being of Indian women between 1949 and 1974. She found that stature improved *very little* during the period, but that differences in height were related to caste and religious membership, suggesting that the benefits of the SET and modernization are not just moderated by socioeconomic status, as several studies have demonstrated, but also more culturally specific and subtle variables; generalization about the benefits of the SET should be undertaken with caution. Importantly, Guntupalli demonstrates that anthropometric data can be used to inform public policy decisions.

In the last 30 years, researchers have confirmed that stature serves as a proxy for adaptation. Stature has long been used by economic historians (Komlos, 1994) to measure wealth, health, and general well-being. It has also gained traction in anthropology, with many studies using stature series to assess health changes with the FET. For example, in *Paleopathology at the Origins of Agriculture* (Cohen and Armelagos, 1984), a number of contributors found a pattern of decreased stature following the Neolithic transition. Mummert et al. (2011) follow-up study, which analyzed additional skeletal samples from around the globe, found the same trend in many of them. Cohen (1989) found the same pattern of declining stature and, by inference, health, during the process of urbanization. Here, Guntupalli and Koepke are among very few scholars to apply stature series data to the explicit question of health trends during and before the SET. Unlike data from the FET, however, stature data from populations preceding and undergoing the SET, including those from the 20th century (Guntupalli), show more ambiguous patterns of changes in health and nutrition.

In addition to applying more traditional methods and data sets to the SET, other contributions to the volume encourage the use of more novel methodologies to tackle some ongoing questions about the SET. Expanding on Sattenspiel (2011), Orbann and colleagues offer a primer on the use of agent-based modeling for testing patterns of influenza transmission in a remote community in Newfoundland hypothetically undergoing the 1918 Spanish influenza pandemic. Agent-based models explicitly model a process within a group of distinct *individuals*, which can include organisms, households, communities, or other distinct units. In its preliminary stages, their model provides compelling results about the pattern of the virus' rapid spread within the community; more advanced versions should grant tremendous insight into the population dynamics and transmission patterns of epidemic infectious diseases within communities preceding and undergoing the SET.

Other contributors tackle issues of variation within the SET, unpacking and critiquing the notion that transition theory adequately models the SET in all contexts in which it was experienced. Perry notes explicitly that epidemiologic transition theory mistakenly presents a neat package of the changes in morbidity, mortality, and fertility that industrialized groups experienced. Barrett et al. (1998) were aware that the theory often masks cultural differences, socioeconomic factors, and environmental variables. Studies in this volume bear out that critique. For instance, Perry's analysis of skeletal data and vital records demonstrates that infant and maternal health in preindustrial London became compromised shortly *before* the decline in mortality from acute epidemic infectious diseases, which complicates the clear trend predicted by epidemiologic transition theory.

Many studies in the volume also focus on single communities or regions and, in doing so, break up the focus on national-level data in studies of the SET, which obscures state-, regional-, and community-level variation in trends. Several of these studies also take the important step of looking at the SET outside of large urban areas or look at urban–rural differences (Budnik; Kitson; Millard et al.; Sattenspiel and Shattuck; Schell), which have dominated the few studies of the SET that have used subnational-level data. For instance, building on earlier work (Sattenspiel and Stoops, 2010), Sattenspiel and Shattuck look for signals of the SET in mortality and longevity trends in Columbia, Missouri, which was a small community in the 19th century. They found that Columbia experienced the same pattern reported for larger cities, but did so late, most notably in the early 1920s. Studies such as this are especially germane as the SET models population health, but in relation to urbanization, modernization, and industrialization. However, in the 19th and early 20th centuries, the overwhelming majority of populations still lived in rural areas or small towns. Importantly, the methods and data source that Sattenspiel and colleagues use are not unique to this community; they used cemetery gravestone data (augmented by death certificates), which is available for most historical communities, large and small. When historical, or even skeletal, data is scarce, incomplete, or unavailable, headstone data represent a grossly underutilized resource for studies of the SET and for expanding such studies into a greatly widened range of time periods and communities.

Degenerative Disease and the SET

The *age of degenerative and man-made diseases* is the final stage in Omran's (1971) original epidemiologic transition model. *Man-made* toxin exposure is one of the components in this stage. However, as noted earlier, this is a neglected area of study for the 19th- and early 20th-century SET in high-income developed nations. Tooth enamel and lead isotopes provided an avenue for Millard and colleagues to examine anthropogenic lead contamination in British archaeological samples, looking for temporal, socioeconomic, and geographical (urban–rural) differences in exposure to the harmful toxicant. Perhaps unsurprisingly, they found clear evidence of anthropogenic lead exposure in industrial era archaeological samples from in and around London. Lead isotopes, which reveal sourcing, showed that rural populations were relatively buffered from the harmful effects of industrialization, which is keeping with the findings of other studies on the SET, including in this volume (i.e., Budnik, Koepke). Intriguingly, though, high status did not exert the typical, expected buffering effect; higher-status individuals displayed higher concentrations, and lead isotopes revealed that exposure came from ores, rather than air pollution. The interpretation is that higher-status households were more exposed than their poorer counterparts through daily contact with lead artifacts, most likely pewter, ceramics, and other high-cost goods. Fitting well with Millard and colleagues' findings, Schell centered his discussion of anthropogenic toxicant exposure, morbidity, and industrialization around concepts of social inequality and the differential apportionment of risk in a given society. Schell drew his evidence from multiple studies of lead exposure, detected through blood lead levels, in marginalized modern communities, specifically indigenous Mohawk communities from around Albany, New York. This raises a key point; even populations that are not clearly exposed to the hazards of industrialization—the communities in Schell's studies live on rural reservations in New York—can be disproportionately harmed by it if they are marginalized. Impoverished areas, whether rural or urban, serve as convenient dumping sources for industrial waste and can be more contaminated than urban industrial areas. These findings have implications not only for studies of environmental quality and degenerative disease during the SET, but also for public health policy. Schell notes that the Mohawk can avoid exposure in their environments, but that this would require changes in behaviors that are key to Mohawk culture and sovereignty. This creates a paradox where classic public health solutions could place their culture at risk.

The SET in Context

In my other contribution to this volume (Zuckerman and Armelagos), I discuss epidemiologic transition theory, the source of the SET, and the controversy surrounding its origin(s). In what is known as the *McKeown thesis*, McKeown (1979) argued that medical advances played a minimal role in the decline of infectious disease mortality. He showed that morbidity and mortality from diseases such as tuberculosis and diphtheria declined long before effective medical interventions, such as antibiotics and immunizations, were developed and implemented (Figure 18.1). Interestingly, this argument and its supporting evidence have been proven incorrect (Szreter, 2002), and critics have focused on his use of evidence for improved nutrition and failure to consider improvements in public health practices (Kunitz, 1991; Schofield et al., 1991; Johansson, 1992; Johnson et al., 2003). However, his insistence

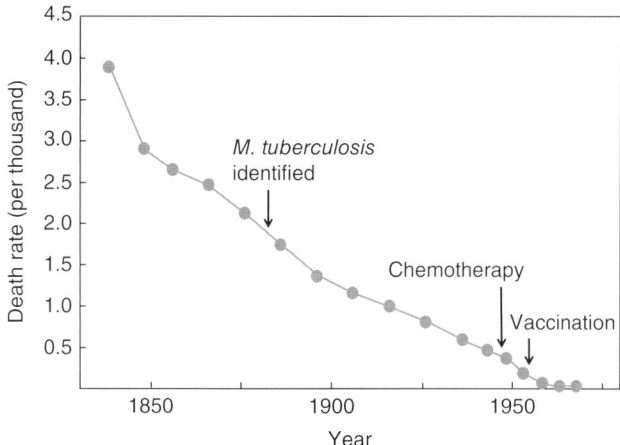

Figure 18.1. Mortality by tuberculosis in England and Wales. (*Source*: E. Gherardi; redrawn from McKeown et al. (1975).)

on the primacy of nutrition and standards of living in improved health continues to enjoy popularity. For instance, Colgrove (2002: 725) states that while McKeown's *empirical foundation* and conclusions were flawed, his thesis that population growth was due to a decline in mortality from infectious disease has remained. He summarizes the McKeown thesis, stating that

> This decline was driven by improved economic conditions that attended the Industrial Revolution, which provided the basis for rising standards of living and, most important, enhanced nutritional status that bolstered resistance to disease. Other variables that may have been operating concurrently—the development of curative medical interventions, institution of sanitary reforms and other public health measures, and a decline in the virulence of infectious organisms—played at most a marginal role in population change. Put another way, the rise in population was due less to human agency in the form of health enhancing measures than to largely invisible economic forces that changed broad social conditions.

As Figure 18.1 demonstrates, there is substance to the argument that medical advances did not cause but instead contributed to already declining mortality and morbidity from infectious disease. Indeed, while advances, such as germ theory, provided for a better understanding of the source of infectious diseases that was essential in instituting measures necessary to control them, other cultural practices and public health interventions preceded them and outweighed them.

Another issue with epidemiologic transition theory specifically comes in relation to the theory of the demographic transition. This theory is a multistage descriptive model of secular declines in total mortality, fertility, and consequent population growth that occurred following the industrial revolution. The first stage is largely preindustrial and is characterized by high-normal and crisis mortality, high fertility, and stationary or slow population growth. The remaining stages involve continued declines in mortality, which started in the mid-1800s in high-income developed nations, and a concomitant rise in population (Thompson, 1929) (Figure 18.2). Omran's (1971) theory of the epidemiologic transition—the SET here—models cause-specific patterns of mortality within the total mortality modeled by the theory of demographic transition. There is a problem with demographic transition theory however, as it assumes that populations during the Paleolithic were at

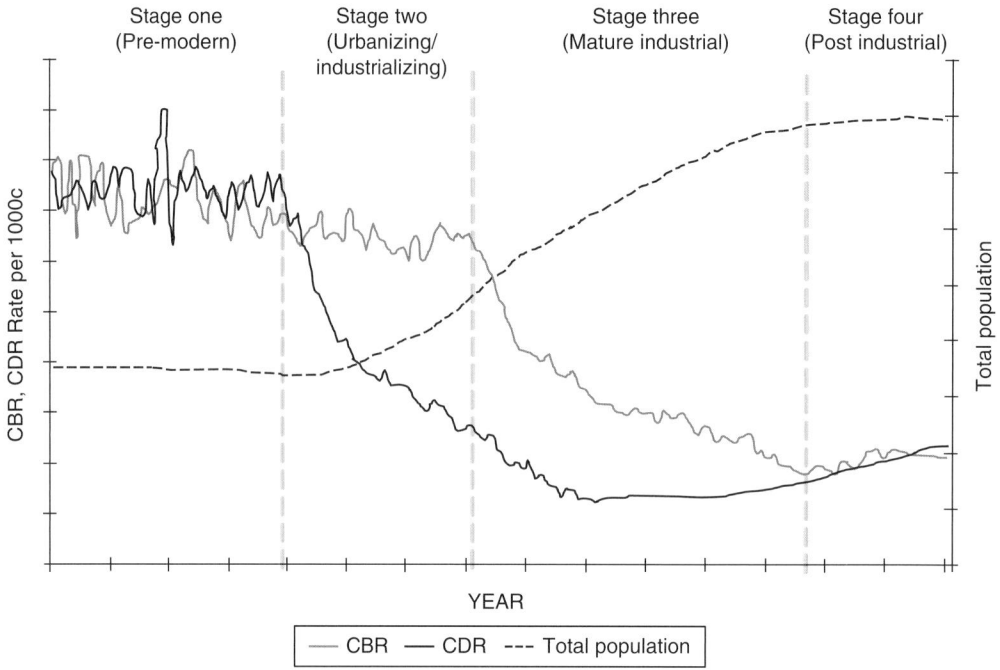

Figure 18.2. The demographic transition model. Stage 1: preindustrial, characterized by high mortality and fertility, stationary to slow population growth; Stage 2: declining normal mortality (noncrisis), rapid increases in population growth; Stage 3: declining mortality, declining fertility (high-income developed nations: mid–late 19th century; LMICs: c.1920); Stage 4: population stability: low and declining mortality, stable to declining fertility, and stable to shrinking population (developed nations); stable to growing population size, low mortality and fertility (LMICs). (*Source*: K. Montgomery.)

maximum fertility. If this was the case, then the increase in population during the FET would not have been possible (Armelagos et al., 1991).

The SET and Novel Environments

One of observations that is apparent from the contributions to this volume is that environmental change creates new opportunities for the emergence of disease. This observation is at the heart of epidemiologic transition theory. In the FET, agriculture fostered a new diseasescape with populations growing in sedentary settings. Consequent environmental contamination, increased population density, and animal domestication fostered new diseases, particularly zoonoses. In the SET, the introduction of public health measures, medical advances, and improved nutrition was responsible for declining infectious disease mortality and rising degenerative disease mortality. In the SET, particularly, major changes in socioeconomic adaptations, as described in this volume, resulted in major changes in patterns of human biological adaptation. There is a distinctive dynamic apparent and echoed in this volume: urbanization (Anroman; Budnik; Koepke; de la Cova), fueled by industrialization (DeWitte; Perry; Kitson; Budnik), creates the emergence or exacerbation of multiple infectious diseases (Anroman), such as tuberculosis (Beemer), influenza (Hallman and Gagnon; Orbann et al.), allergies and autoimmune diseases (Zuckerman and Armelagos), parasites (Reinhard and Pucu de Araújo), and noninfectious pollutants (Schell, Millard et al.). Human growth and development were affected by the SET (Guntupalli; Koepke). Significant factors affected human adaptation and produced changing patterns of mortality and morbidity (Kitson; Budnik; Perry).

Kitson provides a good demonstration of the relationship between disease, mortality, and changing environments. In his contribution, Kitson analyzed demographic developments in a small community, Sedgley, in England between 1700 and 1850 to counter a basic tenant in Omran's original model: that the rapid population growth of the demographic transition occurred due to reduced mortality, which was primarily caused by improved public health and nutritional status. Furthermore, Kitson questioned Omran's treatment of mortality change as an exogenous variable, independent from societies' endogenous factors. Using family reconstitution, he demonstrated that rapid population growth occurred in Sedgley alongside urbanization and industrialization, but that it was also accompanied by increased child mortality. While countering Omran's model, these results follow the pattern suggested by Cohen and Armelagos (1984) and Cohen (1989) in which economic change and declining environmental quality are linked to poor health among children, who are society's most vulnerable members.

Budnik also tackles issues of industrialization, urbanization, environmental change, rural–urban differences, human adaptation, and mortality, with a focus on several large regions of Poland. Importantly, in addition to looking at traditional measures such as crude death rates, life expectancy, and infant death rates, Budnik also used various biological indices to examine opportunities for natural selection operating in the 19th- and 20th-century Poland and detect various populations' ability to buffer themselves and adapt to changing environmental conditions. These measures reveal that rural populations were biologically buffered from the hazards of urbanization and industrialization, in contrast to urban populations, and that natural selection took less of a toll on Polish populations as sanitation, hygiene, and control of infectious diseases improved over time.

In evolution, it is adaptations within the context of novel environments that provide *raw material* for evolution. Human populations can certainly adjust to novel environments, but they will always have competition from pathogens. The winner of the contest of adapting to novel environments? Hands down: the pathogens. This is primarily because of their short generation time, which makes them superbly capable of adapting to changing environments. While human populations have a generation length of approximately 20–30 years, some pathogens have generation times of a few seconds to several hours. Even tuberculosis (*Mycobacterium tuberculosis*), which has a relatively long generation time of 15–20 h, will have reproduced itself more than 12,000 times in the time that it takes an average human to do so.

Where do *we*, as researchers, go from here? For me, the contributions to this volume have forced me to reconsider the complexity of the SET. Barrett and colleagues' (1998) conceptualization of epidemiologic transitions focused on the FET and TET, and these have been subject to most research within anthropology and evolutionary biology. However, Zuckerman and the contributors to this volume force us to transfer our gaze to the SET. We can no longer treat the SET as an afterthought.

Application of the SET

There are a number of unresolved issues with the role of the SET in the human diseasescape. We have eradicated smallpox and rinderpest (cattle plague) (Morens et al., 2011) and are almost free of polio. Rates of malaria and HIV/AIDS are declining overall.

However, as predicted by the theory of the SET, there has been an increase in mortality and morbidity from many degenerative diseases, such as metabolic disease, cancer, and respiratory disease, which account for 60% of all deaths worldwide and 80% of deaths in LMICs (Ash et al., 2012).

Degenerative diseases and NCDs impose a tremendous economic and human cost on modern populations. Globally, NCDs cause more than 35 million deaths annually, with 80% of these deaths occurring in LMICs. While medical interventions have improved the outlook for many, such as those with cardiovascular disease and breast cancer, the resources required for effective treatment are beyond the reach of many in LMICs. Therefore, practices that prevent NCDs must be established (Lutter and Lutter, 2012). Unfortunately, the magnitude of changes necessary to actually change this epidemiologic picture is substantial. According to Ezzati and Riboli (2012: 1484),

> Effective approaches for large-scale NCD prevention include comprehensive tobacco and alcohol control through taxes and regulation of sales and advertising; reducing dietary salt, unhealthy fats, and sugars through regulation and well-designed public education; increasing the consumption of fresh fruits and vegetables, healthy fats, and whole grains by lowering prices and improving availability; and implementing a universal, effective, and equitable primary-care system that reduces NCD risk factors, including cardiometabolic risk factors and infections that are precursors to NCDs, through clinical interventions.

Preventive strategies for NCDs such as that mentioned earlier can be grouped into three levels: primary, secondary, and tertiary. Importantly, findings from the studies of the SET, including those in this volume, can grant key insight into which strategies are likely to be the most effective in reducing morbidity and mortality from degenerative diseases and NCDs. Many of them, such as improving key aspects of nutrition, are very basic, yet have proven to be highly effective over the past two centuries (Tool, 2012).

Primary strategies aim at preventing the development of NCDs; currently, many focus on the developmental origins of health and disease (DOHaD), based on emerging evidence that nutrition and environmental exposures at early stages of development have a critical impact on subsequent risks for NCDs. For instance, basic food and water additives have proven to be extremely effective against many conditions and could be employed globally to cost effectively reduce morbidity and mortality from many conditions. These include adding iodine to salt to prevent goiters and mental retardation (Zimmermann, 2008); fortifying flour, bread, and grains with folic acid to prevent neural tube defects (Jacques et al., 1999); and enriching flour with niacin, a type of B vitamin, to help ward off pellagra. Fluoridating water was a major factor in reducing dental caries and strengthening bone in the U.S. (Ripa, 2008). Additionally, breastfeeding can improve mental abilities in children and greatly reduce later life risk for NCDs in both mothers and children, and represents a lost cost, highly effective primary strategy for improving health. The WHO, for example, recommends that infants be exclusively breastfed for the first 6 months. However, only one-third of infants in LMICs meet this goal, and funding for research and intervention to increase breastfeeding and its duration is limited (Lutter and Lutter, 2012).

Secondary strategies are implemented after an NCD has been diagnosed and are aimed at halting or slowing the progress of the condition. Preventing these events provides the greatest potential for cost-effective reductions in mortality and increased morbidity from NCDs. These strategies include either modification of risk behaviors, such as promoting physical activity and smoking cessation (Houston and Kaufman, 2000), or the use of medications, such as beta-blockers and lipid-lowering drugs. Aspirin, for example, has become an important therapeutic for cardiovascular disease (Cleland et al., 2004), while vaccines are under development for obesity (i.e., *flab jab*) and nicotine addiction (NicVAX) (Trivedi, 2012). However, as with many preventative and therapeutic strategies, patients in many parts of the world, both marginalized and poor individuals in high-income developed countries, and most in LMICs, do not receive the full benefit of these therapies.

Tertiary strategies focus on the management of established, complicated, long-term NCDs and are aimed at preventing further deterioration and maximizing quality of life. They involve strengthening existing health care systems, such as those that provide physical rehabilitation or patient support (Emanuel, 2012). Problematically, health care networks in most nations, both high-income developed and LMICs, suffer from inadequate resources and uneven health care costs. In the U.S., for instance, half of the population underutilizes existing health care resources, while 10% of the population expends nearly two-thirds of all health care costs, largely for the management of NCDs. Problematically, controlling these costs requires effective primary and secondary prevention strategies (Emanuel, 2012).

The burden of NCDs undermines global development efforts (Angell et al., 2012). Effective prevention of NCDs will require both economic and structural changes as well as behavioral changes (Smith, 2012). All of these are likely to be elusive. For instance, overeating, smoking, excessive alcohol consumption, and physical inactivity are known to adversely affect health, but interventions have not been successful. In part, this is because human behavior is cued by environmental stimuli, not filtered by conscious reflection (Ash et al., 2012). Interventions targeting these automatic bases of behaviors have been suggested and await testing of their effectiveness (Marteau et al., 2012).

Conclusion

An Added Burden: Syndemics, the SET, and the TET

The TET involves the emergence of novel infectious diseases and the reemergence of previously controlled infectious diseases, including those that are drug resistant. Barrett et al. (1998) and others have tied this trend to environmental and behavioral changes, namely, environmental degradation from development with increased contact between humans and novel pathogens and vectors; and increased trade, ease of rapid transport, and human mobility. Unprecedented levels of urban development and global climate change are also major drivers of these processes (Luber and McGeehin, 2008). Climate change and increased global temperatures are expected to exacerbate vector-borne diseases in areas where they are currently endemic and establish them in new areas (Morens and Fauci, 2008). Urban living is now the keystone of modern human ecology and plays a crucial role in shaping emerging diseases (McMichael, 2000). The TET demonstrates that while human populations can adapt to the increased biological costs associated with urban living, as they have with the SET, and despite the positive, buffering features of the urban revolution (Dye, 2008), urbanization can still impose a heavy epidemiologic burden. McMichael (2000: 1117), for instance, describes urban centers as the "incubators and gateways for infectious diseases." Indeed, as many developing LMICs that are experiencing the SET and the TET concurrently demonstrate, the positive aspects of urbanization are hampered by the pressures and priorities of globalization. This is especially true for impoverished and marginalized populations within LMICs, who may experience increased morbidity and mortality from NCDs and degenerative disease, while also suffering morbidity and mortality from emerging, reemerging, and drug-resistant infections, such as tuberculosis and HIV/AIDS. In addition to the risks of respiratory and diarrheal disease, according to McMichael, urban poor face the risk of vector-borne infections, pollution, traffic hazards, and urban heat load, which is exacerbated by climate extremes (McMichael, 2000). This fact will be magnified by the fact that the areas of greatest urban growth are occurring in the warmest parts of the world, such as India and Bangladesh. Economic disparity is the primary factor in relation to patterns of disease and urbanization (Patel and Burke, 2009), both between regions and within them; squatters, for instance, or *shadow citizens*, number more than one billion globally and do not reap the health benefits that their richer fellow citizens do (Neuwirth, 2005).

The concurrent SET and TET in LMICs have created the worst of two worlds: increased morbidity and mortality from both NCDs and degenerative diseases and from emerging and reemerging infectious diseases. The unfinished agenda of infectious diseases in many countries combined with aging populations throughout much of the globe, and exacerbated by climate and environmental change, urbanization, and industrialization, will only drive the epidemic of NCDs and infectious conditions (Bygbjerg, 2012). Furthermore, studies of the DOHaD demonstrate that children exposed to infection and malnutrition early in life are more likely to develop NCDs later in life, creating a vicious synergistic dynamic (Bygbjerg, 2012). In a world with limited resources, we tend to intervene with either infectious diseases or NCDs, but the lessons of the SET and TET show that we must intervene with both to improve population health and facilitate human adaptation (Bygbjerg, 2012).

References

Angell SY, Danel I, KM DeCock. 2012. Global indicators and targets for noncommunicable diseases. Science 337(6101):1456–1457.

Armelagos GJ, Harper KN. 2010. Emerging infectious diseases, urbanization and globalization in the time of global warming. In: Cockerham WC, editor. The new Blackwell companion to medical sociology. Hoboken: Wiley Publishing. p 291–311.

Armelagos GJ, Goodman AH, Jacobs K. 1991. The origins of agriculture: population growth during a period of declining health. Popul Environ 13(1):9–22.

Ash C, Kiberstis P, Marshall E, Travis J. 2012. It takes more than an apple a day. Science 337(6101):1466–1467.

Barrett R, Armelagos G. 2013. An Unnatural history of emerging infections. Oxford: Oxford University Press.

Barrett R, Kuzawa C, McDade T, Armelagos G. 1998. Emerging and re-emerging infectious diseases: the third epidemiologic transition. Annu Rev Anthropol 27:247–271.

Bocquet-Appel J-P, Masset C. 1982. Farewell to paleodemography. J Hum Evol 11:321–333.

Boldsen J, Milner G, Konigsberg L, Wood J. 2002. Transition analysis: A new method for estimating age from skeletons. In: Hoppa R, Vaupel J, editors. Paleodemography: Age distributions from skeletal samples. Cambridge: Cambridge University Press. p 73–106.

Bygbjerg IC. 2012. Double burden of noncommunicable and infectious diseases in developing countries. Science 337(6101):1499–1501.

Cleland JG, Findlay I, Jafri S, et al. 2004. The Warfarin/Aspirin Study in Heart failure (WASH): a randomized trial comparing antithrombotic strategies for patients with heart failure. Am Heart J 148(1):157–164.

Cohen MN. 1989. Health and the rise of civilization. New Haven: Yale University Press.

Cohen MN, Armelagos GJ. 1984. Paleopathology at the origins of agriculture. Orlando: Academic Press.

Colgrove J. 2002. The McKeown thesis: a historical controversy and its enduring influence. Am J Public Health 92(5):725–729.

Dye C. 2008. Health and urban living. Science 319:766–769.

Emanuel E. 2012. Prevention and cost control. Science 337(6101):1433.

Ezzati M, Riboli E. 2012. Can noncommunicable diseases be prevented? Lessons from studies of populations and individuals. Science 337(6101):1482–1487.

Farmer P. 1996. Social inequalities and emerging infectious diseases. Emerg Infect Dis 2:259–269.

Greven PJ. 1972. Four generations: Population, land, and family in colonial Andover, Massachusetts, vol. 134. Ithaca: Cornell University Press.

Harper KN, Armelagos GJ. 2010. The changing disease-scape in the third epidemiological transition. Int J Environ Res Public Health 7(2):675–697.

Houston T, Kaufman NJ. 2000. Tobacco control in the 21st century. JAMA 284(6):752–753.

Jacques PF, Selhub J, Bostom AG, Wilson PW, Rosenberg IH. 1999. The effect of folic acid fortification on plasma folate and total homocysteine concentrations. N Engl J Med 340(19):1449–1454.

Johansson SR. 1992. Measuring the cultural inflation of morbidity during the decline in mortality. Health Transit Rev 2(1):78–89.

Johnson RJ, Hurtado A, Merszei J, Rodriguez-Iturbe B, Feng L. 2003. Hypothesis: dysregulation of immunologic balance resulting from hygiene and socioeconomic factors may influence the epidemiology and cause of glomerulonephritis worldwide. Am J Kidney Dis 42(3):575–581.

Komlos J. 1994. Stature, living standards, and economic development: Essays in anthropometric history. Chicago: University of Chicago Press.

Kunitz SJ. 1991. The personal physician and the decline of mortality. In: Schofield R, Reher D, Bideau A, editors. The decline of mortality in Europe. Oxford: Clarendon Press. p 248–262.

Luber G, McGeehin M. 2008. Climate change and extreme heat events. Am J Prev Med 35(5):429–435.

Lutter CK, Lutter R. 2012. Fetal and early childhood undernutrition, mortality, and lifelong health. Science 337(6101):1495–1499.

Marteau TM, Hollands GJ, Fletcher PC. 2012. Changing human behavior to prevent disease: the importance of targeting automatic processes. Science 337(6101):1492–1495.

McKeown T. 1979. The role of medicine: Dream, mirage or nemesis. Princeton: Princeton University Press.

McKeown T, Record R, Turner R. 1975. An interpretation of the decline of mortality in England and Wales during the twentieth century. Popul Stud 29:391–422.

McMichael AJ. 2000. The urban environment and health in a world of increasing globalization: issues for developing countries. Bull World Health Org 78(9):1117–1126.

Morens DM, Fauci AS. 2008. Dengue and hemorrhagic fever: a potential threat to public health in the United States. JAMA 299(2):214–216.

Morens DM, Holmes EC, Davis AS, Taubenberger JK. 2011. Global rinderpest eradication: lessons learned and why humans should celebrate too. J Infect Dis 204(4):502–505.

Mummert A, Esche E, Robinson J, Armelagos GJ. 2011. Stature and robusticity during the agricultural transition: evidence from the bioarchaeological record. Econ Hum Biol 9(3):284–301.

Neuwirth R. 2005. Shadow cities: A billion squatters, a new urban world. New York: Routledge.

Omran A. 1971. The epidemiologic transition. Milbank Mem Fund Q 49:509–538.

Patel RB, Burke TF. 2009. Urbanization—an emerging humanitarian disaster. N Engl J Med 361(8):741–743.

Ripa LW. 2008. A half-century of community water fluoridation in the United States: review and commentary. J Public Health Dent 53(1):17–44.

Sattenspiel L. 2011 Regional patterns of mortality during the 1918 influenza pandemic in Newfoundland. Vaccine 29S:B33–B37.

Sattenspiel L, Stoops M. 2010. Gleaning signals about the past from cemetery data. Am J Phys Anthropol 42:7–21.

Schofield R, Reher D, Bideau A. 1991. The decline of mortality in Europe. Oxford/New York: Clarendon/Oxford University Press.

Singer M. 2009. Introduction to syndemics: A critical systems approach to public and community Health. San Francisco: Jossey-Bass.

Singer M, Clair S. 2003. Syndemics and public health: reconceptualizing disease in bio-social context. Med Anthropol Q 17(4):423–441.

Smith R. 2012. Why a macroeconomic perspective is critical to the prevention of noncommunicable disease. Science 337(6101):1501–1503.

Swedlund AC. 2010. Shadows in the valley: A cultural history of illness, death, and loss in New England, 1840–1916. Amherst: University of Massachusetts Press.

Szreter S. 2002. Rethinking McKeown: the relationship between public health and social change. Am J Public Health 92(5).722–725.

Thompson W. 1929. Population. Am J Sociol 34:959–975.
Tool LE. 2012. Public health measures in disease prevention. Science 337(6101):1468–1470.
Trivedi B. 2012. Chronic disease vaccines need shot in the arm. Science 337(6101):1479–1481.
Turshen M. 1977. The political ecology of disease. Rev Radic Pol Econ 9:45–60.
Wrigley EA. 1990. Continuity, chance and change: The character of the industrial revolution in England. Cambridge: Cambridge University Press.
Wrigley EA, Schofield RS. 1989. The population history of England 1541–1871. Cambridge: Cambridge University Press.
Zimmermann MB. 2008. Research on iodine deficiency and goiter in the 19th and early 20th centuries. J Nutr 138(11):2060–2063.

Chapter 19

The Second Epidemiologic Transition from an Epidemiologist's Perspective

Nancy L. Fleischer and Robert E. McKeown
Department of Epidemiology and Biostatistics, Arnold School of Public Health, University of South Carolina, Columbia, SC

Introduction

In this chapter, we offer an epidemiologist's perspective on the second epidemiologic transition. We review the classic theory of epidemiologic transitions and its shortcomings and discuss how the evolution of epidemiology as a discipline has veered both toward—in social epidemiology—and away from—within molecular and genetic epidemiology—embracing an overarching theory of population health. What has eluded the discipline thus far is practical implementation of a comprehensive, unifying model with both theory and methods to integrate and interpret the vast range of data now available to us. The socio-ecological model (see IOM, 2003) theoretically unites these data, but we believe full implementation is yet to be realized.

Epidemiologic transition theory is rarely emphasized in modern epidemiology (Harper and Armelagos, 2010) or epidemiologic training. However, the classic theory and its extensions can help broaden the epidemiologist's understanding of the ever-complex, multiple dimensions of health and disease over time and, in doing so, contribute toward improving population health. As we highlight here, the perspectives expressed in this volume can move us toward realizing that goal.

Epidemiology: An Evolving Discipline

Epidemiologists sometimes trace the origins of their discipline to the Hippocratic School of ancient Greece (Winkelstein and French, 1972), though there is better justification for seeing our roots in the sanitary movement of the 18th century and the emergence of more refined approaches—such as Snow's investigations of cholera—in the mid-19th century (Johnson, 2006). Susser (1985), however, has cogently argued that epidemiology really emerged as a discipline only in the latter half of the 20th century with the publication of textbooks and the development of refined methods.

Because this volume focuses on novel and creative approaches to studying the second transition, from an epidemiologist's perspective, it is interesting to note transitions within the discipline of epidemiology itself, including its definition, scope, and methods. Though *epidemic* is much older, the modern use of *epidemiology* is typically traced to the 19th century and the London Epidemiological Society. More specifically, John Snow's seminal work on the cholera epidemics in London in the 1850s demonstrates how studying the distribution of events often sheds light on potential determinants. However, Snow's work was much more than just putting dots on maps; his theory was grounded in a range of data and approaches, underscoring that even in the early days of modern epidemiology, a narrow, single-minded focus on one type of data or one method was insufficient for uncovering the underlying causes of the cholera epidemic.

One fascinating indicator of changes in the uses of *epidemiology* and *epidemiologist* is a series of letters to the editors of *The American Journal of Public Health* in 1942 (Editors, 1942) in response to the editors' request for letters "On the subject of what and who, and even why, is an epidemiologist." The published letters reflect a very different discipline from the one we now practice. Wheeler's (1942: 759) insightful, if somewhat whimsical, observation is pertinent to the topic of this essay:

> The rate, commonly used to express...risk, is not a bad shibboleth by which to test the genuineness of one who calls himself an epidemiologist. If he quibbles mightily over the numerator and grabs up anything that offers for a denominator, he is really a clinician at heart. If he accepts anything that offers—be it only reported cases—for the numerator and bickers endlessly over the denominator he is probably a biostatistician. If he frets indefinitely over both numerator and denominator he is more likely to be an epidemiologist.

It is lack of attention to denominators that sometimes undermines the conclusions of those who seek to describe epidemiologic transitions. In several letters, even in the mid-20th century, the focus was primarily, though not exclusively, on infectious diseases and outbreaks, perhaps a reflection of being on the cusp of a transition. Over the past 70 years, much has changed about the discipline: the broadening of subject matter beyond infections and outbreaks, much greater diversity in background and training with less emphasis on the MD and more on the PhD, and increasingly refined study designs and quantitative methods.

Modern epidemiology is the study of the distribution and determinants of health-related states or events in human populations (Porta, 2008); some definitions also emphasize the application of this knowledge to improving health, preventing disease, or informing policy. Epidemiology's scope has expanded to include not only so-called chronic conditions or noncommunicable diseases (NCDs), but also intentional and unintentional injury, mental disorders, well-being, and health-related behaviors and practices from lifestyle to medical interventions and health policy.

In addition to the transition in definition, epidemiology has undergone a methodological transition. The work of Snow, Koch, Pasteur, and others led to the eventual dominance of the germ theory of disease, which in turn led to epidemiology's focus on specific agents of disease: epidemiology reduced to laboratory science. A part of the development Susser (1985) recounts in the latter half of the 20th century had to do with the development of new methods and study designs to focus attention on the increasing threat from NCDs, especially cancer and cardiovascular disease (CVD).

The Epidemiologic Triangle and the Causal Web

Because of the dominance of infectious disease investigations prior to the mid-20th century, epidemiologists developed a simple triangle of factors as a heuristic for investigation. The *epidemiologic triad* (or triangle) (Rockett, 1999; Krieger, 2001a) consisted of characterizing the presumptive causal (typically infectious) agent, the susceptible host, and the environment that contributed to the development or maintenance of the agent, the susceptibility of the host, or likelihood of contact between the two. For vector-borne diseases, a fourth component was added for the vector. This model was quite effective for identifying causal agents, transmission routes, factors that contributed to or impeded transmission, and host characteristics that increased or decreased their susceptibility to infection. Though drawn as a triangle, it was clear that the agent, the host, and their relationship all existed within an environmental context that affected and could be affected by each of them or their interaction (see Anroman, this volume).

It became apparent, however, that even with infectious diseases, causal pathways and associations are typically more complex than this model allowed. As epidemiology began to focus more on NCDs, especially CVD and cancers, a causal web model emerged (MacMahon et al., 1960; Stallones, 1966), which could more completely depict the complex interrelationships and pathways among the multiple factors that interact to increase (or decrease) disease risk. The model was the basis for *risk-factor* epidemiologic investigations of CVD and cancer in the latter half of the 20th century, focusing on specific attributes, characteristics, or behaviors of individuals (Krieger, 1994, 2001a).

The Socio-ecological Model

Even this more complex model, however, failed to capture the multidimensional nature of factors that affect health and disease. In an influential article, Krieger (1994) suggested that in addition to researching relationships within the web, epidemiologists need to look for *the spider*—those larger factors and contexts that influence or create the web itself. This approach, represented by the socio-ecological model for population health (Krieger, 2001a; IOM, 2003), assumes that a broad spectrum of systems and interrelated determinants of health exist, which may act synergistically or antagonistically. It expands the domain of epidemiologic studies vertically, both *upward* to consider broader biologic, mental, behavioral, social, and environmental systems and factors (e.g., policy, culture, and economic environments) and *downward* to the genetic and molecular levels. It also extends it horizontally to consider trends over time. These trends encompass not only the common developmental issues of lifespan epidemiology or changes over the course of longitudinal studies, but also evolving associations among the various levels over time.

A Broader Range of Outcomes Considered

Transitions in epidemiology itself are reflected in both methodological development and the expanding list of outcomes of major concern. The sanitary movement, with its rudimentary comparisons of mortality in association with living conditions, gave way to epidemic investigations (most notably cholera), ushering in the dominance of the germ theory. With greater focus on NCDs in the 20th century and refinement of methods for clinical trials and observational studies, we entered another phase in the development of epidemiology. There is some concern that the incorporation of genomic and other *-omic* data and the focus on molecular epidemiology could portend a return to the near single-minded focus of germ theory. The current interest in *precision medicine* (NRC, 2011) as a means of more a specific molecular nosology of disease may give some credence to that concern, though the National Research Council report calls for integrating broader levels of data. However, the increasing application of epidemiologic methods to a broader range of health-states and events, including injury (both intentional and unintentional), mental disorder, obesity, and well-being, and the increasing realization that health and disease are complex entities with multileveled, multidimensional determinants (Krieger 1994; 2001a; Susser and Susser, 1996a,b; IOM, 2003) suggest that the field is not yet at the point of relinquishing population methods for gene sequencers and other analytic machines.

Social Epidemiology and Complex Systems Approaches

The socio-ecological model has led to increased emphasis on contextual analysis (Diez-Roux, 1998) and, more recently, complex systems thinking (Diez-Roux, 2007, 2011). The growing influence of social epidemiology has been central to these developments (Berkman and Kawachi, 2000; Kawachi and Berkman, 2003; Krieger, 2001a,b; Oakes and Kaufman, 2006). At the same time, epidemiology has also witnessed a greater focus on genetic and epigenetic epidemiology. The goal is an approach that recognizes and encompasses the multidimensional aspects of health, both *vertically* and *horizontally*. In the process, we have come to recognize that health outcomes are multidimensional as well, with impacts not only on organ systems and whole-body functioning, but on psychological well-being, development, family and friend relations, academic or occupational performance, and social interactions.

The concept of epidemiologic transitions provides the breadth of perspective and long-term views that were absent in early models used in epidemiologic investigations of risk factors. The increasing reliance on socio-ecological models now opens up the range of possibilities to include social, cultural, policy, and economic factors (Harper and Armelagos, 2010). The chapter by Zuckerman and Armelagos (this volume) represents this approach from the perspective of anthropology, but with a clear message for epidemiology. It demonstrates that we have much to learn from working in concert with anthropologists who have such profound understanding of the larger scope of epidemiologic research. Of particular interest among social epidemiologists are the ways that these upstream, sometimes global determinants differentially impact vulnerable populations, either because of factors such as race, social standing, and poor education, or from other disadvantages (see Schell, this volume). As Harper and Armelagos (2010: 676) note, "Situating a disease within a particular context using the transition model may provide clues to possible proximate as well as ultimate causes, prevention strategies, and predictions regarding future trends." The result is that epidemiologists have begun to realize that while understanding health-states and events, including disease entities, may require very precise definitions for some types of study, investigations of patterns of health-states may yield more insight into more general and further upstream common determinants and thus potentially have a much larger impact on population health.

The Theory of Epidemiologic Transitions

A Recap of Classic Epidemiologic Transition Theory

Epidemiologic transition theory consists of two interrelated components. The first includes changes in population dynamics and composition, especially age distribution and fertility rates, with shifts from younger to older population distributions. The second involves changes in mortality patterns, encompassing overall death rates, shifts in the relative contributions of major categories of causes of death, declining maternal mortality, and overall increasing life expectancy. In Omran's (1971) classic formulation, the first transition, dating to prehistory and termed the "Age of Pestilence and Famine," was characterized by high and fluctuating mortality rates, variable life expectancy with low average lifespan, and periods of unsustainable population growth. Morbidity and mortality from infectious agents were high, as were infant and maternal mortality, and fertility rates. The second transition, the "Age of Receding Pandemics," involved decreasing mortality and morbidity from infectious agents and a concomitant rise in the NCD burden. The age distribution of the population shifted toward older ages as mortality rates decreased, especially at young ages, and fertility declined; this volume focuses largely on this second component and the relative importance of causes of death. Omran referred to the third transition as the "Age of Degenerative and Man-Made [sic] Diseases." In the classic theory, this involved further increases in life expectancy and declines in fertility, which became more important to population growth as so-called degenerative and anthropogenic diseases largely (but not completely (see Hallman and Gagnon; Orbann et al.; both this volume)) replaced epidemics as major sources of mortality.

Some Proposed Modifications to Transition Theory

Among the early modifications to Omran's theory was Olshansky and Ault's (1986) proposal of a fourth transition, the "Age of Delayed Degenerative Diseases." This involves a continued decline in mortality rates, but with a shift in NCDs to the elderly even as survival increased among them. Olshansky et al. (1998) further proposed a fifth stage to account for the profound impact of HIV/AIDS and other emerging infectious diseases. Graziano (2010) has proposed a different fifth stage, an "Age of Obesity and Inactivity." Graziano suggests that rapid increases in overweight and obesity over the past two decades pose a threat to the supposed pattern of NCDs in ever-older populations described in Olshansky and Ault's fourth stage.

Omran (1971) also attempted to account for deviations from his overall theory by characterizing three models for how the transitions progress. The *Classical* model featured slow progression through various stages from high mortality and fertility to much lower rates of both. According to Omran, this was the experience in Western Europe, with economic development and accompanying advances in sanitation, public health, and medical knowledge being major drivers. Japan provided an example of the *Accelerated*, model which involved the same phases as the Classical model, but with more rapid progression in the trends and transitions of anthropogenic diseases, though many of the same factors were purported to be at work as in the Classical model. The third *Contemporary* model describes the transition in low- and middle-income countries (LMICs) where public health interventions and international aid have contributed to declines in mortality, but fertility rates remain high and the dominance of NCDs had not yet emerged.

Omran (1977) also adapted his theory with a particular focus on the U.S. He hypothesized that the strongest impact of the decline in infectious diseases and resulting reductions in overall mortality and relatively greater role for NCDs would be seen among children more than adults, among women more than men as a result of the declining fertility and decreasing maternal mortality rates, and among whites more than blacks.

Critiques of Epidemiologic Transition Theory

From an epidemiologic perspective, classic transition theory (i.e., Omran, 1971) falls short in a number of areas. Proposed explanations tend to be speculative, lacking reliable data or relying on small samples or data that is unlikely to be generalizable, though some of the chapters in this volume are excellent counterexamples. A number of critics have noted insufficient emphasis on social factors, such as poverty and income inequality, and differences among demographic subgroups, such as by race, sex, or location (Wilkinson, 1994; Gaylin and Kates, 1997; Smallman-Raynor and Phillips, 1999; Pearson, 2003; Armelagos et al., 2005; Sanders et al., 2008; McKeown, 2009); weakness in accounting for emerging and reemerging infectious diseases, particularly HIV/AIDS and treatment-resistant strains of TB and staphylococcus (Gaylin and Kates, 1997; Smallman-Raynor and Phillips, 1999; Armelagos et al., 2005; Sanders et al., 2008; McKeown, 2009); the frequent failure to differentiate between the *proportion* of deaths due to a specific cause or group of causes from the *risk* of dying overall or from some specific cause or class of causes (Pearson, 2003; McKeown, 2009); the oversimple division of noncommunicable and infectious diseases, which ignores the infectious etiology of many chronic conditions; and the relative absence of consideration of morbidity, disability, and quality of life by virtue of the narrow focus on mortality (McKeown, 2009). The latter omission is more striking in light of health researchers and organizations increasingly relying on broader measures of life expectancy, such as disability-adjusted life years (DALYs) (WHO, 2008; 2009; Murray et al., 2013). We address several of these shortcomings further later. For now, the argument here is that the determinants of transitions in patterns of morbidity and mortality, like the etiology of disease, disability, and death, are multidimensional, and so-called upstream social, cultural, and global transitions play a prominent role. This is something that Omran's original theory implicitly recognized but inadequately explored.

Among the more salient critiques of Omran's theory have been those of Caldwell (2001), Mackenbach (1994), and Caselli et al. (2002). Caldwell disputed Omran's focus on mortality rather than morbidity. He also critiqued Omran's emphasis on the role of *ecobiological and socioeconomic factors* in reducing mortality in Western Europe and comparative downplay of public health interventions, medical advances, global economic trends, and improved living conditions. The medical advances cited are directly related to prevention and control of infectious diseases—one could argue in large part because of the victory of the germ theory over miasma, though Caldwell does not—including pasteurization, antiseptics, better sanitation, improved hygiene, and enhanced midwifery training.

Caldwell (2001: 160) also critiqued Omran's three variant transition models because of their failure to recognize "the global nature and historical sequence of the mortality transition as it spread." Instead, Caldwell asserted that each society manifests its own peculiar model; many of the chapters in this volume explore this argument. Mackenbach (1994) also criticized the *Western model* as vague and difficult to verify, especially the Age of Pestilence and Famine, though anthropological research could provide evidence to explore this. He did, however, recognize the theory's utility for examining variation in patterns over time and among countries, as well as for understanding historical patterns and projecting future trends.

Caselli et al. (2002) pointed to deviations from the theoretical model in the modern era. These included dramatic changes in morbidity and mortality after the fall of the Soviet Union, the devastating impact of HIV/AIDS in sub-Saharan Africa, and the greater-than-expected increases in life expectancy in Western countries. Consistent with Caldwell's contention of the specificity of the model to each society, Caselli et al. emphasized the central role of the distinct constellation of historical, cultural, and economic development in the trajectory of the epidemiologic transition in each country.

Challenges for Epidemiology and for Transition Theory

The chapters in this volume demonstrate how focused inquiries on specific conditions can shed light on epidemiologic transition theory. In particular, they demonstrate how diverse approaches and bodies of evidence can be used to respond to and partially remedy these critiques and debates, such as those regarding which specific practices and phenomena provoked the transition and resulted in declining mortality and morbidity from infectious disease (see Reinhard and Pucu de Araújo, this volume). They can also reveal how these varied between different nations as well as within them (see Orbann et al., this volume), as well as how they varied for different segments of populations, such as whites and blacks (see de la Cova, this volume), or women (see Guntupalli, this volume; de Witte this volume) and children (see Perry, this volume; Kitson, this volume). The challenge now is to demonstrate how transition theory thus illuminated can shed light on current public health challenges.

Another part of the challenge is to broaden the frame of reference for epidemiologists. This is most likely to happen working in multidisciplinary teams with joint expertise. Such teams are much more likely to develop innovative approaches to questions and effective applications of findings (Mabry et al., 2008; Sulsky et al., 2012; McKeown, 2013a,b). There are now trends in epidemiology that give us hope that such expansion will be the case, and this volume provides an example of the frame-expansion needed for epidemiologic questions.

The Role of the Epidemiologic Transition in the Education of Epidemiologists

One might argue that epidemiology was transformed during the 20th century by a perception of the second epidemiologic transition, without necessarily relying on the development of that theory by Omran (1971). Ironically, Omran's theory *per se* has not played as large a role in epidemiologic education as one might have expected, in the U.S. at least. Students are taught about changing demographics and changes in the cause of death rankings, but less attention is given to Omran's theory and specific arguments. If one examines citations of his major publications, one sees extensive citation across disciplines, but not the kind of attention from the epidemiologic literature that one might expect for a theory with such wide-ranging influence, not to mention with *epidemiology* in its name.

Possibilities for Epidemiology and Transition Theory

Though epidemiologic transition theory has not generally been taught as such in epidemiology programs in the U.S., epidemiology students are introduced almost immediately to concepts of population dynamics and shifts in major causes of morbidity and mortality, though not with anything approaching the historical reach of Omran's theory or the papers in this collection. A reworked transition theory, such as the one proposed in this volume, could contribute to the efforts of epidemiologists to understand the larger-scale determinants of population health, particularly in regard to challenges such as obesity. For example, Allison and colleagues (Keith et al., 2006) are investigating plausible explanations for the increasing rates of obesity beyond sedentary lifestyles and altered diet. They propose several larger-scale environmental or behavioral factors for consideration, such as endocrine disruptors, sleep deficits, temperature control, decreased smoking, and factors relevant to the epidemiologic transition, including increasing age at first birth, and intergenerational and intrauterine factors. Only a few of these, however, would explain the evidence of increasing weight trends in several animal populations—domestic, laboratory, and feral. They are seeking to understand common contributing factors for both animals and humans (Klimentidis et al., 2011).

Key Differences between Epidemiology's Perspective and Classic Transition Theory

Key differences exist between transition theory and epidemiology. Transition theory looks at large-scale patterns—mortality rates, life expectancy, fertility rates, age distributions, and broad social and cultural contributors to changing patterns—while epidemiology focuses more on specifics—specific diseases, health-states or events, and causal factors. Education in epidemiology attends to mortality trends, including trends in broad categories of causes of death, and possible contributors to these changes. As some of the authors in this volume have noted, however, the period of study tends to be limited to the modern era, due to the recent advent of vital records data in Western Europe and the U.S. So even these efforts rarely exhibit the scope of epidemiologic transition theory. Thus, although some epidemiology students may not have heard of epidemiologic transition theory as such, they will be familiar with many of its components. Perhaps the reason is that epidemiologic training tends to focus on addressing current challenges to public health. There may not be much confidence in what distant historical patterns and past social transitions reveal about current patterns. However, this is a perspective that we and other scholars (Harper and Armelagos, 2010) believe to be a shortsighted. Certainly the little attention to transition theory can also be attributed to epidemiology training's emphasis on specifically defined outcomes and, since the rise of germ theory and the subsequent emphasis on risk-factor epidemiology, on rather specific infectious agents, and other defined proximate exposures or risk factors (vs. the sanitary movement's emphasis on living conditions). With the rise of social epidemiology, we see, once again, due regard for the larger context in which people live and work, including both physical environments and social, cultural, economic, and political domains.

The Theory of Epidemiologic Transitions and Modern Epidemiological Theory and Practice

Epidemiologic Transition Theory as Descriptive Epidemiology

Epidemiologic studies are typically of two general types: descriptive studies and etiologic studies. The former represent a traditional approach. They characterize the distributions or patterns of health-states and events over time or place or in terms of personal characteristics. Etiologic studies focus on understanding determinants, the causes or origins of health-states or events, and therefore focus on very specific health-related outcomes. One might argue that epidemiologic transition theory is an instance of descriptive epidemiology writ large by virtue of examining very long-term secular trends over large geographic areas, with attention to major cultural shifts. After all, descriptive studies do not generally focus on etiology; they are undertaken without the rigorously and specifically defined outcomes of etiologic studies, and many focus on mortality trends. In that regard, the broad outcome categories of epidemiologic transition theory are similar. Further, though classic epidemiologic transition theory (i.e., Omran, 1971) proposed broad, general explanations for the observed changes, the proposals were not generally models of rigorous causal inference but instead tended to be more speculative. This volume demonstrates that more rigorous evidence is being sought and is being used to refine epidemiologic transition theory. It is an indication of the maturation of the theory, which is likely to make it more attractive for consideration by other disciplines, epidemiology included. Millard and colleagues (this volume) represents an example of one such rigorous approach.

Rethinking the Infectious–Chronic Disease Distinction

In addition to a broadening scope and methods in epidemiology—trends to which an enhanced transition theory could contribute—there have also been advances in understanding the nature and etiology of many supposed NCDs. The key underlying insight is that no clear distinction can be drawn between infectious and chronic diseases. Studies of inflammatory processes, common in infections, but also associated with noninfectious origins, are now being investigated for outcomes as disparate as CVD, cancer, diabetes, and preterm delivery. For instance, cervical cancer is associated with certain strains of HPV, gastric and duodenal ulcer are often the result of infection with *Helicobacter pylori*, oral pathogens have been implicated in preterm birth and diabetes, and the list goes on. Failing to appreciate that the traditional distinction between infectious and chronic diseases is no longer valid presents an impediment to applying classic epidemiologic transition theory to the global health trends currently occurring and may call for a modified version of the theory. This is not unrelated to the argument made by Beemer (this volume) that changes in disease definition and precision of nosology present problems of interpretation for disease trends. We see this on a more limited scale with implementation of each new edition of the International Classification of Diseases (ICD). For example, witness the dramatic rise in deaths from Alzheimer's disease during the implementation of ICD-9 as compared to rates observed under previous editions (Murphy et al., 2013).

A Sometimes Overlooked Critical Distinction in Epidemiology

One of the fundamental lessons in epidemiology is that the proportion of deaths from a given cause does not provide an indication of the risk of death from that cause. As Gage (2005; this volume) discusses, this is highly germane to understanding demographic and epidemiologic transition theory as well as the relationship between modern environments and human health. Importantly, these two measures are often conflated and misinterpreted in studies of the second epidemiologic transition (see Holloway et al., 2013 as an example). Using cross-sectional data on suicide, or intentional self-harm, in the U.S. provides an excellent example of this phenomenon.

An Example Illustrating the Distinction

In 2010, there were 4600 deaths from suicide in the U.S. among persons aged 15–24 out of a total 29,551 deaths in that age group, which translates to 15.6% of all deaths. By comparison, out of 765,474 deaths in 2010 among those 85 and older, 968 were from suicide, or 0.13%, a striking contrast. However, the number of deaths by suicide in relation to the population size, an approximation for the risk of death from a specific cause in the same year, yields a rate of 10.5 per 100,000 population among those 15–24 years of age, but 17.6 per 100,000 among those 85 and older. In short, the risk of death from suicide in the U.S. in 2010 was approximately 70% higher among the most elderly compared to the younger age group, which is completely masked by a focus on the percentage of deaths due to this cause (data from Murphy et al., 2013).

Implications of This Distinction for Transition Theory

If we translate this from a comparison of two age groups in the same population in the same year to comparisons of whole populations over time, it becomes clear that the proportion of deaths from specific causes or classes of causes tells us very little about the risk of death from those causes. Indeed, a category of causes may increase in proportion of deaths while at the same time declining in risk, if other causes also decline. That, indeed, is what we observe in some populations. The point is that the decline in deaths from infectious diseases—either an absolute decline in rates or a relative decline as a proportion of deaths—does not mean that the corresponding increase in the proportion of deaths from NCDs entails an increase in the risk of death from NCDs. Further, quite apart from this misinterpretation of two different measures, a decline in deaths from one cause must, of necessity, result in an increase in the proportion of deaths from other causes that decline less rapidly because everyone must die eventually from something. This seemingly trivial observation is, nevertheless, sometimes overlooked in discussions of transition theory (see Holloway et al., 2013), but is critical to understanding the course of epidemiologic transitions of the sort typically associated with the classic version. Deaths from anthropogenic and degenerative causes could increase dramatically as a proportion of all deaths even while the risk of death from those causes was declining (Gage, 2005; this volume) if the risk of death from other causes was declining more rapidly.

The Scope of Classic Transition Theory

The classic transition model focuses primarily on the age distribution of the population (and related life expectancy and fertility rate) and on the distribution of mortality by age and cohort, overall and by major categories of causes. The proposed determinants of the observed (or hypothesized) patterns and of major changes in those patterns are very large scale, dealing with shifts in cultural and societal organization, and tend to be more theoretical. They lack empirical evidence of sufficient scope to warrant the broad claims made. The studies in this volume may contribute by providing evidence and encouraging additional empirical investigations of determinants across settings and time periods.

Age, Period, and Cohort Effects in Epidemiology

When examining trends, epidemiologists often look for age, period, and cohort effects for clues to determinants of the trends. Disentangling the three effects is not straightforward because they are correlated, and there may be more than one alternate explanation for a given pattern. Further, it is possible for combinations of two or three of the effects to be present in a single trend.

Age Effects

Age effects (i.e., developmental effects) are patterns in the occurrence of morbidity or mortality with age within a cohort. For example, a typical mortality curve has a slanted **J** shape with mortality rapidly declining in the first month of life, reaching a fairly stable low point in early childhood, and then increasing exponentially until the oldest ages (see Gage, 2005; this volume). The exact shape of the curve varies with time and location. In the U.S., the Gompertz rule of exponential increase in mortality with age has been found inadequate at oldest ages because of a slowing of the mortality increase (Anderson, 1999). Because transition theory posits declining mortality that is differential across the lifespan, along with declining fertility, the shape of the survival curve as well as the shape of the population distribution by age should look different in different stages of the transition (see DeWitte, this volume). Transition theory could be enhanced by more specific attention to how the theory predicts these changes. Perry's contribution to this volume may be one approach to this issue.

Period Effects

A period effect occurs when an event or change in experience or exposure impacts the mortality or morbidity risk of all or most of a population simultaneously, though not necessarily uniformly. The effect may persist, in which case one sees a noticeable, if not abrupt, change in the curve that is sustained over time, or the effect may be transient, as in the influenza pandemic of 1918, though the impact on the population distribution may last for a generation. Transition theory posits periodic impacts on morbidity and mortality, but could be strengthened by more specific attention to the nature of period effects and how they played out over time, as shown, for example, by Hallman and Gagnon (this volume).

Cohort Effects

Finally, cohort effects (i.e., generational effects) reflect the differing experiences and exposures of different cohorts that impact their risk throughout the lifespan in ways that differ from the cohorts before or after. One of the clearest examples of cohort effects is in rates of cigarette smoking and subsequent rates of tobacco-related diseases and deaths. Transition theory could make substantial contributions to epidemiologic study in the area of cohort effects because of its emphasis on how differing cohorts have been affected by changing large-scale determinants. See Koepke (this volume) for an example that could have bearing on this question.

It may be helpful to think of epidemiologic transitions in terms of wide-ranging, long-term period effects that modified the mortality and to some extent morbidity curves by age over long periods of time in various populations. A lingering question is the extent to which large cohort effects also played a role in some stages of the transition and, if so, how that illuminates our understanding of the underlying, driving forces behind transitions. In addition to changes in exposure characteristic of these period effects, other aspects of the cohort effect include dramatic changes in the shape of age-effect curves, thus producing the demographic shifts that are characteristic of transition theory. Approaching transition theory from the perspective of age–period–cohort analysis as understood in epidemiology is another point of potential collaboration among epidemiologists, anthropologists, demographers, and other social science researchers.

The Theory of Epidemiologic Transitions in a Rapidly Changing World: A Risk Transition in LMICs

The association between economic development and health is not uniformly positive. At least in some areas, development accompanied by globalization has brought with them threats to health associated with greater availability of tobacco products, dietary changes toward more calorie-dense and higher-fat foods, changes in what food is produced and how, and less physical activity in both work and home settings. The result is that, far from being an unalloyed benefit, economic development can be accompanied by various economic, behavioral, cultural, and social transformations that increase disease burden (WHO, 2002; Beaglehole and Yach, 2003; Yach et al., 2004).

The Double Burden in LMICs

An important variant to the classical theory is observed in LMICs that carry a double burden of high rates of emerging NCDs, while infectious disease rates continue at high levels. While NCDs are typically thought to be diseases of more affluent countries and people, LMICs now contribute a larger proportion of CVD globally than do high-income countries (Reddy, 1999). The transition characterized by reduction in the infectious disease burden has not yet occurred even though mortality and fertility rates may be decreasing (Boutayeb, 2006). Further, the rapidly increasing and aging global population and the impact of changing nutrition and sedentary lifestyles are becoming increasingly evident, transforming the character of the epidemiologic transitions that are now taking place. The rate of increase in NCDs in LMICs exceeds past rates of change observed in high-income countries (United Nations, 2007), but also exceeds the rates of enhancements and reforms to health services (Yach et al., 2004) and behavior change (Popkin, 2006).

In recent years, we have seen a global shift in both the prevalence and social patterning of NCD risk factors, creating a *risk transition*. This is particularly evident in rates of tobacco use and obesity in poor countries (WHO, 2002; Yach et al., 2004). In the U.S., there has been considerable emphasis and fairly extensive research (though of uncertain effectiveness) on disparities in health and disease, particularly across racial-ethnic and socioeconomic lines. In addition, we now see emerging evidence of increasing NCD risk among poor people in LMICs, who also suffer inadequate access to health care (Beaglehole and Yach, 2003; Yach et al., 2004).

The implication of these transitions is the prospect of another type of epidemiologic transition in both low- and middle- and high-income countries. In this mode, changing patterns of food consumption and reduced physical activity contribute to increasing risk of NCDs, while communicable diseases continue to burden the poor at disproportionate levels. These patterns will emerge first in urban areas rather than rural areas (Leeder et al., 2004), contributing to yet another disparity that we are only beginning to understand. Population differences in NCDs occur due to variation in demographic effects (e.g., older age structures, migration to urban areas), environmental factors, early childhood programming, and gene frequency and expression (Yusuf et al., 2001b). Global social, cultural, political, and economic changes have caused major changes in the structures of populations, the occupations available, incomes, expenditures, education, diet, and physical activity levels, which have in turn increased chronic disease risk factors, morbidity, and mortality (Yusuf et al., 2001a). Another key issue is that while the global population as a whole is aging, developing countries are aging at a faster rate than developed countries (Mathers and Loncar, 2006; United Nations, 2007). Also troubling is that morbidity and mortality from NCDs are already higher and affect younger populations in developing compared to developed countries (King et al., 1998; Reddy and Yusuf, 1998; Leeder et al., 2004; Strong et al., 2005; Abegunde et al., 2007).

Global Urbanization

One component of globalization is increasing urbanization that is disproportionately concentrated in LMICs. The global shift in the world's population to urban areas is so marked in these countries that by 2030, it is projected that 80% of the world's urban population will reside in LMICs. The plight of these mostly poor urbanites is typically a low priority for urban planning (Greehalgh et al., 2007). The impacts of urban environments on health have been studied in high-income countries, where, apart from gentrification, urbanites are now characterized by ethnic or racial minority persons, with limited access to health care and health-promoting amenities, food deserts, and exposure to environmental toxins (Phillips, 1993; see Schell, this volume). Making healthy choices is often not an option.

Changing Patterns of Morbidity with Socioeconomic Position

In the past, affluence was associated with a higher burden of NCDs. This was because of a joint function of greater risk for infectious diseases and earlier mortality in lower-income groups and greater access to contributors (e.g., tobacco products) to NCDs among the affluent. Studies of the social gradient of chronic disease risk across generations (Cassel et al., 1971) and historical literature reviews (Kaplan and Keil, 1993) have noted a shift over time from higher rates of NCDs among those with higher socioeconomic position (SEP) to higher rates among those of lower SEP. If patterns continue and projections are borne out, LMICs will eventually exhibit the patterns we have already observed in high-income countries: substantial disparities in the occurrence of NCDs, with the poor bearing a disproportionate burden.

Pearson (2003) provides an intriguing and suggestive explanation for some current patterns of NCD morbidity and mortality in people of different SEP. In his model, persons of higher SEP are often early adopters of adverse health behaviors because they possess the means and opportunities to engage in or adopt new products or lifestyles, such as excess food intake, sedentary lifestyles, or smoking. At the same time, and for the same reasons, they are also likely to become better informed—and do so earlier—of the detrimental impact of these products and behaviors. They also possess the means to undertake behavioral change, perhaps even modify their environment, so they also become early adopters of disease prevention and health promotion behaviors, such as better nutrition or smoking cessation. This sequence results in an increase in NCD risk followed by a reversal of the curve with a decline in morbidity and mortality. Lower SEP populations, however, will lag behind this curve and may exhibit increases in NCD occurrence and mortality while the curve for higher SEP groups is decreasing. If one envisions this version of an epidemiologic transition as income-specific epidemiologic curves of morbidity or mortality rates through time, the shape might approximate an inverted U, with the curve for those with higher SEP groups moving ahead of the curve for lower (Pearson, 2003).

Conclusion

We have referred to Omran's proposal as the classic theory of epidemiologic transition. A theory is formulated to answer a question, explain a phenomenon or phenomena, stimulate future research and shape systematic investigation, generate principles or hypotheses, serve as a basis or rationale for action, or provide a coherent framework for a discipline or field of endeavor. The issue for epidemiologists, and public health researchers and practitioners in general, is what question does this theory answer? What phenomena are explained—not merely described—and how does it accommodate the numerous instances of deviation from the model proposed? How is current epidemiologic research motivated and guided in systematic ways? What epidemiologic principles are engendered; what hypotheses suggested? How is public health policy and practice informed by this theory? In what way is epidemiologic practice provided with a more solid grounding, a more coherent vision, and more unified framework by epidemiologic transition theory? We believe there can be positive and fruitful answers to these questions, especially with contributions such as those in this volume, but more fruitful applications of the theory in public health await much greater conversation and collaboration between epidemiologists and (other) social science disciplines.

Epidemiologists bear some responsibility to explore these issues and to seek out the conversations, but the questions raised earlier may also need to be addressed forthrightly so that the possibilities are more clearly focused. One of this volume's strengths is the breadth of approaches and topics addressed. Epidemiologists are increasingly feeling challenged in the Herculean task of encompassing data ranging from the genomic to the personal to the social system to global factors, and accounting for the dynamics of these interrelated domains over time. The contributors here provide valuable examples of how techniques from a range of disciplines can be applied to illuminate perplexing questions over very long time periods.

This volume deals with the second epidemiologic transition, which is but one of the transitions Omran described, all of which fell under the second of his five propositions. The substance of much of the criticism of Omran's work seems to focus on this second proposition and the description of the transitions it included, along with the corollary fifth proposition on models of transition. The other propositions seem to be less problematic. The first, on the role of mortality as a primary driver of population dynamics, seems at first glance to be a truism. However, Omran expanded the proposition to encompass fertility rates and changing age distributions in populations. The third posited that the biggest impact of the transitions was among children and reproductive-age women. The fourth emphasized the role of demographic and socioeconomic transitions in the hypothesized epidemiologic transitions. It seems to us that all three of these propositions continue to play out globally today. All three can be examined and tested much more directly and may have more universal applicability than the specific form or sequence of the transitions described in propositions two and five, though it still seems fruitful to explore the occurrence and the nature of health and disease transitions as they emerge globally today.

The various attempts to modify or expand the theory of epidemiologic transitions are analogous to what Thomas Kuhn (1970) described when proponents of dominant scientific paradigms seek to address anomalies that do not easily fit the traditional model. The increasing accommodation to these anomalies can eventually result in the dominant paradigm becoming unwieldy and merit its replacement by a new paradigm that provides a more elegant explanation for both the original and the newly observed phenomena. It remains to be seen whether these modifications to the epidemiologic transition theory and efforts such as the chapters in this volume will result in a renewed, more fruitful version of Omran's thesis or eventually contribute to its demise and replacement by an even more comprehensive transdisciplinary model. This novel model would address all that we have learned while also accommodating all that we continue to discover through innovative approaches such as those outlined by these researchers, while at the same time raising fertile new questions to explore.

References

Abegunde DO, Mathers CD, Adam T, Ortegon M, Strong K. 2007. The burden and costs of chronic diseases in low-income and middle-income countries. Lancet 370(9603):1929–1938.

Anderson RN. 1999. Method for constructing complete annual U.S. life tables. Vital health statistics, series 2, Data Evaluation and Methods Research, No. 129. Hyattsville: National Center for Health Statistics.

Armelagos GJ, Brown PJ, Turner B. 2005. Evolutionary, historical and political economic perspectives on health and disease. Soc Sci Med 61(4):755–765.

Beaglehole R, Yach D. 2003. Globalisation and the prevention and control of non-communicable disease: the neglected chronic diseases of adults. Lancet 362(9387):903–908.

Berkman LF, Kawachi I, editors. 2000. Social epidemiology. New York: Oxford University Press.

Boutayeb A. 2006. The double burden of communicable and non-communicable diseases in developing countries. Trans R Soc Trop Med Hyg 100(3):191–199.

Caldwell JC. 2001. Population health in transition. Bull World Health Org 79(2):159–160.

Caselli G, Mesle F, Vallin J. 2002. Epidemiologic transition theory exceptions. Genus 9:9–51.

Cassel J, Heyden S, Bartel AG, et al. 1971. Incidence of coronary heart disease by ethnic group, social class, and sex. Arch Intern Med 128(6):901–906.

Diez-Roux AV. 1998. Bringing context back into epidemiology: variables and fallacies in multilevel analysis. Am J Public Health 88(2):216–222.

Diez-Roux AV. 2007. Integrating social and biologic factors in health research: a systems view. Ann Epidemiol 17(7):569–574.

Diez-Roux AV. 2011. Complex systems thinking and current impasses in health disparities research. Am J Public Health 101(9):1627–1634.

Editors. 1942. What and who is an epidemiologist? Am J Public Health 32(April):414–415.

Gage TB. 2005. Are modern environments really bad for us? Revisiting the demographic and epidemiologic transitions. Yearb Phys Anthropol 48:96–117.

Gaylin DS, Kates J. 1997. Refocusing the lens: epidemiologic transition theory, mortality differentials, and the AIDS pandemic. Soc Sci Med 44(5):609–621.

Graziano JM. 2010. Fifth phase of the epidemiologic transition: the age of obesity and inactivity. JAMA 303(3):275–276.

Greenhalgh S, Montgomery M, Segal S, Todaro MP, UN Population Fund. 2007. State of world population 2007: unleashing the potential of urban growth. Popul Dev Rev 33:639–640.

Harper K, Armelagos GJ. 2010. The changing disease-scape in the third epidemiological transition. Int J Environ Res Public Health 7:675–697.

Holloway K, Henneberg R, Lopes M, et al. 2013. Secular trends in tuberculosis during the second epidemiological transition: a Swiss perspective. Adv Anthropol 3:78–90.

Institute of Medicine and Committee on Assuring the Health of the Public in the 21st Century. 2003. The future of the public's health in the 21st century. Washington, DC: National Academy Press.

Johnson S. 2006. The ghost map: The story of London's most terrifying epidemic – and how it changed science, cities and the modern world. New York: Riverhead Books.

Kaplan GA, Keil JE. 1993. Socioeconomic-factors and cardiovascular-disease – a review of the literature. Circulation 88(4):1973–1998.

Kawachi I, Berkman LF, editors. 2003. Neighborhoods and health. New York: Oxford University Press.

Keith S, Redden DK, Katzmarzyk PT, et al. 2006. Putative contributors to the secular increase in obesity: exploring the roads less traveled. Int J Obes 30:1585–1594.

King H, Aubert RE, Herman WH. 1998. Global burden of diabetes, 1995–2025: prevalence, numerical estimates, and projections. Diabetes Care 21(9):1414–1431.

Klimentidis Y, Beasley T, Lin H, et al. 2011. Canaries in the coal mine: a cross-species analysis of the plurality of obesity epidemics. Proc R Soc Biol Sci 278(1712):1626–1632.

Krieger N. 1994. Epidemiology and the web of causation: has anyone seen the spider? Soc Sci Med 39:887–903.

Krieger N. 2001a. A glossary for social epidemiology. J Epidemiol Community Health 55:693–700.

Krieger N. 2001b. Theories for social epidemiology in the 21st century: an ecosocial perspective. Int J Epidemiol 30:668–677.

Kuhn TS. 1970. The structure of scientific revolutions. Chicago: University of Chicago Press.

Leeder S, Raymond S, Greenberg H, Liu H, Esson K. 2004. A race against time: The challenge of cardiovascular disease in developing economies. New York: Columbia University.

Mabry PL, Olster DH, Morgan GD, Abrams D. 2008. Interdisciplinary and systems science to improve population health: a view from the NIH Office of Behavioral and Social Science Research. Am J Prev Med 35(2S):S211–S224.

Mackenbach JP. 1994. The epidemiologic transition theory. J Epidemiol Community Health 48(4):329–331.

MacMahon B, Pugh TF, Ipsen J. 1960. Epidemiologic methods. Boston: Little Brown & Co.

Mathers CD, Loncar D. 2006. Projections of global mortality and burden of disease from 2002 to 2030. PLoS Med 3(11):e442.

McKeown RE. 2009. The epidemiologic transition: changing patterns of mortality and population dynamics. Am J Lifestyle Med 3(1 Suppl):19S–26S.

McKeown RE. 2013a. What's an epidemiologist to do? Epidemiology 23(6):923–926.

McKeown RE. 2013b. Is epidemiology correcting its vision problem? A perspective on our perspective: 2012 presidential address for American College of Epidemiology. Ann Epidemiol 23(10):603–607.

Murphy S, Xu J, Kochanek K. 2013. Deaths: Final data for 2010 National Vital Statistics Reports. Hyattsville: National Center for Health Statistics.

Murray CJ, Abraham J, Ali MK, et al. 2013. The state of US health, 1990–2010: burden of diseases, injuries, and risk factors. JAMA 310:591–608.

National Research Council (NRC), Committee on a Framework for Development of a New Taxonomy of Disease. 2011. Toward precision medicine: Building a knowledge network for biomedical research and a new taxonomy of disease. Washington, DC: National Academies Press.

Oakes JM, Kaufman JS, editors. 2006. Methods in social epidemiology. San Francisco: Jossey-Bass.

Olshansky SJ, Ault AB. 1986. The fourth stage of the epidemiologic transition: the age of delayed degenerative diseases. Milbank Q 64(3):355–391.

Olshansky SJ, Carnes BA, Rogers RG, Smith L. 1998. Emerging infectious diseases: the fifth stage of the epidemiologic transition? World Health Stat Q 51(2–4):207–217.

Omran AR. 1971. The epidemiologic transition. A theory of the epidemiology of population change. Milbank Mem Fund Q 49(4):509–538.

Omran AR. 1977. A century of epidemiologic transition in the United States. Prev Med 6(1):30–51.

Pearson TA. 2003. Education and income: double-edged swords in the epidemiologic transition of cardiovascular disease. Ethn Dis 13(S2):S158–S163.

Phillips DR. 1993. Urbanization and human health. Parasitology 106:S93–S107.

Popkin BM. 2006. Global nutrition dynamics: the world is shifting rapidly toward a diet linked with noncommunicable diseases. Am J Clin Nutr 84(2):289–298.

Porta M. 2008. A dictionary of epidemiology. 5th ed. Oxford: Oxford University Press.

Reddy KS. 1999. Emerging epidemic of cardiovascular disease in the developing countries. Atherosclerosis 144:143.

Reddy KS, Yusuf S. 1998. Emerging epidemic of cardiovascular disease in developing countries. Circulation 97(6):596–601.

Rockett IRH. 1999. Population and health: An introduction to epidemiology. 2nd ed. Washington, DC: Population Bulletin.

Sanders JW, Fuhrer GS, Johnson MD, Riddle MS. 2008. The epidemiological transition: the current status of infectious diseases in the developed world versus the developing world. Sci Prog 91(Pt 1):1–37.

Smallman-Raynor M, Phillips D. 1999. Late stages of epidemiological transition: health status in the developed world. Health Place 5(3):209–222.

Stallones RA. 1966. Prospective epidemiologic studies of cerebrovascular disease. Washington, DC: U.S. Government Printing Office.

Strong K, Mathers C, Leeder S, Beaglehole R. 2005. Preventing chronic diseases: how many lives can we save? Lancet 366(9496):1578–1582.

Sulsky SI, Kreiger N, McKeown RE. 2012. Not continuing along previous lines: exploring how new directions emerge in epidemiologic research. Ann Epidemiol 22(5):369–371.

Susser M. 1985. Epidemiology in the United States after World War II: the revolution of technique. Epidemiol Rev 7:147–177.

Susser M, Susser E. 1996a. Choosing a future for epidemiology: I. Eras and paradigms. Am J Public Health 86(5):668–673.

Susser M, Susser E. 1996b. Choosing a future for epidemiology: II. From black box to Chinese boxes and eco-epidemiology. Am J Public Health 86(5):674–677.

United Nations. 2007. World population ageing. New York: United Nations Department of Economic and Social Affairs.

Wheeler RE. 1942. Letter: what and who is an epidemiologist? Am J Public Health 32(7):759–760.

WHO. 2002. The world health report 2002: Reducing risks, promoting healthy life. Geneva: World Health Organization.

WHO. 2008. The global burden of disease: 2004 update. Geneva: World Health Organization.

WHO. 2009. Global health risks: mortality and burden of disease attributable to selected major risks. Geneva: World Health Organization.

Wilkinson RG. 1994. The epidemiological transition: from material scarcity to social disadvantage? Daedalus 123(4):61–77.

Winkelstein W, French FE, editors. 1972. Basic readings in epidemiology. 3rd ed. New York: MSS Educational Publishing Co.

Yach D, Hawkes C, Gould CL, Hofman KJ. 2004. The global burden of chronic diseases – overcoming impediments to prevention and control. JAMA 291(21):2616–2622.

Yusuf S, Reddy S, Ounpuu S, Anand S. 2001a. Global burden of cardiovascular diseases – part II: variations in cardiovascular disease by specific ethnic groups and geographic regions and prevention strategies. Circulation 104(23):2855–2864.

Yusuf S, Reddy S, Ounpuu S, Anand S. 2001b. Global burden of cardiovascular diseases – part I: general considerations, the epidemiologic transition, risk factors, and impact of urbanization. Circulation 104(22):2746–2753.

Chapter 20
Methodological Perspectives on the Second Epidemiologic Transition: Current and Future Research

Richard H. Steckel
Department of Economics, The Ohio State University, Columbus, OH

Introduction

Insights into the need for current and future research on the second epidemiologic transition can be gained by reviewing the intellectual history of the demographic transition. Here I place the chapters in this volume in perspective relative to the evolution of the huge body of literature on this topic. In particular, the struggles and successes of the concept of the demographic transition point to ways that future study of the second epidemiologic transition may unfold.

Research on the Demographic Transition

By the 1980s, researchers had ultimately discovered the enormous complexity of the demographic transition with respect to triggers or antecedents and the nature of change at the regional, national, or local level. This is not a basis for pessimism but rather a realistic assessment of the considerable potential for research on the recently enunciated epidemiologic aspect of the larger subject.

As a pioneer in demography, Warren Thompson was concerned with inequality and population growth in relation to land resources. His famous article (Thompson, 1929), "Population," never used the term *demographic transition*, but the data he organized on population growth suggested that many countries in the world were moving or had moved from one type of demographic regime to another. Despite Landry's (1934) restatement and labeling of the concept a few years later, the idea remained dormant until long-run population trends became clearer following World War II. Over a decade after the publications by Thompson and by Landry, articles by Davis (1945), Notestein (1945), and others gained traction such that the model emerged as a central concern of demography and later diffused throughout the social sciences as a useful hypothesis for organizing research.

The first point to note is the lengthy period of time—over 30 years—it took for a critical mass of interest, research, and policy to coalesce following a statement of the concept in 1929. In my view, the major stimulus was clear evidence that world population could outrun resources, particularly food. The simple logic of the mathematics of population growth was compelling. Beginning with the publication of *The Greatest Problem* (Lucas, 1960) and into the late 1960s, books such as *The Population Bomb* and organizations like the Club of Rome predicted a dire future for humanity (Ehrlich, 1968). A few scholars such as Simon (1981) criticized the argument, and others doubted the urgency and hype with which it was presented, but the pessimistic predictions were plausible enough and ultimately contributed to greater funding for population research as well as extensive new programs such as the Green Revolution, China's one-child policy, and various efforts to promote fertility control, especially in developing countries.

Study of the demographic transition soon branched into specialties of fertility and mortality. Though not lacking research and policy challenges, mortality was thought to be manageable or at least improvable in the developing world, where most of humanity lived, by applying a suite of inexpensive, modern public health measures, such as water purification, sanitation, and vaccination. Thus, in the second phase of the demographic transition in developing countries, fertility remained high relative to mortality such that world population was scheduled to double from two to four billion between 1930 and 1975, most of which was forecast to occur in poor countries. Research understandably focused on fertility as a way to manage the population problem.

But fertility control was a different matter, involving a complicated mesh of personal choice, culture, labor-market options for women, schooling, child-rearing costs, the technology of birth control, expectations for children in the old-age care of their parents, and so forth. To help understand these issues, in the late 1960s, the Office of Population Research launched the Princeton Fertility Project, which ultimately produced numerous articles and seven books on the historical fertility transition in several European countries. A volume, Coale and Watkins (1986), summarized the findings, reporting that their efforts could not identify widespread, much less specific, universal, socioeconomic triggers—beyond broad background conditions of the desire, the capability, and the cultural legitimacy to limit family size—that led the historical fertility decline in countries that were well developed in the 20th century. Some have criticized the methodology and the conclusions of the project, but most scholars agree that no single stimulus or small, homogeneous collection of stimuli, led to the fertility transitions in these countries (Guinnane, 2011). If a generalization is permitted, voluntary fertility decline was associated with *modernization*, a phenomenon taking numerous specific forms that make modeling and forecasting difficult across countries.

World population growth rates peaked in the 1960s at 2.3% per year, equivalent to a doubling in global population size roughly every 30 years.[1] Since then, the growth rate has fallen sharply to about 1.2% and currently is forecast to drop below 1% in 2020. Some projections anticipate that world population will peak at 10 billion or less in 2100. The success of the Green Revolution and the emergence of other methods of increasing food production using forms of biotechnology, the world's recent slow-down in population growth, and realistic projections of the future have deflated the urgency of the population problem. Some observers say that today population is not even a problem in relation to food supply. In any event, the alarm of the 1960s has drifted to other concerns such as energy, pollution, and climate change.

Will a sense of urgency, as found for the study of the *population explosion* in the 1960s and 1970s, ever emerge for research on epidemiologic transitions? If so, what would the trigger(s) be? In my view, a disease crisis, a dreadful pandemic such as the 1918 Spanish

---

[1] http://esa.un.org/wpp/

influenza pandemic (see Hallman and Gagnon, this volume), would create that urgency. This possibility should be taken seriously because the coevolution of humans and pathogens has a long history punctuated by unpleasant surprises (Lederberg, 1998). Moreover, scholars have identified several nasty protagonists, such as the class of known pathogens that are becoming or have become antibiotic resistant (e.g., tuberculosis), and emerging (or apparently new) diseases such as the Ebola virus and HIV/AIDS (Barrett et al., 1998; Harper and Armelagos, 2010). Pessimists note that international trade and widespread travel are now uniquely high in world history, enabling rapid transmission of pathogens while creating favorable conditions for virulent mutations to occur. One may ask whether weak and ineffective governments of the world will be able to enforce quarantines and other public health measures while providing essential services necessary to avert chaos. Moreover, our world of specialization, global travel, and trade create enormous interdependencies such that bottlenecks of production and distribution could readily appear during a disease crisis. It is plausible to think that many people underestimate our vulnerability on this front and that society underinvests in the study of and adaptations to disease crises.

On the optimistic side, public health officials are alert to these possibilities and will be on guard to apply new technologies of diagnosis and defense. Still, the shocks of the past are disquieting, and therefore it is prudent to invest in the study of epidemiologic transitions. This book is an important effort in that direction.

The Research Agenda

How does this collection of essays enhance our knowledge of epidemiologic transitions, and where should research go from here? First, it is useful to describe concisely what has been accomplished relative to what could and should be done if enough resources were available. Table 20.1 organizes the chapters in this volume by geographic location, time period, and data source. Note that there is widespread, but far from universal, geographic representation, a shortcoming given that most countries have witnessed at least some segment of the epidemiologic transition. Europe (including the UK) and North America are well represented but comprise about one-sixth of world population. The most populous region, Asia, is represented by one study, and there are none for the large continents of Africa and South America. Of course, it is far too much to expect one volume to encompass the globe (or if it did, the effort would be spread very thin), but it remains true that relatively little research on epidemiologic transitions has been done on these parts of the world. This is a pity because scientific investigation thrives on the study of diversity, and the current state of research omits regions that are or were very likely to have been different in their pathogens, modes of transmission, health care facilities, culture, technology, and so

Table 20.1. Classification of papers by geographic location, data source, and time period.

Location (N)	Data sources (N)	Primary time period (N)
UK (5)	Skeletal remains (7)	20th century (4)
Continental Europe (2)	Death registration (5)	19th century (5)
U.S. (6)	Church records (2)	18th century (3)
Canada (2)	Bills of mortality (1)	Pre- and postindustrialization (1)
Asia (1)	Measured stature (1)	Middle Ages to industrialization (3)
Global (1)	Archaeoparasites (1)	Neolithic to present (1)

forth. Therefore, my first recommendation is to undertake many more projects that will diversify geographic areas and time periods of study. This will require considerable funding, but using the framework of cost–benefit analysis, I claim that it would be a good investment for society; lack of preparation for a disease crisis will eventually have disastrous consequences that might be lessened or averted by funding suitable research. Indeed, if I were rewriting my presidential address on "Big Social Science History" (Steckel, 2007), I would include this in my list of proposed projects.

It is clear, however, that the topics for this volume were not chosen from a smorgasbord of options, but were constrained by the availability of evidence. In any emerging field of study, it is wise to pick first the low-hanging fruit, here the time and places where data are readily available. It simply turns out that Europe and North America have the best historical records, including ecclesial series on baptisms and burials, and vital records assembled by governments. But these are far from the only useful sources of evidence. Skeletal remains are widely available around the globe, as can be seen from a reconnaissance undertaken by Steckel, Larsen, and collaborators (Steckel, 2007; Steckel et al., 2009). Moreover, skeletal remains contain information on dental health, degenerative joint disease, and periosteal reactions, which provide useful insights into morbidity, something sorely needed in the study of the epidemiologic transition. Stature series, often collected by military organizations or by governments creating civil registration or identities, are widely available around the world for the past two centuries (Baten and Blum, 2012). Hence, while this volume draws upon the regions and time periods for which stature and skeletal data relevant to the epidemiologic transition have been analyzed for this purpose as of this time, there is no shortage of evidence for additional studies outside of these contexts, and these sources should be further exploited to help control the urge to generalize from small amounts of data. In fact, premature generalizations can be counterproductive if they absorb resources later required to overturn erroneous conclusions. These resources could be used more productively elsewhere.

The chapters in this volume assembled evidence that could be organized over time: a good result because tempo is an essential element of the transition process. Researchers might have been tempted to analyze only data collected at a point in time and then use cross-sectional differences to project what would have happened over time. Cross-sectional evidence can be informative but is most useful in this context when analyzed to understand change over time. For example, cross-sections may reveal changing patterns of inequality in causes of death across social class or occupations as the transition unfolded. One can then inquire about the mechanisms that may have protected certain groups against ill health. Quite likely, the epidemiologic transition was shaped as much by social and economic phenomena as biological ones. At least this is a hypothesis worth exploring.

In the future, I see a need for greater statistical rigor guided by models and by clearly formulated hypotheses. Many of the chapters here adopted what could be called exploratory data analysis or the sifting of evidence in search of patterns that defined or accompanied the epidemiologic transition. This approach is valuable in the early stages of investigation, which is what this volume represents, in which ideas are crystallizing and tentative generalizations are being formulated. As a result of these chapters, however, the field has now advanced to the point where future researchers should be in a position to compare results from different types of evidence. This should be done cautiously because agreement does not necessarily imply confirmation if the sources measure somewhat different aspects of health.

Some of the data analysis employed here is mainly one dimensional, considering change or differences with respect to one variable, such as time, community size, stage of industrialization, age, or sex. This is largely a product of the nascent state of research on the subject. More multidimensional approaches (see Guntupalli; DeWitte; Koepke; Budnik; Perry; de la Cova, all this volume) create opportunities to detect interactions or interdependencies of variables that operate synergistically. For example, did the effect of industrialization on health vary by age and sex and race or ethnicity? Future research should take advantage of multivariate regression models to address these questions.

An additional advantage of these models, at least as invoked by standard statistical packages such as SAS, SPSS, or STATA, is their ease in generating standard errors. It is important to know whether temporal patterns in the data and parameter estimates fall within the bounds of probable sampling error.

Communicable diseases such as smallpox, cholera, and typhoid often receive the spotlight in research on the transition, but prior to the 20th century, poor methods of food preservation led to spoilage, and water supplies in many communities were frequently infested by pathogens, creating less spectacular scenes of death. In fact, it would be useful to know how many illnesses and deaths were caused by these apparently mundane pathogens prior to the rise of the germ theory of disease. In this vein, Fogel (2004) estimates that only a small share (perhaps 5%) of overall mortality was associated with crises prior to the late 19th century. Several of the chapters in this volume do explore these more mundane aspects of the transition (see Reinhard and Pucu de Araújo; Zuckerman and Armelagos, both this volume) and should encourage future research on these issues.

One may also ask whether it is opportune to distill knowledge from empirical patterns found here so as to revise the theory of the epidemiologic transition or to formulate different or related theories that speak to causation. It is too much to ask for a wholesale reworking of the theory, but for the assistance of future researchers, it would be useful to articulate the holes in the theory brought to light by these chapters. For example, the standard enumeration imagines reductions in communicable diseases to have been prime movers in improving health. Yet the chapters by Reinhard and Pucu de Araújo and Zuckerman and Armelagos on parasites and symbiotic microorganisms and by Schell and Millard and colleagues on toxins in the environment indicate that other types of agents were also important.

With regard to understudied topics, where do effective methods of food preservation, such as refrigeration and canning, and water treatment enter the formulation of the theory of the epidemiologic transition? How about draining swamps, which reduced the spread of disease by mosquitoes? In what ways were the identification and treatment of vitamin deficiencies important for health? How did new machines lessen labor demands that improved net nutrition (think of the steel plow, the reaper, and electrification in industry)? Research has yet to clarify how lower transportation costs contributed to the rise of international markets in food that lessened the vulnerability of any particular region to harvest failures (consider that the last harvest crisis in Sweden occurred in the late 1860s). How might the telegraph, invented in the mid-19th century, have led to sharing information that reduced (or possibly exaggerated?) epidemics? What did improved dental care contribute? It is plausible that dozens, if not hundreds, of small technological improvements contributed to the second epidemiologic transition. Clearly, fertility was not the only complex aspect of the demographic transition; researchers are challenged by a plethora of interesting topics for the study of the epidemiologic transition. The challenge will remain finding data that allows them to address these issues.

Consequences

Much research in this volume is devoted to understanding causation, to identifying and measuring the impact of forces that initiated and governed the pace of the epidemiologic transition. This is a worthy agenda and builds the foundation for future, complementary areas of research. I particularly recommend that future work address the consequences of the reduction in communicable disease and of the return of a disease crisis,

Health and Living Standards

For many years, I have introduced the topic of the standard of living by asking students to believe in reincarnation, such that they would be reborn at random to parents in a country they could choose based on two characteristics, which I place on a list. Inevitably, average health (e.g., life expectancy at birth) and per capita income appear on the list, along with inequality, political freedom, and the potential for socioeconomic mobility. Health is clearly a central aspect of the standard of living, to the point that economists now discuss the value of a statistical life (Ashenfelter, 2006). Although it would be a huge job, researchers could attempt to estimate the economic value of the epidemiologic transition. The effort could be made manageable by considering particular countries and time periods. I suspect that the contribution to human welfare, via improved health, has been enormous, possibly as great as measured GDP itself. Large-scale, long-term stature series, such as that compiled by Koepke (this volume), could be used to tackle aspects of this issue.

Education

For now-industrialized countries, the epidemiologic transition roughly doubled life expectancy at birth, from approximately 35–40 years to about 75–80 years. Working life or the length of a career also increased substantially, such that the payback period from investing in education was substantially lengthened. Individuals therefore had a greater incentive to invest in education. One could measure the strength of the incentive by considering evidence for individual countries, creating a laboratory for studying the interaction of health and education (see Guntupalli, this volume).

Epidemics and Interdependence

Even though it lacks a precise definition, globalization has been a catchword for our generation, loosely understood to mean growing interdependence of countries, organizations, and individuals brought together by lower transportation costs, new communication technologies, lower tariffs, and new software that manages the logistics of coordination.

Figure 20.1 provides information on one important aspect of globalization over the past several decades, a growth in world trade relative to Gross Domestic Product (GDP). One can see that results differ little whether GDP is unadjusted (nominal) or adjusted (real) for inflation. Imports as a share of world GDP more than doubled in the 35 years after 1970, from 13% in 1970 to about 29% (in real terms) in 2005. This shift has been so sudden that I suspect few health officials have contemplated its implications for the production and distribution of goods and services. Increasingly, we live in an interdependent world where a sneeze in one country can lead to a cold in another. I am not talking about the biological transmission of pathogens, but rather the implications of widespread illness and mortality of the labor force for the normal operation of the economy and the political system. The fragility of the entire system, even in the absence of a disease crisis, is illustrated by the global scale of the financial crisis that began in 2008. A large pandemic could interrupt trade, exposing these interdependencies. Bottlenecks in the production and distribution of strategic products, ordinarily traded around the world, would impose massive adjustment costs and possibly political chaos, especially in countries where governments are already weak and ineffective. Large countries such as the United States, which produce many types of products, might be less affected, but considerable hardship would ensue.

The data in Figure 20.1 understate the impact of a pandemic because the numbers assume that trade consists of whole finished products rather than a mix of intermediate and final products. Consider the case of automobiles, whose parts such as engines, spark

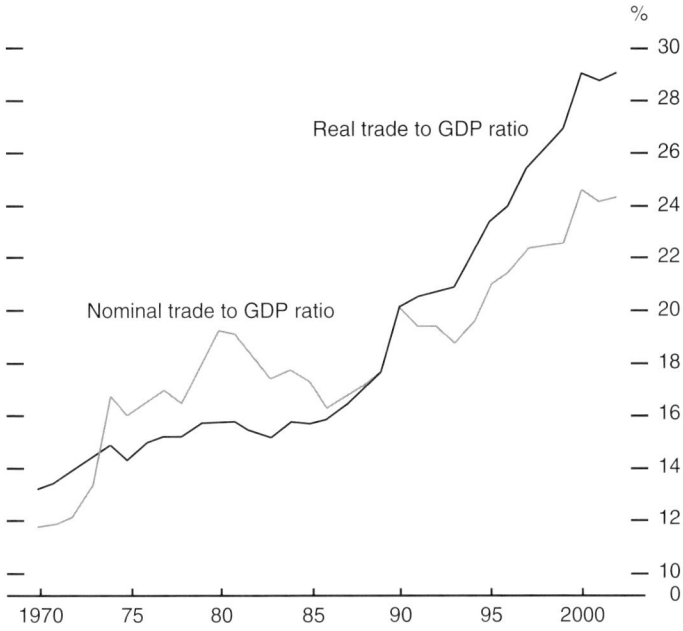

Figure 20.1. World imports as a ratio of world GDP. *Source*: Dean and Sebastia-Barriel (2004: 310).

plugs, brakes, and so forth are made in many different parts of the world but shipped and assembled in one place. If important links in this chain of production are dismembered, it is conceivable that no automobiles would be produced simply because a few strategic parts such as spark plugs were lacking.

Automobiles and other durable goods such as refrigerators and houses last a long time, and so we could muddle through without their production for a few years, but consumables such as food, gasoline, and electricity are another matter. Anyone who doubts the importance of these items merely needs to recall gasoline shortages of the 1970s (or during the aftermath of hurricane Sandy), their discomfort from a power outage that lasted more than a few days, or to a blizzard that disrupted the delivery of food. In the absence of food, it is plausible that many people would riot within a week. Because they lack local production of food and have limited storage capacity, cities would be most vulnerable in this regard. Electricity is essential for operating our computers, which are important for efficient logistics in the production and distribution of products. This does not give either the government or the private economy, already stressed by a pandemic, much time to make accommodations to the loss of key people in the labor force.

Lack of information flows would cramp other forms of specialization, such as outsourcing. How would cloud computing operate without functioning server farms? Could American X-rays be read in India? How about call centers in Thailand, computer programming in China, and data storage in Australia?

Conclusion

In sum, there are enormous economic advantages to specialization, division of labor, and trade. But there are also great vulnerabilities that would be exposed by large-scale disruption of our economic system, costs that I suspect many individuals have not thought about because it has yet to happen on a large scale. Hopefully, study of the consequences of past pandemics will inform preparations for the next one, which some say is sadly inevitable.

References

Ashenfelter O. 2006. Measuring the value of a statistical life: problems and prospects. Econ J 116: C10–C23.
Barrett R, Kuzawa C, McDade T, Armelagos G. 1998. Emerging and re-emerging infectious diseases: the third epidemiologic transition. Ann Rev Anthropol 27:247–271.
Baten J, Blum M. 2012. Growing taller, but unequal: biological well-being in world regions and its determinants, 1810–1989. Economic History of Developing Regions 27: S66–S85.
Coale AJ, Watkins SC. 1986. The decline of fertility in Europe: the revised proceedings of a conference on the Princeton European Fertility Project. Princeton: Princeton University Press.
Davis K. 1945. The world demographic transition. Ann Am Acad Pol Soc Sci 237(43): 1–11.
Dean M, Sebastia-Barriel M. 2004. Why has world trade grown faster than world output? Bank of England Quarterly Bulletin 44: 310–320.
Ehrlich PR. 1968. The population bomb. New York: Ballantine Books.
Fogel RW. 2004. The escape from hunger and premature death, 1700–2100: Europe, America, and the Third World. Cambridge: Cambridge University Press.
Guinnane T. 2011. The historical demographic transition: a guide for economists. J Econ Lit 49(3): 589–614.
Harper K, Armelagos GJ. 2010. The changing disease-scape in the third epidemiological transition. Int J Environ Res Public Health 7: 675–697.
Landry A. 1934. La Révolution Démographique. Paris: Sirey.
Lederberg J. 1998. Emerging infections: an evolutionary perspective. Emerg Infect Dis 4(3): 366–371.
Lucas FL. 1960. The greatest problem, and other essays. London: Cassell.
Notestein F. 1945. Population: the long view. In: Schultz TW, editor. Food for the world. Chicago: University of Chicago Press, p 36–57.
Simon JL. 1981. The ultimate resource. Princeton: Princeton University Press.
Steckel RH. 2007. Big social science history. Soc Sci Hist 31: 1–34.
Steckel RH, Larsen CS, Sciulli PW, Walker PL. 2009. The history of European health project: a history of health in Europe from the late Paleolithic era to the present. Acta Univ Carol Med Monogr 156: 19–25.
Thompson WS. 1929. Population. Am J Sociol 34(6): 959–975.

Chapter 21
The Current State of Knowledge on the Industrial Epidemiologic Transition: Where Do We Go from Here?

Timothy B. Gage
Department of Anthropology and the Department of Epidemiology, University of Albany, State University of New York, Albany, NY

Introduction

The industrial demographic and epidemiologic transitions are sometimes referred to as theories. However, since they are simple empirical descriptions based on the available data, it might be better to call them models. Models are intended to describe general patterns that all (or at least most) industrial demographic and epidemiologic transitions share. The model thus defines the commonalities, which are important on the one hand, but also provides a null hypothesis against which variation from the common pattern can be identified.

A historical perspective differentiates between the industrial *demographic transition*, which describes the decline in mortality and fertility that has accompanied the industrial revolution, and the *epidemiologic transition*, which was developed a half century later and describes the changes in cause of death that accompanied the demographic transition. Many chapters in this volume tend to combine these two processes, as well as others, together under the heading of the *epidemiologic transition*. However, it is useful to distinguish between the various *transitions*.

The term *demographic and epidemiologic transition* used here refers to the trends during the industrial period. This volume refers to this as the second demographic/epidemiologic transition, which distinguishes it from the *first* demographic/epidemiologic transition that occurred earlier as a result of the adoption of agriculture. However, it is possible that other demographic/epidemiologic transitions occurred prehistorically, such as when California Native Americans learned how to process acorns. Further, there is a literature that refers to the postmodern decline in fertility to below replacement, as the second demographic transition (e.g., Lesthaeghe and Neidert, 2006). Future research on this and related subjects should be attentive to preserving and utilizing these distinctions.

The Industrial Demographic Transition

The *demographic transition* is a model describing the decline in mortality and fertility that occurred over the last 300 years. Since this process coincides with the industrial revolution, it is assumed to be the catalyst. Sweden is often used as an example because Sweden was the first western country to carry out reliable censuses and register deaths and is thus the longest series of life tables currently available, using consistent data and methods (Figure 21.1). In the 1750s, mortality was relatively high and varied from year to year. It is assumed that there was a general (baseline) level of mortality, interspersed by years of *crisis* (higher than baseline) mortality (Flinn, 1984). Prior to the transition, crisis mortality was thought to vary spatially as well as temporally. Early in the transition, however, crisis mortality became more organized, so that spatial variation within a year declined, and crisis mortality (yearly variation) may have increased. This process appears to culminate with the influenza pandemic of 1918–1919. Beginning in the 1800s, crisis mortality starts to decline in frequency and amplitude, with the exception of the 1918–1919 influenza epidemic, as does baseline mortality. While dramatic, the decline in crisis mortality is thought to account for less than 6% of the total decline in mortality (Caldwell, 2001). The decline in baseline mortality accelerated after 1850. By 1950, mortality had reached relatively low levels and the rate of decline slowed, although mortality has continued to decline until the present.

Less is known about the preindustrial situation prior to the industrial revolution. Family reconstitution studies going back to 1500 or so suggest that *if anything*, mortality increased in England and Wales prior in the early industrial era (Wrigley and Schofield, 1981; Wrigley et al., 1997; Kitson, this volume). A consensus on this issue has not yet emerged (Caldwell, 2001). It is possible that this increase in England and Wales was the result of greater urbanization and was not shared by less urbanized countries such as Sweden. In any event, mortality in Sweden in the 1750s appears to be similar to that reported in England and Wales in the

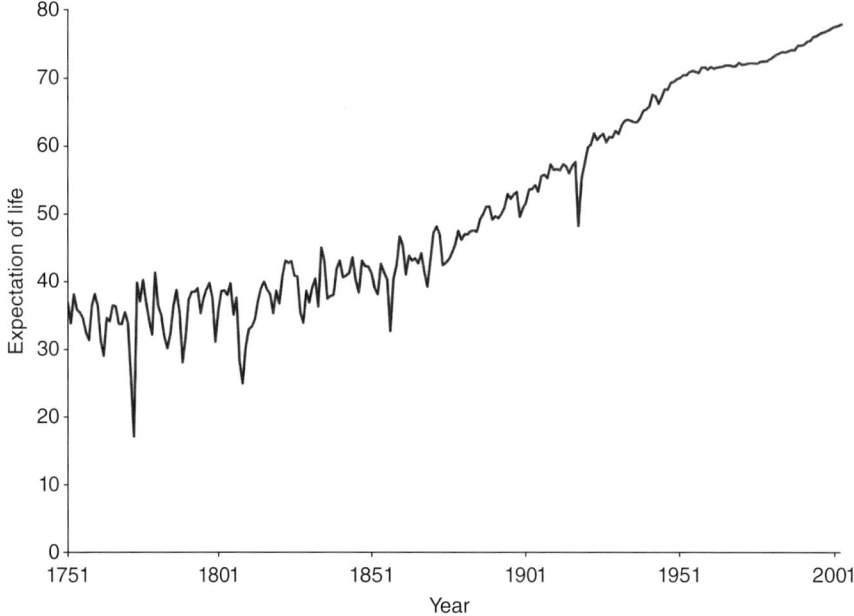

Figure 21.1. Yearly, male, expectation of life 1751–2003 Sweden. *Source*: Human Mortality Database (data downloaded on 1/6/2005). Figure reprinted from Gage (2005).

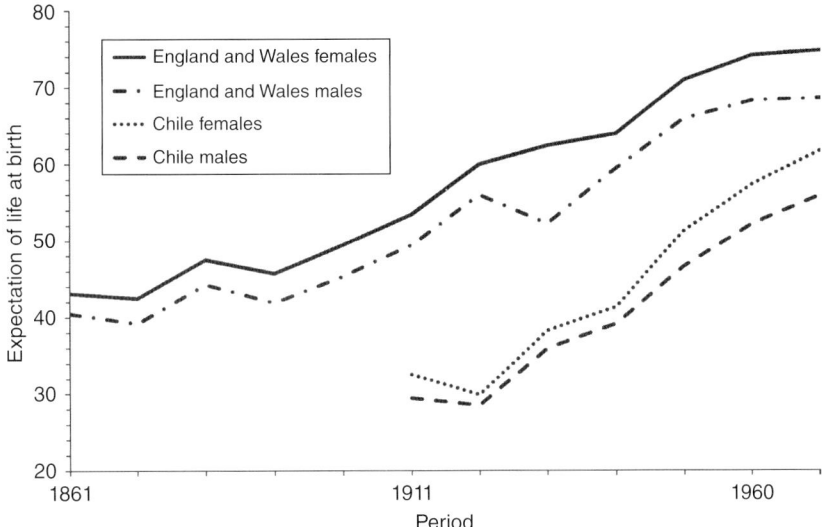

Figure 21.2. Expectation of life at 10-year intervals by sex for England and Wales, and Chile. *Source*: Preston et al. (1972). Figure reprinted from Gage (2005).

1720s based on family reconstitution methods and back projection. More research will be necessary to characterize demographic regimes prior to the beginning of the industrial era.

The general decline in mortality is thought to have occurred in developed nations as well as in low-income, less developed nations, although the timing and details of the decline clearly varied. For example, in Chile (Figure 21.2), the decline appears to have begun about 1920 and occurred at a more rapid rate than in England and Wales or Sweden. Consequently, the variation in mortality among nations (developed and less developed) has tended to converge. Baker (1987) and Gage (1989) have argued that mortality decline in the less developed nations began generally around 1920, which is interesting given the proximity of this date to the 1918–1919 influenza pandemic.

This is a particularly odd history. On the one hand, the decline in crisis mortality corresponds with a transition from a regime of epidemic to endemic infectious disease mortality. This could be due to simple population growth, to improvements in transportation, and/or to the growth of urbanism or all three, that is, effective population density. Judging by the 1918–1919 influenza pandemic, which spread rapidly around the world with very few populations omitted, by 1920, the world's population had become a single population with respect to infectious disease. On the other hand, Fenner (1970) argues on theoretical grounds that due to changing social organization (i.e., increases in effective population density), many new, specifically human infectious diseases emerged, such as measles, as well as a general intensification of infectious disease. It is ironic then that just at the point at which infectious disease is becoming endemic, *normal* mortality really begins to decline in earnest, not only in the now industrialized countries but perhaps on a worldwide basis. This leads to the heretical hypothesis that increasing population density might have something to do with the decline in infectious disease mortality.

Age-Specific Trends during the Demographic Transition

The secular trends in age-specific mortality are a new addition to the demographic transition, largely due to improvements in the quality of data available, such as the Human Mortality

Database (nd). Much of the traditional literature has assumed that the decline in mortality was largely due to declines in infant mortality and that adult mortality has largely been unaffected. Much of this perception can be attributed to the characteristics of expectation of life as a measure of the level of mortality. Expectation of life is much more sensitive to a single death averted during infancy than a single death averted at any older age. Lee and Carter (1992), on the other hand, have argued that relative mortality declined at a constant rate at all ages as total mortality declines; in other words that the difference in log of mortality is a constant (Figure 21.3). As a generalization, this is clearly more accurate than the previous view that all declines in total mortality were due to infant mortality.

Using the age-specific data from Sweden from the Human Mortality Database, I have argued that mortality declined more or less at all ages once normal mortality began to decline c. 1851 in Sweden. Infant mortality and mortality at ages over 40 declined the most in absolute terms (Figure 21.3), while childhood mortality declined the most in relative terms (Figure 21.2). However, the declines in adult mortality began earlier, around 1851, whereas infant and childhood mortality declines did not clearly begin until after 1901. Most of the declines in total mortality in the last 20–30 years have been due to declines in mortality at the older ages. The decline in infant mortality accounts for only about 25% of the decline in lifetime risk that accompanied the demographic transition.

DeWitte, Perry, and Kitson (all this volume) extend the age-specific results to periods leading up to the industrial era; using skeletal and family reconstitution data for England, they all report increases in infant (childhood) mortality and reductions in adult mortality prior to 1750. If expectation of life did not decline or declined only slightly during this period (see earlier), adult risk must have been far larger than the increase in the risk of infant death, since expectation of life is more sensitive to infant deaths. At least in England, the age pattern of mortality appears to have been changing, unless the decline in the expectation of life (increase in mortality) during this period was much larger than the data currently support. It remains to be determined if this is a general pattern or unique to England.

The Decline in Fertility

Fertility is probably more variable among populations than mortality. This is because fertility has a large cultural component. In any event, fertility in 1750 in European populations was relatively high within marriage, although many did not marry. In less developed countries, fertility may have been lower within marriage, but marriage was more universal. Taken together, fertility was similar in both types of population (Leridon, 1977). In any event, the *demographic transition* observes that fertility remained high after mortality began to decline. Figure 21.4 shows the decline in fertility for a developed country and a developing country. In the United States, fertility decline continued into late 1940s and early 1950s when the baby boom interrupted the decline, but has continued to decline since. The decline in fertility began much later in the developing countries, such as Costa Rica, where it occurred in the early 1960s and did so with breathtaking speed. Again, the demographic systems of developed and less developed countries have tended to converge.

Population Growth: How Will the Demographic Transition End?

The industrial demographic transition is not over yet. In those settings in which the decline in mortality preceded fertility, populations grew rapidly. In fact, the growth rate of the human population at least until the 1960s exceeded an exponential rate (von Foerster, 1960). von Foerster's modeling predicted the end of the human species in 2026 on the basis that that is the year his predictions indicated that the human population would reach infinity,

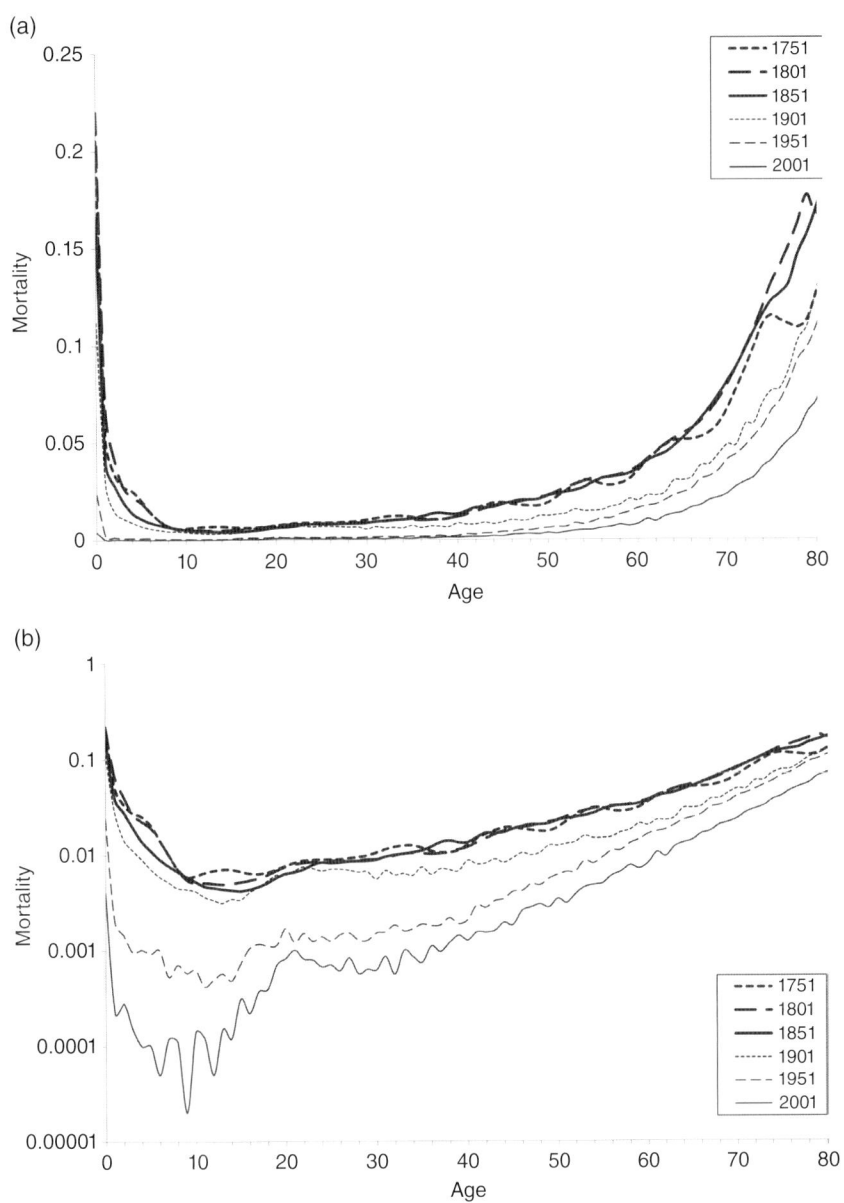

Figure 21.3. Age-specific male mortality (a) and log age-specific male mortality (b) 1751–2001 Sweden. *Source*: Human Mortality Database (data downloaded on 1/6/2005). Figure reprinted from Gage (2005).

which of course is not possible. While this notion seems patently ridiculous, world population was ahead of von Foerster's projections based on United Nations' world population estimates, until the mid-1990s (Figure 21.5). The decline in the growth rate has been quite recent and dramatic. Population growth has been relatively linear since the growth rate peaked in the 1960s (Lam, 2011). Whether mortality and fertility remain low in the future will depend upon whether human populations can maintain fertility at replacement levels and cope

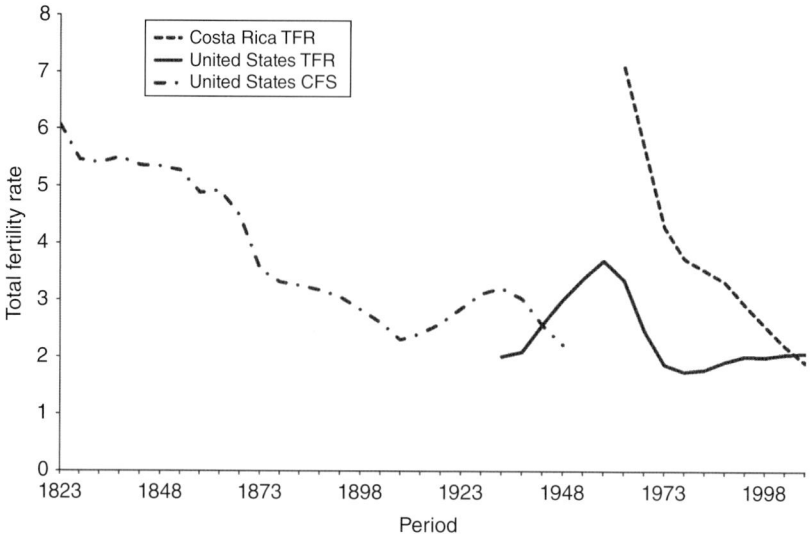

Figure 21.4. The total fertility rate, 1800–1990, United States and Costa Rica. *Source*: United Nations Demographic Yearbooks, various.

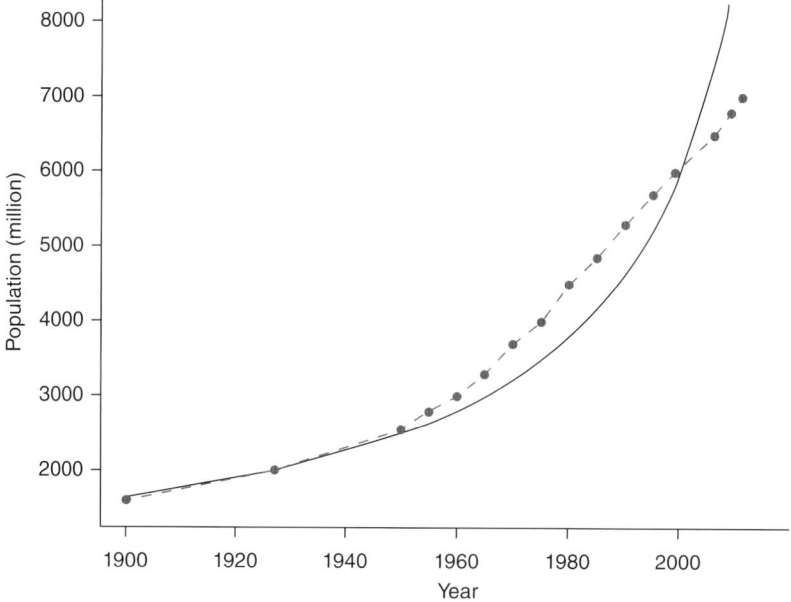

Figure 21.5. Human Population Growth, United Nations estimates (dotted line), and von Foerster et al. (1960) predictions (solid line). *Source*: United Nations, World Population Prospects: The 2010 Revision, and the World at Six Billion 1999.

economically with an age structure in which there are relatively few young people and many old people (see Lam (2011) for an optimistic view and Wood (1998) for a pessimistic view). Clearly, the demographic transition is not yet completed. But von Foerster's predictions suggest that it may not be long in coming.

The Epidemiologic Transition

The epidemiologic transition was developed as a companion piece to the demographic transition and describes the shift from a predominance of infectious disease mortality to a predominance of chronic disease mortality (Omran, 1977). Pretransition populations were subject to high death rates to infectious disease, evident by the periods of crisis mortality (Figure 21.1). During the decline in mortality, the relative frequency of deaths due to infectious disease declined, and the degenerative diseases became relatively more common. At some point, the total infectious deaths and total degenerative deaths reach parity, and the lines cross. The primary goal of Gage (2005) paper was to point out that the observed data concerning the incidence of degenerative deaths is highly biased by advances in medical diagnosis of cause of death. When statistically corrected using a surrogate measure of improvements in diagnosis (i.e., the proportion of deaths assigned uninterpretable causes of death, such as old age), there is evidence that the risk of degenerative deaths also declined with the general decline in mortality, with the exception of cancers, which increased at least until recently (Preston, 1976; Gage, 2005). The degenerative causes declined more slowly, however, than the infectious causes of death, so an epidemiologic transition still occurs. Clearly the risk of degenerative disease mortality did not increase with modernization, so modernization has generally been beneficial both with respect to infectious and degenerative disease mortality.

It is important to note that Omran (1977) was primarily concerned with the number of deaths, while my analysis was concerned with the risk of death. The number of deaths by cause is of great public health utility, such as in determining how many hospital beds will be needed for x or y, etc. On the other hand, risk of death is more useful for understanding the underlying biological and social mechanisms of the transition and thereby addressing the question of whether modernization has been bad for human health.

Several papers herein are concerned with the timing of epidemiologic transition and comparisons among countries of the point in time at which the degenerative diseases superseded the infectious diseases. If the comparisons are made on the basis of the observed number of deaths, then this has utility, particularly in the public health sphere. On the other hand, if an analysis is conducted using risk and standardized age structures (see Gage, 2005), then there is little comparative utility. The point of transition would change with the substitution of a different standard age structure. Alternatively, an analysis based on lifetime risk could be used. However, this would be essentially the same as imposing a standard age structure in which the number of individuals at each age is assumed constant across age. The illustration in Gage (2005) was largely to clarify the process, and the particular year of transition was not meaningful.

The same cautions apply to the interpretation of the observed numbers of deaths. The age structure of a population has an impact upon the point of transition in those observed data as well. To make matters more complicated, the age structure changes across the period of the transition. So the exact point of the observed transition will be influenced by the age-specific risks of death due to the various causes, the details of the age-specific total mortality, and the details of fertility (not to mention migration), all of which are time-dependent variables across the demographic and epidemiologic transitions. Finally, the dynamics of nonstable populations, which are not well understood, could influence the point of transition as well. Identifying the cause of the variation in the point of transition may be complex or counterintuitive. Thus, for example, a later epidemiologic transition in rural villages, compared to more urban villages, could be due to higher fertility and younger age distribution in rural agricultural villages.

The Morbidity Transition or Health Transition

Considerably less is known about the history of morbidity than mortality, largely because there are less reliable data concerning morbidity. It is a common assumption that mortality and morbidity are highly positively correlated and that morbidity follows the demographic and epidemiologic transitions. But this is not necessarily the case (Riley, 1987, 1990; Wood et al., 1992; Robine et al., 2003; Gage, 2005). Much of the literature on the relationship between morbidity and mortality, particularly studies of contemporary populations, involve degenerative disease morbidity at the older ages (Robine et al., 2003; Gage, 2005). But the lack of a necessary empirical correlation between mortality and morbidity with respect to infectious disease is also clear. Riley and colleagues (Riley, 1987; Alter and Riley, 1989) provide historical data indicating the disassociation of mortality and morbidity based on records of the Odd Fellows Society originally published by Watson (1903; cited in Alter and Riley, 1989). This account details the trends in morbidity from 1840 to 1900 in England and Wales, during a period of general decline in mortality. However, the number of sick days, a measure of morbidity, increased over the period. Of course, these data are questionable. But surveys in the modern era indicate that self-reported measures of morbidity have increased in Japan, the U.S., and Britain as mortality has declined, while in Hungary, where mortality increased from the 1960s into the 1980s, self-reported morbidity declined (Riley, 1990). On the other hand, Costa (2000) has shown that there is a long-term decline in chronic disease morbidity from 1910 to the present based on Civil War Veteran data compared to NHANES, a contemporary U.S. nutritional survey.

If mortality and morbidity can be either positively or negatively correlated, they should not be equated or confused. There is a need to clearly distinguish between the demographic and epidemiologic transitions (i.e., the mortality and cause of death transitions, respectively), and the health transition (i.e., morbidity transition). If generalizations can be made at this point, it is thought that early in the epidemiologic transition (or prior to it), incidence tends to be high, but the prevalence of morbidity is low because of the selective effects of mortality, and as the transition proceeds, incidence declines but mortality selection declines faster, so prevalence increases (Myers et al., 2003). More research is needed to determine if this is a reasonable generalization.

Traditionally, the frequency of skeletal lesions has been used as a measure of health and mortality (e.g., Cohen and Armelagos, 1984). However, skeletal lesions, particularly healed skeletal lesions, are best considered to be measures of morbidity. If the nature of the relationship, *negative* or *positive*, between morbidity and mortality depends upon other conditions, skeletal lesions cannot necessarily be used as evidence of mortality transitions (Wood et al., 1992). This does not mean that skeletal lesions are not useful. They could be used to examine health transitions. They can also be used to study the relationship of morbidity and mortality, that is the relative selective mortality of healthy vs. morbid individuals. DeWitte and colleagues (2008, 2010, 2012, this volume) have found that morbid individuals have higher mortality than healthy individuals.

A remaining problem is determining the incidence rates of morbidity in prehistoric populations, or the rate at which healthy people become morbid, using skeletal data. The traditional literature with respect to the agricultural transition assumes that incidence rates increased because the nutritional value of agricultural food stuffs are not as high as the products of hunting and gathering, and that therefore agricultural diets are more likely to be inadequate leading to morbidity. This literature also assumes that infectious diseases increase due to increases in population density, contact with domesticated animals, and the sedentary life styles of agriculturalists. On the other hand, an argument can be made that sedentary lifestyles might significantly reduce the case fatality rate of morbid individuals. However, this is largely hypothetical. What is needed is a way to test this hypothesis.

Other Transitions

The Height Transition

The examination of historical trends in height began as an attempt to empirically quantify McKeown's nutrition or standard-of-living hypothesis (McKeown, 1976; Floud et al., 1990). The problem is that stature is a complex outcome of energy balance, which includes intake, as well as, various expenditures, including responses to infectious disease morbidity (Scrimshaw, 2003; Gage et al., 2012). Consequently, it is not a measure of nutrition that is independent of morbidity or mortality. In any event, Floud and colleagues' initiative has resulted in historical documentation of the *height transition*, which is interesting in its own right. In general, height has increased with the industrial demographic and epidemiologic transitions. But again, the secular trends in height may not be monotonic. For instance, in the U.S., stature declined in the latter half of the 19th century before beginning to increase again (see Floud et al., 2011). It is interesting that this decline in stature occurs at about the same time that baseline mortality really began to decline significantly in the currently Westernized countries. Is it possible that the decline in stature in the U.S. in the late 1800s was due to increases in infectious disease morbidity associated with the early phase of the decline in mortality?

One very important advantage of height is that it is more likely to be available over a longer time span than any other variable. Garofalo (this volume) examines pre- and postmedieval growth in England and Wales. She reports no differences in stature, despite economic stagnation in the postmedieval period, but does note declines in activity levels (energy expenditure) between the two periods. Koepke (this volume) obtains similar results for Europe, as a whole from the 8th century BC to the 18th century AD. In fact, stature increases slightly throughout, although stature may be detrimentally affected during the Roman period. These results suggest that there may have been little overall change in morbidity during these periods or that declines in activity levels were sufficient to counteract any changes in morbidity or changes in any of the other components of energy balance. But what does stature tell us about mortality or morbidity?

There are clear associations of stature and mortality in historical and contemporary populations both at the individual and population levels (British Association for the Advancement of Science, 1883; Dublin et al., 1949; Gage and Zansky, 1995). In osteological populations, DeWitte has shown that stature (at the individual level) can be a risk factor for mortality in London during the Black Death (DeWitte and Hughes-Morey, 2012). Koepke (this volume) provides a particularly interesting analysis of the spatial and temporal distribution of height in Europe from the 8th century BC to the 18th century AD. She finds systematic spatial variation across Europe in stature, which has remained consistent over time. Similarly, Coale and Demeny (1983) report spatial variation in the shape of the mortality curve, which is roughly consistent with Koepke's spatial variation in stature, which persisted over the industrial *demographic transition*. Gage and O'Connor (1994) have found that this shape variation in the human mortality curve is highly correlated with dietary variation. Could the variation in age pattern of mortality and stature be related somehow through diet?

Subnational Transitions

Many of the chapters in this volume examine variation in the timing and dynamics of the demographic transition and the epidemiologic transition within, as well as, between nations. The demographic and epidemiologic transitions are intended as general models, which hopefully describe what all of these transitions have in common, not how they differ.

Nevertheless, studies of the variation in these transitions can be useful for identifying intranational commonalities of the transitions and variation that might contribute to a better understanding of the causes of the transitions. The problem that must be recognized is that intranational demographic analysis is more complicated than national demographic analysis, and the data are more likely to be flawed. For example, at least in Western European countries, intranational migration complicates data quality, analysis, and interpretations (Ruggles, 1992). The exception here is sex differences, which can be addressed at the national level. There is a vast literature on sex differences in all the transitions considered earlier except the height transition during the historic period, which is based largely on male military conscripts.

The most has been written about rural vs. urban differences in mortality in the early phases of the demographic transition. The early Bills of Mortality collected in cities in England and Wales were intended to alert the authorities to epidemics within a city. These data have been used by demographers to show that the number of deaths clearly outstripped the number of baptisms, which hopefully represent most live births. This has led some to the conclusion that cities were demographic sinks and would have disappeared except for continual rural-to-urban migration. This view was probably reinforced by the rather squalid conditions of early urban areas as recounted by observers of the time. It would certainly be valid to conclude that cities, closed to migration, would disappear if deaths outnumbered births. However, these cities were not closed to migration, which like birth is a way of entering the population of interest. The migrant's contribution to the number of deaths must be accounted for. In my view, the sensationalism of this description of the demographics of cities detracts from the science. I have always wondered what could have possessed people to leave rural areas for these deadly urban areas, unless the counterfactual of remaining in the rural areas was in fact more serious. Nevertheless, while the situation has been sensationalized, that does not mean that mortality was lower in urban areas than rural areas.

The conventional generalization of urban–rural differentials is that mortality was higher in urban than rural areas historically, but then shifted to higher rural than urban mortality in more recent times. This latter shift would constitute a rural–urban mortality transition, due perhaps to the availability of modern medicine. However, the generality of this description is in question. Woods (2003), a proponent of higher urban mortality, has recently reviewed the debate concerning historical urban–rural mortality differentials. Historical Chinese and Japanese data suggest that mortality may have been lower in urban areas compared to rural areas in northeastern Asia. It is possible that better urban social institutions (e.g., sanitation) in Asia kept urban mortality in check. However, the quality of these data is questionable. In France, where rural mortality also appears to have exceeded urban mortality historically, Woods argues that this is due to high mortality among urban-born infants who were placed with wet nurses in rural towns, which, in mortality data, would artificially increase infant mortality in towns while reducing it in cities. But of course it is not clear how much of the excess mortality is due to wet nursing, rural towns, and/or the interaction of rural towns and wet nursing. In England, the higher urban mortality is clearer. Woods argues that urban mortality exceeded rural mortality because childhood mortality and perhaps infant mortality were higher in urban areas. Early childhood mortality appears to be correlated with the log of population density, while adult mortality, and to a lesser extent infant mortality, is largely independent of population density (Woods, 2003). This more nuanced view of the rural/urban difference in England is certainly a much more reasonable concept than the original view of urban areas as demographic sinks.

Clearly, additional analysis will be necessary to resolve the historical urban–rural debate. However, if a simple cultural practice such as wet nursing can reverse the underlying rural–urban differential in France, or social institutions reduce urban mortality below rural mortality historically in Asia, how strong can rural–urban density differences be with respect to mortality?

A similar debate concerns higher rural mortality compared to urban mortality, particularly infant and childhood mortality, in contemporary populations. In this case, the argument is over whether there is or is not a rural–urban differential, after accounting for demographic, educational, and socioeconomic, etc., factors. Many studies suggest that after correcting for these factors, rural–urban differences are small or nonexistent. On the other hand, Sastry (1997) using a multilevel modeling approach with the Demographic and Health Survey of Brazil has shown that community-level variables can play a significant role. These results suggest that differences in medical care services, which tend to be better in urban settings, do not have a large impact, but that sanitary water supplies and community influences on individual behavior do. Perhaps community characteristics do make a difference.

Where Do We Go from Here?

Empirical models are only as good as the data, methods, and resulting assumptions upon which they are based. Fortunately, demographic data, even historic and prehistoric data, are becoming more available, and methods are becoming increasingly sophisticated.

More and Creative Use of Data

Even the best understood transition discussed earlier, the demographic mortality transition, is not well documented. What we would like to have in order to address this better is consistently collected and analyzed data from a reasonable sample of high-income, developed and low-income, developing countries. What we have instead are models cobbled together using a variety of countries and types of data and methods for different time periods. I used the data from Sweden earlier because it is the longest continuous series of relatively good mortality data based on standard national demographic methods currently available. Similar data for England and Wales do not have this same time depth. The earlier discussion of earlier periods is addressed with family reconstitution data and back projection methods with data from England and Wales (Wrigley and Schofield, 1981; Wrigley et al., 1997; Kitson, this volume). Data from the rest of Europe, and particularly the less developed countries, are even more sporadic. Budnik's (this volume) analysis of Polish data is therefore an important addition to this literature. Because of the current data limitations, it is important not to overinterpret the results, that is, overly complicate the general description of the transitions. Models are intended to describe the common features, not the unique features. Clearly, more data are needed.

What has surprised me over the last 20–30 years is that data, even historical data, which has to have been collected in the past, are increasingly available. In part, this is due to a movement toward data sharing by researchers, sometimes at the insistence of funding agencies. Anthropology has lagged behind on data sharing, due to the tradition of individual *ownership* of data, and even particular populations, but this is changing, too. Of course, standard national demography is based on administrative data collected by national governments and is available to anyone. The historical demographic data for Europe, particularly England and Wales, followed based on family reconstitution. This is largely due to the efforts of Wrigley and Schofield (1981) and Wrigley and colleagues (1997). However, historical demographic data are no longer restricted to Europe, largely due to the efforts of Lee and colleagues to assemble data for China and Japan (see Dong et al., nd). The Asian data sometimes contain information on migration at the individual level, information that is generally lacking in the West. Hopefully, more *historical* datasets will be rediscovered.

New national administrative data, not traditionally available, have also contributed to our understanding of the industrial demographic and epidemiologic transitions. The opening of historical census data at the individual level, such as the Integrated Public Use Microdata Series, U.S.A (IPUMS), provides a new source of historical data (e.g., Riley, 1990; Elman and Myers, 1997). Other similar projects from other countries include the North Atlantic Population Project and IPUMS International. The use of U.S. Civil War Veterans medical records is another administrative dataset, which has contributed greatly to our understanding of the health transition, (e.g., Costa, 2000). The historical *height transition* studies are based on administrative military records, the stature of conscripts (Floud et al., 1990).

In the less developed countries, national administrative data are generally lacking. Here the World Fertility Survey and later Demographic Health Survey are an important but underutilized data source, particularly across countries. The paper by Sastry (1997) concerning rural–urban differentials and community characteristics indicates the potential for creative use of these data sources. Another rich source of demographic data is the Matlab Health and Socio-Economic Survey, from rural Bangladesh.

Finally, anthropological data have begun to be assembled and made available. In particular, efforts have been made to gather consistent databases of skeletal remains. For instance, two particularly notable initiatives, the Global History of Health Project, which focuses on the Western Hemisphere and Europe, led by Steckel and Rose (2002),[1] and another in England, the Wellcome Osteological Research Database at the Museum of London (see Perry and DeWitte, this volume), are currently under way. These projects are critical to moving paleodemography ahead. In ethnography, the recent addition of Chagnon's Yanomamo data to the Interuniversity Consortium for Political and Social Research's collection is an important step forward.

Overall, there is every reason to believe that demographic data will continue to improve.

New and Improved Methods

The use of more sophisticated methods has also contributed to our understanding of demographic transitions. Some are simply improvements in the application of standard statistics. An example is the inclusion of an important covariate, which reduces misclassification bias, in regressions concerning the secular trends in cardiovascular disease (Preston, 1976; Gage, 2005). This actually reverses the direction of the secular trend. Another example is Sastry (1997), who employed multilevel regression models to examine rural–urban differences in health. This model allows individual-level and community-level covariates to be modeled properly in the same regression.

Over the last few decades, there have also been important improvements in the methods of anthropological demography, in particular, the issue of skeletal age estimation. In the past, in the anthropological literature, the conditional probability of a set of aging criterion given age (from a reference sample) has been assumed to be identical to the conditional probability of age given a set of aging criterion. Bocquet-Appel and Masset (1982) demonstrated that this assumption is incorrect and leads to estimates that are biased toward known-age reference samples. One solution to this problem is the application of Bayes' theorem. However, this was not recognized in the paleodemography literature until 2002 (Hoppa and Vaupel, 2002). In any event, theoretically appropriate aging techniques have only begun to be applied. Only a few

---

[1] http://global.sbs.ohio-state.edu/

examples are currently available, and it is not yet clear how this will affect our understanding of paleodemography (Gage et al., 2012). However, several of the skeletal demography papers here have exploited this new method (e.g., DeWitte; Perry, both this volume).

Other explicitly methodological improvements have included model life table systems for anthropological populations. The application of the Siler model, a parametric survival model, has made it possible to summarize and compare life tables derived from incomplete demographic data (Gage, 1988) and has provided a method for developing empirical model life tables (Gage and Dyke, 1986), as well as methods for summarizing risk into infant, exogenous, and senescent components (Gage, 2005). For instance, O'Connor (1995) used Gage and Dyke's method to summarize the published paleodemographic life tables. These will need to be revised given the recent advances in aging skeletons, particularly if they make a big difference in age estimates. Gurven and Kaplan (2007) employed the same method to provide a summary of extant anthropological population life tables. They also make use of the components of the Siler model to provide a number of alternative measures of mortality other than expectation of life, which better reflect changes in senescent mortality, and aging in these populations in general. DeWitte and colleagues (2008, 2010, 2012, this volume) have used components of the Siler model in the context of morbidity models to examine the effect of previous morbidity (i.e., frailty) on mortality.

Model life tables, such as those provided by Coale and Demeny (1983) or Gurven and Kaplan (2007), are empirical models, much like the demographic and epidemiologic transitions, and describe the general mortality characteristics of the populations they are applied to. Again, however, these are only as good as the data upon which they are based. The Coale and Demeny (1983) models are based almost exclusively on the European experience, post-1850 and with expectations of life at birth above 35 years. Earlier life tables, and higher mortality life tables, were rejected because they had *odd* patterns of mortality, thought to be due to errors in the data. All of the model life tables proposed by Coale and Demeny (1983) with expectations of life lower than 35 years were extrapolated from the results of those life tables with expectations of life above 35 years. However, by the 1850s, industrialization, and urbanization, was well under way in Europe, and it is possible that the high infant mortalities of high mortality life tables in the Coale and Demeny series are due to these issues. Similarly, the Gurven and Kaplan (2007) series for anthropological populations consist of life tables from populations post-1920, which were exposed at least sporadically to modern human infectious disease, maintained endemically by the world population. Clearly, this is a considerably different disease environment than that which prehistoric hunter-gatherers, and even prehistoric agriculturalists, were exposed to. In this regard, use of the Siler model or something similar, which brings none of these assumptions with it, is extremely useful.

There are also indirect demographic methods that do not require standard demographic data. Some such as the empirical model life tables, particularly those by Coale and Demeny (1983), have been widely used, but others have not. Some of these methods could be very useful, particularly in interdisciplinary approaches to the demographic transition and epidemiologic transition. In this regard, the variable R methods (Preston and Coale, 1982), some of which have been applied anthropologically (Gage et al., 1984a,b), are an important advance. These indirect techniques do not require stationary or stable assumptions, but have been underutilized. For example, all of the life tables used by Gurven and Kaplan (2007) were estimated using conventional demographic methods, despite the significant advantages of variable R methods (Gage and Dyke, 1986). Another variable R method proposed by Preston and Coale (1982) requires only age-at-death data, but does not assume that the population is stationary or even stable. This method might be useful for historical age-at-death series but requires good control of year-of-death data. Hence this might not be useful with prehistoric skeletal data. To my knowledge, this method has never been applied. Finally,

Gage (2010) provides a comprehensive discussion of an indirect method that can in theory provide mortality and fertility estimates based on stable population theory and that is compatible with Bayesian age estimation from an age-at-death distribution alone. This might be applicable to paleodemographic populations, although large sample sizes will be needed. Fortunately, movements to assemble large databases from skeletal collections are now well under way.

Conclusion

Our understanding of the demographic and epidemiologic changes that coincided with the industrial revolution is currently based on a rather small and unrepresentative sample of the world population. The entire model is patched together using disparate data and methods. Nevertheless, the trends during the industrial period are reasonably well understood, at least with respect to mortality. The issues with respect to morbidity and perhaps height are not as well established. The conclusion of the demographic transition is not yet understood at all, at least empirically. However, given the recent trends with respect to increases in the volume and type of data, not to mention creative use of data, and improvements in the sophistication of methods, a clearer model of the general trends will emerge. Even an understanding of the conclusion to the demographic transition may not be that far off.

Acknowledgments

I would like to thank Sharon DeWitte for her comments on the manuscript.

References

Alter G, Riley J. 1989. Frailty, sickness and death: models of morbidity and mortality in historical populations. Popul Stud 43:25–45.
Baker PT. 1987. Modernization and human biological responses. In: Harrison GA, Tanner JM, Pilbeam DR, Baker PT, editors. Human biology. New York: Oxford University Press. p 529–543.
Bocquet-Appel JP, Masset CL. 1982. Farewell to paleodemography. J Hum Evol 11:321–333.
British Association for the Advancement of Science. 1883. Final report of the anthropometric committee. London: British Association.
Caldwell JC. 2001. Demographers and the study of mortality. Ann N Y Acad Sci 954:19–34.
Coale AJ, Demeny P. 1983. Regional model life tables and stable populations. New York: Academic Press.
Cohen MN, Armelagos GJ. 1984. Paleo-pathology at the origins of agriculture. Orlando: Academic Press. p 615.
Costa DL. 2000. Understanding the 20th century decline in chronic conditions among older men. Demography 37:53–72.
DeWitte SN, Wood JW. 2008. Selectivity of Black Death mortality with respect to pre-existing health. Proc Natl Acad Sci U S A 105:1436–1441.
DeWitte SN, Bekvalac J. 2010. Oral health and frailty in the Medieval English cemetery of St. Mary Graces. Am J Phys Anthropol 142:341–354.
DeWitte SN, Hughes-Morey G. 2012. Stature and frailty during the Black Death: the effect of stature on risks of epidemic mortality in London, A.D. 1348–1350. J Archaeol Sci 39:1412–1419.
Dong H, Lee J, Campbell C. nd. New sources of global social science: historical panel data from East Asia. http://epc2012.princeton.edu/papers/121125 (accessed October 30, 2012).
Dublin LI, Lotka AJ, Spiegelman M. 1949. Length of life: A study of the life table. New York: The Ronald Press.

Elman C, Myers G. 1997. Age and sex differentials in morbidity at the start of an epidemiological transition: returns from the 1880 U.S. census. Soc Sci Med 45:943–956.

Fenner F. 1970. The effects of changing social organization on the infectious diseases of man. In: Boyden SV, editor. The impact of civilization on the biology of man. Canberra: Australian National University Press. p 48–76.

Flinn MW. 1984. The stabilization of mortality in preindustrial Western Europe. J Eur Econ Hist 3:285–318.

Floud R, Wachter K, Gregory A. 1990. Height, health and history: Nutritional status in the UK, 1750–1980. Cambridge: Cambridge University Press.

Floud R, Fogel RW, Harris B, Hong SC. 2011. The changing body: Health, nutrition, and human development in the western world since 1700. Cambridge: Cambridge University Press.

Gage TB. 1988. Demographic estimation from anthropological data: a review of the non-stable methods. Curr Anthropol 26:644–647.

Gage TB. 1989. Biomathematical approaches to the study of mortality. Yearb Phys Anthropol 32:185–214.

Gage TB. 2005. Are modern environments really bad for us? Revisiting the demographic and epidemiologic transitions. Yearb Phys Anthropol 48:96–117.

Gage TB. 2010. Demographic estimation: indirect techniques for anthropological populations. In: Larsen C, editor. A companion to biological anthropology. New York: Wiley-Blackwell Publishing. p 179–193.

Gage T, Dyke B. 1986. Parameterizing abridged mortality tables: the Siler three-component hazard model. Hum Biol 58:275–291.

Gage TB, O'Connor K. 1994. Nutrition and variation in level and age pattern of mortality. Hum Biol 66:77–103.

Gage TB, Zansky SM. 1995. Anthropometric measures of nutritional status and the level of mortality. Am J Hum Biol 7:679–691.

Gage TB, Dyke B, Riviere PG. 1984a. Estimating mortality from two censuses: an application to the Trio of Surinam. Hum Biol 56:489–502.

Gage TB, Dyke B, Riviere PG. 1984b. Estimating fertility and population dynamics from two censuses: an application to the Trio of Surinam. Hum Biol 56:691–701.

Gage TB, DeWitte SN, Wood JW. 2012. Demography part 1: mortality and migration. In: Stinson S, Bogin B, O'Rourke D, editors. Human biology: An evolutionary and biocultural perspective. Hoboken: Wiley-Blackwell. p 695–756.

Gurven M, Kaplan H. 2007. Longevity among hunter-gatherers: a cross-cultural examination. Popul Dev Rev 33:321–365.

Hoppa RD, Vapel JW. 2002. Paleodemography, age distributions from skeletal samples. Cambridge: Cambridge University Press. p 259.

Human Mortality Database. nd. University of California, Berkeley (USA), and Max Planck Institute for Demographic Research (Germany). Available at: www.mortality.org or www.humanmortality.de (accessed November 6, 2013).

Lam D. 2011. How the world survived the population bomb: lessons from 50 years of extraordinary demographic history. Demography 48:1231–1262.

Lee RD, Carter LR. 1992. Modeling and forecasting U.S. mortality. J Am Stat Assoc 419:659–671.

Leridon D. 1977. Human fertility. Chicago: University of Chicago Press.

Lesthaeghe R, Neidert L. 2006. The second demographic transition in the United States: exception or textbook example? Popul Dev Rev 32:669–698.

McKeown T. 1976. The modern rise of population. New York: Academic Press.

Myers GC, Lamb VL, Agree EM. 2003. Patterns of disability change associated with the epidemiologic transition. In: Robine JM, Jagger C, Mathers CD, Crimmins EM, Suzman RM, editors. Determining health expectances. New York: John Wiley and Sons. p 59–74.

O'Connor KA. 1995. The age pattern of mortality: a micro-analysis of Tipu and a meta-analysis of twenty-nine paleodemographic samples. Ph.D. thesis. Albany: University at Albany, State University of New York.

Omran AR. 1977. Epidemiologic transition in the United States: the health factor in population change. Popul Bull 32:1–41.

Preston SH. 1976. Mortality patterns in national populations. New York: Academic Press.
Preston SH, Coale AJ. 1982. Age structure, growth, attrition and accession: a new synthesis. Popul Index 48:217–259.
Preston SH, Keyfitz N, Schoen R. 1972. Causes of death: Life tables for national populations. New York: Seminar Press.
Riley J. 1987. Disease without death: new sources for a history of sickness. J Interdiscip Hist 42:537–563.
Riley J. 1990. The risk of being sick: morbidity trends in four countries. Popul Dev Rev 16:3:403–432.
Robine JM, Jagger C, Mathers CD, Crimmins EM, Suzman RM. 2003. Determining health expectancies. New York: John Wiley and Sons.
Ruggles S. 1992. Marriage, migration and mortality: correcting sources of bias in English family reconstitutions. Popul Stud 46:507–522.
Sastry N. 1997. What explains rural-urban differentials in child mortality in Brazil? Soc Sci Med 44:989–1002.
Scrimshaw NS. 2003. Historical concepts of interactions, synergism and antagonism between nutrition and infection. J Nutr 133:316s–321s.
Steckel RH, Rose JC. 2002. The backbone of history: Health and nutrition in the western hemisphere. Cambridge: Cambridge University Press.
von Foerster H, Mora PM, Amiot LW. 1960. Doomsday: Friday, 13 November, A.D. 2026. Science 132:1291–1295.
Wood JW. 1998. A theory of preindustrial population dynamics: demography, economy and well-being in Malthusian systems. Curr Anthropol 39:99–135.
Wood JW, Milner GR, Harpending HC, Weiss KM. 1992. The osteological paradox. Curr Anthropol 33:343–370.
Woods R. 2003. Urban-rural mortality differentials: an unresolved debate. Popul Dev Rev 29:29–46.
Wrigley EA, Schofield RS. 1981. The population history of England 1541–1871: A reconstitution. Cambridge: Harvard University Press.
Wrigley EA, Davies RS, Oeppen JE, Schofield RS. 1997. English population history from family reconstitution 1580–1837. Cambridge: Cambridge University Press.

Index

Page numbers in *italics* denote figures, those in **bold** represent tables.

accelerated model of epidemiologic
 transition 357
adaptation 339–52
Aedes aegypti 25, 26–7
African Americans 243–64
 age of death 250, **250**
 dental health 246, **248**, **252**, 253
 emancipation 257–8
 infectious diseases 250–3, **251–2**
 medical care 259
 skeletal data 246–9, **247**, **248**
 skeletal pathologies 250–3, **251–2**
 stature 244–5, **254**
age at death 41
 African Americans 250, **250**
 distributions **39**, 41, 44
 Euro-Americans 250, **250**
 infants 227–9, 232–3, *233*
  gestational weeks *234*
  pathological conditions *235*
age effects in epidemiology 362
age estimation 39, **39**, 49, 227–8, 236, 341, 388–9
age of man-made diseases 151, 344
Age of Pestilence and Famine 237–9
age-specific mortality rates 37, 41–2, 44–5, **45**, 46, *98*, *99*, 379–80
 influenza 123–35
 men *381*

agent-based modeling 105–22
 advantages 118–20
  historic/ethnographic background 118–19
  modeling of historic populations 119
 influenza pandemic (1918) 105–22
  behaviors and movement 114
  disease process 114–15
  model behavior 115–16, *116*
  model structure 113
  sensitivity analyses 116–18, **117**
 scope 107–8
 use of 119–20
agropastoral populations 309–10
Albany, New York 329–34, **331**, **332**, **334**
allergic disease 10, 302–3
 geographic distribution 304–7
American Civil War 245, 256–7
Americas
 archaeoparasitology 327–34, **331**, **332**, **334**
 see also USA
animal reservoirs of disease 18
Antebellum Puzzle 244–5
archaeoparasitology 2, 11, 312–14, 321–36
 Americas 327–34, **331**, **332**, **334**
 Europe 322–7, **324–6**
autoimmune disease 10, 274, 302–3
 geographic distribution 304–7

Bayes' theorem 388–9
beef consumption, and stature 58, 64, *65*
bioarchaeology 2, 36
   infant morbidity/mortality 225–41
Biological State Index 144, 156, 157
bog bodies 314
burials
   Columbia, Missouri 165–7, *166*, *167*, *170*, *172*
   infants 226–7
   London 37–9
   see also skeletal data

Canadian influenza pandemics (1890/1918) 123–35
   death rate *128*, *130*, *132*
   deaths from all causes *131*, *132*
   population size 127–8, **127**
caries, dental **252**
caste system 206, 216–17, *217*
causal web 354–5
cause-of-death reporting 2, 7, 36–7, **94–5**
   India 202
   Poland 145
   reliability and accuracy 87–90, **89**
   skeletal data 9, 36
   see also specific causes
children 9
   environmental hazards 202
   frailty 47–8
   infectious diseases 202
   lead exposure 279–99, **286–91**
      lead isotope ratios **290–1**
      natural levels 285, *292*, 293
      skeletal material 284–5
      sources of exposure 295–6, *295*
      study sites 281–4, *282*
   low- and middle-income countries (LMICs) 202
   malnutrition 24
   mortality *182*, 192–3, **193**
cholera **94**, **95**, 97, 146, 183, 190, 302, 353
   epidemiology 354
   mortality rate *98*
chronic inflammatory disorders (CID) 7–8, 301, 303, 304, 305
   allergic/autoimmune disease 302–3, 304–7
   infectious-chronic disease distinction 360–1
   unequal burden 310–12
classical theory of epidemiologic transitions 180–1, 201, 356–7, 359–60, 362
climate change, nutritional effects 58, 65–6, **71**, **72**
cohort effects 363

Columbia, Missouri 163–77
   burials 165–7, *166*, *167*
      by age group *170*, *172*
   death certification 168, 173
   history and demographics 165–8
   infectious diseases 168, 173
   mortality patterns 171–2
communicable diseases *see* infectious diseases
consumption *see* tuberculosis
contemporary model of epidemiologic transitions 357
coprolites 312–14, 322, 329
Coventry, childhood lead exposure 281, 283, 285
cremains, stature estimation from 64
cribra orbitalia 40, **45**, 46–7
   infants 232, 234, 236
   and iron-deficiency anemia 293
crowd diseases 118, 308
crowd parasites 328–9
Crow's I_m index 143–4, 156
cultivation methods 59, 66, **71**, **72**
   three-field rotation 66
cultural changes 32, 120, 239, 249, 260, 267

dairy consumption, and stature 58, 64, **71**, **72**
death registration 83–90, *85*, 168, 173
   reliability and accuracy 87–90, **89**
   see also cause-of-death reporting
degenerative diseases 310, 344, 348
   mortality rate *97*
demographic transition 4–5, 345–6, *346*, 377, 378–82, *378*, *379*
   age-specific trends 379–80
   fertility decline 380
   population growth 380, 382, *382*
   research on 369–71
dental health
   abscesses **252**
   African Americans 246, **248**, **252**, 253
   caries **252**
   Euro-Americans 246, **248**, **252**, 253
   see also teeth
developed nations
   infectious diseases 303
   life expectancy *379*
   see also individual countries
Developmental Origins of Health and Disease Hypothesis (DOHaD) 7, 348
   influenza pandemics 123–35
diet 74
   beef consumption 58, 64, *65*
   dairy consumption 58, 64, **71**, **72**
   and stature 56–7
   Western 203

dirt 322–7, *324–6*
disability-adjusted life years (DALYs) 358
disease
   crowd diseases 118, 308
   demographic impact 59, 66, **71, 72**
   emergence of 18
   germ theory 6, 163, 180, 340, 345, 354, 355, 358, 360, 373
   see also individual diseases
disease-specific mortality rates
   consumption, tuberculosis and phthisis *96*, 97
   degenerative diseases *97*
Drinker, Elizabeth Sandwith 22, 28–9, 30
dysentery 18, 24, 29–31, **94**, 245

Ebola virus 371
ecobiology 180, 358
edentulism **252**
education 374
enamel hypoplasia 47, **252**, 253
England and Wales
   childhood lead exposure 279–99
   Industrial Revolution *see* Industrial Revolution
   life expectancy 183, *379*
   parish registers 185
   poor relief system 183–4
   population growth 188
   second epidemiologic transition 179–84
   see also London
environmental factors 10–11, 17–34, 35–53, 267–78
   risk allocation 268–74
      lead exposure 269–72, *270*
      organic compounds 272–4
   social stratification 267–8, 274–5
   total environment 17–34
   see also specific locations
epidemics 7–8, 374–5, *375*
   see also individual diseases
epidemiologic transition theory 1–2, 4–11, 139, 302, 356–9, 377, 383
   First Epidemiologic Transition *see* First Epidemiologic Transition
   Second Epidemiologic Transition *see* Second Epidemiologic Transition
   Third Epidemiologic Transition 340, 349–50, 357
   Fourth Epidemiologic Transition 357
   archaeoparasitology 2, 11, 312–14, 321–36
   challenges to 358–9
   classic 180–1, 201, 356–7, 359–60
      scope of 362
   critiques 357–8

   as descriptive epidemiology 360
   future studies 387–90
   and hygiene hypothesis 307–10
      agropastoral populations 309–10
      degenerative disease 310
      Paleolithic baseline 308–9
   infectious-chronic disease distinction 360–1
   low- and middle-income countries (LMICs) 363–5
   modifications 357
   Omran's theory 180–1
   role in education of epidemiologists 359–60
epidemiologic triangle/triad 354–5
epidemiology 353–68
   age effects 362
   causal web 354–5
   cohort effects 363
   complex systems approach 356
   critical distinction 361–2
   history 353–4
   outcomes 355–6
   period effects 362–3
   social 356
   socio-ecological model 355
   see also epidemiologic transition theory
ethnographic background 118–19
Euro-Americans 243–64
   age of death 250, **250**
   dental health 246, **248**, **252**, 253
   infectious diseases 250–3, **251–2**
   skeletal data 246–9, *247*, **248**
   skeletal pathologies 250–3, **251–2**
   stature 244–5, **254**
Europe
   archaeoparasitology 322–7, **324–6**
   England and Wales *see* England and Wales; London
   nutritional studies 55–80
      data acquisition 60–1
      determinants of stature 58–60, 64–7, 70–3, **71, 72**
      estimation of stature 61–4
      temporal trends 67–70, *67*
   Poland 139–61
evolutionary paradigm 339–52

family histories 185–7, **186**
   reconstitutable minority 186, *187*
fecal-borne parasites 312–14, 322–7, *324–6*, 330–1, 332–3, **332, 334**
femur length 61–2
fertility 144, 185, 370, *382*
   decline in 380
   and infant mortality 239

filth-borne parasites 322–7, **324–6**, 329–34, **331**, **332**, **334**
First Epidemiologic Transition 309–10, 340
  Americas 327–9
  Europe 322–7, **324–6**
food
  lead contamination 280–1
  preservation 373
  see also diet
Fourth Epidemiologic Transition 357
frailty 37, 40, 42–4, *43*, 45, **45**, 46–8

gender inequality 60, 67, **71**, **72**
  India 201–23
  mortality rate 44, 45–6, **46**, 48–9
  see also men; women
geochemistry 2
germ theory 6, 163, 180, 340, 345, 354, 355, 358, 360, 373
Gompertz-Makeham mortality model 41, 43, 44

health 268, 374
  inequalities 311
  transition 5, 384
height transition 385
heirloom diseases 303–4
Helicobacter pylori 35, 304
helminths 314, 322
  see also individual species
Henry, Louis 185
historic populations
  Columbia, Missouri 163–77
  Holyoke, Massachusetts 90–9, *93*, **94–5**, *96–9*
  India 201–23
  London 35–53, 183
  Massachusetts 81–101
  Philadelphia 17–34
  Poland 139–61
HIV/AIDS 358, 371
Holyoke, Massachusetts 90–9, *93*, **94–5**, *96–9*
hookworm 256, 257, 328
human behavior 18
  and mortality rates 48–9
hygiene hypothesis 301–20
  allergic and autoimmune disease 302–3, 304–7
  chronic inflammatory disorders 310–12
  and epidemiologic transitions 307–10
    agropastoral populations 309–10
    degenerative disease 310
    Paleolithic baseline 308–9
  heirloom and souvenir microorganisms 303–4
  infectious diseases 310–12
  paleoparasitology 312–14
  see also epidemiologic transition theory

in-migration 243–64
India, gender inequalities 201–23
  anthropometric data 207–10, **208**, *209*, **210**
    adolescent growth 212
    stature and age 212
  caste system 206, 216–17, *217*
  causes of death 202
  childhood place of residence 212
  economic growth 205
  ethnic divisions 206
  life expectancy 203–4, *203*
  macroeconomic data 210–12, *211*, **212**
  mortality biases 212
  political and economic changes 204–5
  population growth 205
  postcolonial period 204–5
  regions **211**
  religious divisions 206
  states *211*
  study methods 212–13
  welfare 205
    by caste 216–17, *217*
    determinants of 218–20, **219**
    levels and trends 213–14
    and religious affiliation 215–16, *216*
    rural-urban differences 217–18, *218*
    and stature 214–15, *214*, *215*
  Western diet 203
industrial demographic transition *see* demographic transition
industrial populations
  age and sex estimation 39–40, **39**
  skeletal data 38–9
Industrial Revolution 179–97, 230, 244, 345
  childhood lead exposure 279–99
  infant morbidity/mortality 237–9
industrialization 8, 184
  health consequences of 2
  and life expectancy *150*
  mortality patterns 35–53
infants
  age definition 226
  lead exposure 272
  see also children
infant morbidity/mortality 47, 146–8, **147**, *148*, 181, 182–3, *182*, 192–4, **193**, *194*
  bioarchaeological studies 225–41
  differential burial 226–7
  Industrial Revolution 237–9
  morbidity patterns 229–30
  mortality profiles
    age-of-death 227–9, 232–3, *233*
    gestational weeks *234*
  pathological conditions *235*

Second Epidemiologic Transition 230–7
  weaning deaths 226, 228–9
infectious diseases 17–34, **248**
  African Americans 250–3, **251–2**
  decline in mortality *see* Second Epidemiologic Transition
  developed nations 303
  Euro-Americans 250–3, **251–2**
  unequal burden 310–12
  United States of America (USA) 17–34, 168, 173
  see also epidemics; pandemics; and individual diseases
infectious-chronic disease distinction, 3, 78, 348–9
influenza pandemics 7–8
  agent-based modeling 105–22
    behaviors and movement 114
    disease process 114–15
    model behavior 115–16, *116*
    model structure 113
    sensitivity analyses 116–18, **117**
  Canada (1890/1918) 123–35
  early exposure and mortality risk 123–35
  Newfoundland and Labrador 105–22
  Russian influenza 125, 126
  SEIR framework *112*
  Spanish flu 123–4, 126, 370–1
interdependence 374–5, *375*
International Classification of Diseases (ICD) 82, 361
iron-deficiency anemia 293

Lactobacillus spp. 309, 310
land *per capita*
  and nutritional status 58, 64–5, **71**, **72**
  temporal changes 65
latrines, sediment analysis 312–14, 322–7, *324–6*, 330–1, 332–3, **332**, **334**
lead exposure 344
  18th/19th centuries 280–1
  children 279–99, **286–91**
    lead isotope ratios **290–1**
    skeletal material 284–5
    sources of exposure 295–6, *295*
    study sites 281–4, *282*
  food contamination 280–1
  health effects 280
  increase in 279
  infants 272
  occupational 281
  risk of 269–72, *270*
lice 308
life expectancy 148–51, 183
  by age and area of residence *149*
  by degree of industrialization *150*
  developed nations *379*
  low- and middle-income countries (LMICs) *379*
  men 203, *378*, *379*
  temporal changes *151*
  women *379*
  see also individual locations
living standards 2, 69, 74, 180, 213, 220–1, 374
London 35–53, 183
  age estimation 39, **39**
  age patterns of mortality 41–2, 44–5, **45**, 46
  age-at-death distributions **39**, 41, 44
  cemeteries 37–9, **231**
  childhood lead exposure 283–5
  frailty patterns 42–4, *43*, 45, **45**, 46–8
  industrial 38–40, **39**
  osteological stress markers 40, **45**, 46–7
  preindustrial 38, 39–40, **39**
  sex differences in mortality 44, 45–6, **46**, 48–9
  sex estimation 39–40
  skeletal data 37–9
    Broadgate Cemetery 38
    Cross Bones Cemetery 38–9
    St. Bride's Lower Churchyard 38
    St. Thomas' Hospital Cemetery 38
  see also England and Wales
low- and middle-income countries (LMICs) 1–2, 4, 202, 340, 342
  double disease burden 363–4
  fertility *382*
  life expectancy *379*
  risk transition 363–5

McKeown, Thomas 1, 56, 33, 71, 79–80, 101, 118, 133, 139, 141, 149, 332–4, 345, 346, 385
McKeown thesis 5–6, 344–6, *345*, *346*
malaria 18
malnutrition 24, **248**
  and stature 56–7
  see also diet
Malthus, Thomas 180–1
'man-made' diseases 10, **97**, 279, 344
Massachusetts 81–101
  causes of death **94–5**
  death registration 83–90
    Registration Act (1842) 84–5, *85*
    reliability and accuracy 87–90, **89**
  mortality rate
    age-specific *98*, *99*
    consumption, tuberculosis and phthisis *96*
    degenerative diseases *97*
    Northampton and Holyoke 90–9, *93*, *96–9*

maw-worm (*Ascaris lumbricoides*) 322–7, **324**, 331, **332**, 333
measles 23–5, 145, 229, 303
  1916 epidemic 111
  mortality from 28, 30, 111
  population density effects 20
men
  life expectancy *203*, *378*
  mortality rate 48
    age-specific *381*
  stature estimation 62–3
mercury poisoning 22–3, 273
methods in epidemiology 369–76
microbial deprivation hypothesis 306
military, spread of smallpox by 21–2
model life tables 389
modernization 180, 184, 370
morbidity 5
  infants 229–30
  and socioeconomic status 364–5
  transition 384
mortality rates 5, 5185
  age-specific *see* age-related mortality rates
  Canada *128–32*
  cause of death *see* cause of death
  children *182*, 192–3, **193**
  disease-specific
    consumption, tuberculosis and phthisis *96*
    degenerative diseases *97*
  gender inequality 44, 45–6, **46**, 48–9
  and industrialization 35–53
  infants *see* infant mortality/morbidity
  Massachusetts *93*
  Poland 143, 145–8, **146**, *147*, **147**, *148*
    changes in 151–6, *152–5*
  tuberculosis *96*, *345*
mummified remains, gut contents 314

natural selection indices 143–4, 156–7, **157**
Newfoundland and Labrador *109*
  history and demographics 108–12
  influenza pandemic 105–22
non-communicable diseases (NCDs) 1, 201, 202, 340, 348, 354
  allergic and autoimmune disease 10, 302–3, 304–7
  degenerative disease 97, 310, 344, 348
  in LMICs 363–4
  management 349
  prevention 348–9
  see also degenerative diseases
Northampton, Massachusetts 90–9, *93*, **94–5**, 96–9
novel environments 346–7
nutritional status *see* diet

old friends hypothesis 306, 309
Omran, Abdel 4, 8, 9, 10, 35, 55, 81, 105, 123, 151, 163, 195, 201, 203, 230, 243, 254, 260, 279, 307, 1334, 1747, 21718
osteological stress markers 40, **45**, 46–7
osteomyelitis 246, **251**, 253, 255
overcrowding 24, 133, 146, 183, 258, 283
  see also population growth

palatine fever 28
paleodemographic signatures 6, 36, 37
paleoimmunology 315
Paleolithic baseline 308–9
paleoparasitology 312–14
pandemics 7–8
  influenza 7–8, 105–22
parish registers 185
pathoecology 321–2
Pediculus humanus 308
period effects 362–3
periosteal lesions 40, **45**, 46–7
Philadelphia, Pennsylvania 17–34
  measles 23–5
  physical environment 19
  smallpox 20–3
  yellow fever 25–30
phthisis *see* tuberculosis
pinworm (*Enterobius vermicularis*) 256, 308, 328–9
pleurisy 23
Poland 139–61
  cause of death 145
  death rates 143
  Eastern Pommerania 141
  Great Poland 141–2
  history and demographics 140–1
  indices of opportunity for natural selection 143–4, 156–7, **157**
  infant mortality 146–8, *147*, *148*
  life expectancy 148–51, *149–51*
  mortality rate 145–8, **146**, *147*, **147**, *148*
    changes in 151–6, *152–5*
  Silesia 142
  smallpox 152, 154
  tuberculosis 155–6, *155*
pollution 275
  history 274–5
  lead 269–72, *270*, 279–99
  mercury 22–3, 273
  organic compounds 272–4
  see also environmental factors
polychlorinated biphenyls 273–4
population growth 18, 181, 188, 205, 370–1, 380, 382, *382*
porotic hyperostosis 244, **248**, **251**, 256

Potential Gross Reproductive Rate 144, 157
precision medicine 355
preindustrial populations
    age and sex estimation 39–40, **39**
    skeletal data 38
privies 332–3
    see also fecal-borne parasites
protein consumption, and stature 58, 64, *65*
pseudo-commensals 304, 308

race 9–10
reconstitutable minority 186, 187
regional variation 8
religious affiliation and welfare 215–16, *216*
religious nonconformity 187
research 369–76
    agenda 371–3, **371**
    demographic transition 369–71
rickets 47, 232, 237, 246, **248**, **251**, 256, 275
risk allocation 268–74
    lead exposure 269–72, *270*
    organic compounds 272–4
risk of death *see* frailty
Romanization, demographic impact of 59–60, 67, **71**, **72**
roundworm 256, 313
    see also maw-worm
Russian influenza 125, 126
    see also influenza pandemics

sanitation 126, 163–4, 173, 312–14
scrub typhus 309
scurvy 47, 237
Second Epidemiologic Transition 1–13, 17, 310, 340–4, 349–50
    agent-based modeling 105–22
    application of 347–9
    causes 5–7, 73–5
    consequences of 373–5, *375*
        education 374
        epidemics and interdependence 374–5, *375*
        health and living standards 374
    context 344–6, *345*, *346*
    current state of knowledge 377–92
    data sources 341–2
    degenerative diseases **97**, 310, 344
    English context 179–84
    environmental factors 10–11
        Europe 55–80
        London 35–53
        Massachusetts 81–101
        Philadelphia 17–34
    epidemics and chronic disease 7–8
    epidemiologist's perspective 353–68
    evolutionary paradigm 339–52
    hygiene hypothesis 301–20
    Industrial Revolution 179–97
    infant morbidity/mortality 230–7
    levels of variation 342–4
    methodological perspective 369–76
    novel environments 346–7
    regional and temporal variation 8
    small towns/cities 163–77
    underrepresented communities 9–10
    see also specific locations and factors
Sedgley, England 179–97
    industrial development 189
    mortality patterns 190–5, *191*
        infants and children 192–3, **193**
    population growth 188–9
    urbanization 189–90
SEIR framework 112–13, *112*
sensitivity analysis 116–18, **117**
sewer systems 330
sex differences *see* gender inequality
sex estimation 39–40
sex hormones, and morbidity/mortality 48
Siler mortality model 41–2, 43
skeletal data 9, 36, 37–9
    African Americans 246–9, **247**, **248**
    age estimation 39, **39**, 49, 227–8, 236, 341, 388–9
    Euro-Americans 246–9, **247**, **248**
    Europe 55–80
    infants 227
    London 37–9, 40, **45**, 46–7
    small towns/cities 163–77
    stature 57
        data acquisition 60–1
        determinants of 58–60, 64–7, 70–3, **71**, **72**
        and diet 56–7
        estimation 61–4
        temporal trends 67–70, *67*
skeletal stress markers *see* osteological stress markers
small towns/cities 163–77
smallpox 19, 20–3, 152, 154
    evolution 22
    inoculation 21
    population density effects 20
    spread by soldiers 21–2
    transmission 21
    treatment 22–3
Snow, John 354
social epidemiology 356
social stratification 267–8, 274–5
socio-ecological model of epidemiology 355

socioeconomic status 9–10, 49, 180, 202
   and morbidity 364–5
   and mortality 358
   and parasitism 333
   urban populations 243–64
souvenir diseases 303–4
Spanish flu 123–4, 126, 370–1
   death rate *128*
   see also influenza pandemics
standard of living *see* living standards
stature 343
   African Americans 244–5, **254**
   and biological welfare 214–15, *214*, *215*
   data acquisition 60–1
   determinants of 58–60, 64–7, 70–3, **71**, **72**
      climate change 58, 65–6
      cultivation methods 59, 66
      dairy and beef consumption 58, 64–5, *65*
      disease profile 59, 66
      gender inequality 60, 67
      land *per capita* 58, 64, *65*
      Romanization 59–60, 67
      urban share 59, 66, *66*
      war and conflict 59, 66
   and diet 56–7
      protein consumption 58, 64, *65*
   estimation 61–4
      females 63–4
      from cremains 64
      males 62–3
   Euro-Americans 244–5, **254**
   Europeans 55–80
   height transition 385
   temporal trends 67–70, *67*
   USA **254**
      Antebellum Puzzle 244–5
   women 201–23
subnational transitions 385–7
syndemics 349–50
syphilis 246, **248**, 253, 257

T cells, regulatory 306–7
tapeworms 313, 315
teeth
   edentulism **252**
   enamel hypoplasia 47, **252**, 253
   lead in tooth enamel *295*
   see also dental health
temporal variation 8
theory of epidemiologic transition *see* classic theory of epidemiologic transitions
Third Epidemiologic Transition 340, 349–50, 357
Thompson, Warren 369

total environment 17–34
transition analysis 49
treponematosis **251**
tuberculosis 18
   Europe 155–6, *155*
   mortality rate *96*, *345*
   USA 81–101, 244, **248**, **251**, 257, 258–9
typhoid 24, 29, **94**, 152, 174, 245, 308, 373
   mortality rate *98*, 152, 164
typhus 24, 309

underrepresented communities 9–10
United States of America (USA)
   Albany, New York 329–34, **331**, **332**, **334**
   American Civil War 245, 256–7
   Antebellum Puzzle 244–5
   Cleveland (Ohio) 249, 256
   dental health 246, **248**, **252**, 253
   epidemiologic transition 327–34, **331**, **332**, **334**
   fertility *382*
   Great Migration 255–6
   historical research 249–50
   industrialization 244
   infectious diseases 250–3, **251–2**
   Massachusetts 81–101
   Philadelphia (Pennsylvania) 17–34
   Reconstruction 245–6, 255–6
   St. Louis (Missouri) 249
   skeletal data
      African Americans 246–9, **247**, **248**
      Columbia, Missouri 163–77
      Euro-Americans 246–9, **247**, **248**
   skeletal pathologies 250–3, **251–2**
   stature **254**
   urbanization and in-migration 243–64
   Washington DC 249
urban sinks 183
urbanization 4, 189–90
   infant morbidity/mortality 237–9
   in low- and middle-income countries (LMICs) 364
   nutritional effects 59, 66, *66*, **71**, **72**
   USA 243–64
Usher mortality model 43, *43*

variola virus 20
vectors 26–7, 355
von Neumann neighbors 115

war and conflict, nutritional effects 59, 66, **71**, **72**
waste 322–7, **324–6**
welfare 205

determinants of 218–20, **219**
and religious affiliation 215–16, *216*
rural-urban differences 217–18, *218*
and socioeconomic status 216–17
and stature 214–15, *214*, *215*
Wellcome Osteological Research Database 6, 9, 230–1, 341, 388
Western diet 203
Western model of epidemiologic transition 358
whipworm (*Trichuris trichiura*) 322–7, **325–6**, **332**, 333
women 9
Indian, stature of 201–23
life expectancy *203*

mortality rate 48
stature estimation 63–4

yellow fever 25–30
disease 27–30
mortality 29–30
origins of 25–6
population density effects 20
vector 26–7
virus 27

zoonoses 303–4